Soybean Production in the Midsouth

Edited by

LARRY G. HEATHERLY
U.S. Department of Agriculture
Agricultural Research Service
Stoneville, Mississippi

HARRY F. HODGES
Mississippi State University
Department of Plant and Soil Sciences
Mississippi State, Mississippi

EDITORIAL COMMITTEE: RICHARD WESLEY AND ALAN BLAINE

SPONSORED BY THE MISSISSIPPI SOYBEAN PROMOTION BOARD

CRC Press
Boca Raton   London   New York   Washington, D.C.

Acquiring Editor:	John Sulzycki
Project Editor:	Carol Whitehead
Marketing Manager:	Becky McEldowney
Cover design:	Dawn Boyd
PrePress:	Greg Cuciak
Manufacturing:	Carol Slatter

Library of Congress Cataloging-in-Publication Data

Soybean production in the midsouth / edited by Larry G. Heatherly and
  Harry F. Hodges.
      p. cm.
  Includes bibliographical references and index.
  ISBN 0-8493-2301-0 (alk. paper)
  1. Soybean--Southern States. I. Heatherly, Larry G. II. Hodges, H. F.
SB205.S7S39 1998
633.3'4'0975--dc21   98-24743
                                                                     CIP

No claim to original U.S. Government works
International Standard Book Number 0-8493-2301-0
Library of Congress Card Number 98-24743
Printed in the United States of America 1 2 3 4 5 6 7 8 9 0
Printed on acid-free paper

Preface

Soybean is a major crop in the midsouthern U.S. Its production provides a significant contribution to the agricultural economy of the region, and it is a major export commodity. Its high value is based on the content of both high-quality oil and protein that provide components for both human and animal food.

Research with soybean production has provided information that, when appropriately applied, results in increased soybean yields, profits, and management options. This research has been supported by public funds through state and national programs, by grower-initiated efforts that provide primarily supplemental funds for research in addition to that supported by public funds, and by private industry primarily for variety and chemical development. Agricultural scientists, through the generous provision of funds from the public and private sectors of society, have acquired an unheralded body of physical, chemical, and biological information that relates to the world around us. Much of it can be applied to understanding the principles of soybean production.

Soybean Production in the Midsouth presents the current status of our knowledge derived from these research efforts. It is our intention to identify information that is particularly relevant to understanding production processes, condense it into a reasonable volume, and present it in a format that growers and their support groups can readily use. Emphasis is placed on describing the most up-to-date knowledge of the physical and biological processes that relate to soybean production. We have striven to identify production practices and bring together diverse information from a variety of disciplines that suggest ways for producers to best utilize our soil and climatic resources.

Soybean yields in the midsouthern U.S. have been lower than those in the upper midwestern U.S. There are several reasons for these lower yields in the Midsouth, and special efforts have been made to focus on information relevant to enhancing soybean production in this environment. The primary audience for this monograph is the soybean producer. We anticipate the information in this book will provide, in consolidated form, a ready reference for understanding major aspects of soybean production. The information will provide the basis for adopting new or different procedures and practices that have not been used previously. We also expect students of agronomy and crop production in general will find the information helpful. Agricultural technicians in both the public and private sectors should use this material as a reference for basic and applied soybean production information.

The editors wish to express our thanks to each of the authors for taking the time and effort to prepare their respective manuscripts, to Richard Wesley and Alan Blaine for carefully reviewing each chapter, and to K. Raja Reddy for the technical help. We especially appreciate the financial support of the Mississippi Soybean Promotion Board.

Foreword

During the past quarter century, information and technology in the agricultural sciences have been developed at an unprecedented pace. Concurrent discoveries and new supporting information in computer sciences, molecular biology, genetic engineering, and complex pesticide chemistry also have been developed. In fact, new technologies in these peripheral areas of crop production have resulted in valuable resources from previously unexplored areas, thus further complicating the decision-making process. Basic and applied crop production and management research is contributing information not only from traditional avenues of experimentation, but also by incorporating these new technologies into proven systems to achieve increased soybean yields, higher profits, and more efficient utilization of resources.

It is difficult to remain current and knowledgeable in the many aspects of science that affect crop production. It sometimes seems impossible to understand the application of new technology to crop production because of its dynamic nature. Soybean producers and their supporting infrastructure of resource personnel have few places to go for a review of current, basic information related to everyday management decisions. This new information related to soybean production is fragmented in reports from several unconnected sources, and organized methods for incorporating it with other known facts are just becoming available. Consolidation of information about current, proven technology is needed so that the "whole picture" of soybean production systems can be found in one source.

This book will partially serve that purpose. All important aspects of soybean production in the midsouthen U.S. are covered in this one volume. Topics range from basic plant–water relations to applied topics such as planting date, row spacing, and variety selection. You, the reader, should find the information you require to produce the best possible soybean crop.

Contributors

Alan Blaine
Mississippi State University Extension
  Service
Department of Plant and Soil Science
Mississippi State University
Mississippi State, Mississippi

Glenn R. Bowers
Department of Agronomy
Crop Soil and Environmental Sciences
Purdue University
West Lafayette, Indiana

Normie Buehring
Mississippi Agricultural and Forestry
  Experiment Station
Verona, Mississippi

Dave Buntin
Georgia Griffin Station
University of Georgia
Experiment, Georgia

Seth M. Dabney
National Sedimentation Laboratory
USDA-ARS
Oxford, Mississippi

James C. Delouche
Department of Plant and Soil Sciences
Mississippi State University
Mississippi State, Mississippi

Ernest H. Flint, Jr.
Mississippi State University Extension
  Service
Kosciusko, Mississippi

Joe Funderburk
North Florida Research and Education
  Center
University of Florida
Quincy, Florida

P. D. Gerard
Experimental Statistics Unit
Mississippi State University
Mississippi State, Mississippi

Larry G. Heatherly
Crop Genetics and Production
  Research Unit
USDA-ARS
Stoneville, Mississippi

G. W. Hergert
West Central Research and Extension
  Center
University of Nebraska
North Platte, Nebraska

Harry F. Hodges
Department of Plant and Soil Sciences
Mississippi State University
Mississippi State, Mississippi

Bennie C. Keith
Department of Plant and Soil Sciences
Mississippi State University
Mississippi State, Mississippi

G. W. Lawrence
Department of Entomology and Plant
  Pathology
Mississippi State University
Mississippi State, Mississippi

K. S. McLean
Department of Agriculture
Northeast Louisiana University
Monroe, Louisiana

Robert McPherson
Coastal Plain Experiment Station
University of Georgia
Tifton, Georgia

Krishna N. Reddy
Southern Weed Science Research Unit
USDA-ARS
Stoneville, Mississippi

K. Raja Reddy
Department of Plant and Soil Sciences
Mississippi State University
Mississippi State, Mississippi

John S. Russin
Louisiana State University Agricultural
  Center
Baton Rouge, Louisiana

Glover B. Triplett
Department of Plant and Soil Sciences
Mississippi State University
Mississippi State, Mississippi

Jac J. Varco
Department of Plant and Soil Sciences
Mississippi State University
Mississippi State, Mississippi

Richard A. Wesley
Application and Production
Technology Research Unit
USDA-ARS
Stoneville, Mississippi

Frank D. Whisler
Department of Plant and Soil Sciences
Mississippi State University
Mississippi State, Mississippi

J. L. Willers
Crop Science Research Laboratory
USDA-ARS
Mississippi State, Mississippi

Bob Williams
Extension Economist (retired)
Mississippi State University Extension Service
Mississippi State University
Mississippi State, Mississippi

Contents

chapter one

Economics of soybean production in Mississippi

Bob Williams

Contents

Introduction

For the past 17 years, many Mississippi farmers have experienced numerous economic struggles with the soybean enterprise. In some years low yields resulted in poor economic conditions. That was the case in 1986, when we averaged only 17 bu/acre. Low prices have caused problems in a number of years. One of the best examples was 1986, when prices averaged less than $5.00/bu and November 1986 futures put in a life-of-contract high of only $5.56/bu. In other years, high production costs have been a major negative factor. That was the case in the early 1980s, as total costs in many sections of the state approached $200/acre, with total costs being defined as total specified expenses plus land, management, and general overhead costs. Because of these numerous poor economic conditions, acreage in the state has declined sharply. In 1980, 3.85 million acres were harvested. By 1996, that figure had declined to only 1.75 million acres, for a decline of more than 2 million acres in the 17-year time period. There have been only two year-to-year increases in acreage during the past 17 years.

Even with the rather poor economic conditions that have existed since 1980, many Mississippi farmers are more optimistic about the economics of the soybean enterprise. Recent statewide yields of more than 30 bu/acre have caused some of that optimism. Fairly good prices in 1993 and in late 1995, as well as in 1996 when November futures moved into the upper one third of the historical price range, have also been factors causing optimism. Also, soybean farmers have controlled production costs fairly well in recent years and that is a plus for the future.

0-8493-2301-0/99/$0.00+$.50
© 1999 by CRC Press LLC

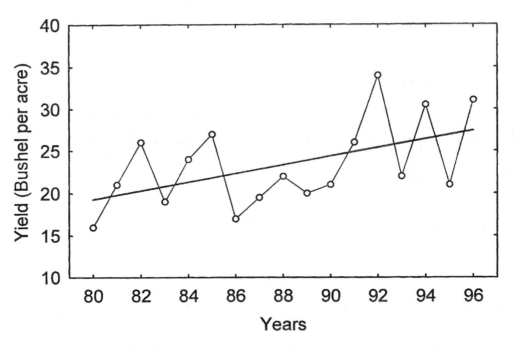

Figure 1.1   Trends in average soybean yields for Mississippi, 1980–1996.

Because of the importance of yields, prices, and cost of production relative to the economics of soybean production, some attention will be given to all three factors in this chapter. However, most of the attention will be directed at prices and how to get a better price for soybeans, and at cost of production with emphasis on controlling costs.

Yield

Since yields play an important role in determining the economics of soybean production, it is necessary to review our most recent yield history. Figure 1.1 shows soybean yields for Mississippi from 1980 through 1996. Yields have certainly been variable but the 17-year trend is upward. In 1980, Mississippi farmers averaged only 16 bu/acre, which is the lowest figure for the 17-year time period. Data from the Mississippi Agricultural Statistics Service indicate the summer of 1980 was extremely hot and dry, and yields reflect that. The highest statewide average yield during this 17-year time period was 34 bu/acre recorded in 1992. The weather conditions in the summer of 1992 prompted one agricultural writer to say that many Mississippi farmers thought they had died and gone to the Midwest. Because of excellent growing conditions and good management practices, Mississippi soybean yields were at an all-time high. Also, in 4 of the past 6 years we have been at or above 26 bu/acre, and we have been above 30 bu/acre in 3 of the past 5 years.

As we look at the trend line for the past 17 years (Figure 1.1), we note that the line is at 19 bu/acre in 1980 and up to about 27 bu/acre in 1996. That is an increase of 0.47 bu/year. That is significant, and over time will add about 850,000 bu annually to Mississippi soybean production without adding any acres. Using a $6.00 selling price, the additional bushels will add about $5.0 million annually to farm gate sales in the state; at a price of $7.00, the additional bushels will add about $6.0 million annually.

As we look at similar data for the U.S., yields have also moved higher. The average annual increase in U.S. soybean yields for the same time period is 0.6 bu. Even though

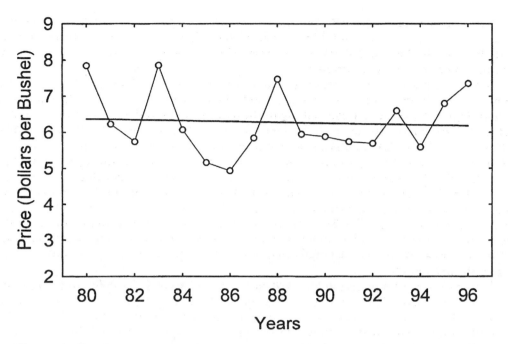

Figure 1.2 Trends in average soybean prices received by Mississippi farmers, 1980–1996.

we are making progress, the gap between Mississippi and U.S. yields continues to widen. Correcting that situation should be a major goal for all parties of the soybean industry.

As the data indicate, we have made significant strides in the past 17 years. However, to stay competitive with other U.S. and world soybean farmers, the annual increase in yields must be at a higher rate. That can be accomplished, and there are several critical success factors that can and will cause that to happen. As an economist, I will not try to tell you what those critical success factors are. However, I will suggest that your county agent, your soybean specialist, and many of the other authors of this book can help you identify those critical success factors for your operation. After you determine what those factors/practices are, and as you become more timely and better at implementing those practices, your soybean yields should improve significantly.

Price

Soybean prices have trended down in recent years (Figure 1.2.) In 1980, trend line prices were at about $6.35/bu. By 1996, the trend line had declined to about $6.15/bu. With 5 of the past 8 years showing prices in the $5.55 to $5.95 range, it is easy to see why the 17-year trendline is declining and why current trendline prices are just above the $6.00 level. The lowest price per bushel received for Mississippi soybeans in the 1980 to 1996 period was $4.93 in 1986, and the highest was $7.85 in 1983. The average price for the 17-year time period was $6.27. The average for the 5-year time period of 1992 through 1996 was $6.39. The lowest November futures price for the 17-year time period was $4.60 recorded on February 17, 1987. The highest November futures price recorded in the 17-year period was $10.46 recorded on June 23, 1988. That is a difference of almost $6.00/bu that happened within essentially a 16-month time period. It certainly gives some validity to the statement that "low prices tend to be self-correcting."

One critical success factor in getting an above-average price for soybeans is the timing of sales. The late Gene Futrell told the story of the Indian Chief who was known far and wide for his rain dances. The key to the Indian Chief's success was timing. In marketing, timing is a critical success factor.

Seasonal trends for November soybean futures are shown in Figures 1.3 and 1.4. They show data for 10-year and 15-year time periods. Figure 1.3 shows data for the entire time period. Figure 1.4 shows similar data with drought years excluded. It should be pointed out that these charts do not include data for 1994, 1995, or 1996. The 1994 November futures chart shows data that is somewhat normal from a seasonal standpoint. The highs occurred in May and June, and the lows occurred in early October during harvest. However, the 1995 November futures chart shows some rather unusual figures. The highs occurred in October and November. That is the first time since 1983 for November futures to record highs during the harvest season. That does not occur often, but it did occur in 1980, 1983, and again in 1995. The 1996 November futures chart was more normal in that the highs occurred in mid-summer and by mid-October prices were down by about $1.50/bu. A near life-of-contract high did occur in mid-September, which is a little unusual in non-drought years.

Figures 1.3 and 1.4 point out how critical it is to price early. As stated in the previous paragraph, pricing at harvest time was a good alternative in only 3 of the past 17 years. However, in most years, pricing in the late spring or early summer is the best alternative. In nondrought years, prices tended to be highest in April, May, and June. The lows normally occur at peak harvest in early- to mid-October, but a rally usually occurs near the end of harvest season in early to mid-November. As beans are sold out of storage in early to mid-February, prices usually decline prior to starting the spring rally. If you like to use call options in your marketing program, give some thought to buying those calls at the harvest-time low or at the February low just prior to the start of the spring rally. If you buy the call and prices advance substantially, the call can be liquidated at a profit.

In short crop years, such as 1980 and 1983, prices tend to peak at harvest. When the short crop years are included in the seasonal charts, the peak tends to come in June and is followed by a lower peak in early to mid-September. The lows still tend to come in October.

Knowing seasonal tendencies can be extremely helpful. In the 1994 season, peaks in the $6.90 to $7.00 range occurred in late May and mid-June. Prices declined sharply after that as the large crop got larger. The lows occurred in early October at about $5.25. In 1995 some fairly decent figures occurred in the May, June, and July time period. As the U.S. crop struggled with late summer/early fall weather problems and as the crops in South America struggled with early season dryness, prices moved to new highs in late October and early November. In 1996, a life-of-contract high for November futures at $8.25 occurred in mid-July. Another good peak occurred in September, but November futures were in the $6.60 to $6.75 range by mid-October/early November.

Another critical success factor in getting an above-average price for soybeans is knowing the basis trends for your area. Basis is defined by Futrell and Wisner (1987) as the difference between a futures price and a cash price. It normally reflects the cost of transportation to the delivery point plus the cost of storage, interest, and insurance until delivery. It is different in various areas of the state and changes as time passes.

Figure 1.5 shows soybean basis at Greenwood, MS for the 1988 through 1993 time period. As the chart indicates, basis is normally wide in June, July, and August. It is normally narrow or most favorable for the producer in December, January, and February. This seems logical since many commercial firms begin preparing for the upcoming harvest season in mid-to-late summer. As farmers utilize on-farm storage in December, January, and February, basis begins to improve as commercial firms try to move grain from on-farm

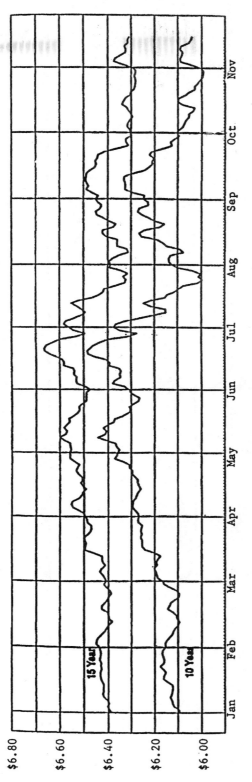

Figure 1.3   November soybean futures prices. (Courtesy of the Stewart-Peterson Group.)

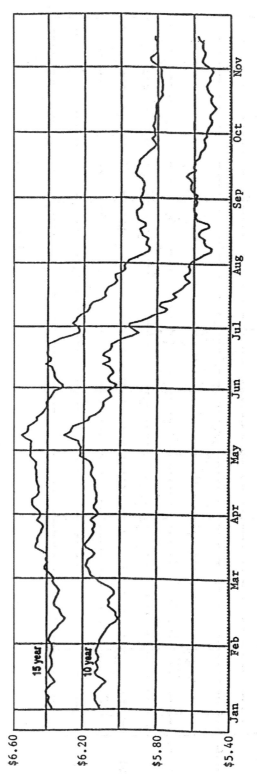

Figure 1.4 November soybean futures prices, drought years excluded. (Courtesy of the Stewart-Peterson Group.)

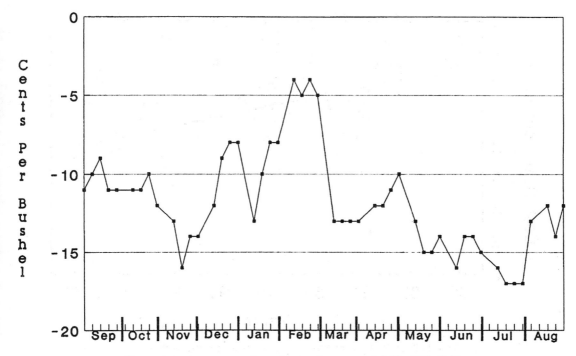

Figure 1.5   Average soybean basis, Greenwood, MS, 1988–1993.

storage to their processing/export facilities. Also, basis tends to become wide when prices are highest (May, June, early July) and narrow when prices are lowest (October through February).

Basis is one area where Mississippi farmers tend to have an advantage over most midwestern U.S. farmers. A recent check of over 100 observations of cash prices at Clarksdale, MS, Des Moines, IA, and an average for 20 U.S. locations shows the cash price at Clarksdale higher about 95% of the time. That is partly true because of the processing facility at Clarksdale, but a look at many other locations in the Mississippi major soybean-producing regions shows somewhat similar results in most situations.

In addition to knowing seasonal trends for November futures and knowing basis trends for your area, producers should consider a number of other critical success factors in getting an above-average price for soybeans. One of those is having an estimate of production costs. A good set of farm records that shows costs and returns by enterprise is extremely valuable. Another one is setting price objectives and then developing the discipline to price soybeans once those objectives are reached. Another critical success factor in getting above-average prices is being familiar with the marketing alternatives that are available in your area. And finally, develop the discipline to review past marketing decisions in the context of helping make better future decisions.

Cost of Production

Total soybean production costs per bushel have declined in Mississippi in recent years. The trend line on Figure 1.6 puts estimated total costs at about $8.50/bu in 1980 and $6.00/bu in 1996. The variation in cost per bushel has been fairly large during the 17-year time period. A number of factors have caused that variation. When statewide average yields have been at or below the 21-bu level, cost per bushel has been, in most years, at

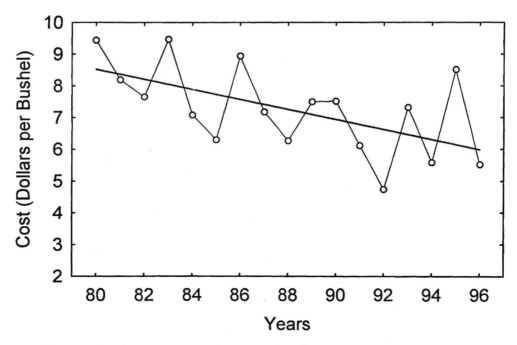

Figure 1.6 Trends in average total Mississippi soybean production costs, 1980–1996.

or above the $7.50 level. When statewide average yields have been at or above the 25-bu level, cost per bushel has been at or below the $6.50 level. In those years when yields have been at or above the 30-bu level, cost per bushel has been at or below the $5.60 level. That occurred in 1992, 1994, and 1996.

In the early 1980s, production costs per acre were relatively high. Farmers were not quite as cost conscious as they are now and land values and rental rates were at an all-time high. As land values and rental rates declined after the 1981 peak, cost per acre began to decline and cost per bushel also trended down. Our figures show cost per acre moving down from 1982 through 1988. Since 1988, costs per acre have moved substantially higher. With the decline in costs per acre in the mid-1980s and because of the higher yields that have been produced in recent years, it is easy to see why the cost per bushel trend line is moving in a favorable direction.

The cost per bushel data were developed using a weighted statewide total cost per acre based on the Mississippi State University soybean budgets, other printed cost data, and statewide average yields as published by the Mississippi Agricultural Statistics Service. Acreage harvested in the Delta and non-Delta regions was the weighting factor used. Keep in mind these figures are estimates and are not meant to be used as absolutes.

The highest cost per bushel was $9.47 in 1983. The next highest was at $9.44 in 1980. The lowest cost per bushel was in 1992 when yields averaged 34 bu in Mississippi and estimated total costs per bushel were only $4.74. The next lowest was $5.52 in 1996, when yields averaged 31 bu in Mississippi.

The average cost per bushel for 1980 through 1996 was $7.26. For 1992 through 1996, cost per bushel averaged $6.34. The average cost per bushel for 1980 through 1996 is substantially above the average price received for the similar time period. The good news is that for the past 5 years, the average price received has been slightly above the average cost per bushel, and the current cost per bushel trend line is about 15¢ below the current price per bushel trend line. These figures tell us, however, that in most of the past 17 years, total costs per bushel have averaged substantially above the price received per bushel. The only major exceptions were 1988, 1992, 1994, and 1996. This explains, to a large extent, why soybean plantings declined from 4 million acres in 1980 to slightly less than 2 million acres in 1996.

What are some of the critical success factors we need to think about as we attempt to move toward the time when costs per bushel will consistently be below the prices received? It is fair to say that as producers do a better job controlling costs, they need a good estimate of production costs. We mentioned this in the section on prices, and we also mentioned the need and value of good enterprise records as we try to estimate cost of production.

Tables 1.1 through 1.4 show the Mississippi State University 1997 estimates for current production costs. Both direct costs and fixed machinery costs are shown for the Delta, the Brown Loam area, the Coastal Plain, and the Black Belt. Total specified expenses (the sum of direct costs and fixed machinery costs) vary from a low of $112/acre in the Delta to a high of $146/acre for the Coastal Plain area. Keep in mind these estimates are presented in an effort to assist you in putting together cost figures for your farm. Your production and marketing decisions should be based on your figures rather than any we present. The cost figures in Tables 1.1 through 1.4 do not include management, general overhead, or land costs. There are a number of decisions that should be based on total costs per acre or total costs per bushel, and these other charges should be added to total specified expenses to get total costs. The data shown on Figure 1.6 includes all costs. The cost estimates shown in Tables 1.1 through 1.4 were published by the Department of Agricultural Economics at Mississippi State University with substantial input from numerous research and extension personnel. These cost estimates are revised annually and are usually available by February 1 each year.

Information about soybean production practices in Mississippi is provided from surveys of soybean producers across the state in 1996. Dr. Stan Spurlock, Agricultural Economist at Mississippi State University, estimated revenues, costs, and net revenues for each producer surveyed. The statistical results from the surveys were computed for five soil resource areas and are presented in Tables 1.5 through 1.9. Almost one half of the surveyed soybean cropland was irrigated in the Upper Delta area. The irrigation component of the production system caused the average fixed machinery cost to be higher in this area than in the other four areas.

Average herbicide cost, the largest direct-cost item in each area, was lower in the Black Belt area than in the other areas. Average fertilizer cost was higher in the Upper Coastal Plain area. A portion of the surveyed farms in each area of the state used custom combining. Average land rental rates varied somewhat from one area to another. In many of the cost categories, there was a wide range of values estimated. This implies that there are "low-cost" farms and "high-cost" farms in a given year. There was also a wide range of soybean yields in the survey. The variability in both revenues and costs leads to a substantial amount of variation in net revenues. Farmers need to analyze their own situations to determine if their revenues are high enough to offset their expenses. If not, alternative production practices should be evaluated.

Another critical success factor in controlling costs is to check prices at more than one input supply firm. Input prices may vary from firm to firm, and small savings add up to significant dollars over long time periods. Where possible, give some consideration to purchasing larger quantities and paying cash. Discounts on volume purchases and cash discounts can also add up to significant savings over time.

Reducing fixed cost per acre and per bushel is another critical success factor. Farming a few more acres or selling some machinery and farming a few less acres or doing custom work for another farmer or having another farmer do custom work for you are ways to reduce fixed costs. Being more timely in the application of specific practices is another

Table 1.1   Estimated Costs per Acre Soybeans, Clay Soil, 10-Row 30" Equipment, Delta Area, Mississippi, 1997

Item	Unit	Price, $	Quantity	Amount, $	Your farm
Direct Expenses					
Custom					
App by air (3 gal)	appl	2.60	0.2500	0.65	_____
Herbicide					
Treflan EC	pt	3.67	2.0000	7.34	_____
Scepter 70 DG	oz	6.30	2.8600	18.02	_____
Blazer	pt	7.21	0.7500	5.41	_____
Fusilade DX	oz	0.87	1.2000	1.04	_____
Insecticide					
Larvin 3.2	oz	0.39	4.5000	1.76	_____
Haul					
Haul soybeans	bu	0.16	30.0000	4.80	_____
Seed/plants					
Soybean seed	lb	0.28	40.0000	11.20	_____
Adjuvant					
Surfactant	pt	0.89	0.2500	0.22	_____
Crop oil (petroleum)	pt	0.83	0.2000	0.17	_____
Operator labor					
Tractors	hour	7.50	0.6010	4.51	_____
Self-propelled equipment	hour	7.50	0.1220	0.92	_____
Hand labor					
Implements	hour	5.87	0.1455	0.85	_____
Unallocated labor	hour	7.50	0.6507	4.88	_____
Diesel fuel					
Tractors	gal	0.82	6.8514	5.62	_____
Self-propelled equipment	gal	0.82	0.6954	0.57	_____
Repair and maintenance					
Implements	acre	3.80	1.0000	3.80	_____
Tractors	acre	3.52	1.0000	3.52	_____
Self-propelled equipment	acre	5.73	1.0000	5.73	_____
Interest on operating capital	acre	3.15	1.0000	3.15	_____
Total direct expenses				84.14	_____
Fixed expenses					
Implements	acre	7.73	1.0000	7.73	_____
Tractors	acre	9.73	1.0000	9.73	_____
Self-propelled equipment	acre	10.54	1.0000	0.54	_____
Total fixed expenses				27.99	_____
Total specified expenses				112.13	_____

Note: Cost of production estimates are based on last year's input prices.

critical success factor in controlling costs. A few examples include being more timely with the application of pesticides and being timely with planting, harvesting, cultivating, and applying water. Implementation of specific low-cost technology can, on some farms, reduce cost per bushel and be a critical success factor. Some of that technology may come from the past, while some may emerge from the good manager who spends time thinking about ways to control costs.

One final critical success factor to consider for controlling and/or reducing costs relates to marginal economic analysis. As each dollar is spent, the expected returns from

Table 1.2 Estimated Costs per Acre Soybeans, 12-Row 20" Equipment, Brown Loam Area, Mississippi, 1997

Item	Unit	Price, $	Quantity	Amount, $	Your farm
Direct expenses					
Custom					
App by air (3 gal)	appl	2.60	0.5000	1.30	_____
Fertilizer					
Lime (Spread)	ton	26.88	0.2500	6.72	_____
Phosphorus (46% P2O5)	cwt	12.30	0.6600	8.12	_____
Potash (60% K2O)	cwt	7.26	1.0000	7.26	_____
Herbicide					
Treflan EC	pt	3.67	1.5000	5.51	_____
Scepter 70 DG	oz	6.30	2.8600	18.02	_____
Storm	pt	8.58	1.5000	12.87	_____
Fusilade DX	oz	0.87	1.2000	1.04	_____
Insecticide					
Larvin 3.2	oz	0.39	9.0000	3.51	_____
Haul					
Haul soybeans	bu	0.16	30.0000	4.80	_____
Seed/plants					
Soybean seed	lb	0.28	40.0000	11.20	_____
Adjuvant					
Surfactant	pt	0.89	0.5000	0.45	_____
Crop oil (petroleum)	pt	0.83	0.2000	0.17	_____
Operator labor					
Tractors	hour	7.50	0.5700	4.28	_____
Self-propelled equipment	hour	7.50	0.1220	0.92	_____
Hand labor					
Implements	hour	5.87	0.2740	1.61	_____
Unallocated labor	hour	7.50	0.6228	4.67	_____
Diesel fuel					
Tractors	gal	0.82	6.4980	5.33	_____
Self-propelled equipment	gal	0.82	0.6954	0.57	_____
Repair and maintenance					
Implements	acre	4.11	1.0000	4.11	_____
Tractors	acre	3.33	1.0000	3.33	_____
Self-propelled equipment	acre	5.73	1.0000	5.73	_____
Interest on operating capital	acre	5.00	1.0000	5.00	_____
Total direct expenses				116.50	_____
Fixed expenses					
Implements	acre	7.99	1.0000	7.99	_____
Tractors	acre	9.23	1.0000	9.23	_____
Self-propelled equipment	acre	10.54	1.0000	10.54	_____
Total fixed expenses				27.76	_____
Total specified expenses				144.26	_____

Note: Cost of production estimates are based on last year's input prices.

that dollar should certainly be greater than a dollar, and dollars should be spent where the greatest "marginal" return is expected. By doing this consistently, chances of making a profit on the crop are greatly increased. As this is done, work these decisions out on paper. As John Ikerd, Extension Economist in Missouri, has said many times, "If it won't

Table 1.3 Estimated Costs per Acre Soybeans, 6-Row 30″ Equipment, Coastal Plain, Mississippi, 1997

Item	Unit	Price, $	Quantity	Amount, $	Your farm
Direct expenses					
Custom					
App by air (3 gal)	appl	2.60	0.5000	1.30	_____
Fertilizer					
Lime (Spread)	ton	26.88	0.2500	6.72	_____
Phosphorus (46% P2O5)	cwt	12.30	0.6600	8.12	_____
Potash (60% K2O)	cwt	7.26	1.0000	7.26	_____
Herbicide					
Treflan EC	pt	3.67	1.0000	3.67	_____
Scepter 70 DG	oz	6.30	2.8600	18.02	_____
Classic	oz	17.43	0.5000	8.72	_____
Fusilade DX	oz	0.87	1.2000	1.04	_____
Insecticide					
Methyl Parathion 4E	pt	3.12	0.5000	1.56	_____
Haul					
Haul soybeans	bu	0.16	25.0000	4.00	_____
Seed/plants					
Soybean seed	lb	0.28	40.0000	11.20	_____
Adjuvant					
Surfactant	pt	0.89	0.5000	0.45	_____
Crop oil (petroleum)	pt	0.83	0.2000	0.17	_____
Operator labor					
Tractors	hour	7.50	1.0130	7.60	_____
Self-propelled equipment	hour	7.50	0.1530	1.15	_____
Hand labor					
Implements	hour	5.87	0.3340	1.96	_____
Unallocated labor	hour	7.50	1.0494	7.87	_____
Diesel fuel					
Tractors	gal	0.82	6.6858	5.48	_____
Self-propelled equipment	gal	0.82	0.8721	0.72	_____
Repair and maintenance					
Implements	acre	3.84	1.0000	3.84	_____
Tractors	acre	3.38	1.0000	3.38	_____
Self-propelled equipment	acre	7.10	1.0000	7.10	_____
Interest on operating capital	acre	4.98	1.0000	4.98	_____
Total direct expenses				116.29	_____
Fixed expenses					
Implements	acre	7.49	1.0000	7.49	_____
Tractors	acre	9.35	1.0000	9.35	_____
Self-propelled equipment	acre	13.05	1.0000	13.05	_____
Total fixed expenses				29.89	_____
Total specified expenses				146.18	_____

Note: Cost of production estimates are based on last year's input prices.

work on paper, it most likely won't work in the field." There may be other critical success factors in controlling costs, but these are some of the ones that may be applicable to a large number of mid-southern U.S. soybean farmers.

Table 1.4   Estimated Costs per Acre Soybeans, 8-Row 30" Equipment, Black Belt Area, Mississippi, 1997

Item	Unit	Price, $	Quantity	Amount, $	Your farm
Direct expenses					
Custom					
App by air (3 gal)	appl	2.60	0.5000	1.30	_____
Fertilizer					
Phosphorus (46% P2O5)	cwt	12.30	0.6600	8.12	_____
Potash (60% K2O)	cwt	7.26	1.0000	7.26	_____
Herbicide					
Treflan EC	pt	3.67	2.0000	7.34	_____
Scepter 70 DG	oz	6.30	2.8600	18.02	_____
Classic	oz	17.43	0.5000	8.72	_____
Fusilade DX	oz	0.87	1.2000	1.04	_____
Insecticide					
Methyl Parathion 4E	pt	3.12	0.5000	1.56	_____
Haul					
Haul soybeans	bu	0.16	28.0000	4.48	_____
Seed/plants					
Soybean seed	lb	0.28	40.0000	11.20	_____
Adjuvant					
Surfactant	pt	0.89	0.5000	0.45	_____
Crop oil (petroleum)	pt	0.83	0.2000	0.17	_____
Operator labor					
Tractors	hour	7.50	0.7170	5.38	_____
Self-propelled equipment	hour	7.50	0.1220	0.92	_____
Hand labor					
Implements	hour	5.87	0.2645	1.55	_____
Unallocated labor	hour	7.50	0.7551	5.66	_____
Diesel fuel					
Tractors	gal	0.82	8.1738	6.70	_____
Self-propelled equipment	gal	0.82	0.6954	0.57	_____
Repair and maintenance					
Implements	acre	3.98	1.0000	3.98	_____
Tractors	acre	4.19	1.0000	4.19	_____
Self-propelled equipment	acre	5.73	1.0000	5.73	_____
Interest on operating capital	acre	4.63	1.0000	4.63	_____
Total direct expenses				108.96	_____
Fixed expenses					
Implements	acre	8.07	1.0000	8.07	_____
Tractors	acre	11.61	1.0000	11.61	_____
Self-propelled equipment	acre	10.54	1.0000	10.54	_____
Total fixed expenses				30.21	_____
Total specified expenses				139.17	_____

Note: Cost of production estimates are based on last year's input prices.

Summary

The economics of soybean production in Mississippi in the past 17 years has varied significantly as the data in Figure 1.7 indicate. For the most part, the situation has been somewhat poor. However, for a number of reasons, the situation is improving. Yields have

Table 1.5   1996 Survey Results from 32 Upper Delta–Area Soybean Producers in Mississippi, 48% Irrigated

Item	Mean	Std. Dev.	Minimum	Maximum
Yield (bu/acre)	36.40	9.49	10.00	55.00
Revenue (@ $7.10/bu)	258.41	67.38	71.00	390.50
Selected direct cost items:				
Herbicides	29.60	17.64	5.51	85.05
Seed	18.41	6.81	8.40	35.40
Repairs and maintenance	16.03	4.52	3.04	25.39
Diesel fuel	10.93	4.14	2.54	16.27
Custom harvest/haul	7.93	4.00	1.60	33.08
Operator labor	7.35	2.40	1.55	12.92
Overhead labor	6.61	2.16	1.39	11.62
Interest on operating capital	3.71	0.90	1.98	6.60
Total Direct cost	107.40	22.15	57.24	153.98
Machinery fixed cost	54.65	15.84	10.50	76.90
Direct + fixed cost	162.06	31.28	81.18	197.03
Land rent	46.03	12.88	20.00	78.00
Total specified cost	208.09	36.73	101.18	257.03
Net revenue	50.32	71.07	−131.80	243.78

Notes: In the "mean" column, total direct cost does not equal the sum of the selected direct cost items listed because several minor cost categories are omitted. A discussion of how to interpret the standard deviation of an item (the "std. dev." column) is presented in another section of this book and is not repeated here. In the "minimum" and "maximum" columns, addition to obtain total costs and subtraction to obtain net revenue are not appropriate since each figure in those two columns could have come from different producers.

Table 1.6   1996 Survey Results from 26 Lower Delta–Area Soybean Producers in Mississippi, 8% Irrigated

Item	Mean	Std. Dev.	Minimum	Maximum
Yield (bu/acre)	27.06	7.38	6.00	48.00
Revenue (@ $7.10/bu)	192.09	52.41	42.60	340.80
Selected direct cost items:				
Herbicides	29.23	12.10	7.34	69.34
Seed	16.07	2.88	3.36	49.65
Repairs and maintenance	12.62	4.32	2.75	25.41
Custom harvest/haul	7.08	5.20	1.12	32.40
Diesel fuel	6.49	4.52	1.90	16.54
Operator labor	4.83	2.52	1.94	12.95
Overhead labor	4.35	2.27	1.74	11.65
Interest on operating capital	3.36	1.23	1.89	6.42
Total direct cost	88.69	24.77	52.78	148.48
Machinery fixed cost	32.26	18.75	6.47	77.62
Direct + fixed cost	120.95	41.89	59.25	215.73
Land rent	39.07	2.70	20.00	60.00
Total specified cost	160.02	42.04	89.25	255.73
Net revenue	32.07	43.36	−120.47	145.56

Notes: In the "mean" column, total direct cost does not equal the sum of the selected direct cost items listed because several minor cost categories are omitted. A discussion of how to interpret the standard deviation of an item (the "std. dev." column) is presented in another section of this book and is not repeated here. In the "minimum" and "maximum" columns, addition to obtain total costs and subtraction to obtain net revenue are not appropriate since each figure in those two columns could have come from different producers.

Table 1.7 1996 Survey Results from 20 Upper Brown Loam–Area Producers in Mississippi, Dryland

Item	Mean	Std. Dev.	Minimum	Maximum
Yield (bu/acre)	31.89	8.95	18.00	45.00
Revenue (@ $7.10/bu)	226.38	63.51	127.80	319.50
Selected direct cost items:				
Herbicides	27.66	14.68	0.00	61.93
Seed	14.86	4.21	8.40	33.60
Repairs and maintenance	12.15	2.76	5.24	15.69
Custom harvest/haul	6.32	5.81	1.60	30.00
Fertilizers	5.43	7.78	0.00	29.85
Diesel fuel	4.91	1.98	0.72	7.46
Operator labor	4.88	1.78	1.15	9.65
Overhead labor	4.39	1.60	1.03	8.69
Interest on operating capital	3.46	1.06	1.27	5.43
Total direct cost	89.37	20.25	49.36	131.36
Machinery fixed cost	25.47	6.36	12.28	32.92
Direct + fixed cost	114.83	22.73	68.94	163.58
Land rent	37.55	8.53	17.00	65.00
Total specified cost	152.38	27.64	105.19	213.58
Net revenue	74.00	75.38	−71.58	191.63

Notes: In the "mean" column, total direct cost does not equal the sum of the selected direct cost items listed because several minor cost categories are omitted. A discussion of how to interpret the standard deviation of an item (the "std. dev." column) is presented in another section of this book and is not repeated here. In the "minimum" and "maximum" columns, addition to obtain total costs and subtraction to obtain net revenue are not appropriate since each figure in those two columns could have come from different producers.

Table 1.8 1996 Survey Results from 20 Coastal Plain–Area Soybean Producers in Mississippi, Dryland

Item	Mean	Std. Dev.	Minimum	Maximum
Yield (bu/acre)	31.92	6.53	20.00	48.00
Revenue (@ $7.10/bu)	226.62	46.37	142.00	340.80
Selected direct cost items:				
Herbicides	30.81	21.80	6.32	83.04
Seed	16.78	3.08	8.96	23.80
Repairs and maintenance	14.31	0.88	12.25	16.88
Fertilizers	13.11	13.90	0.00	35.75
Operator labor	6.41	1.30	4.29	9.98
Diesel fuel	5.90	1.08	3.88	10.56
Overhead labor	5.77	1.17	3.86	8.98
Custom harvest/haul	5.59	1.07	3.20	7.52
Interest on operating capital	4.21	1.52	2.09	7.58
Total direct cost	105.37	30.87	55.57	175.29
Machinery fixed cost	29.88	1.98	25.30	37.29
Direct + fixed cost	135.24	31.04	83.29	203.77
Land rent	32.44	8.55	10.00	50.00
Total specified cost	167.68	29.42	112.46	225.71
Net revenue	58.94	56.67	-71.00	164.29

Notes: In the "mean" column, total direct cost does not equal the sum of the selected direct cost items listed because several minor cost categories are omitted. A discussion of how to interpret the standard deviation of an item (the "std. dev." column) is presented in another section of this book and is not repeated here. In the "minimum" and "maximum" columns, addition to obtain total costs and subtraction to obtain net revenue are not appropriate since each figure in those two columns could have come from different producers.

Table 1.9 1996 Survey Results from 23 Black Belt–Area Soybean Producers in Mississippi, Dryland

Item	Mean	Std. Dev.	Minimum	Maximum
Yield (bu/acre)	31.05	6.23	15.00	45.00
Revenue (@ $7.10/bu)	220.45	44.25	106.50	319.50
Selected direct cost items:				
Herbicides	18.47	6.60	8.58	52.84
Seed	13.96	3.61	9.80	37.80
Repairs and maintenance	11.59	1.80	5.47	16.86
Custom harvest/haul	5.38	1.67	2.40	24.64
Fertilizers	5.05	9.54	0.00	29.46
Operator labor	4.67	1.35	2.45	11.12
Diesel fuel	4.34	1.65	1.85	9.69
Overhead labor	4.20	1.22	2.20	10.00
Interest on operating capital	2.65	0.79	1.56	6.12
Total direct cost	72.19	18.65	52.84	131.17
Machinery fixed cost	24.06	4.24	13.48	37.93
Direct + fixed cost	96.25	22.44	73.08	162.51
Land rent	29.45	0.88	15.00	35.00
Total specified cost	125.70	22.33	102.25	197.13
Net revenue	94.75	51.27	−62.60	211.22

Notes: In the "mean" column, total direct cost does not equal the sum of the selected direct cost items listed because several minor cost categories are omitted. A discussion of how to interpret the standard deviation of an item (the "std. dev." column) is presented in another section of this book and is not repeated here. In the "minimum" and "maximum" columns, addition to obtain total costs and subtraction to obtain net revenue are not appropriate since each figure in those two columns could have come from different producers.

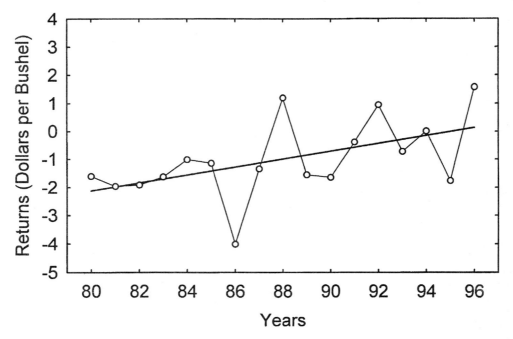

Figure 1.7 Trends in estimated average soybean returns above total costs for Mississippi, 1980 -1996.

trended higher, prices received have trended slightly lower, and total cost per bushel has trended lower. As the gap between price received and total costs per bushel becomes larger, the chances increase for soybean farmers to cover costs and make a profit.

Since 1980, the best year for Mississippi soybean farmers was 1996 (Figure 1.7). We averaged 31 bu/acre, the average price received was $7.10, and the estimated total cost per bushel was $5.52. That provided an average profit above all costs of $1.58/bu or about $50/acre. The poorest year since 1980 was 1986. We averaged only 17 bu/acre, got $4.93/bu, and had an estimated cost per bushel of $8.94. That translates into a loss per bushel of $4.01. That was certainly a difficult year.

Since 1980, the poorest 3 years were 1981, 1982, and 1986. Our best three have been 1988, 1992, and 1996. The good news is the poorest years were in the earlier part of the time period, whereas the best years have been in the latter part of the time period. These numbers indicate that the economic situation for soybeans has improved. While 1995 was a setback for most mid-southern U.S. producers, the situation rebounded in 1996 and we are moving in the right direction at a fairly decent pace.

As we look to the last years of this century, chances are good that 1 and possibly 2 of the 3 years will be good ones. The current supply-and-demand numbers for soybeans and the trends we have discussed indicate that kind of situation. That will put us in position to enter the 21st century to cover all costs consistently and produce a profit. If that kind of situation emerges, it will be extremely positive for the mid-southern U.S. farm economy and the total economy of the soybean-producing states of the region. By paying attention to some of the critical success factors we have mentioned, plus others that might be equally or more important on your farm, the chances of having a more prosperous economic future with soybeans will be increased.

References

Breakeven Analysis for Major Crop Enterprises, Mississippi, Agricultural Economics Department, Co-operative Extension Service/Mississippi State University, 1980–1995.

Crop Production, Annual Summary, National Agricultural Statistics Service, U.S. Department of Agriculture, Washington, D.C., 1980–1996.

Crop Values, National Agricultural Statistics Service, U.S. Department of Agriculture, Washington, D.C., 1980–1996.

Futrell, G. A. and R. N. Wisner, *Marketing for Farmers*, 2nd ed., Doane Information Services, St. Louis, MO, 1987.

Soybeans 1997 Planning Budgets, Agricultural Economics Report No. 78, Mississippi State University, December 1996.

Stewart-Peterson Group, West Bend, WI, 1994.

The Brock Report, Richard A. Brock & Associates, Inc., Milwaukee, Wisconsin, 1994–1995 (published weekly).

Williams, B., *Economics and Marketing Materials, Mississippi Soybeans*, Agricultural Economics Department, Cooperative Extension Service/Mississippi State University, 1995.

chapter two

Soil erosion and soybean production

Glover B. Triplett and Seth M. Dabney

Contents

Introduction

Erosion is a natural process. Geologic forces constantly uplift land from the sea and when this happens erosion begins. Rocks and other parent material weather and disintegrate to release nutrients and to form soil. This soil supports vegetation which covers the land surface. Vegetation slows but does not stop the erosion process. Under these conditions, an equilibrium is established between soil loss and soil formation. This stabilizes the soil surface within the time span of human records. For longer periods of time, the effects of geologic erosion are glaringly obvious. The Grand Canyon is the product of erosion and might be considered the largest gully in the world. In eastern Ohio and Kentucky, West Virginia, and the western Pennsylvania region, several coal seams were formed sequentially during past geologic ages, separated by layers of sediment, and then covered with hundreds of feet of additional sediment. Coal seams can now be found at predictable elevations along hillsides in the region. Their exposure in this manner is a result of erosion removing the sediment covering them, along with the coal seams themselves, in the valleys that were formed. This created the landscape now present in the region. In human terms, geologic erosion is very slow, ranging from 0.01 to 1 ton/acre/year.

Human activities such as agriculture, mining, and construction result in accelerated erosion. Soil movement from accelerated erosion can be 10 to 1000 times as great as from geologic erosion, and far exceeds the capacity of the soil to renew itself. Annual soil loss is influenced by several factors. For tilled soybean production on a 4% slope in Mississippi without cover, soil loss might equal 15 to 20 tons/acre/year. Since an acre inch of soil weighs approximately 150 tons, erosion at this rate would require 7 to 10 years to remove that inch of soil. The soil loss within 1 year at this rate probably could not be detected with research designed to test the loss of crop productivity.

The effects of soil loss are cumulative and serve to reduce the productive potential of the land. Early European immigrants brought farming systems with them from a region where erosion is a lesser problem because rainfall intensity is lower. They cleared land, farmed it until productivity declined, then moved on to new areas. After all, there was plenty of land to the west. Now, there is no unused land to settle and we have to deal with effects of earlier mismanagement. Much of the land area in the southeastern U.S. was cleared, farmed, and then abandoned to trees or converted to grassland after crop productivity declined (Miller et al., 1985). Because of high soybean prices some of this land was brought back into production during the 1970s. Productivity was high initially, but declined rapidly under tilled agriculture. Even applications of plant nutrients did not maintain productivity. Thus, the soil has some capability to recover from mismanagement, but considerable time is required for recovery to take place.

The history of mankind contains numerous examples of land misuse and the eventual consequences. Agriculture is considered to have first developed in the area between the Tigris and Euphrates rivers in the Middle East approximately 11,000 years ago. Yet, there is very little agriculture in that area today. What happened? There has been speculation of a climate change as being part of the problem. Another possible explanation is a reduction in productivity occurred as the soil was degraded through erosion. Soil washed from the hillsides was deposited on the floodplains where traces of villages have been found buried under 20 feet of silt. When erosion from the uplands cut into the subsoil, and these materials became the dominant source of sediment, the productivity of both uplands and alluvial areas declined.

There are areas where agriculture has been sustained for thousands of years. The Nile valley is one such area. In Europe, where erosion is less of a problem than in America, crops have been produced since Roman times. Thus, agriculture need not reduce productivity of the land as long as the soil resource is protected and maintained. Soil loss tolerance levels (*T*) have been established for soils in the U.S. This *T* value is considered the amount of annual soil loss that will not decrease soil productivity and ranges from 2 to 5 tons/acre/year, depending on soil depth and other characteristics. Soil loss tolerance levels for some major Mississippi soils are shown in Table 2.1.

Table 2.1 Erodibility (*K*) Factors and Soil Loss Tolerance Limits (*T*) of Selected Soil Series from Six Land Resource Areas in Mississippi

Soil series	K factor horizons			T values (tons/acre)
	A	B	C	
Brown Loam Resource Area				
Grenada	0.49	0.43	0.37	3
Lexington	0.49	0.43	0.34	3
Loring	0.49	0.43	0.43	3
Memphis	0.49	0.37	0.37	5
Upper Coastal Plains Resource Area				
Cahaba silt loam	0.24	0.28	0.24	5
Cahaba sandy loam	0.17	0.28	0.24	5
Ora	0.28	0.37	0.32	3
Prentiss	0.28	0.24	0.24	3
Ruston	0.28	0.28	0.28	5
Smithdale	0.28	0.24	0.24	5
Sweatman	0.37	0.28	0.28	3
Black Land Prairie Resource Area				
Brooksville	0.37	0.32	0.32	4
Kipling	0.32	0.32	0.32	5
Okolona	0.37	0.32	0.32	4
Vaiden	0.32	0.32	0.32	4
Lower Coastal Plains Resource Area				
Benndale	0.20	0.28	0.32	5
Lucedale	0.24	0.24	0.24	5
McLaurin	0.20	0.20	0.20	5
Ruston	0.28	0.28	0.28	5
Savannah	0.37	0.28	0.24	3
Delta Resource Area				
Bosket	0.24	0.32	0.24	5
Dubbs	0.37	0.37	0.37	5
Dundee	0.37	0.32	0.32	5
Sharkey	0.32	0.28	0.28	5
Interior Flatwoods Resource Area				
Adaton	0.43	0.32	0.32	5
Bude	0.49	0.43	0.37	3
Falkner	0.49	0.43	0.24	4
Longview	0.49	0.43	0.43	5

Costs of erosion

Soil erosion has both on-site and off-site costs. One on-site cost involves decreased crop yield potential. Soils that contain natural pans or compacted layers that restrict root growth and water movement are especially sensitive. These pans may be several feet thick and cannot be eliminated by tillage. On many soils, pans occur at 16 to 20 in. depth and erosion which reduces soil thickness above these pans also reduces the available rooting zone, thereby critically reducing the available water-holding capacity of soil that is explored by roots. Fragipan material commonly has greater density, lower organic matter levels, and lower water-holding capacity than topsoil. Some highly productive soils have been eroded to the point that most of the topsoil has been lost and tillage now is into the pan. When this occurs, profitable crop production may become impossible, depending on commodity price levels.

Two studies in Mississippi serve to illustrate the importance of soil depth and the influence of erosion on soybean productivity. Rhoton (1990) conducted a 3-year study on Grenada soil which has a naturally occurring fragipan at the 16 to 22 in. depth. Sites were identified that had only slight past soil erosion, moderate past erosion, and severe past erosion that had removed half or more of the topsoil. Fertilizer and lime were added so that crop nutrition would not be a limiting factor in determining crop yields. Yields for the slightly eroded sites were 28% greater than for the severely eroded sites during the 3 years of the study. During a dry year, yields were 47% greater for the slightly eroded sites, reflecting the increased water storage available with the deeper soil profile. Soil organic matter levels were positively correlated with increased yield; the slightly eroded sites had greater soil organic matter levels. Soybean yield reduction was greatest (0.9 bu/acre/in. of soil lost) where the soil thickness above the fragipan was between 20 and 24 in. Yields were relatively constant at about 30 bu/acre where soil thickness above the pan was between 8 to 16 in., indicating that roots can penetrate the fragipan to some extent and utilize some of the water stored there.

The other study (Hairston et al., 1989) was on soils in the Black Belt land resource area. These soils are underlain with chalk, which is nearly impervious to root growth and water movement, and which, as such, is considered to be parent material but not yet soil. Over long periods of time, the chalk material does weather slowly to form soil. In this 3-year study there was a direct relationship between soil depth and crop yield. If the soil layer was 3 to 6 in. deep, each inch of soil increased soybean yields by 2 bu. From 6 to 12 in., the increase was 1 bu, and from 12 to 24 in., the increase was 0.5 bu. From 24 to 60 in., the increase was 0.2 bu. The yield response to soil depth in this study was attributed to water available to the crop. As with the study on the Grenada soil (Rhoton, 1990), yields were greater with the same soil depth as organic matter level increased, underscoring the importance of this factor. Note, however, that the reaction to removal of an inch of soil was different on the two soils. Yield loss from soil erosion was more rapid on the Black Belt soil where soil thickness was less than 16 in., whereas it was more rapid on the Grenada soil where the soil thickness was greater than 20 in.

Tillage and erosion effects combine to degrade the soil and reduce potential productivity if soil organic matter becomes depleted. Soil under permanent vegetation develops a surface layer rich in organic matter that decomposes and releases nutrients for plant growth. Since organic matter is constantly being deposited on the surface as plants die or drop their leaves, an equilibrium is established between decomposition and replenishment. A part of the organic matter near the soil surface coats the mineral particles, which forms stable aggregates, and improves the soil structure, creating large particles which are difficult to erode and which resist crust formation and surface sealing. Stable channels or macropores formed by the action of earthworms, roots, and other soil organisms increase the rate of water infiltration. With tillage, all of this changes. The rate of organic matter decomposition is greatly accelerated, releasing nutrients. This was important to early settlers who used little or no fertilizer. However, as tillage also destroys continuous macropores, soil structural stability is reduced, crusting becomes more of a problem, and runoff increases. With increased runoff from a soil surface that has reduced stability, the erosion rate increases and the loss of more of the rainfall as runoff increases the likelihood of drought stress. Since most of the soil organic matter is located at or near the surface, the erosion removes a relatively larger percentage of this valuable layer. Thus, the problem feeds on itself. With continued tilled cropping, organic matter declines, infiltration rates and water-holding capacity are reduced, and productivity decreases.

During the past half-century, we have witnessed dramatic increases in crop productivity. In 1940, for example, the state average corn yield was around 15 bu/acre; in 1994, the yield exceeded 100 bu/acre, or more than a sixfold increase. Factors important in this yield improvement include improved hybrids, better control of pests, higher plant populations, timely planting, and greatly increased application of plant nutrients. Increasing crop productivity made possible through the technology of improved inputs tends to mask the continued degradation of our soil resource. However, the productivity decline witnessed in the 1970s for land brought into production and operated with essentially the same practices underscores the need for wise soil management.

In recent years, research has been directed toward defining the nature of the problem of declining productivity with continued cropping. Also, systems are being developed that maintain soil productivity with annual cropping, as well as restoring productivity for eroded sites. In Georgia, no-tillage with large inputs of organic matter from crop residue and/or cover crops was studied on a site that had been cropped for several decades and consequently had decreased productivity and reduced soil organic matter. After 4 years, crop yields for the untilled treatments were greater than for the tilled areas. Organic matter increased in the soil surface and infiltration rates were greater than for the untilled system. The scientists involved concluded that improved soil water for the crop was a dominant factor in the increased yields (Bruce et al., 1995).

Scientists in Mississippi conducted a similar study with similar results. McGregor et al., (1992) produced soybeans on a Loring soil with a 3 to 4% slope and compared no-tillage and conventional tillage. During the first 2 years of the study, conventional tillage had greater yields than no-tillage. During the fourth through the eighth years, no-tillage soybeans yielded 44% greater than did conventional soybean. A rainfall simulator was used to measure runoff and soil loss during the third, fourth, and seventh years of the study. Such a test allows soil loss comparisons with similar rainfall events which are hard to obtain with natural rainfall. Both conventional and no-tillage plots were lightly tilled prior to the simulator runs. Runoff during the initial run from the conventional plots was 21% greater than that from no-till plots in the third year of the study, 13% greater in the fourth year, and 52% greater in the seventh year. Soil loss during initial runs for these same years was 62, 34, and 350% greater, respectively, from conventional-till plots than from no-till plots. Eroded soils with a conventional-till history were more susceptible to further erosion under additional intensive tillage than soils protected through a no-till management system. These scientists concluded that the loss of productivity progressed as the erosion continued. Additional work will be needed to determine the mechanisms involved in the yield trends reported in these two studies. Nevertheless, the practical implications regarding maintenance of soil productivity are obvious.

Not all the soil eroded on larger fields moves completely off site. Some from steeper slopes moves downslope where it is redeposited. With time, subsoil is exposed on steeper slopes, and productive potential of those spots as well as the overall productivity of the field is reduced. Plant nutrients applied to the soil are often incorporated but remain near the surface. Erosion removes this nutrient-rich soil layer, necessitating the expense of increased fertilizer application.

Off-site costs of erosion are receiving increased attention by society. Sediment originating from tilled fields is deposited in streams, increasing the frequency of dredging required to keep navigable channels open, and increasing the flood hazard of all streams. Increased runoff rates from degraded land aggravate stream channel stability problems so that stream bank erosion becomes a major source of sediment. The capacity of reservoirs for flood control, for water supply, or for power generation is diminished due to deposited sediment. Nutrients attached to soil particles contribute to algae blooms and eutrophication of lakes. Pesticides may also move off site, either dissolved in runoff or attached to soil particles. Concern with these and other off-site problems has prompted the development of regulations governing practices affecting soil loss.

How erosion occurs

Erosion involves three phases: *detachment, transport,* and *deposition* (Troeh et al., 1991). Raindrop impact and flowing water are the major energy sources involved in detachment. Flowing water has energy where it concentrates and can cut channels as it moves downhill. Raindrops from intense thunderstorms have several times the energy contained in the same amount of rain falling slowly over a longer period of time such as is common during the winter season. Raindrops are distributed thoroughly over the land, whereas water flowing downhill concentrates to detach soil in localized areas.

Anything that intercepts raindrops and dissipates their impact energy before they reach the soil surface greatly reduces soil detachment. For sites covered with permanent vegetation, the soil detachment problem is minimized because leaves, stems, other plant parts, and debris may cover the soil surface completely. When a canopy forms during the growth cycle of annual crops, the erosion hazard is greatly reduced. The most critical time for erosion in tilled annual crop production is between primary tillage and canopy formation. This is why mulch cover in the form of vegetative residue is both important and effective in controlling erosion. This cover is present when the crop is planted and bridges the gap between planting and canopy formation. Another factor is that tillage loosens soil particles, greatly increasing the ease of detachment. Where runoff volumes are equal, bare untilled soil resists detachment much better than bare, tilled soil (Van Doren et al., 1984).

During intense storms, falling raindrops may splash soil particles a distance of several feet even on flat surfaces. If the rain is falling on a slope, a greater percentage of the splash movement is downhill, but flowing water is necessary for more than a token amount of transport to occur. Thus, if the rainfall infiltrates into the soil as fast as the rain falls, there will be no runoff and little transport, even though detachment is taking place. During the transport phase, a sorting of soil particles occurs. Sand and coarse silt particles require moving water or raindrop splash to remain in suspension. If the water slows, these particles settle out in a short distance. Clay-sized particles, on the other hand, settle very slowly and even slight water movement is enough to keep them suspended. Therefore, flowing water can transport these particles for miles, and lakes collecting sediment may remain turgid for weeks following a storm. Since much of the cation exchange capacity in soil resides in the clay fraction, soil is degraded by loss of clay. Mulch cover together with good soil aggregation is useful in regulating the transport process. Surface mulch slows runoff so that larger particles often are deposited at or near the site of detachment.

Predicting soil erosion by water

USLE and RUSLE

Until well into the present century, only minimal efforts had been made to study the erosion process, to estimate soil losses, and to evaluate factors important in the overall erosion process. Scientists in Missouri were leaders in soil erosion research with plots established shortly before World War I. By the late 1920s, there was research dealing with erosion at several locations. The dust bowl of the 1930s focused the attention of the nation on the seriousness of the erosion problem and prompted conservationists to initiate a search for solutions. Some 10,000 plot years of data were analyzed by Wischmeier and Smith (1978) and used to develop the Universal Soil Loss Equation (USLE) to predict soil loss from water erosion. This model has recently been updated, refined, and developed into the computer program RUSLE (Revised Universal Soil Loss Equation) that will soon be available in all USDA-NRCS field offices (Renard et al., 1994). The basic USLE formula is

$$A = RKLSCP$$

where A = average annual soil loss,
R = rainfall and runoff factor,
K = soil erodibility factor,
LS = slope length and steepness factors, considered together,
C = cropping-management factor, and
P = erosion control practices factor.

The value of each of the factors is assigned for a given field, soil, slope, crop, tillage method, and set of conservation practices used. These factors are combined in the equation and multiplied to give an estimate of annual soil loss. Since rainfall durations, amounts, and intensities vary widely from year to year, the equation is not precise for predicting losses for individual years; however, it should be viewed as a good prediction mechanism over a 30-year period. A brief description of each factor and its influence on soil loss will help in understanding the application of the prediction equation.

Rainfall
The rainfall factor (R) considers both the total amount of precipitation and the intensity of the storms. Rains associated with frontal passage are commonly much less erosive per inch of precipitation than the intense thunderstorms common in spring and summer. In practice, rainfall amounts, precipitation patterns, and storm-intensity information are collected on a regional basis and values are calculated for large areas. A single number then characterizes rainfall for the zone. In Mississippi, rainfall factors range from 300 in the northern part of the state to 600 along the coast. The larger number in the coastal area reflects hurricane activity which, although infrequent, is highly erosive. In the upper Midwest, rainfall factors may be 100 or less. If we apply these numbers to the potential for erosion, we find that with the same crop, soil, and cultural practices, we can expect from three to almost six times the soil loss in Mississippi as in the upper Midwest. This makes water erosion control much more critical for crop production in this region.

Soil erodibility
The K factor in the USLE is used to indicate the ease of detachment and transport for soil particles. This factor ranges from 0.49 for some of the soils in the Brown Loam region to 0.17 for other soils. Soils with greater clay content and stable aggregates that do not disperse readily with raindrop impact, resist detachment and do not erode readily. Soils with a coarse, porous texture that maintain high infiltration rates resist erosion because there is less runoff to transport the particles even though detachment is taking place. The K factor reflects the erosion rate that occurs when a soil is maintained in a bare fallow condition and crusts are regularly disrupted by tillage. Based on research, K factors in RUSLE vary slightly throughout the year with soils usually being most erodible in the

late winter and least erodible during late summer. Increases in soil organic matter in response to crop rotation or reduced-tillage management are reflected more in rapid changes to the *C* factor than long-term changes in the *K* factor. *K* factors and *T* values for some important Mississippi soils are shown in Table 2.1.

Slope length and steepness

The factor for length and steepness of slopes (*LS*) has a profound effect on erosion and these influences are usually considered together in the *LS* factor. When rainfall starts, water begins to infiltrate into the soil. As the rain continues, the infiltration rate is exceeded and water collects on the soil surface. The force of gravity pulls the excess water down the slope, and the steeper the slope, the faster the water moves. As it moves down the slope and collects, the energy of the flowing water dislodges soil particles and forms rills that continue to deepen. As the slope increases in length, the amount of water that collects and moves becomes greater, increasing soil movement in the process. Rapidly flowing water carries dislodged soil particles in suspension until the rate of flow slows. This might be at the foot slope of the hill, on the floodplain of a stream, or in the ocean at the mouth of a river.

Long and/or steep slopes increase the erosion potential so greatly as to make production of tilled crops impractical, especially in the midsouthern U.S. with our high climatic factor. In USLE, *LS* values were commonly selected from tables; in RUSLE, they are calculated by computer from user-supplied slope length and steepness data. In both these models, slope length is defined as the length water travels until it enters a concentrated flow channel or until deposition begins. Although slope steepness cannot be changed readily for a given field, terraces that intercept water reduce the effective slope length. *LS* values for different slope lengths and steepness are shown in Table 2.2.

Cropping methods

The influence of soil management and cropping methods on water erosion are reflected in the *C* factor. Kind of tillage, time of tillage, implements used, postemergence cultivation, crops planted, time of planting, crop sequence, residue cover of the soil surface, and changes in soil organic matter all affect the *C* factor. *C* values (numerical) are determined as the ratio of erosion resulting from specific management systems to that from the bare fallow treatment (*C* = 1) used to determine the *K* factor. In USLE, tables were the source of *C* values to be substituted in the equation. In RUSLE, "soil loss ratios" (defined the same way as the *C* factor but pertaining to only a period of the year) are calculated from a combination of subfactors including crop canopy and residue cover, and vary throughout the year. *C* factors for selected soybean management systems are shown in Table 2.3 (Wischmeier and Smith, 1978).

Currently, affecting the *C* factor offers the greatest opportunity to control soil loss through selection of different soil management options. Data compiled and used in developing the original USLE came mostly from tilled fields on different slopes, soils, and in several climatic zones and *C* values for most conventionally tilled crops ranged from 0.5 to 0.2. Much of the work in development and evaluation of no-tillage as an erosion control practice took place after the USLE was developed, so more information has been incorporated into the newer RUSLE. Soil with permanent vegetative cover develops an organic layer near the surface that coats soil aggregates and increases their stability. With tilled cropping, the organic matter declines and aggregates become easier to disperse. Rotations containing legumes, as practiced in the 1940s and earlier, not only furnished nitrogen to the grain crops, but also helped reduce erosion both by increasing cover over the cycle of the rotation and by making the soil more resistant to erosion. Research in Mississippi has

Table 2.2   Slope-Effect Table (Topographic Factor, *LS*)

% Slope	Slope length, ft												
	20	40	60	80	100	110	120	130	140	150	160	180	200
0.2	0.05	0.06	0.07	0.08	0.08	0.08	0.09	0.09	0.09	0.09	0.09	0.10	0.10
0.3	0.05	0.07	0.08	0.08	0.09	0.09	0.09	0.09	0.10	0.10	0.10	0.10	0.11
0.4	0.06	0.07	0.08	0.09	0.09	0.10	0.10	0.10	0.10	0.11	0.11	0.11	0.11
0.5	0.06	0.08	0.08	0.09	0.10	0.10	0.10	0.11	0.11	0.11	0.11	0.12	0.12
1.0	0.08	0.10	0.11	0.12	0.13	0.14	0.14	0.14	0.14	0.15	0.15	0.15	0.16
2.0	0.12	0.15	0.17	0.19	0.20	0.21	0.21	0.22	0.22	0.23	0.23	0.24	0.25
3.0	0.18	0.22	0.25	0.27	0.29	0.30	0.30	0.31	0.32	0.32	0.33	0.34	0.35
4.0	0.21	0.28	0.33	0.37	0.40	0.42	0.43	0.44	0.46	0.47	0.48	0.51	0.53
5.0	0.24	0.34	0.41	0.48	0.54	0.56	0.59	0.61	0.63	0.66	0.68	0.72	0.76
6.0	0.30	0.43	0.52	0.60	0.67	0.71	0.74	0.77	0.80	0.82	0.85	0.90	0.95
8.0	0.44	0.63	0.77	0.89	0.99	1.04	1.09	1.13	1.17	1.21	1.25	1.33	1.40
10.0	0.61	0.87	1.06	1.23	1.37	1.44	1.50	1.56	1.62	1.68	1.73	1.84	1.94
12.0	0.81	1.14	1.40	1.61	1.80	1.89	1.98	2.06	2.14	2.21	2.28	2.42	2.55
14.0	1.03	1.45	1.78	2.05	2.29	2.41	2.51	2.62	2.72	2.81	2.90	3.08	3.25
16.0	1.27	1.80	2.20	2.54	2.84	2.98	3.11	3.24	3.36	3.48	3.59	3.81	4.01
18.0	1.54	2.17	2.66	3.07	3.43	3.60	3.76	3.92	4.06	4.21	4.34	4.61	4.86
20.0	1.82	2.58	3.16	3.65	4.08	4.28	4.47	4.65	4.83	5.00	5.16	5.47	5.77
25.0	2.63	3.73	4.56	5.27	5.89	6.18	6.45	6.72	6.97	7.22	7.45	7.90	8.33
30.0	3.56	5.03	6.16	7.11	7.95	8.34	8.71	9.07	9.41	9.74	10.06	10.67	11.25
40.0	5.66	8.00	9.80	11.32	12.65	13.27	13.86	14.43	14.97	15.50	16.01	16.98	17.90
50.0	7.97	11.27	13.81	15.94	17.82	18.69	19.53	20.32	21.09	21.83	22.55	23.91	25.21
60.0	10.35	14.64	17.93	20.71	23.15	24.28	25.36	26.40	27.39	28.36	29.29	31.06	32.74

	Slope length, ft												
	400	500	600	700	800	900	1000	1100	1200	1300	1500	1700	2000
0.2	0.12	0.13	0.14	0.15	0.15	0.16	0.16	0.17	0.17	0.18	0.19	0.19	0.20
0.3	0.13	0.14	0.15	0.16	0.16	0.17	0.18	0.18	0.18	0.19	0.20	0.21	0.22
0.4	0.14	0.15	0.16	0.17	0.17	0.18	0.19	0.19	0.20	0.20	0.21	0.22	0.23
0.5	0.15	0.16	0.17	0.18	0.18	0.19	0.20	0.20	0.21	0.21	0.22	0.23	0.24
1.0	0.20	0.21	0.22	0.23	0.24	0.25	0.26	0.27	0.27	0.28	0.29	0.30	0.32
2.0	0.31	0.33	0.34	0.36	0.38	0.39	0.40	0.41	0.42	0.43	0.45	0.47	0.49
3.0	0.44	0.47	0.49	0.52	0.54	0.56	0.57	0.59	0.61	0.62	0.65	0.67	0.71
4.0	0.70	0.76	0.82	0.87	0.92	0.96	1.01	1.04	1.08	1.12	1.18	1.24	1.33
5.0	1.07	1.20	1.31	1.42	1.52	1.61	1.69	1.78	1.86	1.93	2.07	2.21	2.40
6.0	1.35	1.50	1.65	1.78	1.90	2.02	2.13	2.23	2.33	2.43	2.61	2.77	3.01
8.0	1.98	2.22	2.43	2.62	2.81	2.98	3.14	3.29	3.44	3.58	3.84	4.09	4.44
10.0	2.74	3.06	3.36	3.62	3.87	4.11	4.33	4.54	4.74	4.94	5.30	5.65	6.13
12.0	3.61	4.04	4.42	4.77	5.10	5.41	5.71	5.99	6.25	6.51	6.99	7.44	8.07
14.0	4.59	5.13	5.62	6.07	6.49	6.88	7.26	7.61	7.95	8.27	8.89	9.46	10.26
16.0	5.68	6.35	6.95	7.51	8.03	8.52	8.98	9.42	9.83	10.24	11.00	11.71	12.70
18.0	6.87	7.68	8.41	9.09	9.71	10.30	10.86	11.39	11.90	12.38	13.30	14.16	15.36
20.0	8.16	9.12	9.99	10.79	11.54	12.24	12.90	13.53	14.13	14.71	15.80	16.82	18.24
25.0	11.78	13.17	14.43	15.59	16.66	17.67	18.63	19.54	20.41	21.24	22.82	24.29	26.35
30.0	15.91	17.79	19.48	21.04	22.50	23.86	25.15	26.38	27.55	28.68	30.81	32.80	35.57

shown that with no-till management, C factors can be below 0.01 (see Table 2.4). As with all the USLE factors, the smaller the number, the greater the reduction in predicted erosion. Erosion would be 100 times larger for bare cultivated fallow than from a cropping system with a C factor of 0.01.

Table 2.3 A Comparison of C Factors for USLE Equation for Various Soybean Production Systems in Mississippi

		C-Values
1.	Continuous soybeans, no-till, Rdl (4000+ lb.), WC	0.040
2.	Continuous soybeans, no-till, Rdl (<4000 lb.), WC	0.050
3.	Soybeans no-till, doublecropped with small grain or ryegrass, Rdl (4000+ lb.)	0.060
4.	Soybeans no-till, doublecropped with small grain or ryegrass Rdl (<4000 lb.)	0.070
5.	Soybeans stubble planted, WC, Rdl (3000+ lb.), *maximum* 2 cultivations	0.100
6.	Soybeans stubble planted, WC, Rdl (<3000 lb.), *maximum* 2 cultivations	0.130
7.	Soybeans broadcast or drilled, no fall plowing, Rdl, WC, seeded preparation	0.140
8.	Soybeans broadcast or drilled, no fall plowing, Rdl, WC, seeded preparation, April 15	0.150
9.	Soybeans conventional planted, Rdl (3000+ lb.), no fall plowing, WC, seeded preparation, April 15, *maximum* 2 cultivations	0.180
10.	Soybeans conventional planted, Rdl (<3000 lb.), no fall plowing, WC, seeded preparation, April 15, *maximum* 2 cultivations	0.200
11.	Soybeans conventional planted, Rdl (<3000 lb.), WC, no fall plowing, seeded preparation, April 1, *maximum* 2 cultivations	0.220
12.	Soybeans conventional planted, Rdl (<1500 lb.), WC, no fall plowing, seeded preparation, April 15, *maximum* 2 cultivations	0.280
13.	Soybeans conventional planted, Rdl (<1500 lb.), WC, no fall plowing, seeded preparation, April 1, *maximum* 2 cultivations	0.320
14.	Soybeans conventional planted, no fall plowing, Rdl (<1500 lb.), WC, seedbed preparation, March 15, maximum 2 cultivations	0.380
15.	Soybeans conventional planted, no fall plowing, Rdl (<1500 lb.), WC, seedbed preparation, March 1, maximum 2 cultivations	0.420

WC = volunteer winter cover; Rdl = residue left.
From NCRS.

Table 2.4 C Factors and Annual Soil Loss from Brown Loam Soil Region Plots with 5% Slope

	Conventional-till		No-till		
		Soil loss,		Soil loss,	
Crop	C factor	ton/acre/year	C factor	ton/acre/year	Ref.
Sorghum	0.04	4.2	0.005	0.6	McGregor and Mutchler (1992)
Corn (grain)	0.09	7.2	0.005	0.4	McGregor and Mutchler (1983)
Corn (silage)	0.14	11.2	0.003	0.3	McGregor and Mutchler (1983)
Soybean	0.12	21.1	0.006	1.2	McGregor (1978)
Soybean	0.10	19.6	0.008	1.4	Mutchler and Greer (1984)
Cotton/vetch[a,b]	0.13	9.6	0.010	0.8	Mutchler and McDowell (1990)
Cotton[b]	0.31	31.2	0.053	5.4	Mutchler et al., (1985)

[a] With hairy vetch winter cover crop.

[b] Average of two treatments with differing previous soil management histories.

Practices

Practices, *P*, represent another means to manage soil to reduce erosion loss. These include such techniques as terraces, farming on the contour, buffer strips, and grass hedges (Meyer and Mannering, 1967, Mutchler et al., 1994). Strips of close-growing crops or grass hedges slow water movement and greatly resist erosion because of the closely spaced stems, causing larger soil particles to be deposited the same way as behind terrace impoundments

(Dabney et al., 1993a). These practices complement reduced-tillage management systems as important means of reducing erosion. *P* values of 0.5 to 0.6 are common for many practices, indicating their use would reduce the erosion potential by 40 to 50%.

Applying the USLE

In calculating the erosion potential for a given site, tabular values for the USLE components are multiplied. For example, a producer in Attala County might have a field of Grenada silt loam and want to grow soybeans. He plans to prepare the soil conventionally after March 15 and cultivate twice. The field is on a 4% slope, 400 feet long. No effort will be made to plant on the contour. The tabular values for various components of the USLE (*A* = RKLSCP) are

R = 350
K = 0.43
LS = 0.70
C = 0.38
P = 1.0

In solving the equation for average annual soil loss,

$$A = 350 * 0.43 * 0.70 * 0.38 * 1,$$

or *A* = 40 tons/acre/year, which far exceeds *T* (Table 2.1) for this soil (*T* = 3).
    If the producer plants on the contour (*P* = 0.5), the soil loss potential then becomes

$$A = 350 * 0.43 * 0.70 * 0.38 * 0.5, \text{ or } 20 \text{ tons/acre/year...,}$$

which is still too high.
    If the producer decides to grow doublecrop soybeans following wheat harvest and plant on the contour (*C* = 0.06), the estimated soil loss becomes:

$$A = 350 * 0.43 * 0.70 * 0.06 * 0.5 = 3.2,$$

which is slightly more than the allowable *T* value.
    If another field is selected with the same soil and slope length, but with a 0.5% slope, the *LS* factor becomes 0.15 and the potential soil loss with conventional tillage:

$$A = 350 * 0.43 * 0.15 * 0.38 * 1,$$

or 8.6, which is greater than 2*T* for this soil. Growers likely would not plant on the contour for fields with slopes less than 1%.

Crop management and erosion

Complete elimination of erosion is not an attainable goal since some soil is lost even from undisturbed areas through the action of geologic forces. Realistically, the reduction should be to a level that will permit sustained production without unacceptable off-site impacts over a long period of time. Control to a level that is at or below the tolerance level (*T*) for the soil in question (while producing annual crops) should be attainable in many cases. However, long, steep slopes may need terraces or other erosion-control practices. Any management practice that reduces the rainfall energy striking the soil surface, slows or reduces the flow of water over the surface, or reduces the energy of wind blowing over exposed soil also decreases soil loss.

Perennial crops such as pasture, hay meadows, or trees commonly form dense cover that is present throughout the year and protects the soil from excessive loss. On the other hand, conventionally tilled annual crops have high potential erosion hazards, especially on slopes. The problem intensifies if land preparation begins soon after harvest of a summer crop and the soil remains unprotected over winter. C factors reflect an assessment of the erosion hazard created by various crop production practices.

Tillage

Primary tillage, secondary tillage to prepare a seedbed, postemergence cultivation, implements used, depth of tillage, and amount of residue remaining on the soil surface all affect the erosion potential created by different management systems. Moldboard plowing buries almost all residue, whereas chisel plowing loosens the soil but leaves considerable soil cover. However, multiple passes with the chisel plow, disk, hipper, and field cultivator used in conventional tillage in Mississippi result in residue cover being reduced to less than 5% after planting. Physical structures such as terraces were a primary means used to reduce soil loss from cropping systems that left such low amounts of residue cover. Terraces decrease slope length, slow water movement downslope, and divert it to constructed outlets, thereby decreasing erosion. However, terraces are costly to build and maintain, and old terraces that followed contours closely are inconvenient to use with large-scale modern equipment.

During the past three decades, production systems have been developed that eliminate most or all of the tillage used in the past while maintaining crop yields on most soils. Crop residues, killed winter weeds, and crops planted to produce residue cover the soil surface and greatly reduce the potential for soil loss. These new methods also conserve moisture by increasing infiltration and reducing evaporation (Triplett et al., 1968; Van Doren and Triplett, 1973; Langdale et al., 1992). A key long-term benefit from use of these systems is that crop yields may be maintained at higher levels than with tilled management. These systems seem to be best adapted for better-drained soils located on slopes, or sites where erosion is the greatest problem (Triplett et al., 1996). However, yield reductions from erosion have been reported on some heavier and less well drained soils (Hairston et al., 1984). These findings are consistent with trends reported for other areas of the country (Dick et al., 1991).

Time since the last tillage operation is another variable that can affect the C factor. Where no-tillage is used continuously and the field has not been tilled for several years, soil resists erosion better than does soil on sites tilled during the current season. Soil loss was 90% less for an untilled silt loam soil than for soil that was tilled during the current season (Van Doren et al., 1984). Some soil crusts are resistant to erosion, so the soil loss ratios can change significantly over a period of 6 weeks after tillage (Dabney et al., 1993b). Such differences in soil loss as a function of time since tillage are accounted for only to a limited extent in RUSLE.

The magnitude of soil erosion that occurs from Mississippi crop production has been determined from more than 20 years of research. C factors and erosion rates from a series of studies that compared conventional and no-tillage production systems in Mississippi are summarized in Tables 2.4 and 2.5. Comparison with USLE C values from Table 2.3 indicate that ULSE estimates are within the range of research data reported in Tables 2.4 and 2.5. Three points should be noted from these findings:

1. Erosion is several times greater from sites cropped to cotton than from corn or sorghum. Soybean is intermediate.
2. Use of a cover crop reduces soil erosion for both conventional and no-till systems.

Table 2.5 C Factors and Annual Soil Loss from Soybean Grown on Black Land Prairie Soil Plots with 3% Slope (Hairston et al., 1984)

Tillage system	C factor	Soil loss, ton/acre/year
Conventional + fall chisel	0.31	5.5
Conventional	0.19	3.5
Reduced (no-till plant + cultivate)	0.17	3.0
No-till	0.16	2.9
Reduced-till doublecrop	0.06	1.1

From Hairston, J. E. et al., *J. Soil Water Conserv.*, 39(6), 391–395. With permission.

Table 2.6 C Factors and Annual Soil Loss from Watershed Studies in the Mississippi Delta on Land Graded to 0.2% Slope with Lengths of 550 to 600 ft

Crop	Soil type	Field Size	Years	C factor	Soil loss, ton/acre/year	Ref.
Soybean	Sharkey	39	1979–84	0.31	5.0	Murphree and McGregor (1991)
Cotton	Sharkey	39	1972–77	0.58	8.2	Murphree and Mutchler (1980)
Cotton	Sharkey	39	1973–78	—	6.2	Murphree and Mutchler (1981)
Cotton	Commerce	46	1973-78	—	7.9	Murphree and Mutchler (1981)

3. No-till reduces water erosion more on the wind-deposited brown loam soils (Table 2.4) than on black land prairie (Table 2.5) soils. Since the black land prairie data is quite limited, this conclusion may be modified in the future.

Although the greatest concern for soil loss is on steeper slopes, considerable amounts of soil can be eroded from the relatively flat topography typical in the Mississippi Delta (Table 2.6). In a series of 6-year studies on graded fields (0.2% slope) with rows running up and down the slope (average row lengths of 550 to 600 ft), annual soil erosion from cotton averaged 6 to 8 tons/acre/year. Erosion from soybean planted flat and cultivated on a Sharkey soil averaged 5 tons/acre/year. In natural landscapes, much of the sediment leaving agricultural fields is deposited in ditches and wetlands. Murphree et al., (1985) measured sediment yield (collected from runoff as it left the watershed) from a 640-acre flatland area in Washington County, MS comprising 477 acres of cotton, 86 acres of soybean, 15 acres of wheat, 54 acres of woodland (mostly depressional wetland), and 8 acres of hard-surfaced road and grassed right-of-way. They reported annual sediment yields of only 2.3 tons/acre/year, of which 85% was clay-sized particles. This indicates that up to two thirds of the sediment leaving the fields was trapped in depressional areas, with mainly dispersed clay passing from the area.

Row spacing and postplant cultivation for weed control can affect erosion from cropland. Figure 2.1 summarizes average annual soil loss for several studies in which conventional tillage and no-tillage were compared with that from reduced tillage. In these studies, reduced tillage was defined as no-till planting followed by two row cultivations for weed control. Conventional tillage consisted of chisel plowing followed by disking before planting plus cultivation. Reduced-tillage or stale-seedbed planting retained residue cover on the surface during April and May and greatly reduced annual soil erosion compared with conventional tillage. Nevertheless, row cultivation after crop emergence reduces residue cover, detaches soil particles, and increases potential erosion in June and July if intense rains occur. Dabney et al., (1993b) showed that a single cultivation on sloping brown loam soils increased potential erosion in June and July from 6 to 40 times. Thus, erosion from

Average Annual Soil Loss (t/a), Holly Springs, MS

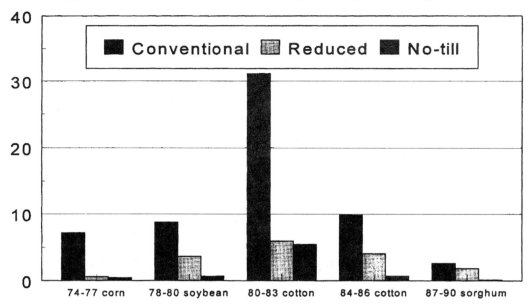

Figure 2.1 Natural-rainfall erosion from conventional-till, reduced-tillage (cultivation after no-till planting), and no-till. Performance of reduced-till relative to no-till depends on occurrence of occasional heavy rains in June or July. Refer to Table 2.4 for references.

reduced tillage can be as low as from no-till in some years, but can be much higher in others (Figure 2.1).

No-till planting of soybean with a grain drill in narrow rows creates more soil disturbance and reduces residue cover compared with planting in wider rows but erosion hazards remain small (Dabney et al., 1993). Crops planted in narrow rows form a canopy more rapidly which protects the soil surface. On some soils, the bare surface between wide soybean rows may crust following rainfall, and this reduces the infiltration rate for subsequent rainfall events. An exception to this is a class of soils called vertisols (Delta or Black Belt clays) that crack when dry. However, if mulch covers 70% or more of the soil surface, erosion is controlled and infiltration rate is maintained, even on soils prone to crusting. Thus, wide rows on crust-prone soils should be cultivated to improve rainfall infiltration and crop growth (Van Doren and Triplett, 1973), but this increases erosion potential if other erosion control measures are not being used. If weeds can be controlled without postemergence cultivation, the best management system for sites with an erosion hazard would be narrow row planting. Alternatively, runoff can be controlled and sediment trapped by a series of parallel grass hedges (Meyer et al., 1995; Dabney et al., 1995).

Crop rotation

Crop rotation can be used to reduce erosion potential. As shown in Table 2.4, some crops create more of an erosion hazard than others. Soils with soybeans and cotton growing on them may have as much as 10 to 100% greater soil loss potential than do soils with corn or sorghum. There are several reasons for this. Neither soybean nor cotton produce a large volume of residue that will cover the soil for an extended period of time. Also, soybean residue decomposes more rapidly than the stalks and leaves of corn. There is experimental evidence that, as a crop, soybean makes the soil more prone to erode. In a study with intense simulated rainstorms, soil loss was 45% greater following soybean in a corn–soybean rotation than for continuous corn at equal amounts of mulch cover (Van Doren et al., 1984).

Soybean rotated with corn and planted without tillage has greater residue cover because the residue provided by the corn persists to control erosion during the soybean crop. This system permits rotation of herbicides, problem grasses are easier to control during the year with soybeans, and some problem broadleaf weeds are easier to control in corn.

Wheat straw also provides extensive, persistent cover, making a wheat–soybean doublecrop system effective in controlling soil loss. However, for this system to function well the straw should not be burned and the soybean crop should be planted no-till. Not all doublecrop systems are as suitable as is soybean following wheat harvested for grain. Edwards (Ned Edwards, personal communication) planted soybean at the Mississippi Brown Loam Station at Raymond, MS following a winter grazing crop and reported greater yields when the soil was tilled. In this case, the yield reductions with no-tillage were possibly caused by compaction from the cattle traffic on the intensively grazed winter crop.

Cover crops

Growing vegetation is often used as a source of soil cover during the winter and spring seasons when summer annual crops are not present. This vegetation can consist of annual weeds or a crop planted specifically to provide cover. Legumes are often seeded as cover preceding corn or sorghum and not only help minimize erosion but also fix nitrogen for use by the grain crop. Legumes, however, would be inappropriate to precede soybeans, since the nitrogen is not needed and management of legumes is often more difficult than that for grass cover crops.

Where a cover crop is used, wheat or some other small grain is ideal to precede soybean. Any crop planted solely to provide cover represents a production expense without the harvested grain to provide a monetary return (Elmore et al., 1992, Heatherly et al., 1993). Thus, a doublecropping system provides both soil cover and the potential for extra income. Where the soil protection of a cover crop is needed to reduce erosion potential on slopes, drainage is usually adequate for wheat to be well adapted. Under these conditions, doublecrop soybean following wheat has been more profitable than either wheat or soybean grown as a sole crop. On a poorly drained Tunica clay, however, a wheat cover crop killed before planting did not increase the soybean yield, reduced net returns, and increased the percent winter ground cover only in years when fall tillage reduced the populations of volunteer winter annual weeds (Elmore et al., 1992).

Wind erosion

Although generally associated with the Great Plains region, wind erosion can be a very serious problem in other parts of the U.S. (Hayes, 1965; Carreker, 1966), particularly on soils with a high percentage of fine sand. As with water erosion, the three phases of wind erosion are *detachment, transport,* and *deposition.* Specific conditions are necessary for wind erosion to occur. The soil surface must be dry and exposed, the wind must have adequate velocity to begin particle movement, and specific-sized particles must be present. Interestingly, loess soils, which were wind deposited, strongly resist wind erosion because few larger particles are present.

As the wind velocity increases on exposed soil, individual particles begin to move. These spin and bounce along, striking the soil surface and providing energy to dislodge other particles until there is an avalanche of movement. Soil wind erosion rates of up to only 1 ton/acre per storm may cause serious abrasive injury to seedling crops (Hayes,

1965). This problem has been aggravated in recent years by the removal of hedgerows, which served as windbreaks, to facilitate installation of center-pivot irrigation systems. Anything that reduces wind velocity at the soil surface serves to control wind erosion. Tillage systems that maintain residue cover on the soil surface are extremely effective. Also, narrow strips of perennial or annual vegetation spaced up to 120 ft apart can be effective (Hagen et al., 1972). Irrigating to moisten the soil surface reduces wind erosion as does tillage, which roughens the soil surface and exposes clods too large for the wind to move. A computer model, Wind Erosion Prediction System (WEPS), has been developed to predict wind erosion hazards (Hagen et al., 1995).

Erosion and economics

Growers are faced with constraints in changing cultural practices. They must ensure the crop is planted on time and at the proper density. Weeds and other pests must be controlled. Soil fertility and productivity must be maintained. They must make a profit. In Mississippi, the greatest early adoption of reduced tillage was for soybeans planted as a second crop following small-grain harvest (doublecropping). Recently, reduced and no-tillage acreage has been increasing for full-season crop production. New herbicides and development of soybean varieties and other crops that tolerate postemergence applications of broad-spectrum herbicides such as glyphosate are making these systems more practical.

Several factors must be considered in the economics of conservation. Both short-term and long-term implications are important. The statement was made earlier that, for a given year, the effect of soil loss could not be easily detected. Thus, if only the current crop is considered, the value of conservation would be low. If a longer-term view is taken, systems that conserve soil and maintain crop productivity are more desirable. Production costs do not vary greatly for seed, fertilizer, tillage, and weed control for different management systems. Thus, profits are driven largely by crop yields. Building structures such as terraces to conserve soil is expensive and the cost must be amortized over the life of the structure and added to the cost of producing the crop. Conservation tillage systems, on the other hand, are very effective in conserving soil and require little additional expense other than purchase of a planter that will operate in untilled soil. On long slopes, however, conservation tillage systems may not adequately control soil loss and terraces may be needed to break the slope length. Grass hedges are also inexpensive relative to terraces. Where hedges can be established from seed, costs may be only $0.03/ft, compared to $2.00/ft for terraces (DeWald et al., 1996).

Pesticide and nutrient movement

Off-site movement of pesticides and fertilizer nutrients applied in crop production is an area of continuing concern. Herbicides dominate the pesticides used by soybean producers, and include products that are applied both preemergence and postemergence. When herbicides are applied, a part of the application may be intercepted by growing plants or crop residue while another portion strikes the soil surface. The herbicide begins degrading as soon as it is applied. Some may be detoxified by photodecomposition, while other portions may be lost by volatilization and in runoff. Some part of the herbicide applied may be attached so tightly to clay or organic matter particles that it is no longer active. Most herbicides are also decomposed by microorganisms. The term *half-life* is used to indicate the amount of time required for one half of the amount initially present to become ineffective. This ranges from a few days to a few weeks, depending on the pesticide in question.

Table 2.7   Herbicide Losses Measured in Runoff (as percent of total applied
to doublecropped soybean), Black Belt Area

Year	DAT[a]	Metribuzin		Metolachlor	
		Conventional tillage	No tillage	Conventional tillage	No tillage
1989	6	1.7	0.7	0.8	0.4
1990	6	1.6	0.5	1.1	0.2
1991	5	0.9	0.7	0.6	0.6
Average		**1.4**	**0.6**	**0.9**	**0.4**

[a] Days after treatment for first runoff event.

From Webster, E. P. and Shaw, D. R. 1996a. Off-site runoff losses of metolachlor and metribuzin applied to differing soybean (*Glycine max*) production systems, *Weed Technol.* 10: 556–564.

Fertilizer may be broadcast onto the soil surface, banded, or, with conventional tillage, incorporated into the soil. Phosphorus is a fertilizer element associated with surface water quality problems because of its contribution to algae blooms. In contrast, nitrate nitrogen is the fertilizer form of most concern in groundwater. If incorporated into the soil, phosphorus is strongly bound to soil and moves primarily attached to sediment. Nitrate nitrogen is not bound to the soil, but moves freely with runoff and drainage waters.

Nationally, considerable research has been directed toward characterizing nutrient and herbicide movement in runoff, determining the presence of these in groundwater, and developing some sense of the hazards posed by this movement. For soil-incorporated herbicides and fertilizers, only a fraction of the total applied is available for movement in runoff. For surface applications, the pesticides and nutrients present on vegetation, crop residue, and soil particles at the surface can be detached and moved in runoff. Leaching of crop residues remaining on the soil surface can contribute soluble nutrients to runoff waters. Thus, while no-till management reduces sediment and total nitrogen and phosphorus losses from soybean cropland, losses of soluble phosphorus may be increased (McDowell and McGregor, 1980; Schreiber et al., 1993). Groundwater and runoff nitrate concentrations have been similar from no-till and conventional tillage management systems.

Movement of pesticides in runoff follows a predictable pattern. Some pesticides are strongly bound to soil while others are not and very small amounts move attached to soil particles. For both types, the highest pesticide concentration is found in the first runoff event following application and declines greatly with subsequent storms (Triplett et al., 1978; Shaw et al., 1992; Webster and Shaw, 1996a). Pesticide concentration is highest when runoff occurs within 1 to 2 days following application. Concentration declines rapidly with time after application, even if no rainfall or runoff occurs. Thus, if 10 days to 2 weeks elapse between application and a runoff event, pesticide concentration and off-site movement will be only a fraction of that expected for a runoff event soon after application. The pesticide may be present at very low levels in runoff throughout the growing season; analytical methods are quite sensitive and can detect parts per billion (ppb) or parts per trillion (ppt) of many common pesticides.

Off-site movement of metribuzin (Sencor, Lexone) and metolachlor (Dual), two common soybean herbicides, was studied over a several-year period on runoff plots at the Black Belt Experiment Station (Webster and Shaw, 1996a). The site was equipped with a rainfall simulator so that rainfall, either natural or created, would occur soon after herbicide application. Results from this study are shown in Table 2.7. For the 3-year period of the study, seasonal losses of both herbicides were below 2% of the total applied. Metribuzin losses were slightly greater than metolachlor losses and losses from conventional tillage systems were slightly more than from no-tillage. It should be noted that in each of the 3 years, runoff did not occur until 5 or 6 days after herbicide application.

Table 2.8 Herbicide Losses Measured in Runoff as
Percent of Total Applied, Black Belt Area

Year	DAT[a]	No-till Doublecrop	Conventional
Metribuzin (Lexone, Sencor), no filter strip			
1991	5	0.5	1.8
1992	2	7.8	1.1
1993	2	11.9	0.9
Average		**6.7**	**1.3**
Metribuzin (Lexone, Sencor), with filter strip			
1991	5	0.4	0.8
1992	2	6.1	0.6
1993	2	1.2	0.7
Average		**2.6**	**0.7**
Metolachlor (Dual), no filter strip			
1991	5	0.3	1.0
1992	2	3.5	1.8
1993	2	4.3	0.6
Average		**2.7**	**1.2**
Metolachlor (Dual), with filter strip			
1991	5	0.3	0.5
1992	2	2.5	1.2
1993	2	0.5	0.3
Average		**1.1**	**0.7**

[a] Days after treatment for first runoff event.

Source: Webster, E. P. and Shaw, D. R. 1996b. Impact of vegetative filter strips on herbicide loss in runoff from soybean (*Glycine max*), *Weed Sci.* 44: 662-671.

In another study at the same location, 6-ft-wide strips of fescue sod were established between the water sample collection area and the area planted to soybean and treated with herbicides (Webster and Shaw, 1996b). Thus, as rainfall collected on the soil surface, the runoff passed over an untreated area covered with grass. Plots without filter strips were also present in the study and metribuzin and metolachlor were the herbicides applied. This study was conducted for 3 years and results are shown in Table 2.8. For 2 years of this study, runoff occurred 2 days after herbicide application and off-site movement was several times greater than in the earlier study, illustrating the principle of greater movement of pesticides in runoff soon after application. Movement of metribuzin was greater than that of metolachlor as in the earlier study, but the tillage effect was reversed. Having a narrow band of sod as a filter strip decreased movement of both herbicides by half or more.

Losses of both metribuzin and metolachlor in runoff and attached to sediment was also monitored on a Grenada–Loring soil complex in northwestern Mississippi (Smith et al., 1995). In this study, the herbicides were applied to watersheds several acres in size. During the first year of the study, a no-tillage watershed was used while in the second through fourth years, both conventional and no-tillage watersheds were treated. In the first year, total seasonal losses of both herbicides was about 4%. In the second year, with rain that created over an inch of runoff falling within 5 days of treatment, losses were in the range of 20% for metribuzin and 10% for metolachlor (Table 2.9). Almost all of the losses were from herbicide dissolved in the runoff, while less than 1% moved attached to the sediment. Losses for the second year are much greater than are commonly reported or than were observed in the third and fourth year when runoff-producing rainfall did not occur for at least 2 weeks.

Table 2.9 Herbicide Losses as Percent of Total Applied,[a] Brown Loam Soil Area

Year	DAT[b]	Metribuzin (Lexone, Sencor)		Metolachlor (Dual)	
		No tillage	Conventional	No tillage	Conventional
1990	6	3.9	—	3.8	—
1991	5	19.5	22.7	8.6	11.0
1992	14	2	1.5	0.8	1.5
1993	25	0.2	0.04	0.1	0.07

[a] Herbicide loss in runoff and sediment was measured, although very little was found attached to sediment.

[b] Days after treatment for first runoff event.

From Smith, S., Jr. et al., *Trans. ASAE*, 38, 1061–1068. With permission.

On an individual field basis, pesticide losses can be as high as 20% of the amount applied, with concentrations as high as 2000 to 3000 ppb, if a large storm creating runoff occurs shortly after application. On a watershed basis, fields are treated over a period of time and large runoff events are infrequent, so the dilution of pesticides by runoff from untreated areas and fields treated at other times reduces the pesticide concentration impact in larger streams. Similarly, soluble phosphorus is readily adsorbed by low-P sediment from eroding stream banks during large storms, thereby reducing the impact of this nutrient on receiving water bodies. Concentrations of metolachlor (Dual) in the range of 2 to 3 parts per million (ppm) are required to kill some fish species, and expected levels are much less, even at the field edge. With dilution by runoff from untreated areas, the effect on nontarget organisms has been minimal. Nevertheless, the best management practice is to apply pesticides and nutrients at times and in ways such that off-site movement is minimized.

Interpretive summary

Rainfall intensities and amounts in Mississippi create an erosion hazard that can remove 10 to 100 tons of soil annually when conventional tillage systems are used to produce crops on sloping sites. This is several times greater than the soil loss tolerance level for our soils, and over a few years it will degrade soil both chemically and physically, thereby reducing its productive potential. Conventional soybean production poses a greater erosion hazard than does most other annual crops, thus intensifying the problem. During the last few years, conservation, stale seedbed, and no-tillage management systems have been developed that maintain mulch cover on the soil and reduce soil loss by 90 to 95% when compared with conventional practices. These systems can reduce soil loss potential below the soil loss tolerance level on many sites and permit annual crop production while maintaining long-term productivity of the soil resource.

References

Bruce, R. R., G. W. Langdale, L. T. West, and W. P. Miller. 1995. Surface soil degradation and soil productivity restoration and maintenance, *SSSA J.* 59:654–660.

Carreker, J. R. 1966. Wind erosion in the southeast, *J. Soil Water Conserv.* 11:86–88.

Dabney, S. M., K. C. McGregor, L. D. Meyer, E. H. Grissinger, and G. R. Foster. 1993a. Vegetative barriers for runoff and sediment control, in Mitchell, J. K., Ed., *Integrated Resources Management and Landscape Modification for Environmental Protection,* American Society of Agricultural Engineers, St. Joseph, MI, 60–70.

Dabney, S. M., C. E. Murphree, and L. D. Meyer. 1993b. Tillage, row spacing and cultivation affect erosion from soybean cropland, *Trans. ASAE* 36:87–94.

Dabney, S. M, L. D. Meyer, W. C. Harmon, C. V. Alonso, and G. R. Foster. 1995. Depositional patterns of sediment trapped by grass hedges, *Trans. ASAE* 38:1719–1729.

DeWald, C. T., S. Bruckerhoff, S. M. Dabney, J. Douglas, J. Henry, J. C. Ritchie, D. Shepherd, and D. Wolf. 1996. Guidelines for the establishment of warm season grass hedges for erosion control, *J. Soil Water Conserv.* 51:16–20.

Dick, W. A., E. L. McCoy, W. M. Edwards, and R. Lal. 1991. Continuous application of no-tillage to Ohio soils, *Agron. J.* 83(1):65–73.

Elmore, C. D., R. A. Wesley, and L. G. Heatherly. 1992. Stale seedbed production of soybeans with a wheat cover crop. *J. Soil Water Conserv.* 47(2):187–190.

Hagen, L. J., E. L. Skidmore, and J. D. Dickerson. 1972. Designing narrow strip barrier systems to control wind erosion. *J. Soil Water Conserv.* 27:269–272.

Hagen, L. J., L. E. Wagner, and J. Tatarko. 1995. Wind erosion prediction system (WEPS), in USDA Wind Erosion Prediction System Technical Description Beta Release 95-08, Wind Erosion Research Unit, USDA-ARS, Manhattan, KS, chap. 1.

Hairston, J. E., J. O. Sanford, J. C. Hayes, and L. L. Reinschmiedt. 1984. Crop yield, soil erosion, and net returns from five tillage systems in the Mississippi Blackland Prairie, *J. Soil Water Conserv.* 39(6):391–395.

Hairston, J. E., J. O. Sanford, F. E. Rhoton, J. G. Miller, and K. B. Gill. 1989. Effects of Soil Depth, Organic Matter, and Rainfall on Soybean Yield in the Mississippi Blackland Prairie, MAFES Tech. Bull.163, 7 pp.

Hayes, W. A. 1965. Wind erosion equation useful in designing northeastern crop protection, *J. Soil Water Conserv.* 10:153–155.

Heatherly, L. G., R. A. Wesley, C. D. Elmore, and S. R. Spurlock. 1993. Net returns from stale seedbed plantings of soybean (*Glycine max*) on clay soil, *Weed Technol.* 7:972–980.

Langdale, G. W., L. T. West, R. R. Bruce, W. P. Miller, and A. W. Thomas. 1992. Restoration of eroded soil with conservation tillage, *Soil Technol.* 5(1):81–90.

McDowell, L. L. and K. C. McGregor. 1980. Nitrogen and phosphorus loss in runoff from no-till soybeans, *Trans. ASAE* 23:643–648.

McGregor, K. C. 1978. C factors for no-till and conventional-till soybean from plot data, *Trans. ASAE* 21:1119–1122.

McGregor, K. C. and C. K. Mutchler. 1983. C factors for no-till and reduced-till corn, *Trans. ASAE* 26:785–788, 794.

McGregor, K. C. and C. K. Mutchler. 1992. Soil loss from conservation tillage for sorghum, *Trans. ASAE* 35:1841–1845.

McGregor, K. C., C. K. Mutchler, and R. F. Cullum. 1992. Soil erosion effects on soybean yields, *Trans. ASAE* 35:1521–1525.

Meyer, L. D. and J. V. Mannering. 1967. Tillage and land modification for water erosion control, in *Tillage for Greater Crop Production*, American Society of Agricultural Engineers, St. Joseph, MI, 58–62.

Meyer, L. D., S. M. Dabney, and W. C. Harmon. 1995. Sediment-trapping effectiveness of stiff-grass hedges, *Trans. ASAE* 38:809–815.

Miller, F. P., W. D. Rasmussen, and L. D. Meyer. 1985. Historical perspective of soil erosion in the United States, in *Soil Erosion and Crop Productivity*, R. F. Follett and B. A. Stewart, Eds., ASA-CSSA-SSSA, 23–48.

Murphree, C. E. and K. C. McGregor. 1991. Runoff and sediment yield from a flatland watershed in soybeans, *Trans. ASAE* 34:407–411.

Murphree, C. E. and C. K. Mutchler. 1980. Cover and management factors for cotton, *Trans. ASAE* 23:585–588, 595.

Murphree, C. E. and C. K. Mutchler. 1981. Sediment yield from a flatland watershed, *Trans. ASAE* 24:966–969.

Murphree, C. E., C. K. Mutchler, and K. C. McGregor. 1985. Sediment yield from a 259-ha flatlands watershed, *Trans. ASAE* 28:1120–1123.

Mutchler, C. K. and J. D. Greer. 1984. Reduced tillage for soybeans, *Trans. ASAE* 27:1364–1369.

Mutchler, C. K. and L. L. McDowell. 1990. Soil loss from cotton with winter cover crops, *Trans. ASAE* 33:432–436.

Mutchler, C. K., L. L. McDowell, and J. D. Greer. 1985. Soil loss from cotton with conservation tillage, *Trans. ASAE* 28:160–163, 168.

Mutchler, C. K., K. C. McGregor, and R. F. Cullum. 1994. Soil loss from contoured ridge-till, *Trans. ASAE* 37:139–142.

Renard, K. G., G. R. Foster, D. Yoder, and D. McCool. 1994. RUSLE revisited: Status, questions, answers, and the future, *J. Soil Water Conserv.* 49:213–220.

Rhoton, F. E. 1990. Soybean yield in response to various depths of erosion on a fragipan soil, *SSSA J.* 54:1073–1079.

Schreiber, J. D., S. Smith, Jr., and R. F. Cullum. 1993. Pesticides and nutrients in southern U.S. shallow ground water and surface runoff, *Water Sci. Technol.* 28:583–588.

Shaw, D.R., C.A. Smith, and J.E. Hairston., 1992. Impact of rainfall and tillage systems on off-site herbicide movement, *Commun. Soil Sci. Plant Anal.* 23(15&16):1843–1858.

Smith, S., Jr., J. D. Schrieber, and R. F. Cullum. 1995. Upland soybean production: surface and shallow groundwater quality as affected by tillage and herbicide use, *Trans. ASAE* 38:1061–1068.

Triplett, G. B., Jr., D. M. Van Doren, Jr., and B. L. Schmidt. 1968. Effect of corn (*Zea mays* L.) stover mulch on no-tillage corn yield and water infiltration, *Agron. J.* 60:236–239.

Triplett, G. B., Jr., B. J. Conner, and W. M. Edwards. 1978. Transport of atrazine and simazine in runoff from conventional and no-tillage watersheds, *J. Environ. Q.* 7:77–84.

Triplett, G. B., Jr., S. M. Dabney, and J. H. Siefker. 1996. Tillage systems for cotton on silty upland soils, *Agron J.* 88:507–512.

Troeh, F. R., J. A. Hobbs, and R. L. Donahue. 1991. *Soil and Water Conservation*, Prentice-Hall, Englewood Cliffs, NJ.

Van Doren, D. M., Jr., and G. B. Triplett, Jr. 1973. Mulch and tillage relationships in corn culture, *SSSA Proc.* 37:766–769.

Van Doren, D. M., Jr., W. C. Moldenhauer, and G. B. Triplett, Jr. 1984. Influence of long-term tillage and crop rotation on water erosion, *SSSA J.* 48:636–640.

Webster, E. P. and Shaw, D. R. 1996a. Off-site runoff losses of metolachlor and metribuzin applied to differing soybean (*Glycine max*) production systems, *Weed Technol.* 10: 556–564.

Webster, E. P. and Shaw, D. R. 1996b. Impact of vegetative filter strips on herbicide loss in runoff from soybean (*Glycine max*), *Weed Sci.* 44: 662–671.

Wischmeier, W. H. and D. D. Smith. 1978. Predicting Rainfall and Erosion Losses, USDA-SEA Agric. Handbook 537.

chapter three

Variety selection, planting date, row spacing, and seeding rate

Larry G. Heatherly, Alan Blaine, Harry F. Hodges,
Richard A. Wesley, and Normie Buehring

Contents

Variety selection

General information

There has been a proliferation of soybean varieties in recent years, and many excellent varieties with diverse characteristics are now available. Because of differing environments, no single variety is superior to all others; however, in most situations there are varieties available that are adapted to specific growing environments.

Decisions regarding the choice of soybean varieties should utilize the information contained in pertinent state variety trials. All states in the midsouthern U.S. conduct variety trials at multiple locations within each state. These trials are usually the most complete yield comparison of available soybean varieties. In addition to yield information, these trials often offer supplemental information that can further aid in variety selection. This includes soil texture of the site on which the test was conducted, planting date, irrigated vs. nonirrigated, date of maturity, plant height at maturity, standability or lodging resistance, and a rating of resistance to diseases and nematodes.

The most important criterion for selecting a variety is yield, but all information about additional traits should be considered. Also, it is best to use multiyear averages of yield whenever this information is available. Select the best varieties from the variety trial with soils as similar as possible to your intended planting situation. Plant these varieties initially on small areas or fields to validate their yield potential on your farm. Remember that these varieties may perform at a lower level on your farm than at the variety trial locations.

State variety trials are an excellent guide, but they are not the only source of varietal information. Company strip tests, variety tests conducted by seed dealers in close proximity to the area to be planted, or a variety test on your or a neighbor's farm are all excellent sources of both observed and printed comparisons. Any variety trial that can be seen and observed during the growing season by a producer is a valuable resource for variety selection information. Many seed companies conduct field days and tours of their variety trials, and these opportunities are also valuable for seeing side-by-side comparisons of many brands of soybean varieties. Seed company literature is the best source of information about the traits of a variety for such things as maturity date and resistance/tolerance to important diseases and nematodes. Varied reputable sources of information regarding soybean variety performance should be used so that as many environments as possible are represented in the decision of which varieties to plant.

A first step in selecting a variety for use in a given operation is determining the texture of the soil that comprises the major portion of a field that will be used for soybean production. The importance of this is discussed in a later section of this chapter. Second, determine the number of acres that will be irrigated since this can affect growth and yield potential. A knowledge of field history — past crops, known disease and nematode problems, surface and internal drainage problems, proneness to flooding, and knowledge of pH and fertility factors — should be considered when selecting a variety. When selecting a variety for a given field, every effort should be made to match the variety with the strengths and weaknesses of that field. Although yield is extremely important, maximum yield potential may have to be sacrificed in order to avoid some potentially devastating problem associated with a field.

In nonproblem fields, choose the highest-yielding varieties, or varieties with a history of consistently high average yields on the soil type that comprises that field. Do not hesitate to try new varieties with a high short-term yield history, but do so on a limited basis until it is conclusively proved that these new varieties will perform at a high level over the long term. Do not abandon older varieties that continue to yield well until yield trials and field experience show a continued advantage for new varieties.

Varietal resistance to pests

Immunity, resistance, and field tolerance of soybean varieties to diseases should be understood. Immunity is the failure of a plant to contract a disease when the causal agent is present. Soybeans, for instance, are immune to infection with southern corn leaf blight because they are not a host plant; that is, this organism will not infect a soybean plant. Soybeans are rarely, if ever, immune to any disease organism for which they serve as a host. If a soybean variety contracts a disease, the causal organism will either cause an adverse reaction in the plant (susceptible) or cause no appreciable reaction (resistance) because of a trait of the host plant. Resistance and field tolerance are the ability of a plant to resist the detrimental effects of a disease even though the disease-causing organism(s) does infect the plant. Resistance usually implies that the plant will not be affected adversely by a single race of a multirace disease, such as phytophthora root rot, while field tolerance indicates that a particular soybean variety will perform at or near maximum potential in the presence of multiple races of the disease organism. In this regard, a variety with field tolerance to a multirace disease such as phytophthora would be preferable to a variety having resistance to race *x* if the field area also has a history of race *y* infestation.

Disease resistance is important when selecting a variety, but not all diseases result in similar losses. A disease such as stem canker can be devastating whenever it occurs, while a disease such as phytophthora root rot may only cause slight yield losses on silt loam and sandy-textured soils. Resistance to the leaf disease frogeye leaf spot is present in many varieties, and should be considered when planting in areas with a known history of infestation since susceptibility will result in yield reduction. Resistance to viruses is becoming an important selection criterion.

There is no known resistance in soybeans to some diseases. Rhizoctonia foliar blight, for example, is an important disease in some areas and infestations will result in yield loss, but resistance has not been identified. Differing levels of susceptibility are known and selection of a less-susceptible variety may lessen the yield loss caused by this disease.

In the variety selection process, verify that a particular disease is prevalent in your area or on your soil type, and contact a plant pathologist to find out the potential yield loss from infestation. If the projected yield loss from infestation by the disease is less than the difference in yield between a higher-yielding susceptible variety and a lower-yielding resistant variety, the susceptible variety should be planted. Varietal resistance to some diseases does not exist, and for these diseases, resistance cannot be a factor in variety selection. Other methods of control of these diseases, such as crop rotation or pesticides, must be used.

Resistance to soybean cyst nematode (SCN) is a varietal trait that is important for plantings on soils with a texture of silt loam or coarser. All available varieties should have SCN resistance information available for them. SCN is not a significant problem for soybeans planted on clay soils; therefore, varietal resistance to this pest on clay soils is not a requirement. Soybean variety selection for SCN-infested fields should be combined with a rotational scheme involving a nonhost plant in order to avoid long-term problems from SCN infestation. Soil sampling and nematode analysis can be conducted to determine the presence of SCN, and the nematode race can be determined by qualified laboratories. Varieties may be resistant to specific races, and knowledge of the predominant races in a field is important for selecting varieties with the proper resistance.

The expansion of soybean production onto sandy, sandy loam, and silt loam soils (traditional cotton soils) in some areas of the midsouthern U.S. has created a need to address the potential damage caused by the reniform nematode. Since soybeans normally have not been grown following cotton, resistance to reniform nematode previously has not been a concern. Varietal resistance and crop rotation are the best cultural practices for control of this pest. Recent screening programs have resulted in a partial list of resistant varieties (Chapter 16), and this list will expand as further screening is conducted. If soybeans are to be planted behind cotton on soils with known reniform infestations, varieties should be selected with reniform resistance in mind. A variety that is tolerant to reniform should be used if possible, and soil samples should be taken in the fall to determine what control strategy should be used the following year. If preferred soybean varieties are not tolerant, or the tolerance of these varieties is unknown, then rotation to a grass crop following cotton is the best alternative method for controlling reniform infestations in future soybean production systems. In fact, 1 year of a grass crop such as corn provides a significant reduction in reniform numbers, and 2 years of a grass crop will virtually eliminate the nematode.

Variety selection related to planting date

Early plantings

Recent research indicates that planting soybeans in some locations in the midsouthern U.S. in April and early May will result in yields and profits that are significantly greater than those obtained from later plantings. When planting in April and early May, varieties within both MG IV and V should be used. Maturity date and plant height at maturity should be considered since these two factors determine harvest time and harvestability. When planted by mid-April, MG IV varieties such as DP 3478 and Dixie 478 that develop normally (optimum rainfall or irrigation) will mature about Sept. 1 to 5, while MG V varieties such as Hutcheson, A 5979, DP 3588, HY 574, and P 9594 that develop normally will mature about Sept. 10 to 15. April plantings of MG IV varieties that are not irrigated may reach harvest maturity 7 to 10 days earlier than those that are irrigated if rain was sufficient during pod-set and pod-filling stages to ensure an adequate fruit load. These varieties reach harvest maturity before damaging populations of defoliating insects develop. MG IV varieties that are planted in early to mid-April and that develop normally can be harvested in early September, while MG V varieties planted at this time and that develop normally can be harvested in mid- to late September. MG IV varieties should be harvested as soon as possible after maturity in order to avoid problems associated with late-emerging weeds and shattering.

Conventional plantings

Varieties within the MG V range of maturity usually result in the highest yield from May and June plantings compared to varieties from earlier or later maturity groups. MG IV varieties can still be planted, but their yields often are slightly to moderately lower than yields from MG V varieties. When planted about May 10 to 15, MG IV varieties such as DP 3478 and Dixie 478 mature about Sept. 15 to 20, while MG V varieties such as Hutcheson, A 5979, DP 3588, HY 574, and P 9594 mature about Sept. 25 to 30.

Doublecrop and ultralate plantings

Soybeans in doublecrop systems are planted from late May (southern parts of region) to mid- and late June (central and northern parts of region) following a winter small grain crop. MG VI or later varieties have been used in these plantings in the past, but planting date studies show an advantage for MG V varieties in the central and northern sections. MG IV varieties may be planted at this late date, but their yields will be slightly to moderately lower than those from MG V varieties. The advantage from planting MG IV varieties behind wheat is the early-October harvest date allows timely planting of wheat in a continuous soybean/wheat doublecrop system or in a rotational doublecrop system such as corn/wheat/soybean. Planting MG VI or later varieties after a winter small grain crop leads to later maturity, and this will result in late October/November harvest in the central and northern sections. This results in the small grain being planted after the optimum time in an unprepared seedbed, or not being planted at all if wet soil results in rutting of the field by the combine during soybean harvest. MG VI and later varieties are recommended for Louisiana plantings made after June 15.

Spring flooding of low-lying areas in the midsouthern U.S. sometimes causes soybean planting to be delayed until July/early August. This can involve significant acreage in some years. In these late plantings, row spacing should be narrowed to 20 in. or less. MG IV and V varieties should be planted at these late dates to avoid the ultralate harvest that will result if MG VI varieties are planted in these situations.

Late plantings of soybeans may be subject to infestations by leaf- and pod-feeding insects. Most defoliation of economic importance to soybeans occurs by insects feeding in late summer/early fall when plants in late plantings are susceptible. Yield losses caused by defoliation are influenced by degree of defoliation and crop growth stage when defoliation occurs. Defoliation studies showed no yield decreases when plants were defoliated of up to 33% of leaf area prior to bloom. However, plants in late plantings will more than likely be setting and filling pods when normal late-season insect infestations occur. Defoliation during this time can reduce yield, especially in irrigated fields or fields that have not experienced appreciable drought stress. Determinate and indeterminate varieties are affected similarly by defoliation. Research has shown that complete defoliation at R5 (beginning seed fill) can reduce yields by 80%, while complete defoliation at R6 (full seed) can reduce yields by 40%. At R7 (beginning maturity), complete defoliation will not significantly reduce yields. Complete defoliation is a drastic condition that may occur in a relatively short period in late plantings because of high densities of one or more species of foliage-feeding insects. Therefore, fields of late-planted soybeans should be scouted closely, and insecticides should be applied to control infestations that may quickly increase to yield-reducing levels, especially in fields with a high yield potential. Late plantings located near corn fields may be especially susceptible to pod feeders and should be scouted closely for infestations of these pests, especially in the southern parts of the region.

Variety selection related to soil texture

Soil texture determines the amount of water available to plants and the speed at which that water will move to roots and be withdrawn by growing plants. Fine-textured soils such as clays have a relatively low amount of water available to plants, and the rate of withdrawal of water by plants is relatively slow. These factors translate into relatively slow growth of soybeans on clay soil with resulting smaller leaves and shorter plants. Conversely, coarse-textured soils such as sandy loams and silt loams have a relatively large amount of water available to plants, give this water up relatively easily, and support rapid growth that translates into larger leaves and taller plants. Therefore, a soybean variety that will grow 30 to 36 in. tall on clay (fine-textured) soil may grow too tall and lodge on a coarser-textured soil. A variety that only gets 18 to 24 in. tall on a silt loam (coarser-textured) soil probably will be too short when grown on clay. It is generally considered that mature plants ranging from 20 to 40 in. tall are acceptable if the proper row spacing was chosen to ensure full canopy development on the low end and that the proper seeding rate was chosen to ensure no lodging on the upper end.

Pest prevalence can be affected by soil texture. For example, phythophthora root rot is a prevalent disease on soybeans grown on clay soils, but may never appear as a significant problem on coarse-textured soils. Soybean cyst nematode, on the other hand, is a significant pest on silt loam and coarser-textured soils, but not on clays. Thus, selection of varieties with resistance to pests should be tempered with knowledge of the soil on which the soybeans will be grown and the likely occurrence of particular pests on these soils.

Selection of Roundup Ready® varieties

Roundup Ready (RR) soybean varieties will probably result in less expense for herbicide weed control, but because of the technology fee, the total cost for weed control may be the same as that for conventional varieties. Therefore, yield potential of RR varieties should be the same as that of conventional varieties in order to realize economic benefit from their use. RR varieties will be preferred for those sites that have problem weeds that conventional weed management systems may not control. In fields where morningglories, teaweed (prickly sida), and hemp sesbania are problematical, additional herbicides may be required to supplement the RR technology. Thus, in these instances additional production costs will be incurred.

Other considerations

Tolerance to herbicides should be considered in variety selection. Some varieties are sensitive to metribuzin, the active ingredient in such herbicides as Sencor, Lexone, and Canopy. Other varieties are sensitive to the active ingredient in Authority Broadleaf. Some varieties are sensitive to the sulfonylurea class of herbicides such as Classic, and some companies have sulfonylurea-tolerant soybean (STS) varieties to counter this. Before selecting a variety for a given environment, read both seed company literature and herbicide labels to determine if the desired variety is sensitive to any of the herbicides that will be used.

Planting date

Early Soybean Production System (ESPS)

Generally, plantings of MG IV and V soybean varieties made in April and early May vs. later result in higher yields and greater net returns in most years. These early plantings, especially of MG IV varieties, are more likely to avoid the late July through early September droughts that are common in the midsouthern U.S. Most of the acreage in the region can be planted in April if a stale seedbed or no-till planting system is used. Plantings made this early will be subject to *Pythium* injury, and should be treated with Apron XL to prevent significant emergence failure and stand loss. Details about ESPS management are presented in Chapter 8.

Desired maturity date

Selection of variety and planting date should be made with desired maturity and harvest dates in mind. A total farm harvest schedule involving all crops should be planned at the beginning of each crop season. Soybean varieties and planting dates should be chosen to match the harvest capacity of the farm equipment and labor complements. April plantings of MG IV varieties that are not irrigated may reach harvest maturity in mid- to late August, and this could conflict with corn and/or rice harvest. Mid- and late April plantings of MG V varieties will reach harvest maturity after mid-September, and this may conflict with cotton harvest. MG V varieties often will be harvested too late to obtain any early delivery price bonus that may be available. In order to realize the total benefits from early planting of MG IV varieties, custom harvesting of some of the crop mix may be desirable.

Irrigated vs. nonirrigated

Over the long term, both irrigated and nonirrigated soybeans produce greater yields and net returns when planted in April and early May vs. later. However, since drought avoidance is the main objective of April planting of soybeans, nonirrigated acres probably should be planted early at the expense of irrigated acres if both are in the farm mix and there is not enough time to plant all acres in April and early May. Irrigation will overcome the effects of drought and most of the effects of later planting, but only early planting offers any real hope for overcoming the effects of drought when irrigation is not an option. Irrigated MG V varieties usually require one or two more irrigations than MG IV varieties planted at a comparable time.

Row spacing

Yield

Soybeans grown in narrow rows (20 in. or less) will sometimes, but not always, yield more than the same varieties grown in wide rows (30 to 40 in.). The yield advantage of narrow rows will be more pronounced when yield levels are relatively high, and in June and later plantings of all varieties. Tall-statured MG V soybeans planted in April and May in wide rows and not irrigated may be more economical to produce since yields from the two row spacings will often be similar and cost of production in wide row systems can be less, especially if cultivation is used in conjunction with herbicides banded over the row.

It is generally assumed that soybeans grown in narrow rows will form a quicker canopy, intercept essentially all available light, and therefore aid in soil moisture conservation because the resulting shade theoretically reduces evaporation. Soybeans grown in narrow rows often form a quicker canopy, but by this time the evaporation component of evapotranspiration is negligible and most of the water used by soybeans after full canopy development is through transpiration. Soybeans grown in narrow rows will likely use more early-season water, especially if stands are optimum and uniform. This means that dryland plantings grown in narrow rows may deplete scarce soil moisture more rapidly and thus suffer more severely from drought later in the season. Narrow-row irrigated soybeans may require more irrigation water.

Choice of row spacing should not be based solely on the presumption that narrow-row soybean systems will yield more than wide-row systems. The majority of row-spacing research in the midsouthern U.S. has shown only a slight yield advantage for narrow rows in conventional plantings (MG V and later varieties planted in May or later). This small yield advantage must be measured against the economics of each system, since production costs in narrow-row systems can be greater.

Management decisions related to row spacing

The row spacing for soybean production should be matched with that used for other crops in the total farm operation. For example, corn and cotton may be grown in 30-in.-wide rows; thus, soybeans rotated with these crops should be planted in this row spacing so that the same equipment can be used for all crops. This is a proper agronomic and economic management decision.

Soybeans grown in narrow rows (20 in. or less in width) usually are not cultivated, or cannot be cultivated effectively for weed control. This dictates that all weed management practices must use broadcast applications of herbicides. If cultivation is desired as a weed management option, row spacing should be 30 in. or wider, and herbicides should be applied on a band over the drill or underneath the canopy later in the season. It is not economically feasible to both cultivate and apply herbicides broadcast in wide-row soybeans.

If soybeans are planted in April on soils that have poor surface and internal drainage, a bed is desirable. Row spacing will have to be 30 in. or wider because individual beds cannot be effectively formed, maintained, and planted in rows narrower than this. If narrow-row planting is desired on beds in these environments, a wider bed capable of supporting several rows must be constructed. Recent equipment developments allow this to be done, and a management system of narrow rows planted on wide beds is now possible.

Row spacing and plant type

Varieties that are tall-statured (30 to 40 in.) and that branch the entire length of the stalk are adapted to wide rows (30 to 40 in.) because they will usually form a complete canopy. Varieties with these traits fall into MG V and higher maturity groups when grown in the midsouthern U.S. Many MG IV varieties that are planted on a large acreage in the Midsouth are short statured (especially when grown on clayey soils), branch only at the lower nodes, and possess a narrow, upright profile. Some early MG V varieties are inherently short, especially when planted in April. Varieties in these categories will form incomplete canopies and require full-season weed management if grown in wide rows.

Wide rows will be inappropriate for plantings of MG IV varieties made on soils such as clays that support slow growth and for plantings of MG IV and V varieties that are made late in the season when growth potential is minimal. A silty or sandy soil usually produces taller plants as a result of more available soil water, but branching of MG IV varieties may not be enhanced by this improved soil moisture environment. Thus, an increase in plant height will improve canopy development potential, but the lack of branching is a significant factor in less canopy development in the MG IV varieties regardless of plant height. Short-statured varieties and/or varieties that do not branch profusely have the most potential for higher yield from rows narrower than 30 in.

Row spacing and irrigation

Yield advantage for narrow vs. wide rows has usually occurred in irrigated plantings or plantings that receive optimum rainfall. This is because drought stress in dryland plantings often overcomes many of the subtle advantages that may exist for a particular row spacing. Also, the canopy structure provided by soybeans in narrow rows is more suitable for maximum interception of sunlight and this should result in maximum photosynthate being produced in the absence of drought stress. This photosynthate, with adequate rainfall and/or proper irrigation management through seed fill, should be translocated to the seed and result in maximum seed size and yield. Thus, any planting that is to be irrigated should be grown in narrow rows.

Seeding rate

It is important to plant the number of seeds that will achieve the desired number of plants per acre in a uniform stand of soybeans. Data collected in the midsouthern U.S. indicate that a density of 80,000 to 120,000 uniformly distributed plants per acre is adequate for

Table 3.1 Number of Seeds per Foot of Row Resulting from, and Cost per Acre for Six Seeding Rates of Soybean of Varied Seed Size

Seed size, no/lb	Seeding rate, seeds/acre	No. seeds/ft of row with row spacing, in.					$/acre when cost/50 lb of seed[a] is			
		7	15	20	30	40	$10	$15	$20	$25
2800	80,000	1.1	2.3	3.1	4.6	6.1	5.71	8.57	11.43	14.29
	100,000	1.3	2.9	3.8	5.7	7.7	7.14	10.71	14.29	17.86
	120,000	1.6	3.4	4.6	6.9	9.2	8.57	12.86	17.14	21.43
	140,000	1.9	4.0	5.4	8.0	10.7	10.00	15.00	20.00	25.00
	160,000	2.1	4.6	6.1	9.2	12.2	11.43	17.14	22.86	28.57
	180,000	2.4	5.2	6.9	10.3	13.8	12.86	19.29	25.71	32.14
3000	80,000						5.33	8.00	10.67	13.33
	100,000						6.67	10.00	13.33	16.67
	120,000	Seeds/ft of row at this seed					8.00	12.00	16.00	20.00
	140,000	size same as for above					9.33	14.00	18.67	23.33
	160,000						10.67	16.00	21.33	26.67
	180,000						12.00	18.00	24.00	30.00
3200	80,000						5.00	7.50	10.00	12.50
	100,000						6.25	9.38	12.50	15.63
	120,000	Seeds/ft of row at this seed					7.50	11.25	15.00	18.75
	140,000	size same as for above					8.75	13.13	17.50	21.88
	160,000						10.00	15.00	20.00	25.00
	180,000						11.25	16.88	22.50	28.13
3400	80,000						4.71	7.06	9.41	11.76
	100,000						5.88	8.82	11.76	14.71
	120,000	Seeds/ft of row at this seed					7.06	10.59	14.12	17.65
	140,000	size same as for above					8.24	12.35	16.47	20.59
	160,000						9.41	14.12	18.82	23.53
	180,000						10.59	15.88	21.18	26.47
3600	80,000						4.44	6.67	8.89	11.11
	100,000						5.56	8.33	11.11	13.89
	120,000	Seeds/ft of row at this seed					6.67	10.00	13.33	16.67
	140,000	size same as for above					7.78	11.67	15.56	19.44
	160,000						8.89	13.33	17.78	22.22
	180,000						10.00	15.00	20.00	25.00

[a] Price per 50 lb bag of seed equates to public ($10), private ($15), and RR ($20–25) varieties. Public varieties are sold in 60-lb bags, but converted to price per 50 lb. of seed ($12/60-lb bag = $10/50 lb) in this table.

maximum yield. Plant populations that fall below this range can be tolerated with little or no yield loss if the stand is uniform and the variety branches profusely. Plant populations higher than this range result in plants that do not produce pods if water is limited during the growing season. Seeding rates, assuming 100% germination and emergence, to achieve populations of 80,000 to 180,000 plants per acre in various row spacings are shown in Table 3.1.

The quality of seed planted, the capability of the soybean planter, and the conditions following planting determine the final stand from any seeding rate. All of these factors should be favorable before planting even the highest quality seed. Planters should be equipped to penetrate the soil to the desired planting depth, and have attachments that will ensure placement of seed at a uniform depth and complete closure of the seed trench. Conditions following planting are difficult to control, but proper closure of the seed trench is a first step since this will prevent rapid drying of the zone around the planted seed.

Adequate surface drainage is required on nearly level soils to prevent prolonged soil saturation and oxygen deprivation in the seed zone.

Seeding rate and cost

Cost differences among the seeding rates necessary to attain the 80,000 and 120,000 optimum population are relatively small. The data in Table 3.1 show the costs for varieties with various size seeds planted at six seeding rates per acre. These costs are based on seed costs ranging from $10 to $25/50 lb of seed. For example, planting 120,000 seeds/acre of a variety that has 3000 seeds/lb will result in 40 lb of seed being planted (120,000 divided by 3000 = 40) on an acre. This translates to a cost of $8/acre for seed that costs $10/50 lb, which is derived from 40 lb of seed per acre divided by 50 lb of seed at $10/50 lb of seed. All other calculations for Table 3.1 values were derived in a like manner.

If a final population of 100,000 plants/acre is desired, and the germination percentage of the seed lot is the only factor considered, the number of seeds to be planted will be 100,000 divided by the germination percentage of the seed lot to be planted. For example, if the germination is 90%, then seeds to be planted would be 100,000 divided by 0.90, or about 111,000 seeds/acre. If a seed lot has 3000 seeds/lb, the amount of seed to be planted will be 111,000 divided by 3000, or 37 lb/acre. If the seed cost $25/50 lb bag, then the cost per acre for this seeding rate will be $18.50/acre (37 divided by 50, or 0.74; this figure is multiplied by $25).

Field emergence rarely equals the seeding rate multiplied by the germination percentage. Often, an unknown number of germinable seeds will not emerge because of poor seed–soil contact, inclement weather, disease, saturated soil, etc. For example, if it is assumed that only 80% of germinable seeds will emerge, the seeding rate should be increased to achieve the desired final stand. By using the above numbers from the seed lot with 90% germination, the 37 lb of seed per acre should be divided by 0.80 to obtain the number of pounds to be planted, which in this case will be 46 lb. This 46 lb of seed planted per acre will cost $23.00 when using seed that costs $25/50 lb (46 lb divided by 50 lb equals 0.92; multiply this 0.92 by the $25/50 lb of seed). A planter should be set to plant 138,000 seeds/acre to adjust for this 90% germination and 80% assumed emergence. This is derived from using the desired final stand of 100,000 plants/acre divided by 0.90 (to correct for less than 100% germination) which is 111,000 seeds/acre. To correct for only 80% emergence, divide this 111,000 by 0.80 to get 138,000 seeds/acre to be planted. From Table 3.1, the planter should be set to drop about 8 seeds/ft in a 30-in. row to achieve the final stand in this example.

Seeding rate for marginal planting conditions

Planting in marginal conditions is sometimes deemed necessary and is a management decision. Conditions that are considered marginal are planting in dry soil and hoping for rain, planting in a rough cloddy seedbed, planting low-quality seed, and planting after mid-June. Increased seeding rates for late plantings may be necessary to obtain a greater number of main stems to compensate for reduced branching and shorter plants.

If the decision is made to increase the seeding rate to compensate for any of these "less than ideal" field conditions that may adversely affect germination and emergence of planted seed, an additional cost will be incurred. If seeds that cost $10/50 lb of seed have 3000 seeds/lb, have a 90% germination, are planted to obtain 100,000 plants/acre, and seeding rate is increased by 20% to compensate for marginal field conditions, then this extra 20% of planted seed will result in an increased cost of $1.48/acre ($7.41 vs. $8.89). If seeds that cost $25/50 lb of seed have 3000 seeds/lb, have a 90% germination, are planted to obtain 100,000 plants/acre, and seeding rate is increased by 20% to compensate for marginal field conditions, then this extra 20% of planted seed will result in an increased cost of $3.70/acre ($18.50 vs. $22.20). If emergence of germinable seeds in this overplanted situation is 100% instead of being reduced by the anticipated 20%, the stand of 120,000 emerged plants/acre will still be an acceptable population. These calculations indicate that slight overplanting to ensure an adequate stand of soybeans is not an expensive proposition. It is important to remember, however, that increasing the seeding rate will not ensure that adequate or uniform stands will be obtained under marginal conditions.

Tangible and intangible costs associated with stand failure and subsequent replanting will be high and may dictate that planting should not be made in marginal conditions. Therefore, planting under marginal conditions is not advocated because of this high risk. However, if it is decided that planting must be conducted under such conditions, using the proper equipment is an absolute necessity. A planter must be equipped to plant at a uniform depth, and this will usually involve disk openers with attached depth bands and wide-spaced depth-control wheels. There should be adequate downpressure on each row or drill unit to ensure that seeds in each row or drill are planted at the desired depth. The seed furrow or slit should be completely closed to ensure optimum seed–soil contact and to prevent rapid drying of the seed trench. A properly equipped and adjusted planter will ensure the best possible stand in a given set of soil conditions. Using a planter with inadequate soil penetration, depth control, and seed slit closing capabilities will generally result in marginal stands even under ideal planting conditions or when using increased seeding rates in less than ideal conditions.

chapter four

Nutrition and fertility requirements

Jac J. Varco

Contents

Soybean nutrient requirements

Soybean is a major row crop in both the U.S. and Mississippi. Acreage in the U.S. in 1995 was 61.2 million with an average yield of 34.9 bu/acre, while 1.8 million acres were harvested in Mississippi with an average yield of 21 bu/acre (Gregory and Corley, 1996). Nutrients removed in the harvested seed at these yield levels are equivalent to approximately 110 lb nitrogen (N), 10 lb phosphorus (P), and 35 lb potassium (K) per acre for the U.S. and 66 lb N, 6 lb P, and 20 lb K per acre for Mississippi. Additionally, other nutrients such as calcium (Ca), magnesium (Mg), and sulfur (S) as well as micronutrients are removed to a lesser degree. Although soybean obtains N through biological fixation, a considerable amount of soil N is also taken up by the crop. From a fertility standpoint, it is apparent that to sustain or augment the current level of soybean productivity, replenishment of soil nutrients is a necessity. This chapter deals with specific fertility needs and management strategies that will help ensure sustainable soybean production.

In 1993, 21% of the 2 million acres of soybean planted in Mississippi received some fertilizer, with 10% receiving N, 20% P, and 21% K (ERS-NASS, 1994). Average nutrient application rates were 16 lb N, 17 lb P, and 50 lb K per acre on the acreage receiving fertilizer. Based on this survey, nutrients applied to the soil are less than removal rates across all major soybean-producing states due primarily to the limited acreage receiving any fertilizer. For the major soybean-producing states, Georgia at 69% and North Carolina at 62% have the greatest proportion of soybean acreage receiving fertilizer, which is likely a reflection of an inherently low soil nutrient-supplying potential.

The greatest demand by soybean for nutrients occurs from flowering through pod fill. During this period Hanway and Weber (1971) found accumulation rates of 4.0 lb N, 0.36 lb P, and 1.34 lb K/acre/day. In North Carolina, Henderson and Kamprath (1970) noted maximum accumulation rates of 6.9 lb N, 0.37 lb P, and 4.1 lb K/acre/day for determinate soybean. On a Chalmers silt loam soil in Indiana, Barber (1978) noted increasing nutrient demand rates through 92 days of growth for P with a peak of 0.56 lb/acre/day, and through 78 days for K with a peak of 2.56 lb/acre/day. For maximum yields, soil nutrient supply must be sustained through pod fill.

Macronutrients

Nitrogen

Nitrogen nutrition

Nitrogen is required in the greatest quantity of all plant nutrients absorbed from the soil. It is present in all amino acids, which are the building blocks of protein, nucleic acids, and chlorophyll (Jones et al., 1991). A deficit of N causes slowed growth, and older leaves become chlorotic and senesce prematurely due to translocation of N compounds to growing points. Conversely, excess N can delay maturity, cause excessive vegetative growth and water use, and greater shoot growth relative to root growth.

Soybean belongs to the Leguminosae family and has the capacity to acquire N from air through biological fixation as well as to utilize soil-derived N. Nitrogen-fixing bacteria known as *Bradyrhizobium japonicum* infect soybean roots with subsequent formation of nodules from modified root tissue. The bacteria reside within the nodules and when effective (pinkish red color inside) form a symbiotic relationship with soybean plants. In essence, bacteria utilize plant photosynthates in exchange for providing organic N compounds, which the plant uses in protein synthesis. Soybean also has the capacity to absorb

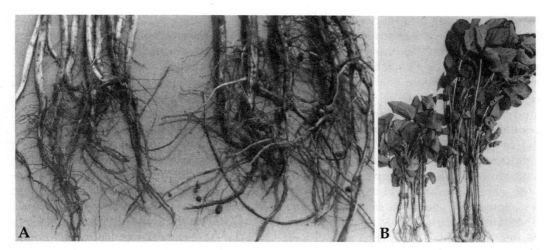

Figure 4.1   (A) Field-grown soybean plant roots showing no effective nodulation on the left and effective nodulation on the right. (B) Early plant height and color differences between uninoculated plants on the left and nodulated plants on the right.

and utilize ammonium and nitrate. These two forms of N are supplied naturally by soil processes and can be augmented through organic and inorganic fertilization.

Typically, well-inoculated soybean can meet N demands through biological fixation and uptake of soil N. Harper et al. (1989) showed the majority (73 to 78%) of plant N during early growth was from soil supplies, while the remainder was derived from biological fixation. The quantity and proportion of N derived from fixation after this period reached a maximum by flowering. Fixation was noticeably decreased during a drought period with conventional tillage as compared to no-tillage, and presumably resulted from decreased availability of soil moisture. Hardy et al. (1973) showed that the period of N fixation corresponded well with maturity group. Zapato et al. (1987) noted a rapid increase in fixation beginning at flowering and continuing through physiological maturity. Overall, 47% of the N assimilated was derived from biological fixation. However, total N derived through biological fixation is highly variable and reported values have been from as low as 25% to as high as 75%.

Nitrogen deficiency symptoms in soybean include short or stunted plants having a pale green color or uniformly chlorotic older leaves in a more-advanced stage of growth (Sinclair, 1993). Assessment of leaf tissue N concentration can be made to confirm suspected N deficiency. The most fully developed leaf sampled prior to pod set should contain a minimum of 4.0% total N (Jones et al., 1991). If N deficiency is confirmed, the cause could be related to poor or no inoculation (Figure 4.1), low available soil N, moisture stress, molybdenum (Mo) deficiency, excessive soil acidity, or low Ca, and possibly soybean cyst nematode infection (Sinclair, 1993).

Fertilizer N response

Soybean yield response to N fertilization has been varied. In Arkansas, across seven locations, N rates of 20 and 40 lb/acre did not increase soybean yield (Maples and Keogh, 1969). Vegetative growth was stimulated, but only on two soils having a pH of 5.3 to 5.5, which suggests that inoculation may have been poor on these soils. Yield was decreased at one site due to greater Johnsongrass (*Sorghum halepense* [L.] Pers.) competition as a result of fertilizer N application. Long et al. (1965) reported no response to N rates of 30 and 60 lb/acre at various locations in Tennessee with differing soil types. A 2.5 bu/acre yield response to 20 lb N per acre on a Sharkey clay was reported by Pettiet (1971), but was attributed to high soil temperatures and dry soil conditions. In Kentucky on a Maury silt loam soil, Brevedan et al. (1978) reported a 22 to 32% yield increase with an application of 150 lb N per acre to soybean at initial bloom. Yield increases over the course of the study were attributed to greater vegetative growth, pod retention, and seed size. These results emphasized the importance of a high N supply during flowering and pod set.

In an effort to elucidate fertilizer N effects on N_2-fixation and yield, studies have been conducted comparing nodulating and nonnodulating soybean isolines. A nonnodulating soybean isoline is not compatible with N-fixing bacteria and does not fix N, but is otherwise genetically similar to its nodulating isoline. Bhangoo and Albritton (1976) found on a Calloway silt loam soil in Arkansas maximum yield at an N rate of 200 lb/acre for a nodulating soybean, while for a nonnodulating isoline, yield increased with up to 400 lb N per acre. Total N uptake without fertilizer for the nodulating isoline averaged 225 lb/acre for the 3-year period, while the nonnodulating isoline averaged 123 lb/acre. For the nonfertilized nodulating isoline, sources of N included biological fixation and soil, while it can be assumed for the nonfertilized nonnodulating isoline that uptake was attributable only to soil N. Biological N fixation decreased linearly with a corresponding increase in fertilizer N rate. Also, total N uptake was greater for the nodulating isoline across N rates, but at a rate of 400 lb/acre there was little difference between the nodulating and nonnodulating isolines.

In eastern Nebraska on a Sharpsburg silty clay loam, Deibert et al. (1979) found a yield response to fertilizer N only for a nonnodulating isoline. The greatest response occurred when fertilizer N was applied at flowering rather than at planting. This corresponds to the period of peak N demand by developing soybean seed. For a nodulating isoline, N applied at rates greater than 40 lb/acre at planting reduced the quantity of N fixed, while no rate of applied fertilizer N reduced fixation when applied at full bloom. Thus, when fertilizer N is applied during peak demand, there is less chance to reduce the amount fixed by the plant.

Conservation tillage and doublecropping practices may alter fertilizer N requirements due to lowered soil N supply. Bharati et al. (1986), working in Iowa on a Nicollet–Webster soil complex with varying tillage intensities, found increased plant height, lodging, and nodes per plant, but no yield response to N rates up to 240 lb/acre. It was concluded that reductions in tillage did not increase the need for fertilizer N. Doublecropping wheat (*Triticum aestivum* L.) and soybean may present special problems with N management. Planting soybean no-till into wheat stubble or in a prepared seedbed following wheat may result in depressed N availability due to microbial immobilization of available soil N during decomposition of wheat residues. Hairston et al. (1987) working on an Okolona silty clay soil in Mississippi found that 25 lb N per acre increased soybean yield only when planted no-till into wheat stubble and a lack of benefit when straw was removed, burned, or incorporated. Also, net returns were increased the greatest in response to N fertilization for the no-till treatment.

Unpredictability in soybean yield response to fertilizer N is likely related to many factors. Some of these would include N fertilization repressing N fixation, variability in soil N-supplying capacity, soil water availability, and environmental conditions in general. Bhangoo and Albritton (1976) recommended fertilizer N rates of 50 to 100 lb/acre to minimize any decrease in N fixation. It is likely that any conditions that result in low soil N availability or that negatively impact nodulation and fixation would increase the probability of a fertilizer N response.

Phosphorus

Phosphorus nutrition

Phosphorus, although required in much lower quantities than either N or K, is very critical to rapid growth and proper development. The most noteworthy functions of P in plants include energy storage and transfer, membrane function, and genetic transfer (Marschner, 1995). On a whole-plant scale, P is critical to root system and plant canopy development, and seed production. Phosphorus content of harvested soybean seed is approximately 0.30 to 0.35 lb/bu harvested. Therefore, maintenance of available soil P would require at least this range returned to the soil each year. Before flowering, the greatest proportion of plant P accumulates in leaves relative to stems and petioles. During pod fill, P is translocated from stems, leaves, and petioles to developing seed (Karlen et al., 1982a). Also, during periods of drought, irrigated soybean plants are able to maintain greater vegetative tissue P levels than nonirrigated ones.

Soybean plants express P-deficiency symptoms as stunted growth, small leaflets, and delayed flowering and maturity (Sinclair, 1993). A minimal tissue P level at flowering of 0.31% or greater was suggested by Bell et al. (1995). A range of 0.26 to 0.5% for the most recently matured leaves prior to pod set is considered sufficient (Jones et al., 1991).

Soil test phosphorus and fertilizer response

Generally, soil test levels of 20 lb/acre or greater result in 100% of potential yield (Hanway and Olsen, 1980). In Mississippi, little to no response is expected at soil test levels greater than 36 lb P per acre (Hoover, 1968). This is in agreement with results found in North Carolina (Kamprath and Miller, 1958) and Alabama (Rouse, 1968). More recently, Sabbe and Mahler (1991) reported a soil test level of 40 lb P per acre produced 95% of maximum yield in Arkansas. Some variation in optimal soil test P levels from state to state is expected due to differing soil extraction procedures. Lins et al. (1985) suggest that optimal soil test P level is dependent on clay content; however, textural determination is not a routine soil test practice.

Soybean requires a constant supply of available P to maintain rapid growth and development. Root length of Williams soybean was found to increase through 90 days of growth and the proportion of roots in the 0 to 6 in. soil depth increased with time (Barber, 1978). Soybean P uptake was less than for corn during early growth, but was greater later in the season. These observations indicate that it would be best to augment soil P within a greater volume of the root zone rather than banding fertilizer at planting. In a study by Ham et al. (1973) in Minnesota on three soils ranging from low to high soil test P, the greatest yield response on a low testing soil under dry conditions occurred with broadcast fertilizer. When rainfall was not limited, a combination of starter and broadcast fertilizer resulted in the greatest yield increase on this soil. No fertilizer response was observed when soil test P was rated as high.

Phosphorus fertilization has been shown to increase both the number of nodules and their weight as well as the number of pods per plant (Jones et al., 1977). Response of soybean to fertilizer P has not been as great as for crops such as corn (deMooy et al., 1973). When a yield response occurs on soils having low P fertility, it is only to rates of around 26 lb P per acre or less (deMooy et al., 1973; Jones et al., 1977; Hairston et al., 1990). Jones et al. (1977) found a response to greater P rates only when K was applied simultaneously (Figure 4.2). Limited response to fertilizer P occurred at 19 locations in Arkansas and a rate of only 13 lb P per acre was sufficient to produce maximum yield (Maples and Keogh, 1969). It has been suggested that soybean grown in rotation with well-fertilized crops such as corn or wheat requires minimal supplemental fertilizer P to optimize yields economically (deMooy et al., 1973).

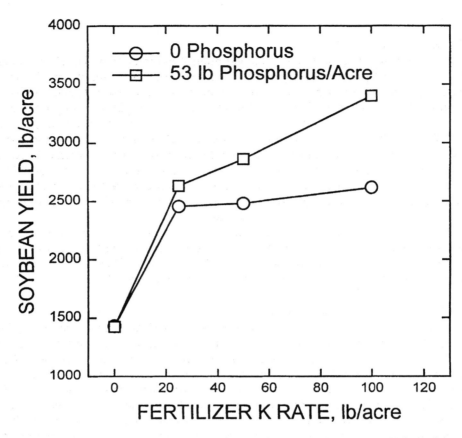

Figure 4.2 Dependency of fertilizer potassium response on phosphorus rate. (Adapted from Jones et al., 1977.)

        Jones et al. (1995) evaluated P fertilization rates and placement methods for no-till and conventional-till soybean at Verona and Brooksville, MS. At the Verona location, the soil was a Leeper silty clay loam having a high soil test P level (158 lb/acre). At Brooksville, the soil was a silty clay loam of the same type and tested low in P (18 lb/acre). Both soils had a pH of 8.0 and belong to the Blackland Prairie land resource area. At Verona, soybean yield responded to fertilization only with no-till. Surface banding at an application rate of 0–45–45 (N–P–K, lb/acre) produced maximum yield with no-till. Due to the experimental design, P and K responses could not be separated at either location. Leaf tissue samples (recently matured trifoliolates) obtained at early flowering showed adequate tissue P for all treatments including the nonfertilized check. Soil test P levels increased less for the 0 to 6 in. depth than for the 0 to 2 in. depth with no-till, suggesting accumulation near the soil surface due to no disturbance or mixing from tillage. Yield increased with P and K fertilization for both tillage systems at Brooksville. Although soil test P was low for this site, tissue levels were within the sufficiency range. As with the Verona location, the response was most likely related to K application since tissue K levels for the check treatment were below the sufficiency range.
        Hairston et al. (1990) suggested positional unavailability of nutrients with no-till and found subsurface banded P and K resulted in greater no-till soybean yields only on fine-textured Okolona silty clay and Memphis silt loam soils. Initial P levels were low for these two soils. The greater response to fertilization with no-till suggests that rooting may not

Table 4.1   Soil Test P Categories and Recommended Fertilizer
Rate for Soybeans

Soil test category	Soil test phosphorus, lb/acre	Recommended phosphorus rate,[a] lb/acre
Very low	0–18	52
Low	19–36	26
Medium	37–72	13
High	73–144	0
Very high	144+	0

[a] Multiply values times 2.29 to convert to fertilizer phosphate rate.

Table 4.2   Soil Test P Results by Land Resource Area for
Samples Tested by the Mississippi State Soil Testing
Laboratory (for the period 1 July 1995 to 30 June 1996)

Mississippi land resource area	Soil test category, %			
	Very low	Low	Medium	High
Upper Coastal Plain	3	9	29	59
Lower Coastal Plain	1	5	38	56
Loess Belt	4	6	31	59
Delta	1	3	23	73
State average[a]	**2**	**5**	**26**	**68**

[a] Variable number of samples from each land resource area.

From McCarty, W. et al., MCES, Mississippi State University, Mississippi State, 1996. With permission.

be as extensive in fine-textured soils. This effect would lessen nutrient uptake, and, therefore, soil test levels may need to be greater with no-till.

Soil test P categories for Mississippi and recommended fertilizer P rates are shown in Table 4.1. The recommended rate at the medium soil test level is essentially a maintenance fertilization rate. The probability of a yield response to fertilizer application is low at the medium soil test level, but becomes greater with the low and very low soil test categories. Soil test P levels for soybean production in Mississippi generally are adequate (Table 4.2). However, certain circumstances, such as exposed acid subsoil or chalk, may result in a P deficiency.

Potassium

Potassium nutrition

Potassium, known commonly as potash, is required by soybean at about 30 to 50% of the quantity of N. Potassium is readily translocated from older to newer plant tissue to meet the demand of the crop if the soil supply is inadequate. It is important in enzyme reactions, protein synthesis, and water use efficiency. It functions also in drought tolerance, fruit and seed quality, and susceptibility to frost and disease resistance. A seed content of 1.85% K would result in harvesting about 1.0 lb K per bushel.

Soybean has been shown to respond to K by increasing pods per plant, root nodules, and total weight of nodules (Jones et al., 1977; Bharati et al., 1986). Coale and Grove (1990) noted increased branch production and pod and seed dry weight under high soil K availability. Hallmark and Barber (1981) found greater root surface area with increasing K supply, but only when soil bulk density was high. Increased root density with low K

Table 4.3   Soil Test K Categories Used by the Mississippi State
                 Soil Testing Laboratory

Soil test category	Cation exchange capacity[a]			
	<7	7 to 14	15 to 25	>25
Very low	0–50	0–60	0–70	0–80
Low	51–110	61–140	71–160	81–180
Medium	111–160	141–190	161–210	181–240
High	161–280	191–335	211–370	241–420
Very high	280+	335+	370+	420+

[a] Soil test potassium lb/acre.

fertility relative to high K fertility suggests a compensatory mechanism by which soybean increases nutrient absorption potential. Greater K uptake can be expected with irrigation, primarily as a result of greater dry matter production and possibly greater K concentration in plant parts (Karlen et al., 1982b).

Potassium deficiency symptoms are visible as irregular yellow mottling along edges of older leaflets (Sinclair, 1993). With progression, necrosis and cupping downward of leaf margins can occur. Wrinkled and deformed seed can result from insufficient K supply during seed formation. A range in tissue concentration of 1.7 to 2.5% for the most recently matured leaves prior to pod set is considered sufficient (Jones et al., 1991).

Recently, an atypical K deficiency has been found in the South and has been discussed by Snyder and Ashlock (1996). Normally, K deficiency begins in older leaves as it is translocated to strong sink areas including new shoots, pods, and seeds. Late-season K deficiency symptoms have been expressed in the middle to upper canopy on soils testing less than 100 lb K per acre. Many growers have been switching to early maturing varieties which can continue to produce vegetative growth during flowering and pod filling. Thus, new vegetative growth competes with developing pods, which have a high demand for K.

Soil test potassium and fertilizer response

Determining soil K availability is highly complex due to soil mineralogical variability. Most soil-testing facilities focus on extracting K adsorbed to the soil in exchangeable form as well as the water-soluble fraction. However, minerals that release K at variable rates exist in the soil and their K status is not currently measured by routine soil tests. Early research with soybean showed the greatest response to soil test levels occurring within the range of 50 to 150 lb/acre (Bray, 1945). To achieve 95% of maximum yield, a soil test of 200 lb K per acre was recommended. More recently, Sabbe and Miller (1991) determined that a soil test K level of 140 lb/acre was necessary to produce 95% of maximum yield in Arkansas.

Soil test K categories for Mississippi are shown in Table 4.3. The underlying philosophy in this categorization is that greater soil test K levels are required with increasing cation exchange capacity. This agrees with the philosophy of maintaining a percentage of the cation exchange capacity saturated with K and that increasing levels of Ca and Mg could negatively affect K uptake.

With grouping of soil test results by major land resource areas, the Delta is the only one with minimal samples testing low in K (Table 4.4). This is a reflection of the high clay–cation exchange capacity soils typically used in the Delta for soybean production. These soils contain large reserves of K due to a high content of K-bearing minerals. Mineralogically, these soils also have a great capacity to retain added fertilizer K. The soil test data for samples analyzed at the Mississippi State Soil Testing Lab for 1995–1996 suggest that significant acreage outside of the Delta would respond favorably to K fertilization.

Table 4.4   Soil Test K Results by Land Resource Area for
Samples Tested by the Mississippi State Soil Testing Laboratory
(for the period 1 July 1995 through 30 June 1996)

Mississippi Land Resource Area	Soil test category, %			
	Very low	Low	Medium	High
Upper Coastal Plain	1	19	23	58
Lower Coastal Plain	2	24	31	43
Loess Belt	0	26	30	43
Delta	0	3	8	89
State average[a]	<1	14	18	68

[a] Variable number of samples from each land resource area.

From McCarty, W. et al., MCES, Mississippi State University, Mississippi
State, 1996. With permission.

Table 4.5   Recommended K Fertilization Rates for
Soybeans Based on Soil Testing Category

Soil test category	Recommended K rate[a], lb/acre
Very low	100
Low	50
Medium	50
High	0
Very high	0

[a] Multiply values times 1.2 to convert to fertilizer potash rate.

Current Mississippi soil test recommendations relative to soil test categories are shown in Table 4.5.

Response to K fertilization depends greatly on soil available K and mineralogy, as well as on other factors limiting crop growth such as rainfall and availability of other essential plant nutrients. In Virginia on a Davidson clay loam soil testing low in P and K, there was a distinct dependence of K response on P rate (Figure 4.2). At fertilizer K rates greater than 25 lb/acre, yield was limited by insufficient P. In Iowa, Bharati et al. (1986) found a response to fertilizer K at a rate of 150 lb/acre for 2 years on a Nicollet–Webster soil complex which initially tested near 200 lb K per acre. The yield increase was attributed to an increase in number of pods per plant. Also, the unfertilized check plots declined to a medium soil test level after two crops. Even with an initially high K rating and an ending medium rating for the nonfertilized treatment, leaf K was below sufficiency.

The greatest probability of a yield response to fertilizer K is on soils testing very low to medium in available K, and generally low application rates satisfy requirements. In Alabama, on soils testing low to medium in K, the greatest yield increase occurred for a 25 lb K per acre application when averaged across six locations and 4 years (Evans et al., 1986). Long et al. (1965) noted a 3-year average yield increase of 5.6 bu/acre to 33 lb K per acre on a Grenada soil initially testing very low in K (55 lb/acre). Keogh and Maples (1970) in Arkansas noted a response to K fertilizer rates of 50 lb/acre only when soil test levels were less than 80 lb/acre.

Secondary macronutrients calcium, magnesium, and sulfur

Calcium plays a critical role in membrane stability and cell integrity (Marschner, 1995). It is necessary to have a constant supply of Ca throughout the soybean growth cycle because it is not readily translocated within the plant. Plant tissue susceptible to a deficiency is found mainly at growing points such as terminal buds and root tips. Plants selectively absorb Ca, although large quantities of available Ca exist in the soil. Soybean has a relatively high requirement for Ca compared with corn and sorghum. Growth, however, would more likely be slowed due to lowered biological N_2-fixation with declining Ca availability (Albrecht, 1932). Absorption of soil Ca is dependent on soil moisture content as shown by Karlen et al. (1982b) who found greater accumulation in soybean leaves when irrigated relative to nonirrigated treatments.

Magnesium is a component of chlorophyll, the site of photosynthesis. It is also required in the formation of protein in plants. A deficiency of Mg is expressed as unthrifty plants with yellowing or interveinal chlorosis of older leaves with progression to a necrosis of affected areas (Grundon, 1987; Sinclair, 1993). Like Ca, it is added to soils when dolomitic sources of lime are used. Availability needs to be high during pod fill when demand is the greatest. Karlen et al. (1982b) found no difference in Mg accumulation between irrigated and nonirrigated soybean suggesting soil moisture supply has a limited influence on uptake.

Research in Mississippi has shown a 5 bu/acre yield increase from application of 60 lb Mg per acre to a Griffith clay loam soil (Cervellini, 1980). Leaf tissue analysis indicated a possible Mg deficiency with the nonfertilized check plots having slightly below the minimal sufficiency level of 0.26%. Fertilization increased tissue Mg to 0.39%. Based on yearly summaries of soil samples analyzed by the MCES Soil Testing Lab, the Lower Coastal Plain has the greatest need for Mg, while 100% of samples from the Delta were high in Mg. Delta soils are naturally high in Mg and will likely not require supplemental Mg fertilization, except on highly leached coarse sands with low CEC. Prior to liming, the soil Mg level should be evaluated so that a decision can be made whether calcitic (primarily $CaCO_3$) or dolomitic [$CaMg(CO_3)_2$] lime should be used.

Sulfur is essential for the synthesis of two amino acids, and, therefore, when deficient can result in an accumulation of sugars and nitrates. Soybean plants deficient in S exhibit yellowing of leaves similar to an N deficiency, but the symptoms are found in newer growth rather than older leaves. Another characteristic is elongated stems which can be thin and hard (Sinclair, 1993).

Plant-available S is derived primarily from the decomposition of plant residues and soil organic matter. Geographic location relative to industry and coal-burning power plants determines atmospheric deposition of S to farmland in rainfall and particulates. Historically, when ordinary superphosphate which contains 12% S was in common usage, S was inadvertently applied to many soils. Kamprath and Jones (1986) summarized S fertility research in the southeastern U.S. and found that a response to fertilization occurred for only two of nine sites. A positive yield response occurred on soils having available S levels of 4 ppm or less, while nonresponsive soils had 8 ppm available S and an accumulation of subsoil S within 8 in. of the surface. In the Blackbelt region of Mississippi, Cervellini (1980) found a 5 bu/acre yield response to 20 lb S per acre on a Griffith clay loam soil. With reduced S emissions, response to fertilizer S may become more common.

Micronutrients

Iron

Iron (Fe) is essential for the synthesis of chlorophyll and for biological N fixation as well as many metabolic functions in plants. It is not readily translocated in plants; therefore, deficiency symptoms are first observed in new growth. Typically, interveinal areas of young leaves will exhibit yellowing or chlorosis. Eventually, brown necrotic spots may form along leaf margins and plant growth can be slowed. Minimum leaf tissue Fe necessary for optimal growth is 50 ppm (Jones et al., 1991).

Iron deficiency or "Fe chlorosis" in soybean is most common on soils with alkaline pH (>7.0) containing calcium carbonate. Thus, the term *lime-induced chlorosis* is sometimes used to describe this condition. With increasing pH and the presence of calcium carbonate, the availability of Fe is drastically reduced. Soybean varieties differ in their susceptibility to Fe deficiency, and thus Fe absorption is believed to be genetically controlled (deMooy et al., 1973) . The rootstock is the controlling factor since Fe-inefficient tops grafted to Fe-efficient rootstock show no sign of chlorosis (Brown et al., 1958). It appears that tolerant varieties increase availability of Fe in the root zone through root secretion of various compounds (Brown and Jolley, 1989).

In Mississippi, Udoh and Nelson (1983) evaluated soybean varieties for Fe efficiency under greenhouse and field conditions. The degree of Fe efficiency was categorized as efficient, intermediate efficiency, and inefficient. Similar trends were present for greenhouse and field evaluations although there were some exceptions. Herbicides such as trifluralin, which can limit rooting if misapplied, can also induce chlorosis. Udoh and Nelson (1986), working on Sumter and Okolona soil types, found Fe-efficient varieties to be more resistant to trifluralin-induced chlorosis. Herbicide effects were greater on a Sumter soil with a pH of 8.2 than on an Okolona soil with a pH of 6.9.

Research on soybean response to Fe fertilization in Mississippi is limited. Cervellini (1980) worked with various soils of the Mississippi Blackland Prairie with known incidences of chlorotic soybean. A combination of field and greenhouse studies that included evaluation of foliar combinations of Fe and manganese (Mn) solutions were performed. Under greenhouse conditions, foliar-applied Fe alleviated chlorosis and increased dry matter production. However, on a Sumter soil under field conditions, a combination of Fe and Mn chelates was necessary to correct deficiency symptoms.

Manganese

Manganese plays a role in reduction-oxidation reactions in plants, enzyme activity, and photosynthesis (Marschner, 1995). It is not considered to be readily translocated from older tissue to new developing leaves. Kluthcouski and Nelson (1979) concluded that Mn is immobile with low foliar concentrations, whereas limited mobility can occur at high leaf concentrations. Manganese deficiency symptoms resemble Fe deficiency with interveinal chlorosis of young leaves, except veins maintain their green color. Eventually, brown necrotic spots develop and leaves may senesce prematurely. The critical or minimal leaf concentration is within a range of 15 to 20 ppm (Jones et al., 1991). In a southern regional study on Mn requirements of soybean, Anderson and Mortvedt (1982) concluded that the intensity of visual deficiency symptoms is indicative of the Mn status of the plant.

Generally, Mn availability decreases as soil pH increases. In the South, response to Mn fertilization is most common on highly weathered, coarse-textured soils of the Coastal Plain, which have been limed to a pH greater than 6.0 (Parker, 1982). States where a response occurred included Florida, Georgia, North Carolina, South Carolina, and Virginia. Generally, the Gulf Coastal Plain contains inherently greater levels of Mn than the Atlantic Coastal Plain (Adams, 1984), and thus Mn toxicity on acid soils of the Gulf Coastal Plain is a concern. Soils with a pH greater than 7.0 have not been studied extensively in Mississippi for Mn availability. Cervellini (1980) noted a 3.2 bu/acre soybean seed yield increase in the Blackbelt on a Griffith soil having a pH of 8.2 and containing free calcium carbonate. Manganese deficiencies can be induced on calcareous soils when chelated Fe is applied to soybean, especially during early growth when temperatures are cool (Moraghan, 1985). During periods of flooding, soil Mn availability increases with a subsequent increase in tissue concentration in developing leaves only (Kluthcouski and Nelson, 1979). Therefore, it is critical to sample the most recently matured trifoliolates to determine the current Mn status of the plant.

Critical soil levels of Mn are difficult to determine because of many environmental factors influencing availability. Also, Nelson (1977) noted that water-soluble Mn is variable depending on soil-handling methods and storage which can cause unreliability in testing.

Copper

Copper is required in many enzyme systems in plants and severe deficiencies interfere with pollen formation and fertilization, grain development, and lignification which produces leaf distortion and bending and twisting of stems. Copper deficiency does not appear to be a problem in the South. Leaf tissue levels of Cu should be within a range of 10 to 30 ppm (Jones et al., 1991).

Zinc

Zinc is required by plants, most notably in enzyme reactions. Zinc-deficiency typically results in plants that are stunted with chlorotic mottling of interveinal areas of young developing leaves. The symptoms become progressively worse in older leaves. Also, a wrinkled appearance due to raised interveinal areas has been described by Okhi (1977). In soybean, a deficiency of Zn can reduce flowering and cause abnormal and slow development of pods. The critical leaf concentration for soybean is 15 ppm (Okhi, 1977).

Zinc deficiency can be a problem on high-pH or alkaline soils low in organic matter. Research in Mississippi on calcareous soils of the Blackland Prairie has been inconclusive regarding the occurrence of this deficiency (Cervellini, 1980). Problems with Zn deficiency could also arise where topsoil depth is reduced by erosion or land-forming operations since most available Zn is associated with organic matter concentrated in the surface soil. Also, excessive available P can induce Zn deficiency (Paulsen and Rotimi, 1968).

Boron

Boron is important in cell membrane and hormonal functions and influences root growth, pollination, and development of terminal buds (Marschner, 1995). Soybean is considered sensitive to B toxicity, but little information is available on B requirements. The minimal tissue B level for optimal growth is suggested at about 21 ppm.

Deficiencies of B are most likely to occur on deep sandy or silty soils with very low organic matter content. Recent research in Georgia was summarized by Gascho (1993). Response to B applied alone or with N during reproduction was found primarily on a deep sandy soil that had little capacity to retain the borate ion. In a regional study conducted in the Midwest, B foliarly applied at 0.25 lb/acre at initial flowering increased yield an average of 1.4 bu/acre across 29 sites tested (Oplinger et al., 1993). Cotton requires more B than soybean, and recent research in Mississippi has shown that adequate quantities are supplied to cotton on Delta soils (Heithholt, 1994). Also, irrigation water generally contains B; thus, fertilization should not be necessary.

Molybdenum

The most important plant function of molybdenum (Mo) is its role in N metabolism. It is found in both nitrate reductase and nitrogenase enzyme systems (Marschner, 1995). Nitrate reductase is required by plants in the utilization of nitrate absorbed from the soil. Nitrogenase is essential for the fixation of N from air by *Bradyrhizobia* bacteria. Deficiencies of Mo resemble N deficiency because of its effect on N utilization. Plants deficient in Mo are stunted and show signs of leaf chlorosis. Molybdenum is required in the least amount relative to all other plant nutrients. Only 1 to 5 ppm is necessary for normal growth (Jones et al., 1991). Seeds can be highly variable in Mo content and serve as an important source during early plant development.

Availability of Mo decreases with decreasing soil pH or increasing acidity. This is one of the primary factors involved in determining pH or liming requirements for soybean. Early research in Georgia and Mississippi showed that Mo fertilization can be very effective on acid soils, but the magnitude of the response decreases to zero with increasing pH (Parker and Harris, 1962; Anthony, 1967). In a regional study conducted in the South, response to application of 0.39 lb Mo per acre was found when soil pH was less than 5.7 (Boswell, 1982). Similar to the results of Parker and Harris (1962) and Anthony (1967), response to applied Mo became progressively less as pH was increased to 5.7 by liming. Yields were optimized within a pH range of 6.0 to 6.5. This study failed to find a strong relationship between available soil Mo and yield.

Due to the low requirement of plants for Mo, only very low rates of fertilizer are necessary. Molybdenum salts can be added to fertilizer blends or can be mixed directly with the seed prior to planting. Additionally, Mo can be foliar applied, but Parker and Harris (1962) found seed treatment to be superior. Although Mo fertilization is an option, maintaining soil pH greater than 5.7 through lime management is a sound practice due to other problems associated with acid soils such as aluminum (Al) and Mn toxicity. Research to date shows that Mo fertilization is not necessary when soil pH is maintained greater than 6.0, but at lower pH levels an economic response is likely.

Soil pH management

Soil Acidity

By nature, soils become more acid with time and degree of weathering. Minerals within the soil that buffer acidity can eventually become depleted. Problems arise for soybean when soil pH declines below 6.0. Availability of some of the essential nutrients decreases, while availability of other elements increases to the point of toxicity.

Manganese toxicity

Excess soil Mn availability results in toxicity. Manganese generally does not inhibit root growth, and thus can continue to accumulate in aerial plant components. Symptoms include interveinal chlorosis of younger leaves, leaf crinkling, and stunting. Older leaves develop red-brown necrotic lesions on their upper surface and in their veins (Grundon, 1987). Soybean leaf tissue data from Louisiana on Falaya and Grenada silt loams and from Mississippi on a Bude silt loam and a Prentiss fine sandy loam showed decreasing Mn concentrations with increasing pH (Woodruff and Jones, 1982). Anderson and Mortvedt (1982) concluded that there was not a clearly defined tissue Mn concentration at which toxicity occurs in the South. This is likely related to a variation in environmental conditions. For example, in a dry year in Louisiana, leaf Mn concentration averaged 618 ppm Mn and did not adversely affect yield. Additionally, varieties differ in their susceptibility to Mn toxicity (Carter et al., 1975). Jones and Nelson (1978) working in Mississippi concluded that toxic Mn effects were eliminated and yields were optimized when tissue levels were less than 200 ppm and soil pH was greater than 5.5.

Aluminum toxicity

Aluminum toxicity is more acute than Mn toxicity due to inhibition of root growth (Marschner, 1995). Elongation of root tips is inhibited and they appear unhealthy with a dark brown color. Aluminum toxicity also interferes with Ca, Mg, and P uptake. Subsoils are commonly more acid with a greater proportion of the cation exchange complex saturated with Al than in the plow layer. Thus, in addition to affecting nutrient uptake, plants growing under conditions of toxic Al are more susceptible to drought stress due to limited rooting. Also, organic matter is known to complex and reduce toxic Al effects. This effect is most active in surface horizons where organic matter is concentrated.

Liming

Liming acid soils decreases toxic levels of Al and Mn, increases availability of nutrients such as P, Ca, Mg, and Mo, and increases microbial activity, most notably N-fixing bacteria. Thus, soybean response to liming can be due to numerous factors. In a study on Coastal Plain soils in Alabama, the critical soil pH for soybean ranged from 5.0 to 5.7 (Rogers et al., 1973). As discussed previously, Jones and Nelson (1978) in Mississippi found optimal growth when soil pH was 5.5 or greater due to alleviation of Mn toxicity.

Liming has multiple-year benefits which must be considered when determining profitability. Jones and Nelson (1983) reported a 6 bu/acre yield increase across a 5-year period on a Bude silt loam soil in Mississippi from a single 3 ton/acre lime application. Similar results were found on a Prentiss fine sandy loam with a yield response occurring 4 of the 5 years and an average yield increase of 7 bu/acre for the 5-year period from a one-time-only application of 2.5 ton/acre. Soil pH for the Bude soil decreased from 5.1 to 4.9 in 5 years for the unlimed treatment and 6.4 to 5.4 when limed. For the Prentiss soil, pH values decreased from 5.1 to 4.8 with no lime and 6.8 to 5.8 when limed. Long et al. (1965) showed similar residual benefits to a single lime application.

Concern about subsoil acidity has prompted an evaluation of whether deep placement of lime is beneficial. Jones and Nelson (1982) tested lime placement and concluded from a greenhouse study that liming subsoil was not necessary if the topsoil was adequately limed. Tupper et al. (1987) evaluated deep placement (6 to 15 in.) of lime in a vertical band in the row using a plow with parabolic shanks modified with a fertilizer hopper to deliver lime in the furrow. Although both the Dundee and Forestdale silt loam soils used in this study had pH levels greater than 6.0 in the 0 to 5 in. depth, surface-applied lime with incorporation resulted in a similar 3-year average yield response to deep-placed lime. Direct comparisons are difficult due to a 4000 lb/acre lime rate with broadcasting and a 1500 lb/acre rate for deep placement, with both applied in 2 consecutive years. Broadcast-applied lime at this rate resulted in an excessive pH (>7.0) after the third growing season, and resulted in a pH increase to a depth of 15 in. Deep placement resulted in a slight increase in pH for the 0 to 10 in. depth and a decrease for the 10 to 15 in. depth. It is noteworthy that in-row subsoiling without liming resulted in a more acid surface (<6.0) and a trend of reduced yield. Therefore, with deep plowing, lime may be necessary due to the mixing of topsoil with acid subsoil.

With respect to soil samples analyzed by the Mississippi State Soil Testing Lab for 1 June 1995 through 30 July 1996, there is a significant acreage in all of the major land resources that could benefit from liming (Table 4.6). In considering liming materials and rates, factors such as cost, available soil Mg, and desired interval before lime is required again should be considered. In general, Coastal Plain and acid soils in the Blackland Prairie would benefit the most from Mg-containing dolomitic limestone, while Loess Belt and Delta soils primarily require calcitic limestones. Adams (1975) showed a response to dolomitic lime compared to calcitic only on a Benndale sandy loam and a Hartsells fine sandy loam having low available Mg.

Use of more costly fluid or pelletized lime at lower rates than standard agricultural lime does not provide an equal capacity to neutralize soil acidity (Snyder et al., 1996). Additionally, more-frequent liming would be necessary with these liming materials, and therefore the costs over time would be greater.

Table 4.6 Soil pH Results by Land Resource Areas in Mississippi for Samples Tested by the Mississippi State Soil Testing Laboratory (for the period 1 July 1995 through 30 June 1996)

Mississippi Land Resource Area	Soil pH range, %				
	<5	5 to 5.4	5.5 to 5.9	6 to 7	>7
Upper Coastal Plain	6	19	33	27	16
Lower Coastal Plain	7	24	39	29	1
Loess Belt	9	26	40	21	5
Delta	2	13	30	46	9
State average[a]	5	18	33	34	10

[a] Variable number of samples from each land resource area.

From McCarty, W. et al., MCES, Mississippi State University, Mississippi State, 1996.

References

Adams, F. 1975. Field experiments with Magnesium in Alabama–Cotton, Corn, Soybeans, Peanuts, Auburn Univ. (Ala.) Agric. Exp. Stn. Bull. 472.

Adams, F. 1984. Crop response to lime in the southern states, in F. Adams, Ed., *Soil Acidity and Liming.* Agron. Monogr. 12, 2nd ed. ASA, CSSA, SSSA, Madison, WI, 211–265.

Albrecht, W. A. 1932. Calcium and hydrogen on concentration in the growth and inoculation of soybeans, *J. Am. Soc. Agron.* 24:793–806.

Anderson, O. E., and J. J. Mortvedt, Eds. 1982. Soybeans: Diagnosis and correction of Mn and Mo Problems, Southern Cooperative Series Bull. 281. University of Georgia, Athens, 1.

Anthony, J. L. 1967. Fertilizing Soybeans in the Hill Section of Mississippi, Miss. Agric. Exp. Sta. Bull. 743.

Barber, S. A. 1978. Growth and nutrient uptake of soybean roots under field conditions, *Agron. J.* 70:457–461.

Bell, P. F., W. B. Hallmark, W. E. Sabbe, and D. G. Dombeck. 1995. Diagnosing nutrient deficiencies in soybean, using M-DRIS and critical nutrient level procedures, *Agron. J.* 87:859–865.

Bhangoo, M. S. and D. J. Albritton. 1976. Nodulating and non-nodulating Lee soybean isolines response to applied nitrogen, *Agron. J.* 68:642–645.

Bharati, M. P., D. K. Whigham, and R. D. Voss. 1986. Soybean response to tillage and nitrogen, phosphorus, and potassium fertilization, *Agron. J.* 78:947–950.

Boswell, F. C. 1982. Soil-plant molybdenum relationships, in O. E. Anderson and J. J. Mortvedt, Eds., Soybeans: Diagnosis and Correction of Manganese and Molybdenum Problems, Southern Cooperative Series Bull. 281, University of Georgia, Athens, 43–47.

Bray, R. H. 1945. Soil-plant relations: II. Balanced fertilizer use through soil tests for potassium and phosphorus, *Soil Sci.* 60:463–473.

Brevedan, R. E., D. B. Egli, and J. E. Leggett. 1978. Influence of N nutrition on flower and pod abortion and yield of soybeans, *Agron. J.* 70:81–84.

Brown, J. C. and V. D. Jolley. 1989. Plant metabolic responses to iron-deficiency stress, *BioScience* 39:546–551.

Brown, J. C., R. S. Holmes, and L. O. Tiffin. 1958. Iron chlorosis in soybeans as related to genotype of rootstock, *Soil Sci.* 86:75–82.

Carter, O. G., I. A. Rose, and P. F. Reading. 1975. Variation in susceptibility to manganese toxicity in 30 soybean genotypes, *Crop Sci.* 15:730–732.

Cervenelli, G. da Silva. 1980. Fertility factors involved in the growth and chlorosis of soybeans [*Glycine max* (L.) Merr.] in the Blackland Prairie land resource area, M.S. thesis, Mississippi State University, Mississippi State.

Coale, F. J. and J. H. Grove. 1990. Root distribution and shoot development in no-till full-season and doublecrop soybean, *Agron. J.* 82:606–612.

Deibert, E. J., M. Bijeriego, and R. A. Olsen. 1979. Utilization of ^{15}N fertilizer by nodulating and non-nodulating soybean isolines, *Agron. J.* 71:717–723.

deMooy, C. J., J. Pesek, and E. Spaldon. 1973a. Mineral nutrition, in B. E. Caldwell, Ed., *Soybeans: Improvement, Production, and Uses*, Agron. Mongr. 16, ASA, Madison, WI, 267–352.

deMooy, C. J., J. L. Young, and J. D. Kapp. 1973. Comparative response of soybeans and corn to phosphorus and potassium, *Agron. J.* 65:851–855.

ERS-NASS. 1994. Agricultural resources and environmental indicators. USDA-ERS Natural Resources and Environment Division, Agric. Hdbk., 705, Washington, D.C.

Evans, C. E., C. H. Burmester, F. Adams, J. T. Cope, Jr., and J. Odom. 1986. Soil Fertility Requirements for Soybeans, Auburn Univ. (Ala.) Agric. Exp. Sta. Res. Rpt. Series No. 4.

Gascho, G. J. 1993. Boron and nitrogen applications to soybeans: Foliar and through sprinkler irrigation, in L. S. Murphy, Ed., in *Foliar Fertilization of Soybeans and Cotton, Proc. Symp. PPI/FAR*, Cincinnati, OH, 8 Nov. 1993. PPI/FAR Spec. Publ. 1993-1, 17–33.

Gregory, T. L. and S. D. Corley. 1996. Mississippi Ag Report 96-17:1–4, USDA-NASS-MASS-MDAC, Jackson, MS.

Grundon, N. J. 1987. Hungry Crops: A Guide to Mineral Deficiencies in Field Crops, Queensland Government, Brisbane, Australia.

Hairston, J. E., J. O. Sanford, D. F. Pope, and D. A. Horneck. 1987. Soybean-wheat doublecropping: implications from straw management and supplemental nitrogen, *Agron. J.* 79:281–286.

Hairston, J. E., W. F. Jones, P. I. McConnaughey, L. K. Marshall, and K. B. Gill. 1990. Tillage and fertilizer management effects on soybean growth and yield on three Mississippi soils, *J. Prod. Agric.* 3:317–323.

Hallmark, W. B. and S. A. Barber. 1981. Root growth and morphology, nutrient uptake, and nutrient status of soybeans as affected by soil K and bulk density, *Agron. J.* 73:779–782.

Ham, G. E., W. W. Nelson, S. D. Evans, and R. D. Frazier. 1973. Influence of fertilizer placement on yield response of soybeans, *Agron. J.* 65:81–84.

Hanway, J. J. and R. A. Olsen. 1980. Phosphate nutrition of corn, sorghum, soybeans, and small grains, in F. E. Khasawneh et al., Eds., *The Role of P in Agriculture, Proc. Symp. ASA, CSSA, and SSSA*, Muscle Shoals, AL, 1–3 June 1976, ASA, CSSA, and SSSA, Madison, WI, 681–692.

Hanway, J. J. and C. R. Weber. 1971. N, P, and K percentages in soybean [*Glycine max* (L.)] plants, *Agron. J.* 63:406–408.

Hardy, R. W. F., R. C. Burns, and R. D. Holston. 1973. Applications of the acetylene-ethylene assay for measurement of nitrogen fixation, *Soil Biol. Biochem.* 5:47–81.

Harper, L.A., J. E. Giddens, G. W. Langdale, and R. R. Sharpe. 1989. Environmental effects on nitrogen dynamics in soybean under conservation and clean tillage systems, *Agron. J.* 81:623–631.

Heitholt, J. J. 1994. Supplemental B, boll retention percentage, ovary carbohydrates, and lint yield in modern cotton genotypes, *Agron. J.* 86:492–497.

Henderson, J. B. and E. J. Kamprath. 1970. Nutrient and Dry Matter Accumulation by Soybeans, N.C. Agric. Exp. Sta. Tech. Bull., 197.

Hoover, C. D. 1968. Research on crops, soils and seed technology, *Miss. Farm Res.* 31(6):3–5.

Jones, G. D., J. A. Lutz, Jr., and T. J. Smith. 1977. Effects of phosphorus and potassium on soybean nodules and seed yield, *Agron. J.* 69:1003–1006.

Jones, J. B., Jr., B. Wolf, and H. A. Mills. 1991. *Plant Analysis Handbook*, Micro-Macro Publishing, Athens, GA.

Jones, W. F. and L. E. Nelson. 1978. Response of field-grown soybeans to lime, *Commun. Soil Sci. Plant Anal.* 9:607–614.

Jones, W. F. and L. E. Nelson. 1982. The effect of subsoil manganese on growth of turnips and soybeans in the greenhouse, *Commun. Soil Sci. Plant Anal.* 13:113–126.

Jones, W. F. and L.E. Nelson. 1983. The Influence of Lime on the Yield of Soybeans. Miss. Agric. For. Exp. Stn. Inf. Sh. 1313.

Jones, W. F., G. E. Jackson, and C. A. Siregar. 1995. Evaluation of Rates and Methods of P and K Application in a Conservation Tillage System, Miss. Agric. For. Exp. Sta. Bull. 1037:1–19.

Kamprath, E. J. and U. S. Jones. 1986. Plant response to sulfur in the southeastern United States, in M. A. Tabatabai, Ed., *Sulfur in Agriculture*, Agron. Monogr. 27, ASA, CSSA, and SSSA, Madison, WI, 323–343.

Kamprath, E. J. and E. V. Miller. 1958. Soybean yields as a function of the soil phosphorus level, *Soil Sci. Soc. Am. Proc.* 22:317–319.

Karlen, D. L., P. G. Hunt, and T. A. Matheny. 1982a. Accumulation and distribution of P, Fe, Mn, and Zn by selected determinate soybean cultivars grown with and without irrigation, *Agron. J.* 74:297–303.

Karlen, D. L., P. G. Hunt, and T. A. Matheny. 1982b. Accumulation and distribution of K, Ca, and Mg by selected determinate soybean cultivars grown with and without irrigation, *Agron. J.* 74:347–354.

Keogh, J. L. and R. Maples. 1970. Soybean Fertilization: Timing and Placement of Phosphorus and Potassium, Arkansas Agric. Exp. Stn. Report Series 185.

Kluthcouski, J. and L. E. Nelson. 1979. Variations in the manganese concentrations in soybean trifoliolates, *Commun. Soil Sci. Plant Anal.* 10:1299–1310.

Lins, I. D. G., F. R. Cox, and J. J. Nicholaides, III. 1985. Optimizing phosphorus fertilization rates for soybeans grown on oxisols and associated entisols, *Soil Sci. Soc. Am. J.* 49:1457–1460.

Long, O. H., J. R. Overton, E. W. Counce, and T. McCutchen. 1965. Lime and Fertilizer Experiments on Soybeans, Univ. Tenn. Agric. Exp. Stn. Bull. 391.

Maples, R. and J. L. Keogh. 1969. Soybean Fertilization Experiments, Ark. Agric. Exp. Stn. Report Series 178.

Marschner, H. 1995. *Mineral Nutrition of Higher Plants*, 2nd ed., Academic Press, San Diego, CA.

McCarty, W., K. Crouse, and L. B. Wampler. 1996. Summarization of Soil Test Data by Selected Crops, MCES, Mississippi State University, Mississippi State.

Moraghan, J. T. 1985. Manganese deficiency in soybeans as affected by FeEDDHA and low soil temperature, *Soil Sci. Soc. Am. J.* 49:1584–1586.

Nelson, L. E. 1977. Changes in water-soluble Mn due to soil sample preparation and storage, *Commun. Soil Sci. Plant Anal.* 8:479–487.

Ohki, K. 1977. Critical zinc levels related to early growth and develop of determinate soybeans, *Agron. J.* 69:969–973.

Oplinger, E. S., R. G. Hoeft, J. W. Johnson, and P. W. Tracy. 1993. Boron fertilization of soybean: a regional summary, in L. S. Murphy, Ed., *Foliar Fertilization of Soybeans and Cotton, Proc. Symposium PPI/FAR*, Cincinnati, OH, 8 Nov. 1993, PPI/FAR Spec. Publ. 1993-1, Norcross, GA, 7–16.

Parker, M. B. 1982. Soil–plant manganese relationships, in O. E. Anderson and J. J. Mortvedt, Eds., Soybeans: Diagnosis and Correction of Manganese and Molybdenum Problems, Southern Co-operative Series Bull. 281. University of Georgia, Athens.

Parker, M. B. and H. B. Harris. 1962. Soybean response to molybdenum and lime and the relationship between yield and chemical composition, *Agron. J.* 54:480–483.

Paulsen, G. M. and O. A. Rotimi, 1968. Phosphorus–zinc interaction in two soybean varieties differing in sensitivity to phosphorus nutrition, *Soil Sci. Soc. Am.* 32:73–76.

Pettiet, J. V. 1971. Are Nitrogen Fertilizers Needed for Soybean? Miss. Agric. For. Exp. Stn. Inf. Sheet 1147.

Rogers, H. T., F. Adams, and D. L. Thurlow. 1973. Lime Needs of Soybeans on Alabama Soils, Auburn Univ. (Ala.) Agric. Exp. Stn. Bull. 452.

Rouse, R. D. 1968. Soil Testing Theory and Calibration for Cotton, Corn, Soybeans, and Coastal Bermudagrass, Auburn Univ. (Ala.) Agric. Exp. Sta. Bull. 375.

Sabbe, W. E. and R. J. Mahler. 1991. Soybean Response to Phosphorus and Potassium Fertilizer in Arkansas, Ark. Agric. Exp. Stn. Bull. 928.

Sinclair, J. B. 1993. Soybeans, in W. F. Bennett, Ed., *Nutrient Deficiencies and Toxicities in Crop Plants*, APS Press, St. Paul, MN, 99–103.

Snyder, C. and L. O. Ashlock. 1996. Late-season potassium deficiency symptoms in southern soybeans, *Better Crops Plant Food* 80:10–11.

Snyder, C. S., J. H. Muir, and G. M. Lessman. 1996. Spring-applied ag lime can provide immediate soybean response, *Better Crops Plant Food* 80:3–5.

Tupper, G. R., H. C. Pringle, III, M. W. Ebelhar, and J. G. Hamill. 1987. Soybean Yield and Economic Response to Broadcast-Incorporated and Deep Band Placement of Lime on Low pH Soils, Miss. Agric. For. Exp. Stn. Bull. 950.

Udoh, D. J. and L. E. Nelson. 1983. Evaluation of Differences in Cultivar Susceptibility to Iron Chlorosis, Miss. Agric. For. Exp. Stn. Res. Highlights. 46.

Udoh, D. J. and L. E. Nelson. 1986. Trifluralin-induced chlorosis in soybeans [*Glycine max* (L.) Merr.] grown in clayey, high pH soils, *Plant Soil* 96:175–184.

Woodruff, J. R. and W. F. Jones. 1982. Crop response to manganese, in O. E. Anderson and J. J. Mortvedt, Eds., *Soybeans: Diagnosis and Correction of Manganese and Molybdenum Problems*, Southern Cooperative Series Bull. 281. University of Georgia, Athens, 11–17.

Zapata, F., S. K. A. Danso, G. Hardarson, and M. Fried. 1987. Time course of nitrogen fixation in field-grown soybean using nitrogen-15 methodology, *Agron. J.* 79:172–176.

chapter five

Root zone associations and their influence on soybean production

Ernest H. Flint, Jr.

Contents

Introduction

When one turns a shovel full of soil in a field, the most apparent living organisms include plant roots and large insects. Soil contains an array of additional organisms that influence crop growth and development. Many soil organisms are microscopic but important nonetheless. Smaller animal forms aid in cycling organic matter and nutrients while others are parasites that feed on plant roots. Some fungi act as saprophytes that facilitate the decay of crop residue and the release of nutrients. Other fungi such as *Pythium* and *Rhizoctonia* cause plant diseases that damage or destroy crop plants. Another group of fungi, mycorrhizae, colonize plant roots and carry on a mutually beneficial or symbiotic relationship with the host plant by acting as auxiliary roots. This association extends the effective rooting zone of the plant, aiding in the uptake of water and nutrients. These fungi utilize energy from the plant for growth and reproduction. They serve as binding threads that assist in the creation of soil aggregates, thereby promoting plant root development (Miller and Jastrow, 1992a). Overall, the association between these fungi and their host plants is beneficial to both the plants and the fungi (G. B. Triplett, 1995, personal communication).

The ultimate role of all plants is to combine environmental components into substances that are utilized as food by other organisms. The most important role of the aboveground portion of a plant is to gather energy via photosynthesis. The primary role of the belowground portion of the plant is to gather raw materials required for plant function and development. The ability of plant roots to obtain mineral nutrients and water from soil depends on their ability to explore the soil.

Extent of effective rooting

Uptake of nutrients and water by plants is dependent on root proliferation, and on the ability of nutrients and water to move to plant roots (Hillel, 1982). Water and nutrients have a natural tendency to diffuse evenly through the soil by mass flow and capillary movement. Rate of movement is influenced by density, degree of compaction, and texture of the soil. Movement of water and nutrients is much slower in clay soil than in sand or silt soils because of the small size of pores through which they must pass, and also because some nutrients carry an electrical charge and are attracted to the clay particles.

Size of soil particles or texture is the most important factor influencing soil porosity. When a high percentage of clay is present, pores are smaller than when sand and silt predominate. Also, the type of clay dictates the expansion characteristics (Hillel, 1982). Organic matter affects porosity and rate of diffusion of air, water, and minerals by filling soil pores with a medium less resistant to diffusion than the mineral fraction. Air and water are usually more abundant in soils with good organic matter levels, and roots move through the soil more readily when organic matter is plentiful.

Soybean plants produce roots according to the constraints of the soil into which the seeds are planted. Coarse-textured soils that are adequately aerated, well drained, and contain sufficient levels of nutrients and water allow plants to build extensive root systems. These plants resist drought and produce good yields under a wide range of field conditions. Soils that are poorly aerated or that have physical or chemical barriers to root penetration often produce plants with poor drought tolerance. Yields from plants grown on such soils are usually low.

Barriers to root penetration are caused not only by physical soil conditions such as compaction by equipment, but also by concentrations of natural and applied chemicals. Acid soil (below about 5.8 pH) is a form of chemical barrier to soybean root penetration (Scott and Aldrich, 1970). Some soils have surface pH levels that are acceptable, but low subsoil pH prevents roots from entering the subsoil. When roots cannot reach water contained in the subsoil, drought tolerance of the crop is reduced. While low pH is likely the most common chemical barrier to root penetration, other types of barriers are also possible. Sodium and potassium salts may concentrate in some soils and limit root growth, and herbicides leached from the surface may reduce root development.

One method for evaluating the effect of chemical barriers on plant roots is to remove an occasional plant from the soil and inspect it for rooting depth and abnormal development. When compared with a well-developed root, the restricted root will have a short taproot and branch roots may be abnormally developed. Regular soil testing of both topsoil and subsoil is a vital source of information for deciding whether root-growth barriers are chemical or physical. When root growth is restricted by a physical barrier, such as a natural fragipan or a traffic pan caused by use of heavy implements, deep tillage may be the short-term answer. Reduced tillage, controlled traffic patterns, or avoidance of implements that cause compaction e.g., a disk harrow, can reduce this problem. Often, however, deep tillage is selected before the actual cause is determined. The problem must be accurately identified before a profitable solution can be found.

Mycorrhizae

What Are Mycorrhizae?

During the 1880s, several workers found fungi present on plant roots without apparent disease. In 1885, Frank named these fungus–root associations *mycorrhizae,* a word formed from the Latin words meaning fungus and root (Frank, 1885). Mycorrhizal fungi are present in all arable soils and have been found colonizing nearly all crop and weed plants. They are affected by soil management practices and crop rotation (Hendrix et al., 1995).

Mycorrhizae are fungi that associate with plant roots and increase their absorptive capacity. The plant and fungus are joined at the cellular level and carry on a two-way exchange of nutrients and water. Mycorrhizae have positive and negative effects. Some types of mycorrhizal fungi, while not pathogenic, have a negative final effect on the host plant. Others are about equal in positive and negative aspects, but most are beneficial to crops. They are active in the absorption and transport of nutrients and water from soil to plants, especially during drought periods (Busse and Ellis, 1985). Their potential impact is discounted by some soil scientists because many of their effects were previously attributed to physical and chemical processes. Those who have studied mycorrhizae have shown that these organisms are only another part of the complex system involved in plant/soil relationships (Miller and Jastrow, 1992b).

Between 1950 and 1960, studies were begun with fungi that were found in association with crop plant roots. These fungi are today referred to collectively as vesicular-arbuscular-mycorrhizal or VAM fungi because they develop vesicles for reserve food storage (used during stress periods), and arbuscules which act as exchange sites for nutrients between the fungus and the host plant. The arbuscules are bundles of coiled fungal strands which develop within root cells. Membranes permit exchange of nutrients between plant cells and the fungus. VAM fungi are the type found in association with soybean roots and most other crops. Roots that are infected by one or more of these fungi are considered to be "colonized." Several strains of VAM fungi are capable of colonizing soybean roots, with some being more beneficial to the plant than others under different conditions (Hendrix et al., 1995).

Role and function of mycorrhizae

Effect on uptake of water

The expanding body of knowledge on mycorrhizae has caused much speculation about their role in plant processes. The most concise statement about the relationship between these fungi and the plants they colonize was presented in 1975 by Pirozynski and Malloch.

They proposed that the association of beneficial fungi and land plants is the primary source of nutrition on Earth, and that plants and fungi have been associated since the adaptation of plants to dry land. A statement as broad as this requires further definition. Let us look at some of the evidence that might suggest the importance of beneficial fungi. Of greatest importance to the growth of most plants, particularly soybeans, is the availability of water. Until recently, the story of how mycorrhizae are involved in water uptake has been very sketchy, but a few studies have provided important pieces of the puzzle.

One of the first studies to describe changes in plant–water relations related to mycorrhizae was done with soybeans. Safir et al. (1972) innoculated sterile soil with a fungus known to produce mycorrhizae and then planted soybeans. After 30 days, mycorrhizal plants showed growth stimulation and differences in water uptake. Water uptake values in mycorrhizal plants were 70% higher than in plants without mycorrhizal root systems. They also observed that mycorrhizal soybean plants recovered from wilting more rapidly than plants without mycorrhizae.

VAM have been shown to enhance the development of nonirrigated soybeans by improving drought tolerance. In a 1985 study, Busse and Ellis stated that the higher yields of colonized plants may have been the result of either improved phosphorus nutrition or of more extensive soil water extraction. One part of their study showed that soil surrounding plants infected with VAM was consistently lower in water content from the surface to a depth of 18 in. They suggested that mycorrhizal root systems were capable of absorbing a greater portion of soil water than nonmycorrhizal roots. Their conclusions stated that drought-stressed mycorrhizal soybean plants yielded 10% more than drought-stressed nonmycorrhizal plants because of improved pod retention and seed fill.

Effect on nutrient uptake

As mycorrhizal strands proliferate and explore the soil, the fungi transport nutrients and water to plants in quantities much greater than roots can acquire alone (Powell and Bagyaraj, 1984). The key nutrient in studies of VAM is phosphorus because of its effect on cell membrane permeability and its role in energy transfer. When plant roots are able to acquire adequate phosphorus, root cell exudation or leakage is minimized. Mycorrhizae are then less likely to colonize roots. Conversely, when phosphorus level is low, more exudate is released into the soil, causing mycorrhizal development to increase.

Along with phosphorus, fungal hyphae absorb other mobile and immobile nutrients and transport them to the plant. Mobile nutrients such as nitrogen and potassium likely would have found their way to the root eventually. Immobile ones like phosphorus and copper that might not have been located by the unaided plant root are supplied to plants in greater quantities when roots are mycorrhizal.

Influence of applied fertilizers on mycorrhizae

Unlike natural environments such as prairie or forest, crop production environments require the addition of nutrients because large quantities are removed by the crop and must be replaced to maintain high production levels. In general, the addition of phosphate fertilizer reduces soluble carbohydrate exudate from roots (Graham et al., 1981). Because of the negative influence of phosphorus on mycorrhizal development, care should be exercised when selecting fertilizer rates and combinations. In long-term fertilizer studies done in Minnesota, highly beneficial mycorrhizae formed by *Glomus* fungal species increased significantly in plots fertilized annually with a complete mixture containing roughly 2.5 times more elemental potassium (K) than phosphorus (P) (Johnson, 1991).

Soil management practices and attributes that affect mycorrhizae

Soil history

Soils utilized for the production of soybeans are variable, and have a varying history of use. This history of use includes tillage practices, erosion control, drainage, fertilization, and cropping sequence. These factors, along with the soil type, determine how well the soil may serve as a medium for root growth and colonization by mycorrhizal fungi (Tisdall and Oades, 1980).

Tillage

Disturbance by tillage breaks down the fungal network built by cover crops and other fungal hosts (Read and Birch, 1988). When this fungal network is disrupted by tillage, colonization is much slower and less extensive. Beneficial effects of fungal networks are reduced under intensive tillage (Thompson, 1987). The improved mycorrhizal colonization in no-till crops has been described in wheat (Yocum et al., 1985), edible beans (Mulligan et al., 1985), and corn (Anderson et al., 1987). These studies showed a higher incidence of mycorrhizal colonization, increased nutrient uptake, and improved yields when soils were not disturbed than when a conventional tillage system was used.

Cover crops and native vegetation

The benefits of cover cropping (planted or native) to soil fertility have long been known, but when mycorrhizal colonization is considered, cover crops become even more important. The increase in organic matter, although temporary, brought about by cover cropping is very important for raising the level of mycorrhizal activity. Soil organic matter is a fraction of the soil mass along with sand, silt, or clay. Soil testing laboratories report an organic matter percentage just as they would report the percentage of sand, silt, or clay from a textural analysis. While this is a good index for comparing one soil to another, a more appropriate way to think of organic matter is to consider the level of organic activity in the soil, or the level of life in the soil mass. Organic matter is really a dynamic concept rather than a static one. When you look at it this way, it is easier to understand how cultural practices influence organic matter in soil.

Choice of cover crop is important since some species may act as companion plants for the warm-season crop. The companion plant maintains the fungus in an active state through the winter, allowing the summer crop to benefit from an active hyphal network as soon as its roots penetrate the soil (Johnson et al., 1991; 1992). One example of a companion plant relationship is that both wheat and soybeans are colonized by the same species of *Glomus* mycorrhizal fungi.

Fallowing.

Fallowing has been used for centuries as a way of managing insects, weeds, and plant diseases. However, fallowing is harmful to mycorrhizae since there are no plants for the fungus to colonize during the fallow period. The active fungal network is destroyed and the fungi must rebuild from spores (the least effective means of survival). Thus, the beneficial effects of mycorrhizae are delayed (Harinikumar and Bagyaraj, 1988). This negative effect of fallowing on mycorrhizae may appear as a direct contradiction to studies that have shown crop yield increases after fallow. When soil is fallowed, organic matter is destroyed, releasing bound nutrients that may produce a plant growth response. Yield increases after fallow are often due to the absence of negative factors rather than a direct positive influence (Crookston et al., 1988).

Soil fertility

Virtually every factor influencing soil fertility affects mycorrhizal fungi and subsequent development of fungal networks. However, the most significant factor is phosphorus (Abbott and Robson, 1984). Mycorrhizal fungi absorb phosphorus and other nutrients along with the water that suspends them. Therefore, both nutrients and water are extracted from the soil more efficiently by mycorrhizal root systems than by an unaided plant root system. Likewise, soybean plants colonized by VAM are more adequately supplied with micronutrients such as copper, zinc, and boron than are nonmycorrhizal plants (Pacovsky et al., 1986).

Soil physical characteristics

The physical characteristics of soil directly influence the development of the host plant root system and mycorrhizae since they control the availability of air and water (Mulligan et al., 1985). Both roots and mycorrhizae require oxygen and water for metabolic processes and survival. When soil is low in oxygen because of compaction or saturation with water, neither roots nor fungi can thrive. Likewise, if soil becomes dry enough to inhibit the extraction of water, neither roots nor mycorrhizae can survive (Saif, 1983).

Interactions with mycorrhizae

Mycorrhizae and soil characteristics

As mycorrhizal fungi become more established in soil, they create an increasing network of hyphae throughout the soil mass. As soil becomes increasingly proliferated by the fungal network, aggregation is increased significantly by the hyphal strands linking soil particles together into microaggregates and later into macroaggregates. As soil aggregates become more stable, they are less vulnerable to breakdown by water and less prone to erosion (Miller and Jastrow, 1990). As fungal hyphae extend through the soil and die at the end of a crop year, soil micropores are created. These pores enhance water and air infiltration, thus improving the soil as a medium for the development of both plant roots and mycorrhizae (Miller and Jastrow, 1992b).

Mycorrhizae and nodulating bacteria

Of particular importance in a soybean production system is the relationship between mycorrhizae and *Rhizobium* bacteria which fix atmospheric nitrogen for use by the crop. In general, soybean roots that are more thoroughly colonized by VAM fungi are more heavily nodulated by nitrogen-fixing bacteria. Studies by Bethlenfalvay et al. (1985) showed that numbers of nodules were somewhat lower in VAM plants; however, another study by Pacovsky et al. (1986) showed that even though numbers of nodules were lower, the size and activity of the nodules present were greatly increased. Soybean plants have been shown to utilize nutrients more efficiently when both nitrogen-fixing *Rhizobium* bacteria and a mycorrhizal fungus were present. Photosynthetic nutrient-use efficiency was also found to be significantly higher in leaves of these plants (Pacovsky et al., 1986; Brown et al., 1988).

The following yield data (Table 5.1) are taken from a 1988 study by Young et al. (1988) in which soybeans were grown without *Rhizobium* and mycorrhizae, colonized alone by *Rhizobium* or by mycorrhizae, and colonized by both *Rhizobium* and mycorrhizae. Yields are the average of six test locations.

Mycorrhizae and nematodes

It appears that nematodes are less damaging to mycorrhizal soybean plants. The symptoms of nematode infection are reduced as indicated by numbers of cysts, young nematodes,

Table 5.1   Inoculation Response of Soybean to
Rhizobium and Mycorrhizae

Type inoculation	Yield (bu/acre)
None	26.8
Rhizobium alone	31.3
Mycorrhizae alone	28.8
Rhizobium and mycorrhizae	34.0

and eggs associated with mycorrhizal plants and soil around their roots (Hussey and Roncadori, 1982; Ingram, 1988). The influence of mycorrhizae on plants has been shown to be very positive, and many of these effects can be duplicated by simply adding phosphorus to the soil (MacGuidwin et al., 1985). However, mycorrhizal fungi have been shown to provide tolerance to nematode damage even when phosphorus has been eliminated as a factor. The influence of mycorrhizae on nematodes has been attributed to physiological changes in the roots (Cooper and Grandison, 1986; 1987).

Mycorrhizae and plant root diseases

Disease suppression by VAM can be attributed to several mechanisms. One of the more apparent ways mycorrhizae reduce root diseases is by improving the efficiency of nutrient uptake, and thereby producing a more vigorous plant. Such plants are better able to withstand stresses that would otherwise make them more vulnerable to attack by disease organisms (Linderman, 1992). The mechanisms by which mycorrhizae may alter the effects of diseases on plant roots include the possibility for morphological alteration of the roots themselves, thereby causing the disease difficulty in attacking the plant. Physiological changes are also a possibility since VAM help satisfy the immediate nutrient needs of the root tissue (Balthruschat and Schonbeck, 1975). Although both of the above-mentioned mechanisms may be involved, the preferred explanation is the alteration of physiological aspects of soil/root relationships. One way of understanding this concept is by looking at a forest soil in which plant roots and a wide diversity of microscopic plants and animals coexist as a finely tuned environment with much of the nutrient flow directed toward trees. Nutrients and water are simultaneously extracted from soil and recycled within a natural system. Outside influences come from animals and humans, but even these factors are incorporated into the system. All of this is taking place in agricultural fields, forests, pastures, greenhouses, lawns, or anywhere plants grow.

Summary

Unlike many practices that are offered to farmers by the research community, the utilization of mycorrhizae for increasing yields does not involve increased monetary inputs. These organisms are naturally occurring in all agricultural soils and only require good soil management and a good soil fertility program. Proven agronomic practices favor the development of mycorrhizal root systems just as they favor good plant development. The only special considerations may be the avoidance of excessive amounts of soil phosphorus, since this element suppresses the development of beneficial fungi, and a reduction in soil disturbance through the use of minimum tillage practices.

Our knowledge of the complex system described in this chapter is at best very limited. These are natural systems that operate well without intervention by humans; our entry into them often causes damage to the fine balance already in place. Improved fungal strains may be developed in future research, enabling farmers to increase yields further through soil inoculation. At this time, however, introduced mycorrhizae have not been proved any more beneficial than naturally occurring fungi. Much study remains to be conducted in the areas of plant nutrition, the balance of nutrients in agricultural systems, and the interaction of plants with both symbiotic and pathogenic organisms.

References

Abbott, L. K. and A. D. Robson. 1984. The effect of mycorrhizae on plant growth, in C. L. Powell and D. J. Bagyaraj, Ed., *VA Mycorrhiza*, CRC Press, Boca Raton, FL, 113–130.

Anderson, E. L., P. D. Millner, and H. M. Kunishi. 1987. Maize root length density and mycorrhizal infection as influenced by tillage and soil phosphorus, *J. Plant Nutr.* 10:1349–1356.

Balthruschat, H. and F. Schonbeck. 1975. Studies on the influence of endotrophic mycorrhizae on the infection of tobacco by *Thielaviopsis basicola*, *Phytopathol. Z.* 84:172–188.

Bethlenfalvay, G. J., M. S. Brown, and A. E. Stafford. 1985. Glycine-Glomus-Rhizobium symbiosis: II. Antagonistic effects between mycorrhizal colonization and nodulation, *Plant Physiol.* 79:1054–1058.

Brown, M. S., and G. J. Bethlenfalvay. 1988. The Glycine-Glomus-Rhizobium symbiosis: VII. Photosynthetic nutrient-use efficiency in nodulated, mycorrhizal soybeans, *Plant Physiol.* 86:1292–1297.

Busse, M. D. and J. R. Ellis. 1985. Vesicular-arbuscular mycorrhizal (*Glomus fasciculatum*) influence on soybean drought tolerance in high-phosphorus soil, *Can J. Bot.* 63:2290–2294.

Cooper, K. M. and G. S. Grandison. 1986. Interaction of vesicular-arbuscular mycorrhizal fungi and root-knot nematode on cultivars of tomato and white clover susceptible to *Meloidogyne hapla*, *Ann. Appl. Biol.* 108:555–565.

Cooper, K. M. and G. S. Grandison. 1987. Effects of vesicular-arbuscular mycorrhizal fungi on infection of tamarillo (*Cyphomandra betacea*) by *Meloidogyne incognita* in fumigated soil, *Plant Dis.* 71:1101–1106.

Crookston, K. R., J. E. Kurle, and W. E Lueschen. 1988. Relative ability of soybean, fallow, and triacontanol to alleviate yield reductions associated with growing corn continuously, *Crop Sci.* 28:145–147.

Frank, A. B. 1885. Uber die auf Wurzelsymbiose beruhende. Ernahrung gewisser. Baume durch unterirdische Pilze, *Ber. Dtsch. Bot. Ges.* 3:128.

Graham, J. H., R. T. Leonard, and J. A. Menge. 1981. Membrane-mediated decrease in root exudation responsible for phosphorus inhibition of vesicular-arbuscular mycorrhizae formation, *Plant Physiol.* 68:548–552.

Harinikumar, K. M. and D. J. Bagyaraj. 1988. Effect of crop rotation on native vesicular-arbuscular mycorrhizal propagules in soil, *Plant Soil* 110:77–80.

Hendrix, J. W., B. Z. Guo, and Z.-Q. An. 1995. Divergence of mycorrhizal fungal communities in crop production systems, *Plant Soil* 170:131–140.

Hillel, D. 1982. *Introduction to Soil Physics*, Academic Press, San Diego, CA.

Hussey, R. S. and R. W. Roncadori. 1982. Vesicular-arbuscular mycorrhizae may limit nematode activity and improve plant growth, *Plant Dis.* 66:9–14.

Ingram, R. E. 1988. Interactions between nematodes and VA mycorrhizae, *Agric. Environ.* 24:169–182.

Johnson, N. C. 1991. Plant and Soil Regulation of Mycorrhizae in Natural and Agricultural Ecosystems, Ph.D. dissertation, University of Minnesota, Minneapolis, MN, *Diss. Abstr.* 92–07792.

Johnson, N. C., F. L. Pfleger, R. K. Crookston, S. R. Simmons, and P. J. Copeland. 1991. Vesicular-arbuscular mycorrhizas respond to corn and soybean cropping history, *New Phytol.* 117:657–663.

Johnson, N. C., P. J. Copeland, R. K. Crookston, and F. L. Pfleger. 1992. Mycorrhizae: a possible explanation for yield decline associated with continuous cropping of corn and soybean, *Agron. J.* 84:387–390.

Linderman, R. G. 1992. *VA Mycorrhizae and Soil Microbial Interactions*, in *Mycorrhizae in Sustainable Agriculture*, American Society of Agronomy Special Publication No. 54. Madison, WI, 45–70.

MacGuidwin, A. E., G. W. Bird, and G. R. Safir. 1985. Influence of *Glomus fasciculatum* on *Meloidogyne hapla* infecting *Allium cepa*, *J. Nematol.* 17:389–395.

Miller, R. M. and J. D. Jastrow. 1990. Hierarchy of root and mycorrhizal fungal interactions with soil aggregation, *Soil Biol. Biochem.* 22:579–584.

Miller, R. M. and J. D. Jastrow. 1992a. The application of VA mycorrhizae to ecosystem restoration and reclamation, in M.F. Allen, Ed., *Mycorrhizal Functioning*, Routledge, Chapman and Hall, New York, 438–467.

Miller, R. M. and J. D. Jastrow, 1992b. The role of mycorrhizal fungi in soil conservation, in *Mycorrhizae in Sustainable Agriculture*, American Society of Agronomy Special Publication No. 54, Madison, WI, 29–44.

Mulligan, M. F. and A. J. M. Smucker, and G. F. Safir. 1985. Tillage modification of dry edible bean root colonization by VAM fungi, *Agron. J.* 77:140–144.

Pacovsky, R. S., G. Fuller, and A. E. Stafford. 1986. Nutrient and growth interactions in soybeans colonized with *Glomus fasciculatum* and *Rhizobium japonicum*, *Plant Soil* 92:37–45.

Pirozynski, K. A. and D. W. Malloch, 1975. The origin of land plants: a matter of mycotrophism, *BioSystems* 6:153.

Powell, C. L. and J. D. Bagyaraj. 1984. *VA Mycorrhiza*, CRC Press. Boca Raton, FL.

Read, J. B. and P. Birch. 1988. The effects and implications of disturbance of mycorrhizal mycelial systems, *Proc. R. Soc. Edinb.* 94B:13–24.

Safir, G. R., J. S. Boyer, and J. W. Geredmann. 1972. Nutrient status and mycorrhizal enhancement of water transport in soybean, *Plant Physiol.* 49:700–703.

Saif, S. R. 1983. The influence of soil aeration on the efficiency of vesicular-arbuscular mycorrhizae: II. Effect of soil oxygen on growth and mineral uptake in *Euphorbium odoratum* L., *Sorghum bicolor* (L.) Moench and *Guizotia abyssinica* (L.f.) Cass. inoculated with vesicular-arbuscular mycorrhizal fungi, *New Phytol.* 95:405–417.

Scott, W. O. and S. R. Aldrich. 1970. Modern soybean production, *Farm Q.*, Cinncinnati, OH.

Thompson, J. P. 1987. Decline of vesicular-arbuscular mycorrhizae in long fallow disorder of field crops and its expression in phosphorus deficiency of sunflower, *Aust. J. Agric. Res.* 38:1494–1499.

Tisdall, J. M. and J. M. Oades. 1980. The effect of crop rotation on aggregation in a red-brown earth, *Aust. J. Soil Res.* 18:423–433.

Yocum, D. H., H. J. Larsen, and M. G. Boosalis. 1985. The effects of tillage treatments and a fallow season on VA mycorrhizae of winter wheat, in *Proc. 6th North Am. Conf. on Mycorrhizae*, Bend, OR, 25–29 June 1984, Forest Res. Lab., Corvallis, OR, 297.

Young, C. C., T. C. Juang, and C. C. Chao. 1988. Effects of *Rhizobium* and vesicular-arbuscular mycorrhiza inoculations on nodulation, symbiotic nitrogen fixation, and soybean yield in subtropical-tropical fields, *Biol. Fertil. Soils* 6:165–169.

chapter six

Tillage systems for soybean production

Richard A. Wesley

Contents

Introduction

Early humans were nomadic in nature and obtained their food from naturally occurring plants and animals. Once permanent settlements were established, it was necessary to obtain food from more limited areas. Desired food plants were grown by hand, which involved pulling all competing vegetation from the food-producing area. Eventually, all desired food plants were grown in selected areas by the complete removal of all existing vegetation with crude stone or wooden implements; this is recognized as the beginning of soil tillage. With the advent of metals for making tools and the use of draft animals, it became feasible to till larger areas, thus providing food for large concentrations of people. Approximately 6000 years ago, farmers in the Nile valley developed the plow by hitching an ox to a hoe, which was the beginning of power farming.

At an early stage in agriculture, row cropping was developed for ease of controlling weeds and preventing soil crusting. This system became more important with the adaptation of power-drawn tools to tractors in the 20th century. Weed control became the major function of tillage and "clean" or "complete" tillage became the accepted production practice. This level of tillage became known as "conventional" tillage. However, it exposed the soil to wind and water erosion.

In the 1930s, after several drought years, erosion problems associated with conventional tillage led to the development of tillage equipment that left residues of previous crops on the soil surface to prevent wind erosion and to conserve soil moisture. As tractors became more available, many new tillage implements were developed. In many instances this led to excessive tillage and soil compaction that often offset the benefits of extra tillage.

A variety of tillage implements, together with the development of new herbicides, fostered new strategies for weed management. These developments, along with the economic pressure to reduce production costs, brought about a critical examination of tillage practices that had been considered essential. This examination led to the development and adoption of conservation tillage (reduced- or no-tillage) systems that reduced soil losses from wind and water erosion, and conserved labor, energy, and soil moisture (Hairston et al., 1984a; Johnson, 1987). The adaptability of these production systems to specific situations may be dependent on the availability of herbicides to control specific weeds (Webber et al., 1987). As new herbicides are developed, a more diversified array of weed problems can be controlled, thus increasing the probability of success of these systems.

Crop production systems

Conventional tillage

In 1972, conventional tillage was practiced on 86% of the total soybean acreage in the U.S., but decreased to 40% in 1995. However, conventional or clean-tillage systems are still the predominant cropping systems used on most of the row crop area in the midsouthern US.

In the midsouthern U.S., conventional tillage inputs are utilized to control weeds, incorporate fertilizers and herbicides, and to prepare a clean seedbed that allows good seed–soil contact for prompt and uniform plant emergence. Production inputs to this system vary from region to region but basically include both pre- and postplant tillage and pre- and postemergence herbicides. Preplant tillage is generally accomplished with one or two passes with either a chisel plow or disk harrow to destroy weeds and incorporate residue from the previous crop. A disk harrow or field cultivator is then utilized to incorporate preemergence herbicides. On clayey soils, spring-tooth or spike-tooth harrows are used to level and smooth the soil prior to planting.

Postplant tillage for soybean planted in 20- to 40-in.-wide rows in conventional production systems involves one to three passes with a row crop cultivator as needed for weed control. Weeds in narrow-row soybeans (rows spaced less than 20 in. apart) are generally controlled with timely application of pre- and postemergence herbicides.

A significant portion of the acreage in the Mississippi River alluvial floodplain contains a high percentage of clay and is capable of holding large amounts of water. If excess rainfall occurs in the spring, the soils dry very slowly and are not receptive to timely tillage necessary for timely plantings. However, if conventional tillage is conducted in the spring, the resulting soil aggregates may remain too large and hard and the soil surface will be too rough and dry for planting. The only remedy is to receive rainfall in sufficient quantity and intensity to cause breakdown of the tillage-induced clods. If timely rainfall is not received, planting of soybean on these soils may be delayed beyond the optimum planting date and thus result in lower than optimum yields (Heatherly, 1988; Heatherly and Elmore, 1983; 1986).

The lack of sufficient rainfall after preplant tillage in the spring can delay planting until late May or June. Research in Mississippi indicates this delay in planting MG V and VII soybeans did not affect the yield of nonirrigated soybeans but reduced the yield of irrigated soybeans (Heatherly and Elmore, 1986). To assure timely planting on these soils, a new production system (stale seedbed) was developed in 1986 (Heatherly and Elmore, 1983; 1986). It involves a primary tillage operation of disking or chisel plowing after soybean harvest. Winter rains settle and smooth the tilled soil. Thus, timely planting can be achieved in a stale seedbed without spring tillage. A burndown herbicide is generally utilized in the spring prior to planting to control the winter vegetation.

Reduced tillage

Numerous factors have influenced the adoption of reduced-tillage systems for soybeans. These factors are the reduction of soil and water runoff; the reduction of soil compaction from reduced equipment traffic; conservation of soil moisture; and timely planting.

A major reason for the increases in the reduced-till production concept is the time savings at planting. The reduced-till system deletes preplant tillage inputs and allows farmers to plant soybeans earlier than in conventional systems. However, the reduced-tillage system includes postplant tillage (cultivation) as needed to assist in weed control.

Preplant tillage inputs in conventional production systems assist in controlling weeds and provide a smooth seedbed for planting soybean. As tillage operations are deleted from the crop production system, weed infestations and shifts in weed species may occur. As these changes occur, chemical weed control must be utilized in a timely manner in lieu of tillage throughout the growing season. Weed populations present at planting must be controlled with a burndown herbicide to ensure maximum crop potential. In the mid-southern U.S., pre- and postemergence herbicides with residual activity are often required in the absence of tillage (Heatherly and Elmore, 1983). If weeds are adequately controlled for the first 5 weeks after soybean emergence, they are no longer competitive for soil moisture and nutrients and do not reduce crop yield potential.

No tillage

In no-till production systems, all primary tillage for seedbed preparation and secondary tillage (cultivation) for weed control are deleted. Thus, an increased demand is placed on surface-applied and foliar herbicides for weed control. In no-till systems, certain perennial weeds such as johnsongrass, bermuda grass, redvine, and purple and yellow nutsedge tend to increase in the absence of tillage. Therefore, weed management systems become more complex and critical to the success of the no-till system. Herbicides required often include a burndown herbicide, one or two preemergence herbicides, and one or more postemergence herbicides.

Herbicides selected for weed control in no-till systems are determined by the existing vegetation at planting time, weed history of crops grown previously in the field, and weeds that emerge during the growing season. Herbicides selected may vary from year to year based on the crops grown and cropping practices employed the previous year.

Environmental effects of tillage systems

General

Crop production practices directly impact our environment. Tillage practices, whether conventional, reduced, or no-till, are responsible for a significant portion of the soil loss and sediment in our surface waters. Changing a tillage practice can have more than one impact on the environment. A reduction in tillage may decrease the level of some pollutants while increasing others.

Soil erosion

Soil erosion is a complex process that involves the detachment and transport of soil particles by raindrop impact and runoff. Sediment losses are generally higher during the months of March through July because of the intensive tillage, little to no ground cover, and pattern of rainfall. However, the erosion rate of a given field is a function of vegetative cover, climatological and topographical factors, and the susceptibility of different soils to erosion. One of the most important factors affecting erosion is one that producers can control: the cropping and land management practices utilized for crop production.

Soil loss and runoff from conventionally tilled fields are, in most cases, significantly higher than from fields with reduced tillage. This is attributed to the destruction of ground cover by tillage operations for seedbed preparation. However, for soils subject to surface sealing, conventional tillage that leaves the surface rough decreases runoff velocity and erosion for the first few rainstorms (Hairston et al., 1984a).

Until recently, soil erosion in the Mississippi Valley area was not considered to be a serious problem because most slopes are less than 2%. However, recent measurements made in the Mississippi Delta indicate otherwise. On a 40-acre watershed with Sharkey clay soil graded to a 0.2% slope, a 2-year average sediment yield rate of 11.7 ton/acre/year was recorded for continuous cotton culture in 1972 and 1973 (Murphree et al., 1976). In a later study (1979–1984) on the same watershed, the annual sediment yield from conventionally tilled soybeans averaged 4.5 ton/acre/year (Murphree and McGregor, 1991). Tillage during the cotton study was more intensive than during the soybean study and contributed to the higher sediment yields during the cotton study.

Reduced and no-till systems that leave part or all of the previous year's crop residues on or near the soil surface protect the soil against raindrop impact and increase surface roughness, thereby reducing runoff velocity and soil detachment. The goal of conservation tillage systems is to maintain crop residues on 30% or more of the soil surface. A 3-year study was conducted on erosion plots located near Holly Springs, MS. Soils on the plots were predominantly a Providence silt loam with 5% slope. In this study, the highest amount of crop residue was maintained by a no-till soybean system and resulted in an annual soil loss of only 0.6 ton/acre/year (Mutchler and Greer, 1984). Cultivation in the reduced-tillage soybean system during the growing season significantly decreased residue cover and, in turn, increased soil loss to 3.3 ton/acre/year. Soil loss from a soybean–wheat doublecrop system where soybeans were cultivated during the growing season averaged 1.1 ton/acre/year. In a conventionally tilled soybean system used as the check, tillage virtually destroyed all ground cover each year and resulted in a soil loss of 7.9 ton/acre/year. Ground cover in the conventionally tilled and reduced-tillage systems averaged 1.7 and 2.0 ton/acre, whereas the no-till and doublecrop systems averaged 3.9 and 3.6 ton/acre, respectively.

In another study in north Mississippi on a highly erodible loessial soil, McDowell and McGregor (1980) reported soil losses from no-till soybean were only about 1% of that from conventionally tilled soybean (0.13 vs. 12.9 ton/acre/year). Additional studies near Brooksville, MS on a Blackland Prairie soil with 3% slope show the annual loss from a soybean–wheat doublecrop system averaged 1.0 ton/acre/year compared to 3.1 ton/acre/year from conventionally tilled monocrop soybean (Hairston et al., 1984a).

Pesticides and sediments

The application of pesticides to agricultural fields or crops inevitably leads to pesticide contamination of runoff water from fields, unless the runoff does not occur during the lifetime of the pesticide residue. Most pesticides show a strong affinity for soil and sediment surfaces. Sediment in runoff water generally has much higher concentrations of pesticides than the associated water carrying the sediment. Thus, tillage practices that reduce erosion generally reduce pesticides in runoff water.

In reduced-tillage systems, herbicide usage is often increased because it replaces certain tillage inputs. This increase in herbicide usage also increases the potential for herbicide losses in runoff, unless soil and water losses are controlled. Broadcast applications of herbicides result in greater potential losses, on an area basis, than band applications. Thus, band application of herbicides should be utilized in reduced-tillage systems where cultivation is utilized to assist with weed control.

Conservation tillage practices greatly reduce soil and water losses. In a Mississippi study, Mutchler and Greer (1984) reported 42% of the rainfall received during the crop production season on the conventionally tilled soybean plots left the field as runoff, whereas only 26 to 30% of the rainfall received on the reduced, no-till, and doublecrop plots was lost to runoff. Thus, runoff was reduced by as much as 7 acre-in./year by the conservation tillage treatments. It was also noted that runoff from all monoculture cropping systems was approximately twice as large in the harvest-to-planting period (winter) as it was in the planting-to-harvest period (summer).

Soil compaction

Compaction of soil occurs whenever there are external forces acting on the soil. Compaction, whether natural or artificial, alters the soil condition and affects the growth of plants. In particular, machinery wheel traffic may cause substantial compaction and alter the soil structure considerably. The soil porosity is reduced by wheel traffic, whereas the density and strength of the soil are increased. These changes are a function of the soil moisture content, pressure applied to the soil by the wheels, and number of passes by machines, and adversely affect the content and movement of air, water, heat, and nutrients in the soil. The reduction in soil porosity reduces the free exchange of air between the root zone and the surface environment. The increase in soil density and strength often reduces root development.

Machinery operations carried out on wet fields can result in very dense subsoil which reduces the permeability of the soil. In addition, the increase in bulk density results in a change in the pore size distribution (decreases the percentage of large pores), which then affects the flow characteristics of water through clay soil (Douglas and McKyes, 1978). A dense soil is detrimental to soil-plant-water relationships because it drains slowly, has a low water storage capacity, and a high suction requirement for plants to extract water. On soft clay soils, the soil structure was significantly damaged by a pressure of 14.2 psi (Kuipers, 1963). These impervious subsoils restrict root penetration and growth, and thereby increase water runoff, erosion, and nutrient loss, and often significantly reduce crop yields.

The best approach to determine whether soil compaction is a limiting factor in a particular field is to observe the crop. If root growth is severely restricted, the crop will suffer from pronounced drought stress and subsequent yield reduction. This is especially true on coarse-textured soils that hold less available water in the effective rooting zone compared with finer-textured soils. Plant symptoms of compaction include slow emergence or poor stands, uneven growth especially in wheel tracks, stunted growth, nutrient-deficiency symptoms (soil compaction can cause poor nodulation in soybean), excessive or early wilting during drought stress, and abnormal rooting patterns.

Because many of these plant symptoms are common to problems other than compaction, the best way to identify soil compaction problems is to use a shovel. Dig up a few plants in the area you suspect may have a compaction problem. Look at the soil structure. Is it cloddy and hard? Are there any large pores in the soil to increase rooting and water infiltration? Most importantly, look at the roots. Does the taproot turn sideways? Is there a proliferation of side-branching roots where a taproot has stopped growing? Are the roots misshapen or abnormally thickened? These symptoms are good indicators that compaction is great enough to be a problem.

A cone penetrometer provides a standard and uniform method of characterizing the penetration resistance of soils. It is a hand-held device consisting of a stainless steel cone, a graduated driving shaft, and a dial readout. The cone is pushed into the soil at a uniform rate. The force required to press the cone through the soil is a measure of the soil strength and is called the "cone index."

During the last two decades, researchers have attempted to reduce or eliminate the effects of compaction on crop growth and yield. In 1969, a controlled traffic system for cotton production was developed. Cotton grown in fields where the wheel traffic was confined to established traffic lanes produced significantly higher seed cotton yields than a conventional system (Cooper et al., 1969; Dumus et al., 1975). In 1980, deep tillage was applied only in the production zone between the traffic lanes and significantly increased cotton yields (Williford, 1980).

Previous research in Florida indicated subsoiling under the row of soybeans increased root density below the plow pan and increased soybean yields 11.0 bu/acre compared with control plots (Rhoads, 1978). In Georgia, a 2-year study with four soybean varieties indicated that subsoiling increased average yield by 16% (Parker et al., 1976). Recent research in Florida on an Arredondo fine sand indicated in-row subsoiling of soybeans increased yields of conventionally tilled and no-till plots 35 and 48%, respectively (Vazquez et al., 1989).

In West Tennessee, chisel plowing or subsoiling consistently and substantially reduced surface compaction of silt loam soils with fragipans located 18 to 24 in. below the surface (Tyler and McCutchen, 1980). Penetrometer readings on Calloway, Grenada, and Henry silt loam soils consistently indicated less resistance where chiseling or subsoiling was performed compared with disking only. However, soybean yields were not increased by chiseling or subsoiling under the row on these soils. The fragipans on all three soils supplied adequate water to the crop. Thus, water was not limiting soybean yields.

Another study in West Tennessee compared no-tillage to shallow and deep tillage for soybean production on soils where compaction had developed over time as a result of surface tillage (Tyler and Overton, 1982). Treatments were disking 4 in. deep; chisel plowing 10 in. deep; moldboard plowing 10 in. deep; between-row subsoiling 10 in. deep; under-row subsoiling 10 in. deep; and no-tillage in a chemically killed winter wheat crop. The soil was a Lexington silt loam that had been in continuous crop production for 70 years with annual moldboard plowing and disking. Over the 3-year study, yields from all treatments were statistically similar.

A recent study in Mississippi evaluated the effects of three deep tillage implements in a controlled traffic system on soybean yield on a Tunica clay soil (Wesley and Smith, 1991; Wesley et al., 1994). All deep tillage was performed in the row direction in the fall when the soil profile was relatively dry. In a sprinkler-irrigated environment, soybean yields from all deep-tilled and conventional systems were similar and averaged

Table 6.1 Yield of Soybeans Grown in a Deep Tillage-Controlled Traffic Study with and without Irrigation on Tunica Clay near Stoneville, MS (1987–1991)

Irrigation treatment	Tillage treatment[a]	Crop year, bu/acre					
		1987	1988	1989	1990	1991	Avg.
Irrigated	DT1	55	53	35	32	57	46
	DT2	57	52	32	30	57	46
	DT3	57	50	35	29	57	46
	C	55	50	36	28	55	45
	Avg	56	51	35	30	57	46
Nonirrigated	DT1	48	41	42	26	56	43
	DT2	48	43	42	25	57	43
	DT3	44	36	41	25	58	41
	C	27	17	40	14	45	29
	Avg	42	34	41	23	54	39
LSD (0.05)[b]							
Compare treatments within irrigation levels		5	5	3	5	5	5
Compare treatments across irrigation levels		6	9	3	5	8	5
Compare irrigation levels		5	8	3	3	7	4

[a] Tillage treatments: DT1 = triplex subsoiler with 1 shank; DT2 = parabolic subsoiler with 2 shanks; DT3 = parabolic subsoiler with 3 shanks; C = disked check.

[b] Significant differences occur at the 0.05 probability level when differences in treatment means equal or exceed the LSD values shown.

Table 6.2 Total Precipitation and Supplemental Water from Irrigation during the May through September Period for Soybean Grown on Tunica Clay near Stoneville, MS (1987–1991)[a]

	Crop year, in.				
	1987	1988	1989	1990	1991
Rainfall	17.5	10.2	28.2	11.1	15.3
Irrigation	12.1	14.5	4.5	7.8	8.1
Total Water	29.6	24.7	32.7	18.9	23.4

[a] The long-term (50-year) average rainfall for the region during the May through September period is 18.7 in.

46.0 bu/acre (Table 6.1). Conversely, in nonirrigated environments all deep-tillage systems produced similar yields that averaged 43.0 bu/acre, and this was significantly greater than that produced by a conventional disked soybean system (29.0 bu/acre). Over the study period (1986 through 1991) deep tillage in the fall in a controlled traffic system increased the yield of nonirrigated soybeans by 48% above that produced by a conventional production system.

Data from the Mississippi study indicate soybean yield response to deep tillage in nonirrigated production systems was a function of growing season precipitation. Over the study period, rainfall at the test site ranged from 10.2 to 28.2 in. (Table 6.2). Yield data indicate the response to deep tillage was greater in the drier growing seasons. Deep tillage in the fall of 1987 increased yields 150% in 1988 (43 vs. 17 bu/acre) when growing season rainfall was only 10.2 in., and increased yields 78% in 1987 (48 vs. 27 bu/acre) when growing season rainfall approximated the long-term average for the region (18.7 in.). In 1989 when May through September rainfall approached 28.0 in., yields from the deep-tilled and conventional treatments were similar. The yield increases resulted from increased water infiltration and storage, improved internal water availability, movement, and drainage, and an overall enhanced soil structure that increased nutrient availability and allowed deeper rooting.

Response of soybean to tillage systems

Agronomic

Differences in crop yields between conventional and no-tillage systems have been documented by numerous researchers. Yields from no-tillage have generally been comparable to or greater than yields from conventional tillage on coarse- and medium-textured soils. However, the opposite has been documented on fine-textured, poorly drained soils (Dick and Van Doren, 1985). Soybeans have produced lower seed yields with no-tillage on poorly drained soils in both the Midwest (Dick et al., 1986a,b) and the Midsouth (Hairston et al., 1984a).

In some instances, plant diseases have been more severe in no-till systems and thus resulted in lower yields (Sanford et al., 1984; Hairston et al., 1988). Soil compaction also has been shown to be more severe in no-till systems where all primary tillage has been deleted (Hairston et al., 1984b). This compaction has been related to soil texture and subsequent root growth.

Nonirrigated studies in Mississippi on an Okolona silty clay and a Memphis silt loam indicated yields from no-till systems averaged 3 to 4 bu/acre less than those from conventional tilled systems (Hairston et al., 1990). It was concluded that the silty clay soil had problems meeting the water demands of soybeans at critical growth stages. This shrinking/swelling soil becomes more compact as it dries and then forms cracks, both of which reduce the ability of the soil to provide sufficient water to meet the demand of the crop. In the same study, differences in seed yield of no-till and conventional-till systems on a Prentiss very fine sandy loam were not significant.

A long-term study (1984 to 1991) was conducted on a Loring silt loam soil with a 3 to 4% slope (McGregor et al., 1992). In 1983, all plots were tilled uniformly preceding planting of soybeans. Beginning in 1984, no-till soybeans were grown on one plot of each pair to provide a minimal loss of productivity caused by erosion, and conventional-tilled soybeans were grown in the other plot to allow continuous loss of productivity due to excessive erosion. Conventional-tilled soybean yields were 12 and 3% greater than no-till yields in 1984 and 1985, and equaled no-till yields in 1986. However, no-till soybean yields were 19, 70, 19, 72, and 40% greater than conventional-till soybean yields during 1987 through 1991, respectively. During the last 5 years of the study, no-till soybean yields averaged 44% greater than conventional-till soybean yields. Thus, the developed trend of lower yields for conventional-till compared to no-till indicated an adverse effect of excessive erosion on soil productivity in the conventional-till soybean plots.

Economic

Economic data from the long-term deep tillage study on Tunica clay in Mississippi are summarized in Table 6.3 (Wesley et al., 1994). Data indicate the yield from nonirrigated conventional production systems averaged 29 bu/acre. Gross income, specified production costs, and net returns above specified costs for this system were $172, $124, and $48/acre, respectively. When supplemental irrigation was utilized in a conventional production system, yields averaged 45 bu/acre; gross income, specified costs, and net returns

Table 6.3 Summary of Soybean Yield, Gross Income, Specified Costs, and Net Returns
for Conventional Tilled and Deep-Tilled Treatments with and without Irrigation
on Tunica Clay near Stoneville, MS (1987–1991)

Irrigation treatment[a]	Tillage treatment[b]	Soybean yield, bu/acre	Gross income, $/acre	Specified costs, $/acre	Net returns, $/acre
NI	C	29	172	124	48
I	C	45	278	195	83
NI	DT2	43	265	136	129
I	DT2	46	282	204	79

[a] Irrigation treatments are NI (nonirrigated) and I (irrigated) with a lateral-move sprinkler system.

[b] Tillage treatments are C (conventional disked check) and DT2 (deep tilled with a parabolic subsoiler with two shanks).

averaged $278, $195, and $83/acre, respectively. In a nonirrigated production system that included deep tillage in the fall when the soil profile was relatively dry, yields from the deep tillage system averaged 43 bu/acre, and gross income, specified costs, and net returns averaged $265, $136, and $129/acre, respectively. Thus, net returns were enhanced considerably by fall deep tillage in nonirrigated production systems on Tunica clay.

Soybean enterprise budgets are prepared annually by researchers and extension personnel. These budgets reflect current production practices and input levels based on survey data provided by producers in various soybean production regions of the state. A sample enterprise budget compiled by the Mississippi State University Department of Agricultural Economics (1995) for soybean produced on clayey soil in the Delta is presented in Table 6.4. This budget reflects the field operations utilized, equipment size required to perform the specified field operations, and direct and fixed costs.

The estimated inputs for conventional production of soybean are listed in Table 6.4. Tillage inputs include a pass with a disk harrow for seedbed preparation and incorporation of residue, two passes with a field cultivator to apply and incorporate preemergence herbicides, and an early and a late cultivation. Estimated costs for these tillage inputs total $32.66/acre (Table 6.4) and this amounts to 25% of the total specified costs of production ($131.89/acre). Similarly, herbicides utilized include both preemergence (broadcast) and postemergence (band) applications for grass and broadleaf weeds and totaled $31.68, or 25% of the total specified costs. When cultivation is used during the growing season, serious consideration should be given to banding all preemergent herbicides to reduce their overall cost. In no-till budgets, all tillage is deleted and weed control is accomplished totally with herbicides. Typical budgets for various soybean production regions show herbicide costs for no-till soybean production systems average between $40.00 and $55.00/acre, or approximately 35 to 40% of the total specified costs of production.

Future farming concepts

Before the industrial revolution, farmers had first-hand knowledge of the physical and biological variability of each of their agricultural fields. Because of this knowledge, farmers would automatically hoe more in weedy areas and spread less fertilizer in the fertile areas.

In the industrial age, large-scale mechanization of farm operations dictated that all fields be treated uniformly without regard to the variability in soil types, fertility, or weed pressure. All production inputs — fertilizer, herbicides, pesticides, etc. — are applied at a constant rate across the entire field area. The end result ranged from insufficient to excessive application of fertilizers, pesticides, and other materials, less than optimum yields, and potential environmental pollution problems.

Table 6.4   Estimated Costs per Acre for Field Operations Soybeans, Clayey Soil, 8-row Equipment, 38-in. Rows, Delta Area, Mississippi, 1995

Operation/ operating input	Size/ unit	Direct cost, $						Direct cost to date, $	Fixed cost, $	Total cost, $
		Op input	Fuel	R&M	Labor	Inter	Total			
Disk harrow	28 ft	—	0.79	2.46	2.14	0.29	5.68	5.68	5.41	11.09
Field cult + inc	33.5 ft	—	0.40	1.43	1.07	0.13	3.03	8.71	3.51	6.54
Treflan	lb ai	7.55	—	—	—	0.35	7.90	16.61	—	7.90
Scepter DG	lb ai	17.31	—	—	—	0.80	18.11	34.72	—	18.11
Field cult	33.5 ft	—	0.32	1.00	0.86	0.10	2.27	36.99	2.47	4.74
Plant	8-row	—	0.40	1.02	2.14	0.16	3.72	40.71	2.57	6.29
Soybean seed	lb	9.20	—	—	—	0.42	9.62	50.33	—	9.62
Spray TR MT (band)	27 ft	—	0.55	0.67	1.50	0.10	2.82	53.16	1.81	4.63
Blazer	lb ai	5.31	—	—	—	0.20	5.52	58.67	—	5.52
Surfactant	pt	0.22	—	—	—	0.01	0.23	58.90	—	0.23
Cultivate — early	8-row	—	0.63	1.09	1.71	0.13	3.57	62.47	2.78	6.34
Spot spray — TR mount	27 ft	—	0.55	0.67	1.50	0.08	2.80	65.27	1.81	4.61
Fusilade DX	lb ai	1.13	—	—	—	0.03	1.16	66.43	—	1.16
Crop oil	pt	0.16	—	—	—	0.00	0.16	66.59	—	0.16
Cultivate — Late	8-row	—	0.40	0.68	1.07	0.07	2.21	68.80	1.74	3.95
App ins by air (3 gal)	appl	1.13	—	—	—	0.03	1.15	69.96	—	1.15
Larvin	lb ai	3.32	—	—	—	0.08	3.40	73.35	—	3.40
Combine soybeans	20 ft	—	0.64	10.89	2.14	0.10	13.78	87.14	18.63	32.41
Haul soybeans	bu	4.00	—	—	—	0.03	4.03	91.17	—	4.03
Totals		49.33	4.68	19.92	14.12	3.13	91.17		40.72	131.89

Note: Cost of production estimates are based on last year's input prices.

The agricultural community is at present facing a new challenge — precision farming methods. The new management systems currently being developed allow farmers to quantify yields and attempt to manage field variability. The integration of Global Positioning Systems (GPS) with automated ground sensors and ground sampling enables farmers to collect spatially referenced data sets of numerous variables that can affect crop production. Software programs are being developed to correlate the specific location of an operating farm truck or implement with previously acquired data sets to build Geographical Information System (GIS) knowledge bases.

Variable-rate applicators are also being developed that may be controlled based on field position and application rates compiled in the GPS/GIS system. The GPS-equipped sprayer may be programmed to adjust spray rates for areas with high insect pressure or sandier soils, and rapidly make on-the-go changes in pesticide or fertilizer application rates. As new technologies are developed and refined, farmers may be able to vary inputs to crop production, including seeding rates, fertilizer, herbicides, pesticides, tillage, and even tillage depths.

References

Cooper, A. W., A. C. Trouse, Jr., and W. T. Dumas. 1969. Controlled traffic in row crop production, in *Proc. 17th Int. Congress of Agricultural Engineering*, Section III, Theme 1:1–6, C.I.G.R., Baden-Baden, Germany.

Dick, W. A. and D. M. VanDoren, Jr., 1985. Continuous tillage and rotation combination effects on corn, soybean, and oat yields, *Agron. J.* 77:459–465.

Dick, W. A., D. M. VanDoren, Jr., G. B. Triplett, Jr., and J. E. Henry. 1986a. Influence of Long-Term Tillage and Rotation Combinations on Crop Yields and Selected Soil Parameters. I. Results Obtained from a Mollic Ochraqualf Soil, Ohio Agric. Res. Dev. Ctr. Res. Bull. 1180.

Dick, W. A., D. M. VanDoren, Jr., G. B. Triplett, Jr., and J. E. Henry. 1986b. Influence of Long-Term Tillage and Rotation Combinations on Crop Yields and Selected Soil Parameters. II. Results Obtained from a Typic Fragiudalf Soil, Ohio Res. Dev. Ctr. Res. Bull. 1181.

Douglas, E. and E. McKyes. 1978. Compaction effects on the hydraulic conductivity of a clay soil, *Soil Sci. Am.* 125(6):278–282.

Dumas, W. T., A. C. Trouse, Jr., L. A. Smith, F. A. Krummer, and W. R. Gill. 1975. Traffic control as a means of influencing cotton yield by reducing soil compaction, ASAE Paper No. 75-1050, ASAE, St. Joseph, MI.

Hairston, J. E., J. O. Sanford, J. C. Hayes, and L. L. Reinschmiedt. 1984a. Crop yield, soil erosion and net returns from five tillage systems in the Mississippi Blackland Prairie, *J. Soil Water Conserv.* 39:391–395.

Hairston, J. E., J. O. Sanford, P. K. McConnaughey, and D. A. Horneck. 1984b. Influence of primary tillage on compactness of Black Belt soils, in *Proc. 7th Southeast No-Tillage Systems Conference,* Headland, AL, 10 July, Alabama Agricultural Experiment Station, Auburn, 185–187.

Hairston, J. E., L. K. Marshall, J. O. Sanford, and P. K. McConnaughey. 1988. Influence of Tillage, Fumigation, Row Location and Fertilizer Placement on Soybean Growth and Yield, Miss. Agric. For. Exp. Stn. Res. Rep. 13(12).

Hairston, J. E., W. F. Jones, P. K. McConnaughey, L. K. Marshall, and K.B . Gill. 1990. Tillage and fertilizer management effects on soybean growth and yield on three Mississippi soils, *J. Prod. Agric.* 3:317–323.

Heatherly, L. G. 1988. Planting date, row spacing, and irrigation effects on soybean grown on clay soil, *Agron. J.* 80:227–231.

Heatherly, L. G. and C. D. Elmore. 1983. Response of soybeans to planting in untilled, weedy seedbed on clay soil, *Weed Sci.* 31:93–99.

Heatherly, L. G. and C. D. Elmore. 1986. Irrigation and planting date effects on soybean grown on clay soil, *Agron. J.* 78:575–580.

Johnson, R. R. 1987. Crop management, in J.R. Wilcox, Ed., *Soybeans: Improvement, Production and Uses,* 2nd. ed., Agronomy Monogr. 16, ASA, CSSA, and SSSA, Madison, WI, 355–390.

Kuipers, H. 1963. Drukverdeling onder Landbouwvoertuigen [Stress distribution under farm vehicles], *Landb-Mechanisatic.*

McDowell, L. L. and K. C. McGregor. 1980. Nitrogen and phosphorus losses in runoff from no-till soybeans, *Trans. ASAE* 23:643–648.

McGregor, K. C., C. K. Mutchler, and R. F. Cullum. 1992. Soil erosion effects on soybean yields, *Trans. ASAE* 35:1521–1525.

Murphree, C. E. and K. C. McGregor. 1991. Runoff and sediment yield from a flatland watershed in soybeans, *Trans. ASAE* 34:407–411.

Murphree, C. E., C. K. Mutchler, and L. L. McDowell. 1976. Sediment yields from a Mississippi delta watershed, in *Proc. Third Federal Inter-agency Sedimentation Conference,* Denver, CO, 1-99 to 1-109.

Mutchler, C. K. and J. D. Greer. 1984. Reduced tillage for soybeans, *Trans. ASAE* 27:1364–1369.

Parker, M. B., N. A. Minton, D. C. Brooks, and C. E. Perry. 1976. Soybean Response to Subsoiling and a Nematicide, Georgia Agric. Exp. Stn. Res. Bull. 181, 1–2.

Rhoads, F. M. 1978. Response of soybeans to subsoiling in North Florida, *Soil Crop Sci. Soc. Fla. Proc.* 37:151–154.

Sanford, J. O., J. E. Hairston, L. L. Reinschmiedt, J. C. Hayes, and P. K. McConnaughey. 1984. Effects of tillage on soybean yields, net returns and incidence of stem canker on Blackland Prairie soil in Mississippi, in *Proc. 7th Southeast No-Tillage Systems Conference,* Headland, AL, 10 July, Alabama Agric. Exp. Stn. Auburn, AL, 118–119.

Soybean Budgets Committee. 1994. Soybean Planning Budgets, 1995, Mississippi Agric. For. Exp. Stn. Ag. Econ. Report 66.

Tyler, D. D. and T. C. McCutchen. 1980. The effect of three tillage methods on soybean grown on silt loam soils with fragipans, *Tenn. Farm Home Sci.* 114:23–26.

Tyler, D. D. and J. R. Overton. 1982. Effects of no-tillage and deep seedbed preparations on soybeans, *Tenn. Farm Home Sci.* 121:2–4.

Vazquez, L., D. L. Meyre, R. N. Gallaher, E. A. Handon, and K. M. Portier. 1989. Soil compaction associated with tillage treatments for soybean, *Soil Tillage Res.* 13:35–45.

Webber, III, C. L., H. D. Kerr, and M. R. Gebhardt. 1987. Interrelationships of tillage and weed control for soybean production, *Weed Sci.* 35:830–836.

Wesley, R. A. and L. A. Smith. 1991. Response of soybean to deep tillage with controlled traffic on clay soil, *Trans. ASAE* 34(1):113–119.

Wesley, R. A., L. A. Smith, and S. R. Spurlock. 1994. Fall deep tillage of clay: agronomic and economic benefits to soybean, Miss. Agric. For. Exp. Stn. Bull. 1015, 8 pp.

Williford, J. R. 1980. A controlled-traffic system for crop production, *Trans. ASAE* 23(1):65–70.

chapter seven

The stale seedbed planting system

Larry G. Heatherly

Contents

Introduction

The Mississippi River Delta occupies about 20 million acres from the Boot Heel of Missouri to New Orleans, LA. About 9.6 million acres of this area consist of shrink-swell clay soils. A significant acreage of shrink-swell clays also occurs in the Blackland areas of Mississippi and Texas. Soybeans are the major crop planted on these soils, usually following tillage that may include chisel plowing, seedbed preparation with a disk- or spring-tined harrow, and harrow incorporation of herbicides. Frequent rainfall in late winter and early spring often results in wet soil through April and into early May when planting should occur on these soils. This wet soil will not allow seedbed preparation tillage until just prior to planting. Seedbed tillage conducted on these wet soils will result in a cloddy soil surface with insufficient moisture for seed germination. This can delay planting (Heatherly and Elmore, 1983; Heatherly et al., 1990) until significant rain replenishes soil moisture and smooths the soil surface. If seeds are planted in this tilled seedbed, poor stand establishment

Table 7.1 Effect of Preplant Tillage, Wheat Cover Crop, and Planting Date (PD) on Average Yields (bu/acre) and Net Returns ($/acre) from Irrigated Leflore Soybeans Planted in a Stale Seedbed at Stoneville, MS (1986–1987)

Preplant tillage	PD[a]					
	Early	Late	Avg.	Early	Late	Avg.
	bu/acre			$/acre		
Fall disk	48	38	43 a	94	36	65 b
Spring disk	48	39	43 a	104	49	76 ab
Prepared seedbed	50	39	44 a	119	58	89 a
None after harvest	49	40	44 a	108	55	82 a
Fall disk + Wheat	48	40	44 a	66	26	46 c
Average	48 a	39 b		98 a	45 b	

[a] Early = May 6, 1986 and May 5, 1987; Late = June 16, 1986 and May 28, 1987.

Average values for yield or net returns that are followed by the same letter are not significantly different.

Source: Heatherly et al., 1990.

will result. Delayed planting results in reduced yields (Tables 7.1 and 7.2) and lower net returns (Table 7.1), even if planting is delayed from early May to only late May.

An alternative to preparing a conventional seedbed on these soils is to plant in a stale or untilled seedbed (Heatherly and Elmore, 1983; Elmore and Heatherly, 1988). This assures earlier planting and better utilization of the early growing season to realize optimum plant growth and yield potential from irrigated and ESPS (Early Soybean Production System) plantings (Heatherly et al., 1990; Heatherly, 1996). However, use of the stale seedbed system for mid-May through June soybean plantings will not overcome the lower yields usually associated with nonirrigated soybeans in these later plantings (Heatherly and Elmore, 1983; Elmore and Heatherly, 1988; Bruff and Shaw, 1992a,b; Oliver et al., 1993; Heatherly et al., 1994; Hydrick and Shaw, 1995).

Definition

A stale seedbed is described as "a seedbed that has received no seedbed preparation tillage just prior to planting. It may or may not have been tilled since harvest of the preceding crop. Any tillage conducted in the fall, winter, or early spring will have occurred sufficiently ahead of the intended planting time to allow the seedbed to settle or become stale. A crop is planted in this unprepared seedbed, and weeds present before or at planting are killed with herbicides" (Heatherly, 1995). The stale seedbed planting system does not preclude tillage since it is a minimum or reduced tillage concept rather than a no-till concept. It is not confined to clay soils; however, their physical properties, their normally saturated or nearly saturated condition in the spring, and benefits from early planting that are usually precluded by seedbed preparation tillage have stimulated more interest in adapting this concept to these soils.

Objectives of stale seedbed system

1. *Avoid planting delays due to conventional seedbed preparation tillage.* Timely planting is important to maximize yields of both nonirrigated and irrigated soybeans in the midsouthern U.S.
2. *Fewer tillage operations.* Fewer tillage operations should result in reduced equipment inventory, lower tillage costs, savings of energy, and more efficient use of time and equipment.

Table 7.2 Yield (bu/acre) of Nonirrigated (NI) and Irrigated (I) Soybeans at Stoneville, MS

Year	Date of planting	Variety	Yield NI	Yield I
1992	Apr. 15	RA 452	42	62
	May 27	RA 452	32	45
	Apr. 15	P 9501	37	62
	May 27	P 9501	30	45
	Apr. 15	P 9592	56	61
	May 27	P 9592	35	43
	Apr. 15	A 5979	53	64
	May 27	A 5979	33	44
1994	Apr. 21	RA 452	39	50
	May 13	RA 452	32	48
	Apr. 21	DP 3499	31	46
	May 13	DP 3499	28	41
	Apr. 21	P 9592	37	53
	May 13	P 9592	33	46
	Apr. 21	A 5979	39	51
	May 13	A 5979	34	50
1995	Apr. 18	DP 3478	43	66
	May 9	DP 3478	30	54
	Apr. 18	P 9501	36	62
	May 9	P 9501	24	—
	Apr. 18	RA 452	34	60
	May 9	RA 452	24	54
	Apr. 18	P 9592	29	57
	May 9	P 9592	28	—
	Apr. 18	DP 3589	32	52
	May 9	DP 3589	26	56
	Apr. 18	A 5979	26	57
	May 9	A 5979	21	58
	Apr. 18	Hutcheson	36	60
	May 9	Hutcheson	28	—

From Heatherly, L. G., *Crop Sci.* 36, 1000–1006, 1996. With permission.

3. *Ensure maximum early-planted soybean acreage.* When the stale seedbed system is used, planting should be the first field operation conducted in the spring, except for a possible application of a preplant burndown herbicide. This will encourage early planting.
4. *Increase management options allowed by timely planting.* Planting earlier allows use of earlier-maturing soybean varieties and crops such as corn, escape from the earliest flushes of some summer annual weeds, earlier harvest when soil is dry, fall tillage of dry soil, and rotation with other early-planted crops such as corn.

Stale seedbed system guidelines

The following points describe the concept and general methodology of the system.

Use of tillage

Shrink-swell clay, silty clay, and clay loam soils do not require tillage to remove physical barriers such as hardpans. Therefore, tillage should be used only for incorporating crop

Table 7.3 Effect of Fall Tillage on Yield (bu/acre) of
Nonirrigated (NI) and Irrigated (I) Soybeans Grown
on Tunica Clay at Stoneville, MS, (1987–1989)

Irrigation	Fall tillage[a]	1987	1988	1989
I	Subsoiled	56	52	34
	Disked	55	50	36
NI	Subsoiled	47	40	41
	Disked	27	17	40

[a] Oct. 1 1986 (dry), Oct. 19 1987 (dry), Dec. 19 1988 (wet).
Planted on May 19 1987, May 16 1988, and May 12 1989.

From Wesley, R. A. and Smith, L. A., *Trans. Am. Soc. Agric.
Eng.* 34, 113–119, 1991. With permission.

residue, weed control, smoothing of the soil surface, and remedying compaction and
rutting in the upper soil profile caused by equipment during the preceding season. Recent
research (Wesley and Smith, 1991) indicates that deep (12 to 15 in.) tillage of clays may
be beneficial for nonirrigated soybean production if it is conducted in late summer or
early fall when the soil is dry to the depth of tillage or deeper (Table 7.3). However, the
dry soil requirement for this deep tillage will occur only if the preceding crop is harvested
early (late summer/early fall), as is the case with corn, grain sorghum, and ESPS plantings.
When a crop is harvested early, a wide window of opportunity for deep tillage of the dry
soil will occur.

Time of tillage

Any tillage that occurs after harvest of the preceding crop should be conducted no later
than 4 to 6 weeks in advance of planting the next crop to allow settling of the seedbed.
Seedbeds may remain idle following harvest, or they may be disked in the fall, late winter,
or early spring to smooth the soil surface or destroy old rows and machinery tracks, and
then left undisturbed until planting (Heatherly et al., 1990; 1992a). All of these stale
seedbed options resulted in yield and net returns that were similar to those resulting from
a prepared seedbed (Table 7.1) given the same planting date. Weed seedlings that will
have emerged since harvest of the preceding crop or since the last tillage operation can
be killed with a preplant burndown herbicide, and the crop can be planted into the stale
seedbed with the dead weed residue remaining on the soil surface. It is critical that 4 to
6 weeks remains after the last tillage operation to allow the seedbed to weather and become
smooth enough to plant at the optimum time so that potentially yield-reducing planting
delays do not occur. Cultivation for weed control after soybean emergence is optional.

Preplant or no preplant tillage

The tillage described above is not necessary unless machinery ruts are present or the
herbicide to be used requires incorporation. However, incorporation tillage exposes the
soil to more evaporation and, if conducted near intended planting time, can delay planting
until after rains occur. The availability of pre- and postemergent grass herbicides reduces
dependence on incorporated herbicides in fields where only annual grass predominates.
If there are no machinery ruts and/or an incorporated herbicide will not be used, soybeans
can be planted in old rows from the previous year with no preplant tillage. Yield and net
returns from stale seedbed plantings with no preplant tillage have been found to be similar
to those from stale seedbeds where fall or spring tillage was conducted (Tables 7.1 and 7.4).

Table 7.4 Effect of Weed Control Treatment (WTRT) on Average (1988-1989) Yields (bu/acre) and Net Returns ($/acre) from Nonirrigated (NI) and Irrigated (I) A 5980 Soybeans Planted in a Stale Seedbed at Stoneville, MS

	Yield			Net returns		
WTRT[a]	I	NI	Avg.[c]	I	NI	Avg.[c]
Burndown	46	20	33 b	168	43	106 b
Burndown + PRE	52	31	41 a	191	101	146 a
Burndown + POST	51	32	41 a	183	81	132 a
Burndown + PRE + POST	50	35	42 a	179	114	146 a
Burndown + PRE + POST[b]	50	30	40 a	182	85	134 a
Average	50 a	30 b		181 a	85 b	

[a] Burndown = Roundup; PRE = preemergent Sencor/Lexone; POST = postemergent broadleaf and grass herbicides. All treatments were cultivated.

[b] No preplant tillage; all other treatments were disked in the fall.

[c] Average values for yield or net returns that are followed by the same letter are not significantly different.

From Heatherly, L. G. et al. *Weed Technol.* 7, 972–980, 1993. With permission.

Planting without tillage

Clayey soils are often too wet in early spring for chiseling or disking. If tillage is conducted on these soils when the surface is dry enough to support equipment, a cloddy soil surface will result, thereby necessitating additional tillage. However, planting can occur on these clay soils when they are too wet for effective preplant tillage. This facilitates early planting. Impressions created by tractor tires during planting will be in the row middles, thus providing a furrow for drainage and irrigation. Excellent control of planting depth can be achieved in the smooth, noncloddy seedbed, and moisture for seed germination and emergence will usually be present at the depth of optimum seed placement.

Preplant weed management

Existing vegetation must be dead or killed at planting (Elmore and Heatherly, 1988; Bruff and Shaw, 1992a,b; Heatherly et al., 1994; Lanie et al., 1994a; Hydrick and Shaw, 1995). This can be accomplished with numerous herbicides. If existing weeds are not killed at planting, yields (Tables 7.5 through 7.7) and net returns (Table 7.5) will be reduced, even with postemergent cultivation. The use rate of burndown herbicides is critical for achieving complete weed kill (Lanie et al., 1993; Hydrick and Shaw, 1994; Lanie et al., 1994a) and subsequent maximum yield potential (Tables 7.6 and 7.8). Knowledge of the weed species that are present and their size is critical for selecting the proper herbicide and rate needed for control.

Weed management after planting

Herbicides with either soil activity or soil and foliar activity can be applied at or after planting to improve weed control (Heatherly et al., 1992a; Lanie et al., 1994b). Use of pre- and postemergent herbicides in addition to a preplant burndown herbicide results in increased yield (Tables 7.4 and 7.7) and net returns (Table 7.4) when highly competitive weeds such as pigweed and common cocklebur appear after crop emergence. When such highly competitive weeds do not appear after crop emergence, the use of pre- and postemergent herbicides may improve weed control (Lanie et al., 1994b) but may not improve

Table 7.5   Effect of Weed Management System in Stale
Seedbed Plantings on Average (1987–1990) Yields
(bu/acre) and Net Returns ($/acre) from Irrigated (I)
and Nonirrigated (NI) A 5980 Soybeans Planted from
Early May to Early June on Clay Soil at Stoneville, MS

Weed management	Yield[b]		Net returns[b]	
system[a]	I	NI	I	NI
1	38 a	20 d	69 a	8 c
2	34 b	20 d	70 a	35 a
3	28 c	17 d	32 b	16 c
Average	33	19	57	20

[a] 1 = burndown and pre- and postemergent herbicides used;
2 = burndown of Roundup used to kill weeds at planting;
3 = burndown of paraquat used to desiccate, but not kill,
weeds at planting. 40-in. rows, cultivated.

[b] Individual values for yield and net returns within each column that are followed by the same letter are not significantly
different.

From Heatherly, L. G. et al. *Weed Technol.* 8, 69–76, 1994. With
permission.

Table 7.6   Effect of Burndown Herbicide
and Rate (lb/acre) on Average Yield
(bu/acre) of Nonirrigated Stale Seedbed
Soybeans Planted on a Silty Clay Loam
Soil at Baton Rouge, LA (1991–1992)

Herbicide	Rate	Yield
Paraquat	0.31	10
Paraquat	0.62	23
Paraquat	0.94	31
Roundup	0.50	19
Roundup	0.75	32
Roundup	1.00	33
Nontreated		5
$LSD_{0.05}$		12

30-in. rows, cultivated.

From Lanie, A. J. et al. *Weed Technol.* 8, 159–164,
1994. With permission.

yield (Tables 7.5 and 7.9) and may result in reduced net returns (Table 7.5) from stale
seedbed plantings of soybeans in years when dryland yield potential is low. The effectiveness of pre- and postemergent herbicides following application of burndown herbicides in stale seedbed soybean plantings depends on the rate of burndown herbicide used
(Table 7.8), and weed size at burndown application (Lanie et al., 1993). If existing weeds
are not killed with burndown herbicides at planting, then application of pre- and postemergent herbicides will not be effective (Oliver et al., 1993).

There are indications that the weed spectrum may change with continued use of the
stale seedbed system. Therefore, tillage and chemical weed control options and crop
rotations must remain flexible on a year-to-year basis. This system should be used with
caution in fields where highly competitive annual or perennial weeds are a recurring
problem.

Table 7.7   Yield (bu/acre) of Stale Seedbed Plantings of Soybeans with
Various Burndown and Selective Herbicide Combinations (lb ai/acre),
Brooksville and Newton, MS (1991–1992)

Burndown herbicides		Selective herbicides		
Herbicide	Rate	Time/herbicide	Rate	Yield
Roundup	0.38	PRE Sencor/Lexone	0.38	9.9
		PRE Sencor/Lexone + Classic	0.32 + 0.05	11.6
		PRE Scepter	0.13	9.4
		POST Scepter	0.13	14.3
		None	—	10.1
Paraquat	0.63	PRE Sencor/Lexone	0.38	13.0
		PRE Sencor/Lexone + Classic	0.32 + 0.05	13.5
		PRE Scepter	0.13	6.5
		POST Scepter	0.13	15.9
		None	—	9.2
None	—	PRE Sencor/Lexone	0.38	14.4
		PRE Sencor/Lexone + Classic	0.32 + 0.05	9.2
		PRE Scepter	0.13	5.9
		POST Scepter	0.13	8.9
		None	—	5.5
$LSD_{0.05}$				3.6

30-in. row spacing, cultivated.

From Hydrick, D. E. and Shaw, D. R. *Weed Technol.* 9, 158–165, 1995. With permission.

Table 7.8   Yield of Stale Seedbed Plantings of Soybeans with Various Burndown and
Burndown + Residual Herbicide Combinations at Baton Rouge, LA (1991–1992)

		Residual herbicide treatment, bu/acre			
Burndown	Rate, lb ai/acre	Sencor/Lexone + Dual + Classic	Scepter + Prowl	Sencor/Lexone + Dual	None
Paraquat	0.11	37.2	22.6	30.2	17.8
	0.63	45.5	27.8	29.3	23.3
	0.94	47.3	25.6	42.5	22.9
Roundup	0.75	29.7	23.0	25.4	23.9
	1.00	32.7	30.3	34.3	32.0
	1.50	46.1	37.3	44.4	31.4
$LSD_{0.05}$			10.0		

30-in. rows, cultivated.

From Lanie, A. J. et al. *Weed Technol.* 7, 960–965, 1993. With permission.

Cover crops

Cover crops have numerous benefits in row crop production, and can be used effectively
in the stale seedbed system (Griffin and Dabney, 1990; Elmore et al., 1992; Heatherly et al.,
1993). The cover crop can be killed at the appropriate time with burndown herbicides
(Griffin and Dabney, 1990; Elmore et al., 1992). However, using wheat as a winter cover
crop may result in lower net returns (Table 7.1) due to the expense incurred in establishing
the wheat cover with no resulting soybean yield increase. The potential long-term benefits
(erosion control, increases in soil organic matter, and increases in soil productivity) resulting from use of cover crops in the stale seedbed system have not been measured.

Table 7.9   Effect of Burndown Herbicide Applied Alone and in
Combination with Residual Herbicides on Average Yield of Nonirrigated
Stale Seedbed Plantings of Soybeans at Baton Rouge, LA (1989–1990)

Herbicide(s)	Rate, lb ai/acre	Yield, bu/acre
Paraquat	0.38	18 b
Paraquat + Sencor/Lexone + Classic	0.38 + 0.32 + 0.05	22 ab
Paraquat + Scepter	0.38 + 0.13	21 ab
Paraquat + Sencor/Lexone	0.38 + 0.38	23 a
Roundup	1.00	19 ab
Roundup + Sencor/Lexone + Classic	1.00 + 0.32 + 0.05	22 ab
Roundup + Scepter	1.00 + 0.13	22 ab
Roundup + Sencor/Lexone	1.00 + 0.38	22 ab

From Lanie, A. J. et al. *Weed Technol.* 8, 17–22, 1994b. With permission.

Overall weed management

The stale seedbed planting system with its inherent absence of conventional seedbed preparation tillage has generated new interest in the use of (1) early-preplant incorporated, preplant surface-applied, and at-planting surface-applied herbicides, (2) burndown and residual herbicide tank mixes, (3) varying rates of burndown herbicides, (4) sequential preplant and at-planting herbicide combinations, (5) pre- and postemergent broadleaf and grass herbicide combinations applied separately from burndown herbicides, and (6) total postemergent weed control programs. Many of the above possibilities have been investigated. In many cases, small yield differences have resulted from the different weed control systems. The most practical weed control program should use cost and return as the deciding factor. A weed control system will be practical and economical only if it results in a monetary return that exceeds its cost. The significance of yield differences due to different weed control strategies should use net returns from a treatment as the deciding factor of its value.

Other management decisions

Once the stale seedbed planting system is adopted, all management decisions regarding subsoiling, preplant tillage, row spacing, irrigation, and weed control should be made to support and perpetuate its use. Each factor can be utilized as conditions allow or dictate, but should not in and of themselves be considered as necessary inputs if they detract from the early planting goal of the stale seedbed planting system.

Precepts for the stale seedbed planting system

1. Seedbed preparation is not required for clayey soils. A planter with double-disk openers and accompanying depth bands is required for effective planting and depth control in the smooth, untilled soil surface. An effective seed furrow closing system on the planter is required for sealing the trench that is created by the slicing action of the double-disk openers. In cases where large amounts of surface residue from a previous crop or dead weed residue resulting from a preplant burndown application occur, special coulters preceding the disk openers may be required. This need will depend on the size and amount of weeds present at the time of the burndown application.

2. No tillage, or fall or late-winter tillage can be used in a stale seedbed planting system.

3. Planting is usually the first tractor-driven operation that occurs on clayey soils in the spring. Tillage operations in the spring can irreparably rut many soils and create soil surface conditions that prevent planting until after a rain. The planting operation may also make ruts, but these can be avoided by delaying planting until the soil will support the tractor and planter. Minor ruts from the planting operation can be tolerated because they will be in the between-row areas and actually provide benefit by creating drainage furrows. The use of track-type tractors for stale seedbed plantings will reduce the rutting dilemma usually associated with early planting on the clayey soils.

4. Vegetation present at planting must be killed with a burndown herbicide.

5. Additional pre- and postemergent herbicides may be needed in irrigated production systems, but the preplant burndown herbicide and postemergent cultivation may be sufficient for weed management in dryland systems using wide rows. Narrow-row production systems may require postemergent weed control in the absence of cultivation.

6. Adequate surface drainage is required, especially for ESPS plantings. This requirement also is especially critical for March and April plantings of corn and grain sorghum that may be grown in rotation with soybeans. Fall bedding is an option for use in rotation systems.

7. Tillage is always an option in the stale seedbed planting system, but only if needed to remedy a specific problem that results in an economic benefit. The stale seedbed planting system does not preclude the use of tillage at any time. Rather, it relegates tillage to those times that will not result in delayed planting.

8. Subsoiling dry soil in the fall leaves the soil surface rough. Winter rains may not sufficiently smooth the surface to allow early planting, so the benefit from subsoiling could be reduced. Thus, the recommended approach (Richard A. Wesley, 1996, personal communication) to ensure early planting when fall subsoiling is used in conjunction with the stale seedbed system is to conduct any surface-smoothing operation in the fall as soon as possible after subsoiling. The ultimate goal is to reap the benefits of both fall subsoiling and stale seedbed planting on a consistent basis year in and year out.

Combinations with other management practices

The stale seedbed planting system for clayey soils is an integral part of the ESPS (Chapter 8), which consists of earlier-than-normal planting of soybeans (Regan, 1991; Bowers, 1995; Heatherly, 1996). Use of the ESPS results in increased yield and profit potential from early-maturing soybeans that are planted in April and harvested in mid-August to mid-September when soil is usually dry. Spring tillage to prepare a seedbed often precludes the advantages of early planting inherent in the ESPS; however, the early or no preplant tillage prescribed in the stale seedbed planting system always allows the earliest possible planting, even if deep tillage was performed the previous fall. The early and limited preplant tillage guidelines of the stale seedbed system also allow the early planting of such crops as corn and grain sorghum that may be grown in rotation with soybeans on clayey soils (Heatherly et al., 1992b; Wesley et al., 1994; 1996).

Acknowledgments

I sincerely appreciate the information provided by Dr. Jim Griffin and Dr. David Shaw.

References

Bowers, G. R. 1995. An early season production system for drought avoidance, *J. Prod. Agric.* 8:112–119.

Bruff, S. A. and D. R. Shaw. 1992a. Early season herbicide applications for weed control in stale seedbed soybean (*Glycine max*), *Weed Technol.* 6:36–44.

Bruff, S. A. and D. R. Shaw. 1992b. Tank-mix combinations for weed control in stale seedbed soybean, *Weed Technol.* 6:45–51.

Elmore, C. D. and L. G. Heatherly. 1988. Planting system and weed control effects on soybean grown on clay soil, *Agron. J.* 80:818–821.

Elmore, C. D., R. A. Wesley, and L. G. Heatherly. 1992. Stale seedbed production of soybeans with a wheat cover crop, *J. Soil Water Conserv.* 74:187–190.

Griffin, J. L. and S. M. Dabney. 1990. Preplant-postemergence herbicides for legume cover-crop control in minimum tillage systems, *Weed Technol.* 4:332–336.

Heatherly, L. G. 1995. Stale seedbed soybean: agronomic and economic implications, *Proc. Southern Weed Sci. Soc.* p. 69.

Heatherly, L. G. 1996. Yield and germinability of seed from irrigated and nonirrigated early- and late-planted MG IV and V soybean, *Crop Sci.* 36:1000–1006.

Heatherly, L. G. and C. D. Elmore. 1983. Response of soybeans (*Glycine max*) to planting in untilled, weedy seedbed on clay soil, *Weed Sci.* 31:93–99.

Heatherly, L. G., C. D. Elmore, and R. A. Wesley. 1990. Weed control and soybean response to preplant tillage and planting time, *Soil Tillage Res.* 17:199–210.

Heatherly, L. G., C. D. Elmore, and R. A. Wesley. 1992a. Weed control for soybean (*Glycine max*) planted in a stale or undisturbed seedbed on clay soil, *Weed Technol.* 6:119–124.

Heatherly, L. G., R. A. Wesley, and C. D. Elmore. 1992b. Cropping systems for clay soil: irrigated and nonirrigated soybean rotated with corn and sorghum, *J. Prod. Agric.* 5:248–253.

Heatherly, L. G., R. A. Wesley, C. D. Elmore, and S. R. Spurlock. 1993. Net returns from stale seedbed plantings of soybean (*Glycine max*) on clay soil, *Weed Technol.* 7:972–980.

Heatherly, L. G., C. D. Elmore, and S. R. Spurlock. 1994. Effect of irrigation and weed control treatment on yield and net return from soybean, *Weed Technol.* 8:69–76.

Hydrick, D. E. and D. R. Shaw. 1994. Sequential herbicide applications in stale seedbed soybean (*Glycine max*), *Weed Technol.* 8:684–688.

Hydrick, D. E. and D. R. Shaw. 1995. Non-selective and selective herbicide combinations in stale seedbed soybean (*Glycine max*), *Weed Technol.* 9:158–165.

Lanie, A. J., J. L. Griffin, D. B. Reynolds, and P. R. Vidrine. 1993. Influence of residual herbicides on rate of paraquat and glyphosate in stale seedbed soybean (*Glycine max*), *Weed Technol.* 7:960–965.

Lanie, A. J., J. L. Griffin, P. R. Vidrine, and D. B. Reynolds. 1994a. Weed control with non-selective herbicides in soybean (*Glycine max*) stale seedbed culture, *Weed Technol.* 8:159–164.

Lanie, A. J., J. L. Griffin, P. R. Vidrine, and D. B. Reynolds. 1994b. Herbicide combinations for soybean (*Glycine max*) planted in stale seedbed, *Weed Technol.* 8:17–22.

Oliver, L. R., T. E. Klingaman, M. McClelland, and R. C. Bozsa. 1993. Herbicide systems in stale seedbed soybean (*Glycine max*) production, *Weed Technol.* 7:816–823.

Regan, J. B. 1991. The Early Soybean Production System — A Risk Management Tool in the Southeastern United States, DowElanco Communication Resources 248-00-004.

Wesley, R. A. and L. A. Smith. 1991. Response of soybean to deep tillage with controlled traffic in clay soil, *Trans. Am. Soc. Agric. Eng.* 34:113–119.

Wesley, R. A., L. G. Heatherly, C. D. Elmore, and S. R. Spurlock. 1994. Net returns from eight irrigated cropping systems on clay soil, *J. Prod. Agric.* 7:109–115.

Wesley, R. A., L. G. Heatherly, C. D. Elmore, and S. R. Spurlock. 1996. Net returns from eight non-irrigated cropping systems on clay soil, *J. Prod. Agric.* 8:514–520.

chapter eight

Early soybean production system (ESPS)

Larry G. Heatherly

Contents

Introduction

The conventional soybean production system involves planting Maturity Group (MG) V, VI, and VII varieties in May and June in the midsouthern U.S., and has resulted in low and static yields during the 1970s and 1980s (Table 8.1). Drought that usually occurs in this region from mid-July through mid-September apparently is the primary cause for this dilemma, since commonly grown MG V, VI, and VII varieties are in high-water-demanding reproductive stages during this hot and dry period (Table 8.2). Thus, they are most susceptible to the yield-limiting conditions imposed by drought stress and concurrent high temperatures.

Moisture deficits that result from decreasing rainfall and increasing evaporative demand (pan evaporation) typically increase or become more negative from April through August at Stoneville, MS (lat. 33°26′) (Table 8.3). The Mississippi state average soybean

0-8493-2301-0/99/$0.00+$.50
© 1999 by CRC Press LLC

Table 8.1 Soybean Planted Acres (% of total)
and Average Yield (bu/acre) in Mississippi
from 1971 through 1997

Period	Planted acres (% of total)			Avg. yield
	May 1–5	May 15–20	June 5–10	
1971–80	2	20	63	21.5
1981–91	3	20	55	22.0
1992	2	26	59	34.0
1993	1	5	33	22.0
1994	24	59	89	31.0
1995	20	48	78	21.0
1996	29	76	95	31.0
1997	23	62	82	31.0

Sources: Anonymous, 1980, 1985, 1991, 1995a, 1995b, 1996, 1997.

Table 8.2 Dates of Developmental Stages for MG IV, V, VI, and VII of Irrigated Soybean Varieties When Planted at Indicated Times at Stoneville, MS

Variety (MG)	Planting date	Dates and days after planting (DAP) of selected stages				
		Beg. bloom	Beg. podset	Beg. seedfill	Full seed[a]	Maturity[a]
DP 3478 (IV)	Apr. 18	May 29 (41)	June 19	July 10	Aug. 10 (114)	Sep. 11 (146)
	May 9	June 12 (34)	June 30	July 28	Aug. 24 (107)	Sep. 18 (132)
P 9501 (IV)	Apr. 18	June 5 (48)	June 22	July 14	Aug. 21 (125)	Sep. 15 (150)
	May 9	June 19 (41)	July 10	Aug. 4	Aug. 28 (111)	Sep. 18 (132)
	May 27	July 5 (39)	July 22	Aug. 10	Sep. 7 (103)	Sep. 28 (124)
RA 452 (IV)	Apr. 18	June 12 (55)	June 30	July 24	Aug. 24 (128)	Sep. 15 (150)
	May 9	June 26 (48)	July 24	Aug. 10	Sep. 1 (115)	Sep. 22 (136)
	May 27	July 12 (46)	July 27	Aug. 14	Sep. 4 (100)	Sep. 18 (114)
A 5979 (V)	Apr. 18	June 19 (62)	July 3	July 28	Aug. 28 (132)	Sep. 18 (153)
	May 9	June 30 (52)	July 17	Aug. 7	Sep. 5 (119)	Sep. 27 (141)
	May 27	July 14 (48)	July 30	Aug. 14	Sep. 8 (104)	Sep. 25 (121)
P 9592 (V)	Apr. 18	June 19 (62)	July 17	July 31	Sep. 5 (140)	Sep. 22 (157)
	May 9	June 30 (52)	July 28	Aug. 14	Sep. 15 (129)	Oct. 2 (146)
	May 27	July 16 (50)	Aug. 5	Aug. 19	Sep. 14 (110)	Oct. 5 (131)
Tracy M (VI)	May 12	July 14 (63)	Aug. 11	Aug. 23	Sep. 12 (123)	Oct. 7 (148)
	June 4	July 29 (55)	Aug. 20	Sep. 1	Sep. 17 (105)	—
	July 8	Aug. 17 (40)	Aug. 31	Sep. 10	Sep. 23 (77)	—
Braxton (VII)	May 12	July 25 (74)	Aug. 16	Aug. 28	Sep. 16 (127)	Oct. 15 (156)
	June 4	Aug. 1 (58)	Aug. 22	Sep. 7	Sep. 24 (112)	—
	July 8	Aug. 23 (46)	Sep. 2	Sep. 14	Sep. 30 (84)	—

[a] Full seed stage may not be reached and maturity may occur earlier for nonirrigated soybeans.

yield for the years 1970 through 1993 (Table 8.1) was significantly and negatively correlated with July plus August pan evaporation ($r = -0.72$) and also with the total moisture deficit (rainfall minus pan evaporation) in July and August ($r = -0.69$) at Stoneville. This simply means that high soybean yields are associated with years in which evaporation is low relative to rain in comparison to low yield in years when evaporation is high relative to rain. This moisture deficit results in increasing drought stress for plants during a normal growing season and is especially detrimental to soybean varieties that are usually planted in the conventional production system. For example, Pioneer Brand 9592 (MG V) planted on May 27 began setting pods on August 5 and began filling seeds on August 19 (Table 8.2).

Table 8.3 Summary of 30-year Average Temperature (°F) and Rainfall and Pan Evaporation (in.) for Growing Season Months, Stoneville, MS (1964–1993)

Month	Air temperature		Rain	Pan evap.	Difference
	Max	Min			
Apr.	74	53	5.4	6.1	–0.7
May	82	62	5.0	7.7	–2.7
June	90	69	3.7	8.5	–4.8
July	91	72	3.7	8.2	–4.5
Aug.	90	70	2.3	7.3	–5.0
Sept.	85	63	3.4	5.8	–2.4

Source: Boykin, D. L. et al., *MAFES Tech. Bull.*, 201, 1995. With permission.

Tracy M (MG VI) planted on May 12 began setting pods on August 11 and began filling seeds on August 23. Dates for the same stages of Braxton (MG VII) planted on May 12 were August 16 and August 28, respectively. Thus, planting MG V, VI, and VII varieties in May and June resulted in pod set and seed fill during periods when moisture deficits and subsequent drought stress are usually the most pronounced.

Yield of soybeans from conventional production systems

The data in Table 8.4 show nonirrigated (NI) and surface-irrigated (I) soybean yields from research at Stoneville for the 1979 through 1990 period. These data show that planting MG V, VI, and VII varieties in May and June was a high-risk enterprise during this period. In many years, NI yields were below 20 bu/acre and only infrequently exceeded 25 bu/acre. Even though there was usually a large response to irrigation in dry years, the NI yield was so low in many years that even this large response to irrigation resulted in only relatively modest yields of irrigated soybeans. Also, up to seven furrow irrigations of 2.5 to 3.0 in. each were required to attain the highest irrigated yields in the driest years. Irrigated yields ranged from 29.7 bu/acre (Bedford planted on May 8, 1980) to 54.3 bu/acre (A 5980 planted on May 25, 1988), but the frequency of I yields exceeding 50 bu/acre was low. The frequency of low yield from the NI plantings confirms the low-yield plateau shown for the entire state in Table 8.1.

Rationale for ESPS

An apparent remedy for this problem is to modify production practices or provide irrigation to avoid the more severe drought periods. Crop water deficits in the midsouthern U.S. usually begin to develop in June and continue into September (Table 8.3). Planting early-maturing varieties in April so that their critical reproductive development coincides with periods of adequate soil moisture and more prevalent rainfall appears to have merit. This early soybean production system (ESPS) is the key to drought avoidance. DP 3478 (MG IV — relative maturity, or RM, 4.7) and Pioneer Brand 9501 (MG IV — RM 5.1) planted on April 18 began blooming on May 29 (41 DAP) and June 5 (48 DAP), respectively (Table 8.2). Both varieties began setting pods in mid-June and began filling seeds in mid-July. Full seed stage was reached on August 10 by DP 3478 and on August 21 by Pioneer Brand 9501. A 5979 (MG V) planted on April 18 started blooming on June 19 (62 DAP), started setting pods and filling seeds on July 3 and July 28, respectively, and reached full seed stage on August 28. Obviously, planting early-maturing varieties in April results in

Table 8.4 Yield (bu/acre) of NI and I Conventional
Soybean Production System Soybeans, and Number
of Furrow or Flood Irrigations (No.) Applied
to Irrigated Soybeans at Stoneville, MS (1979–1990)

Year	Date of planting	Variety	Yield NI	Yield I	No.
1979	June 13	Bedford	40.9	39.7	2
		Tracy	50.1	50.2	2
		Bragg	47.1	53.4	1
1980	May 8	Bedford	10.9	29.7	6
		Tracy	17.1	41.8	7
		Bragg	19.6	52.9	7
1981	May 13	Bedford	14.6	41.3	3
		Bragg	15.3	48.7	4
1982	May 12	Bedford	14.5	33.4	3
		Braxton	15.0	40.4	4
1984	May 14	Braxton	20.2	52.0	5
1985	May 2	Braxton	23.8	42.8	6
1986	May 15	Braxton	1.5	38.6	7
1986	June 3	Sharkey	5.6	43.9	3
1987	May 11	Sharkey	10.5	40.0	3
1988	May 16	Sharkey	33.9	39.8	3
1987	June 8	A5980	13.6	38.9	4
1987	May 6	Leflore	16.4	43.2	6
1988	May 25	A 5980	39.3	54.3	4
		Leflore	32.9	45.9	4
1989	May 8	A 5980	39.8	41.2	2
		Leflore	26.5	32.0	2
1990	May 2	A 5980	19.0	44.3	6
		Leflore	15.9	49.5	6

Sources: Heatherly, 1983, 1988; Heatherly and Elmore, 1986;
Heatherly and Pringle, 1991; Heatherly and Spurlock, 1993;
Heatherly et al., 1994.

reproductive development occurring earlier in the calendar year when rainfall is more prevalent and plant water use is less. Thus, the effective length of the drought period is usually reduced, and much of its detrimental effects are avoided.

Perceived constraints for ESPS

Freeze/frost possibilities are useful for assessing potential early-planting dates for soybeans. The data in Table 8.5 (Boykin et al., 1995) give the probabilities of freeze/frost at Stoneville, MS (lat. 33°26′) for selected March and April dates. Air temperatures in the table were measured at the 5-ft level. Ground-level temperatures may run as much as 3 to 5°F colder at that time of year; therefore, users can generally expect frost at 36°F, a light freeze at 32°F, and a moderate freeze at 28°F. The data indicate that only a 20% chance of frost at Stoneville exists on April 8, and only a 10% chance of frost exists on April 13. The possibility of a light freeze (32°F) is virtually gone by April 3 (10% chance). Thus, April planting of soybeans at the Stoneville latitude is a low-risk proposition from an early-season temperature standpoint. At Jackson, TN (lat. 35°37′), there is a 50% chance that a freeze will occur after April 6, and a 10% chance that one will occur after about April 15

Table 8.5 Probability of Freeze/Frost Occurring after a Spring Date Based on Minimum Air Temperatures, Stoneville, MS (1964–1993)

Prob. level	Spring date threshold temperatures		
	28°	32°	36°
0.35	Mar. 4	Mar. 19	Apr. 2
0.20	Mar. 13	Mar. 27	Apr. 8
0.10	Mar. 22	Apr. 3	Apr. 13
0.05	Mar. 29	Apr. 9	Apr. 18
0.01	Apr. 13	Apr. 21	Apr. 27

Source: Boykin, D. L. et al., *MAFES Tech. Bull.*, 201, 1995. With permission.

(Anonymous, 1995c). Thus, even at this more northerly latitude, April planting can safely occur from an air temperature standpoint.

Days suitable for fieldwork dictate the acreage that can be planted using the ESPS. According to Spurlock et al. (1995), there is an 80% chance of having at least 7.5 days suitable for fieldwork in Mississippi from April 9 to May 1. Thus, the acreage that can be planted between April 9 and May 1 in 8 out of 10 years will be a farmer's daily planting rate × 7.5 days (assuming no seedbed preparation). Jan Deregt of Hollandale, MS (personal communication, November 17, 1995) plants and applies preemergent herbicides at the rate of 200 to 250 acres/day on clayey soils using two planters that travel at 6.5 mph. For him, this translates into 1500 to 1875 soybean acres that can be planted in April in 8 out of 10 years. This example indicates that it is possible to plant a sizeable acreage in most years in the region using the ESPS, especially when no-till or stale seedbeds are used.

ESPS results

Recent reports indicate that the ESPS does indeed have merit for improving the yield potential of midsouthern U.S. soybean acres. Bowers (1995) conducted 3 years (1986 through 1988) of studies at two northeast Texas locations — Blossom (lat. 33°33′) and Hooks (lat. 33°38′) (Table 8.6). Two facts are obvious from this report: (1) early-maturing varieties planted in April yielded more than conventional varieties planted in May, and (2) early-maturing varieties planted in May yielded as much as or more than conventional varieties planted in May. Results from studies conducted in Arkansas in 1987 and published by Regan (1991) corroborate the earlier finding of Bowers. Heatherly conducted studies at Stoneville in 1992 and 1994 through 1997 (Table 8.7) and determined that (1) yields from April plantings of NI MG IV and V soybean varieties equaled or exceeded yields from NI May plantings; (2) MG V varieties planted in April and grown without irrigation always yielded as much as MG IV varieties planted in April and not irrigated; (3) MG V varieties planted in April and grown with irrigation yielded as much as or more than MG IV varieties planted in April and irrigated in 3 of the 5 years; (4) yields of MG IV and V varieties planted in April and grown with irrigation yielded as much as or more than the same varieties planted in May and irrigated; and (5) MG IV varieties planted in May and grown without irrigation yielded as much as MG V varieties planted in May and not irrigated in 4 of the 5 years. Subsequent analysis of the 5 years of Stoneville data showed that net returns from the I and NI April plantings always exceeded those obtained from the I and NI May plantings. This resulted from a combination of lower costs for inputs applied to the April plantings, higher yields from the April plantings, and a higher price received for early delivery of soybeans harvested from the April plantings. In 1995,

Table 8.6 Yield (bu/acre) of MG III through VII
Soybean Varieties Planted in April and May
at Blossom and Hooks, TX in 1986, 1987, and 1988

Planting date[a]	Variety (MG)	Year 1986	1987	1988
	Blossom			
Apr.	Williams 82 (III)	44.0	42.3	22.6
	Crawford (IV)	25.6	26.6	34.5
	Forrest (V)	7.9	17.5	37.6
	Leflore (VI)	4.3	7.8	23.3
	Bragg (VII)	3.2	5.3	16.1
May	Williams 82	15.0	13.8	—
	Crawford	14.3	14.1	—
	Forrest	12.9	12.8	—
	Leflore	10.7	4.3	—
	Bragg	3.8	2.2	—
	Hooks			
Apr.	Williams 82	54.7	24.0	35.6
	Crawford	48.2	31.5	47.9
	Forrest	36.8	11.3	46.8
	Leflore	42.6	6.6	40.5
	Bragg	27.5	11.2	38.0
May	Williams 82	36.0	25.2	—
	Crawford	32.0	10.8	—
	Forrest	43.8	8.1	—
	Leflore	36.5	15.9	—
	Bragg	28.5	20.6	—

[a] Blossom: April 16 and May 15, 1986; April 17 and May 12, 1987; April 22 and May 6, 1988. Hooks: April 17 and May 14, 1986; April 15 and May 11, 1987; April 21 and May 7, 1988.

Source: Bowers, G. R. 1995. *J. Prod. Agric.* 8, 112–119. With permission.

a Mississippi Delta farmer averaged 51.5 bu/acre on 3154 ESPS (MG IV and V varieties) acres that were planted from mid-April to early May and irrigated; in 1997, this same farmer had an average yield exceeding 56 bu/acre. The short form of the above results is that in most years, MG IV and V soybean varieties that are planted in April and grown with or without irrigation will provide greater opportunity for success compared with conventional plantings made in May.

The risk of planting early-maturing soybean varieties after April does not appear to be any higher than using the conventional system (Table 8.4) on nonirrigated and irrigated soils. Results from studies conducted by Bowers (1995) (Table 8.6) and Heatherly (1996) (Table 8.7) and reported by Regan (1991) (Table 8.8) indicate that yields from May plantings of MG III and IV soybean varieties were almost always as high as those from May plantings of later varieties. Thus, the planting window for the ESPS appears to be open into mid-May; however, the majority of the available information indicates that yield potential from April plantings is higher than that from later plantings.

Management inputs for ESPS plantings

Early-maturing varieties planted in the ESPS may respond favorably to narrow-row culture. At Bossier City, LA (lat. 32°25′), an 8 to 10 bu/acre average increase in yield of MG III

Table 8.7 Yield (bu/acre) of NI and I ESPS Soybeans Planted in April (DOP1) and May (DOP2) at Stoneville, MS

Year	Variety	Nonirrigated		Irrigated	
		DOP1	DOP2	DOP1	DOP2
1992	RA 452	42.3	32.4	62.2	45.2
	P 9501	36.9	29.8	61.9	45.0
	P 9592	56.4	34.9	61.2	43.2
	A 5979	52.9	33.2	64.2	43.7
1994	RA 452	39.4	32.1	50.0	48.3
	DP 3499	31.2	27.8	46.4	41.2
	P 9592	37.1	33.0	53.0	45.7
	A 5979	38.6	33.7	51.2	50.1
1995	RA 452	34.3	24.0	59.6	53.7
	DP 3478	43.2	30.3	66.1	53.9
	DP 3589	31.5	25.7	52.2	56.1
	A 5979	25.9	20.9	57.2	57.9
1996	D 478	30.6	30.2	57.7	58.7
	DP 3478	32.3	29.0	57.1	52.3
	DP 3588	42.8	42.2	58.5	50.0
	Hutcheson	45.2	45.2	62.5	61.2
1997	D 478	28.8	28.8	57.9	55.8
	DP 3478	30.0	30.4	62.6	61.8
	DP 3588	34.5	31.5	46.8	56.3
	Hutcheson	36.0	33.3	53.9	63.1
Average		37.5	31.4	57.0	52.6

DOP1 and DOP2: April 15 and May 27, 1992; April 21 and May 13, 1994; April 18 and May 9, 1995; April 30 and May 15, 1996; April 9 and May 12, 1997.

Source: Heatherly, L. G., *Crop. Sci.* 36, 1000–1006, 1996, and unpublished data. With permission.

Table 8.8 Yield (bu/acre) of Soybeans as Affected by Maturity Group and Planting Date in Arkansas (1987)

Variety	Planting date	
	Apr. 7–20	May 12–14
Hempstead County		
MG III & IV	26.8	20.0
MG V & later	15.2	7.9
Lafayette County		
MG III & IV	26.4	9.7
MG V & later	15.2	20.3
NEREC, Keiser		
MG III & IV	46.4	47.6
MG V & later	45.3	34.2

Each value is the average of six varieties.

Source: Regan, J. B. 1991. DowElanco, Indianapolis, IN.

and IV soybean varieties was associated with 10-in. row spacings vs. the same varieties grown in 40-in. rows during 1988, 1990, and 1991 (Table 8.9). Savoy et al. (1992) planted Williams 82 in 14- and 40-in. rows at College Station, TX (lat. 30°35′) on April 20 in 1988

Table 8.9 Yield (bu/acre) of Early-Maturing Soybean
Varieties When Planted in April in Two Row Spacings
at Bossier City, LA (1988, 1990–1991)

Variety (MG)	Row spacing	
	10 in.	40 in.
1988		
Williams 82 (III)	35.2	26.3
AgriPro 4321 (IV)	54.2	41.6
Crawford (IV)	53.6	48.4
P 9442 (IV)	56.7	42.0
1990		
Williams 82	27.1	19.0
Crawford	40.2	32.1
P 9442	26.3	22.4
RA 452 (IV)	38.6	24.8
1991		
Williams 82	35.2	26.3
AgriPro 4321	54.2	41.6
Crawford	53.6	48.4
P 9442	56.7	42.0

Sources: Rabb and Frazier, 1989; Rabb, Micinski, and Colyer,
1990; Rabb and Ryan, unpublished.

Table 8.10 Yield (bu/acre) of I and NI, April 20–Planted
Williams 82 Soybeans as Affected by Row Spacing
at College Station, TX (1988–1989)

Irrigation	Row spacing			
	1988		1989	
	14 in.	40 in.	14 in.	40 in.
I	56.9	53.4	65.7	62.0
NI	50.2	50.4	65.4	60.4

Source: Savoy, B. R. et al., 1992, *Agron. J.* 84, 394–398. With permission.

and 1989 (Table 8.10). They measured small, nonsignificant differences in yield between
the two row spacings. However, the trend was toward higher yields from the narrow
rows. These limited data indicate that yield of early-maturing varieties that are planted
in April may benefit from narrow-row culture.

Other row-spacing considerations lead to the conclusion that early plantings will
benefit from narrow-row culture. MG V determinate varieties planted early may be short
(18 to 24 in.) at maturity (Heatherly, 1996). Reduced height due to early planting is greatest
with determinate varieties, and this may result in an incomplete canopy in wide rows. It
is also generally accepted that early-maturing MG III and IV indeterminate soybean varieties produce a much narrower plant type or structure than do MG V and later determinate
types. This narrow plant growth habit is more suited to narrow rows. Maturity will
commence in August if early-maturing varieties are planted early. Leaves start to shed
and the canopy begins to open in August and early September. High temperatures and
moist soil from either timely rainfall or irrigation during this time may promote a flush
of weeds. These late-season weed infestations can interfere with harvest and contaminate

harvested seed if not controlled. Earlier and more complete canopy closure at the beginning of the growing season and later opening of the canopy at the end of the season result from narrow-row culture. This may offer increased impedance to late-season weed infestations in narrow rows compared with wide rows.

Disease problems associated with ESPS plantings should be similar to those encountered with more traditional planting dates in the midsouthern U.S. There is no documentation of disease problems associated with only ESPS soybeans or unique to only ESPS plantings. The importance of high-quality, disease-free seed, or seed treated for protection against diseases such as *Pythium* spp. that are prevalent in cool, wet soils may be greater for soybeans planted at the earlier dates compared with the more traditional planting dates. However, this is conjecture based on weather summaries for Stoneville, MS (Boykin et al., 1995). For April, average minimum air temperature at Stoneville is 53°F, average minimum 2-in. soil temperature is 58°F, and average rainfall is 5.4 in. For May, average minimum air temperature at Stoneville is 62°F, average minimum 2-in. soil temperature is 67°F, and average rainfall is 5.0. Thus, the cooler air and soil temperatures during April are documented; their possible association with or involvement in the possibility of increased disease problems for April-planted soybean has not been documented. Further work and facts are needed in this area.

Early-planted, early-maturing varieties respond favorably to supplemental irrigation in years with drought stress (Heatherly, 1993). Results of additional work by Heatherly (1996) are shown in Table 8.7. Averaged across 5 years (1992, 1994 through 1997), April-planted MG IV and V soybean varieties yielded 19.5 bu/acre more when irrigated. This is similar to the 21.2 bu/acre average yield increase achieved from irrigating May plantings of the same varieties. Yields of MG IV and V varieties that were planted in April and irrigated (Table 8.7) were higher than yields of MG V, VI, and VII varieties that were planted in May and irrigated in earlier studies (Table 8.4). Significantly, over the long term, fewer irrigations were necessary and subsequently less irrigation water was applied to the ESPS plantings in dry years at Stoneville. Thus, irrigation of ESPS plantings at Stoneville resulted in higher yields with less irrigation water, thereby increasing water use efficiency. Savoy et al. (1992) measured no significant yield effect from irrigating April-planted Williams 82 at College Station, TX in 1988 and 1989 (Table 8.10), but their NI yields averaged 50.3 (1988) and 62.9 (1989) bu/acre, which indicated no drought stress occurred at that location.

Weed control in ESPS

There are no reports of weed control measures for ESPS plantings being different from those required for the conventional production systems. Points to be considered for weed control in the ESPS are as follows.

1. ESPS plantings will be made sufficiently early in the season so that one burndown herbicide application may be sufficient to control emerged weeds. The 1997 stale seedbed budget (Spurlock, 1996) for conventional soybean plantings in Mississippi includes two burndown herbicide applications — one in February and one in May. This split burndown application has its advantages for May and later plantings. The early burndown can utilize a lower herbicide rate because weeds will be small. The second can also use a relatively low rate since it will be necessary to kill only weeds that have emerged since the first burndown application. Using the ESPS, it is possible that the February application can be delayed until the April planting date, although a higher herbicide rate may be required. Another option is to apply the February burndown application, and tank-mix a burndown herbicide with preemergence herbicides for application during the April planting operation. The May application will be unnecessary if planting occurs in April.

2. ESPS plantings should be made in narrow rows, which may preclude effective cultivation of middles for weed control after soybean emergence. If so, all weed control would have to be done with broadcast application of herbicides. Scouting for weed infestations, proper species identification, and proper herbicide selection will become more important without cultivation as a supplemental weed control tool. Effective weed control depends on matching the herbicide to the weed species and on applying the proper herbicide at the appropriate time. Postemergent herbicides should be applied to the smallest possible size of the targeted weed(s).

3. Preliminary work indicates that combinations of preemergence and postemergence broadleaf and grass herbicides should be used for weed control in the ESPS in a manner similar to that used in conventional soybean production systems.

Preliminary consideration has been given to and current research efforts are addressing the use of a preharvest desiccant in ESPS plantings. This consideration has been deemed advantageous since early-maturing varieties planted early have an earlier open canopy. Ramifications of this early maturity have been addressed already. Until definitive research has produced results sufficient for use in making recommendations about the economic use of a preharvest desiccant, the following thoughts are presented.

1. A preharvest desiccant will not be needed if

    a. Weeds present at maturity emerged late in the growing season and their size will not interfere with harvest;
    b. The species present are small-statured annual grasses, small-statured, small-stemmed broadleafs such as teaweed, and/or small-stemmed perennial vines whose presence will not reduce harvest efficiency;
    c. The weeds present have not produced mature seeds that will contaminate the grain;
    d. The desiccant cannot be applied sufficiently ahead of harvest so as to ensure that the weeds are dry at harvest (this may be the case in the high-temperature, low-humidity conditions of late August and early September when the time between maturity of soybean, or 95% mature pod color, and harvest maturity may be as little as 5 to 7 days and the period between desiccant application and harvest must be 7-15 days);
    e. The weed vegetation that is cut by the combine during soybean harvest is returned to the field with the soybean residue so that no foreign matter enters the grain sample; and
    f. If row spacing and variety selection were compatible (i.e., the row spacing was sufficiently narrow to allow the selected variety to form a canopy) and an effective herbicide program was in place, desiccation will not be needed since any weeds present at maturity will be underdeveloped.

2. It is economical to control only those weed infestations that will result in foreign matter in the harvested grain in an amount sufficient to cause dockage.

3. Effective control of early-season weeds will result in fewer significant weed infestations late in the season when the soybean canopy opens at maturity. Late-emerging weeds will be of much less significance at harvest compared with those that are present but underdeveloped beneath the soybean canopy during the growing season. Few weeds have the capacity to emerge and reach a problematical size in the short time between soybean leaf senescence and harvest.

Other advantages from using the ESPS

Traditionally, full-season (MG V, VI, and VII) soybeans have been planted during May and June and harvested in October and November. Rains frequently delay harvest, and wet soils result in harvesting inefficiencies and rutting of fields, especially during November. From 1971 through 1991, only 16% of the Mississippi soybean crop had been harvested by October 10, and less than 50% of the crop had been harvested by October 30 (Anonymous, 1980, 1985, 1991, 1995a). Plantings of MG IV and V varieties made in April to early May mature and reach harvest seed moisture content between mid-August and the end of September, resulting in earlier harvest. This early harvesting and marketing can result in a higher price received (Regan, 1991). When significant acres of soybeans had been planted before May 1 in Mississippi (Table 8.1), 40% of the Mississippi crop had been harvested by October 9 in 1994, 46% by October 9 in 1995, 45% by October 7 in 1996, and 50% by October 12 in 1997. In 1995, when 53% of the Mississippi crop had been planted by May 22, 76% of the acreage had been harvested by October 22. This earlier harvest when soil is dry allows effective fall subsoiling, and this results in greater yields from nonirrigated soybeans planted the following year (Wesley and Smith, 1991). Earlier harvest and fall land preparation also eliminate some weeds before they reach maturity, as well as ensure that the next year's crop can be planted early.

Use of the ESPS should allow avoidance of late-season foliage-feeding insect infestations. The soybean budgets for Mississippi (Spurlock, 1996) list a planned insecticide application to conventional May and June plantings during the month of August. The 1997 ESPS budget does not list this application, since these early plantings of early-maturing varieties will be nearing maturity when usual outbreaks of foliage-feeding insects occur late in the season. It is being assumed that the maturing foliage of these varieties will not be susceptible to significant damage from these pests. The effect of the ESPS on infestation and subsequent damage to seed by the southern green stink bug is unknown.

Shattering

Shattering (opening of pods and releasing seeds before harvest) of early-planted, early-maturing soybean varieties that mature in August or early September in the midsouthern U.S. is often perceived as a potential problem, and there are instances where this has occurred. This shattering is related to the very rapid drying of pods and seeds during this hottest, least humid time of the year. It is generally accepted that shattering is the result of early-maturing varieties (varieties developed in the midwestern U.S.) being grown outside their area of adaptation in the midsouthern U.S. This theory is logical but unproven. Not all early-planted, early-maturing varieties shatter when planted in this region. It is likely that some varieties in all maturity groups will shatter under these conditions. This has not been tested, nor have early-maturing varieties been tested for shattering under late-planted conditions. Until the development and release of early-maturing soybean varieties that are adapted to midsouthern conditions, the only sure way of avoiding this loss is to harvest as soon as the seed moisture is appropriate. This simple solution to reducing the effect of the perceived problem of shattering in ESPS plantings should be achieved easily. According to Spurlock et al. (1995), weather and soil moisture conditions during the anticipated harvest time (August 15 to September 30) allow an 80% chance of at least 29 of these 47 days (62%) being suitable for fieldwork in Mississippi.

Table 8.11   Germination (%) of Harvested
Seed from MG III and IV Soybean Varieties
Planted in April[a] in Arkansas (1989–1990)

Variety	Year	
	1989	1990
MG III		
Williams 82	23	39
A 3966	18	28
Pella 86	22	22
S42-30	38	34
MG IV		
HY 401	37	49
Competitor	29	50
Crawford	40	50
Coker 614	40	43

[a] 1989 planting dates — April 6–11 at three locations; 1990 planting dates — April 11–25 at
three locations.

Source: Mayhew, W. L. and Caviness, C. E. 1994,
Agron. J. 86, 16–19. With permission.

ESPS and seed quality

Seed quality (germination, discoloration, shriveling, etc.) of harvested seed has become a matter of concern when using the ESPS concept. Mayhew and Caviness (1994) grew four MG III and four MG IV April-planted soybean varieties under NI conditions in 1989 and 1990 in Arkansas (lat. 34 to 36°N). Average seed germination for MG III and IV varieties was 28 and 42%, respectively (Table 8.11). Germination percentage was significantly and negatively correlated ($r = -0.72$) with infection with *Phomopsis longicolla*. They did not grow MG V varieties in this study, so it is impossible to say that seed of early plantings of only early-maturing varieties are susceptible to low germination. Heatherly (1993, 1996) measured significantly higher germination of harvested seed from some MG IV and V varieties that were irrigated at Stoneville, MS (Tables 8.12 and 8.13), but this improvement was not always sufficient to impart acceptable levels of germination. Heatherly (1996) also determined that harvested seed of early-planted MG V soybean varieties may also be low in germination, that harvested seed of early-planted MG IV varieties were not always low in germination, and that planting MG IV and V varieties in May and irrigating will almost always ensure seed with the highest germination percentage (Table 8.13). These limited data indicate that early planting of any variety may result in harvested seed with low germination, but the reason(s) are unknown. The low germination, however, may be overcome with later planting, although lower yields can be expected (Tables 8.6 and 8.7). The conclusion drawn from current knowledge is that farmers should not save seed harvested from ESPS plantings for future plantings. Seed for ESPS plantings should be gotten from reputable sources whose seed production was conducted in locations with environments known to produce quality, germinable seeds.

Several studies have used visual ratings of harvested seed as an index of quality of seed harvested from early-planted, early-maturing soybean varieties. These ratings are usually based on percentage of shriveled seed, seed discoloration, etc. However, these ratings have never been related to dockage or reduced value of marketed seed. Thus, they are of little value in assessing quality of seed from ESPS plantings. Since the final determinant

Table 8.12 Germination (%) of Harvested Seed from MG IV Soybean Varieties Grown under NI and I Conditions at Stoneville, MS (1990–1991)

		Year	
Variety	Irrigation	1990	1991
Avery	NI	15	—
	I	34	—
Crawford	NI	44	67
	I	44	85
FFR 464	NI	44	—
	I	72	—
RA 452	NI	—	62
	I	—	89

Planting dates: April 25, 1990 and May 16, 1991.

Source: Heatherly, L. G. 1993, *Crop Sci.* 33, 777–781. With permission.

Table 8.13 Germination (%) of Harvested Seed of MG IV and V Soybean Varieties Planted on Two Dates and Grown under NI and I Conditions at Stoneville, MS (1992–1995)

	Planting date[a]			
	Early		Late	
Variety (MG)	NI	I	NI	I
1992				
RA 452 (IV)	70	68	91	91
P 9501 (IV)	14	36	84	82
P 9592 (V)	73	68	98	98
A 5979 (V)	69	77	93	98
1993				
RA 452	82	—	92	96
DP 3499 (IV)	93	—	88	98
P 9592	93	—	92	97
A 5979	92	—	91	98
1994				
RA 452	95	90	97	96
DP 3499	89	94	92	97
P 9592	94	91	96	98
A 5979	89	89	96	96
1995				
RA 452	38	77	33	94
DP 3478	72	24	71	59
DP 3589	85	93	81	96
A 5979	58	95	72	95

[a] April 15 and May 27, 1992; April 29 and June 3, 1993 (NI), and May 21, 1993 (I); April 21 and May 13, 1994; April 18 and May 9, 1995.

Source: Heatherly, L. G. 1996, *Crop Sci.* 36, 1000–1006. With permission.

of quality of soybeans sold for grain is the price received at the elevator, only those quality factors relating to the price received should be used for determining quality. Any criteria used for assessing quality of seed sold for grain must be related to the value of the sold grain; that is, if shriveling, discoloration, etc. do not result in a reduced price when seed are sold, then they are not accurate measures of economic seed quality.

Combination of ESPS and stale seedbed

Early planting, and not just early planting of early maturing soybean varieties, is stressed in the ESPS concept. The stale seedbed planting system mentioned in Chapter 7 is the choice to ensure this early-planting component, especially on the millions of acres of clay soils in the Midsouth. Primary tillage during the fall followed by use of a burndown herbicide prior to or at planting provides a favorable environment for planting early. Tillage to destroy weeds and incorporate herbicides prior to planting is discouraged since it will dry the topsoil and possibly delay planting until after the next rainfall. Thus, the combination of the stale seedbed planting system and the ESPS should be synergistic; that is, the ESPS is a way of ensuring maximum benefit from early planting allowed by the stale seedbed system, and the stale seedbed system is the only dependable method of ensuring that the ESPS can be used on the maximum acres in most seasons.

Summary of present knowledge

1. In the midsouthern U.S., planting early-maturing (MG III, IV, and V) soybean varieties in April results in maximum nonirrigated and irrigated yields.
2. Weather conditions in April allow a significant acreage of ESPS plantings in the region.
3. ESPS plantings should be made in narrow rows to assure highest yield and maximum early-season canopy development.
4. Damage from late-season foliage-feeding insect infestations will likely be avoided by ESPS plantings.
5. Germinability of seed harvested from ESPS plantings in the midsouthern U.S. may be low. This can be overcome by planting seed fields later in the season (May and June) and irrigating, but yield will be somewhat sacrificed.
6. ESPS plantings will be harvested earlier than conventional soybean production system plantings, and will usually receive a higher price if marketed when harvested.
7. ESPS plantings respond significantly to irrigation, but less irrigation water will be required than for conventional soybean production system plantings.

Summary of unknowns

1. The need for unique or novel weed control measures in ESPS plantings.
2. Specific requirements for preharvest desiccants.
3. Susceptibility to disease infestations unique to the ESPS.
4. Susceptibility to insect (such as southern green stink bug) infestations that may be altered by the different environment provided by the ESPS.
5. Advantage of determinancy or indeterminancy for ESPS plantings, and optimum canopy structure for ESPS plantings.
6. Causes of poor germinability of seed harvested from ESPS plantings.
7. Need for additional protective measures against disease infestations for planted seed and emerged seedlings in ESPS plantings.

8. Effect of discolored and/or shriveled seed harvested from ESPS plantings on marketed grain price.
9. The effect of the ESPS on infestation and subsequent damage to seed by the southern green stink bug.

Acknowledgments

I thank Stan Spurlock, Debbie Boykin, Glenn Bowers, and Jim Rabb for the valuable information they provided. I thank Jeff Tyler for the valuable conversation time and insights.

References

Anonymous. 1980. Mississippi Weather and Crop Report, 1971–1980, Miss. Crop & Livestock Rep. Serv., Jackson, MS.

Anonymous. 1985. Mississippi Weather and Crop Report, 1981–1985, Miss. Crop & Livestock Rep. Serv., Jackson, MS.

Anonymous. 1991. Mississippi Weather and Crop Report, 1986–1991, Miss. Agric. Stat. Serv., Jackson, MS.

Anonymous. 1995a. Mississippi Weather and Crop Report, 1990–1995, Miss. Agric. Stat. Serv., Jackson, MS.

Anonymous. 1995b. Mississippi Agriculture, 1970–1994, Agric. Econ. Dept., Mississippi State University, Starkville.

Anonymous. 1995c. Tennessee Weekly Weather and Crop Bulletin, April 2, 1995, Tenn. Agric. Stat. Serv., Nashville, TN.

Anonymous. 1996. Mississippi Weekly Weather Crop Report, May 7, 1996, Miss. Agric. Stat. Serv., Jackson, MS.

Anonymous. 1997. Mississippi Weekly Weather Crop Report, May 5, 1997, Miss. Agric. Stat. Serv., Jackson, MS.

Bowers, G. R. 1995. An early season production system for drought avoidance, *J. Prod. Agric.* 8:112–119.

Boykin, D. L., R. R. Carle, C. D. Ranney, and R. Shanklin. 1995. Weather data summary for 1964–1993, Stoneville, *MS. MAFES Tech. Bull.*, 201.

Heatherly, L. G. 1983. Response of soybean cultivars to irrigation of a clay soil, *Agron. J.* 75:859–864.

Heatherly, L. G. 1988. Planting date, row spacing, and irrigation effects on soybean grown on clay soil, *Agron. J.* 80:227–231.

Heatherly, L. G. 1993. Drought stress and irrigation effects on germination of harvested soybean seed, *Crop Sci.* 33:777–781.

Heatherly, L. G. 1996. Yield and germination of harvested seed from irrigated and nonirrigated early and late planted MG IV and V soybean, *Crop Sci.* 36:1000–1006.

Heatherly, L. G. and C. D. Elmore. 1986. Irrigation and planting date effects on soybeans grown on clay soil, *Agron. J.* 78:576–580.

Heatherly, L. G. and H. C. Pringle, III. 1991. Soybean cultivars' response to flood irrigation of clay soil, *Agron. J.* 83:231–237.

Heatherly, L. G. and S. R. Spurlock. 1993. Timing of furrow irrigation termination for determinate soybean on clay soil, *Agron. J.* 85:1103–1108.

Heatherly, L. G., C. D. Elmore, and S. R. Spurlock. 1994. Effect of irrigation and weed control treatment on yield and net return from soybean (*Glycine max*), *Weed Technol.* 8:69–76.

Mayhew, W. L. and C. E. Caviness. 1994. Seed quality and yield of early-planted, short-season soybean genotypes, *Agron. J.* 86:16–19.

Rabb, J. L. and J. S. Frazier. 1989. Very early maturing soybeans for north Louisiana, *La. Agric.* 32:11,21.

Rabb, J. L., S. Micinski, and P. D. Colyer. 1990. Evaluation of Early-Maturing Soybeans for Louisiana. Progress Rept. to the Louisiana Soybean & Grain Res. & Prom. Board.

Regan, J. B. 1991. The Early Soybean Production System: A Risk Management Tool in the Southeastern United States, DowElanco, Indianapolis, IN.

Savoy, B. R., J. T. Cothren, and C. R. Shumway. 1992. Early-season production systems utilizing indeterminate soybean, *Agron. J.* 84:394–398.

Spurlock, S. R. 1996. 1997 Planning Budgets: Soybeans. Miss. State. Univ. Dept. of Agric. Econ. Rept. 78.

Spurlock, S. R., N. W. Buehring, and D. F. Caillavet. 1995. Days suitable for fieldwork in Mississippi, *Miss. Agric. & For. Expt. Station Bull.,* 1026.

Wesley, R. A. and L. A. Smith. 1991. Response of soybean to deep tillage with controlled traffic in clay soil, *Trans. Am. Soc. Agric. Eng.* 34:113–119.

chapter nine

Soybean irrigation

Larry G. Heatherly

Contents

Background

From 1971 through 1991, almost 80% of the Mississippi soybean crop was planted after May 20, and average yields for the state were in the low 20s (bu/acre) during this period (Table 9.1). Moisture deficits (rainfall minus pan evaporation) occur due to decreasing rainfall and increasing evapotranspiration as the growing season progresses. Moisture deficits become more negative from April through August at Stoneville, MS (latitude

0-8493-2301-0/99/$0.00+$.50
© 1999 by CRC Press LLC

Table 9.1 Soybean Planting Progress (% of total acreage
planted) and Average Yield (bu/acre) in Mississippi
from 1971 through 1995

| | Planting progress by date | | | Avg. |
Period	Apr 30–May 3	May 15–20	June 5–10	yield
1971–80	2	20	63	21.5
1981–91	3	20	55	22.0
1992	2	26	59	34.0
1993	1	5	33	22.0
1994	24	59	89	31.0
1995	20	48	78	21.0
1996	29	76	95	31.0
1997	23	62	82	31.0

Sources: Anonymous, 1995, 1996, 1997.

Table 9.2 Summary of 30-year Average Temperature
(°F), Rainfall, and Pan Evaporation (in.) for Growing
Season Months, Stoneville, MS (1964 through 1993)

| | Air temp. | | | | |
Month	Max	Min	Rain	Evap.	Diff.
Apr.	74	53	5.4	6.1	–0.7
May	82	62	5.0	7.7	–2.7
June	90	69	3.7	8.5	–4.8
July	91	72	3.7	8.2	–4.5
Aug.	90	70	2.3	7.3	–5.0
Sep.	85	63	3.4	5.8	–2.4

Source: Boykin, D. L. et al., 1995, *MAFES Tech. Bull.*, 201. With
permission.

33°26′) (Table 9.2). This leads to serious drought stress during nearly every growing season. Stoneville weather and state yield data for selected "good" and "bad" yield years in Mississippi (Table 9.3) show how high and low July plus August moisture deficits have been associated with subsequent "low" and "high" state average soybean yields. As a matter of fact, the Mississippi state average soybean yield for the years 1970 through 1993 was significantly and negatively correlated with July plus August pan evaporation ($r = -0.72$) and with the total moisture deficit (rainfall minus pan evaporation) in July and August ($r = -0.69$) at Stoneville. This moisture deficit results in greater drought stress for soybeans that are planted in conventional production systems in the midsouthern U.S.

Soybean development and irrigation

The moisture status of plants is a function of soil water supply, evaporative demand of the atmosphere, and the capacity of the soil to release water. In the field, significant water deficits develop on hot sunny days even in well-watered plants. As water evaporates (transpires) from the leaves, the moisture tensions that develop increase the rate of water uptake from the soil (see Chapter 18 for more discussion of this process). If roots cannot absorb water rapidly enough, plant water tension increases. These tensions become growth-limiting under these conditions. In fact, one of the most sensitive indicators of soil water deficit is leaf growth (Reicosky and Heatherly, 1990). This sensitivity is so great that

Table 9.3   Summary of July and August Average Temperature (°F), Rainfall, and Pan Evaporation (in.) for 1970 through 1993 and for Selected Years at Stoneville, MS, and Mississippi State Average Soybean Yields (bu/acre) (1970 through 1993 and for selected years)

Month, year	Air temp.		Rain	Pan evap.	Diff.	July+Aug. pan evap.	Avg. yield
	Max	Min					
July 1970–1993	91	72	3.7	8.2	–4.5	15.5	22.3
Aug. 1970–1993	90	70	2.3	7.3	–5.0		
July 1979	90	73	6.3	7.1	–0.8	13.6	29.0
Aug. 1979	89	69	2.8	6.5	–3.7		
July 1980	96	74	1.5	9.6	–8.1	18.1	16.0
Aug. 1980	94	72	1.4	8.5	–7.1		
July 1982	91	72	2.4	7.8	–5.4	14.4	26.0
Aug. 1982	90	72	1.5	6.6	–5.1		
July 1985	89	71	1.6	7.6	–6.0	14.7	27.0
Aug. 1985	90	70	4.4	7.1	–2.7		
July 1986	94	74	2.1	9.2	–7.1	17.7	17.0
Aug. 1986	91	68	0.9	8.5	–7.6		
July 1992	90	73	4.1	7.5	–3.4	14.0	34.0
Aug. 1992	86	66	4.5	6.5	–2.0		
July 1994	90	72	11.6	6.5	5.1	13.4	31.0
Aug. 1994	91	67	0.5	6.9	–6.4		
July 1995	91	73	5.8	8.4	–2.6	17.0	21.0
Aug. 1995	95	74	1.4	8.6	–7.5		

Sources: Boykin, D. L. et al., *MAFES Tech. Bull.*, 201, 1995; and National Oceanic and Atmospheric Administration, Mid-south Agric. Weather Center, Stoneville, MS.

most growth occurs at night when atmospheric demand and subsequent transpiration decreases. Leaf tension becomes less at night as water from the soil is translocated to the leaves. In well-watered soils, leaf tension returns to growth-promoting levels very quickly after darkness. In drier soils, leaf water tension recovers much more slowly, and thus provides less time for recovery and growth during the dark hours. Since pod and seed growth, which are also sensitive to plant water deficits, occur later in the season when soil moisture and rainfall are at the lowest seasonal levels, the potential for significant reductions in their growth and development are great. The extent of this reduction depends on the longevity of drought (Hodges and Heatherly, 1983).

A maturity group (MG) V variety planted on May 15 at Stoneville, MS will begin setting pods (R3) on about July 30 and begin filling seeds (R5) on about August 13 (Table 9.4). A MG VI variety planted on May 15 will begin to set pods on about August 15 and begin to fill seeds on about August 27. Obviously, the dates for these critical stages of development occur during the time of greatest drought stress in a normal year (Table 9.2). This coincides with the time of greatest plant leaf area accumulation and subsequently the greatest potential transpiration of water from plants, the time soil moisture is usually at its lowest, and the time of greatest atmospheric demand for water (Hodges and Heatherly, 1983). Only alleviation of drought will allow soybeans at these critical developmental periods to grow most rapidly; however, long-term weather data indicate that receiving needed rainfall during this time is unlikely.

Stages of reproductive development are an important component of soybean irrigation. Irrigation may be timed to coincide with the beginning or ending of a stage or stages of development. Predicted reproductive stages of MG V, VI, and VII soybean varieties planted on May 1 and 15 and June 1 and 15 at Stoneville, MS are given in Table 9.4. These predicted dates are derived from the equations shown in Table 9.5, and these equations are based on data that were collected each year at Stoneville from 1976 through 1995. An

Table 9.4 Predicted Dates of Reproductive Stages for MG V, VI, and VII Soybeans at Stoneville, MS, (1978 through 1995).

Date of		Reproductive stage[a]			
planting	MG	R1	R3	R5	R6
April 15	V	June 15	July 12	July 28	Aug. 27
May 1	V	June 27	July 21	Aug. 6	Sep. 2
	VI	July 8	Aug. 10	Aug. 23	Sep. 19
	VII	July 19	Aug. 15	Aug. 28	Sep. 19
May 15	V	July 7	July 30	Aug. 13	Sep. 7
	VI	July 17	Aug. 15	Aug. 27	Sep. 20
	VII	July 26	Aug. 19	Sep. 1	Sep. 22
June 1	V	July 20	Aug. 8	Aug. 22	Sep. 13
	VI	July 28	Aug. 20	Sep. 1	Sep. 21
	VII	Aug. 3	Aug. 24	Sep. 5	Sep. 26
June 15	V	July 31	Aug. 17	Aug. 29	Sep. 18
	VI	Aug. 6	Aug. 25	Sep. 5	Sep. 23
	VII	Aug. 10	Aug. 27	Sep. 9	Sep. 28

[a] R1 = beginning bloom; R3 = beginning pod set; R5 = beginning seed fill; R6 = full seed.

Table 9.5 Prediction Equations for Determining Estimated Dates of Reproductive Stages of MG V, VI, and VII Soybeans at Stoneville, MS

Stage	MG	Prediction equation[a]	*N*	R^2	CV(%)
R1	V	R1DAP = 87.1 – (0.25 × JULDOP)	170	0.68	5.7
	VI	R1DAP = 112.7 – (0.36 × JULDOP)	122	0.76	5.9
	VII	R1DAP = 141.3 – (0.51 × JULDOP)	43	0.91	5.6
R3	V	R3DAP = 132.3 – (0.42 × JULDOP)	164	0.77	5.3
	VI	R3DAP = 183.2 – (0.67 × JULDOP)	119	0.85	5.7
	VII	R3DAP = 196.0 – (0.74 × JULDOP)	42	0.93	5.1
R5	V	R5DAP = 155.9 – (0.49 × JULDOP)	162	0.82	4.5
	VI	R5DAP = 199.2 – (0.70 × JULDOP)	114	0.81	5.6
	VII	R5DAP = 210.4 – (0.75 × JULDOP)	38	0.91	5.1
R6	V	R6DAP = 202.3 – (0.64 × JULDOP)	161	0.81	4.7
	VI	R6DAP = 251.8 – (0.91 × JULDOP)	101	0.86	4.3
	VII	R6DAP = 239.0 – (0.80 × JULDOP)	28	0.93	3.9

[a] R1 = beginning bloom; R3 = beginning pod set; R5 = beginning seed fill; R6 = full seed; DAP = days after planting; JULDOP = Julian date of planting (Jan. 1 = 1, May 1 = 122, June 1 = 153, Dec. 31 = 365, etc.); *N* = number of data points used to develop equation; R^2 = percentage of variability among the data points explained by the equation.

example calculation is as follows. Pioneer Brand 9592 (MG V) was planted on May 1, or Julian date 122. The equation for predicting beginning bloom or R1 is R1DAP (days after planting of R1) = 87.1 – (0.25 × JULDOP, or Julian date of planting which is 122 for May 1). Therefore, R1 or beginning bloom occurs at [87.1 – (0.25 × 122)] = 57 days after planting. The 57 days after the May 1 planting date is added to the Julian date of planting (JULDOP) of 122 to estimate day 179 as the Julian date of beginning bloom of Pioneer Brand 9592 planted on May 1. This Julian date of 179 translates to June 27. Thus, Pioneer Brand 9592 or any MG V of similar maturity planted on May 1 is estimated to begin blooming on June 27. Estimated date(s) of other stages of MG V, VI, and VII soybean varieties are calculated the same way using the appropriate equation for this location found in Table 9.5.

Management of irrigation

It is widely thought that crops adapt to drought stress and become capable of withstanding drought. There is no evidence to support this view, when it is considered on the basis of producing an economic yield. The limited adaptation that does occur only increases the ability of the plant to survive during drought. This may be a valuable mechanism for a desert shrub, but it is of little value where production of a profitable seed yield is important for a crop such as soybeans.

The advantages from irrigating soybeans in the midsouthern U.S. are well documented (Reicosky and Heatherly, 1990). These results indicate that irrigation of soybeans significantly increases yields by overcoming drought; thus, the remainder of this chapter will be based on this fact. No further presentation of yield response to irrigation will be made. The effectiveness of irrigation in alleviating the effects of drought on soybeans in the region is accepted. If an irrigation system is in place, then it should be used since the ownership or fixed costs associated with the equipment will exist regardless of whether or not the system is used. The question, then, is not whether to irrigate soybeans for significant yield enhancement, but how to do it properly for maximum effect (i.e., maximum yield and economic returns).

Scheduling irrigation

Sound irrigation management requires the addition of the right amount of water at the right time. Irrigation scheduling, a key element of proper management, is the accurate forecasting of water application time and amount for economic yield enhancement. Factors that affect irrigation amount and frequency are determined by the amount of water applied by the previous irrigation (minus runoff), effective rainfall (amount that got into the soil), and estimated water use by the soybean crop since the previous irrigation. Rainfall measurements at the field site can be made easily, and well capacities or system outputs and efficiencies can be measured and/or calculated. Crop water use can be estimated by using pan evaporation numbers from the nearest weather station since actual evapotranspiration during the R1 to R6 period closely resembles pan evaporation (Reicosky and Heatherly, 1990). Estimates of water use based on pan evaporation can be combined with estimates of water supplied by irrigation and rainfall to predict the soil water deficit in the effective rooting zone.

Tensiometers and gypsum blocks have long been available for monitoring soil water status. These instruments require calibration, and labor for installation, reading, and maintenance. They should be placed in representative soil types for a given field. Tensiometers measure soil water tension with great accuracy in the range of 0 to 75 to 80 cbar, which is the range of readily available water in most soils. Previous research at Stoneville, MS indicates that tensiometers provide the best results when placed 12-in. deep in clay soils and 6 in. deep in silt loam and sandy loam soils (Heatherly, 1984; Heatherly and Sciumbato, 1986). Gypsum blocks have a wider effective range than a tensiometer, but the values outside the 60 to 100 cbar range are meaningless for proper irrigation scheduling. When tensiometers are used, care must be exercised to ensure that soil at the depth of measurement does not dry to a level below the measuring capability of the tensiometer. If it does, the tensiometer will lose water and require recalibration before future readings can be used.

Tensiometers have been used to schedule irrigations for soybeans in research studies at Stoneville, MS for years. This experience plus concurrent location weather data and measurements of the amount of water applied at each irrigation have resulted in the following practical approach to scheduling irrigation for soybeans in the region. Pan evaporation in the region typically will be about 0.25 in. of water per day during the R1 to R6 (July to August) period of soybean development. Thus, in the absence of rain, about 3.0 in. of water (net applied to soil) will be needed about every 12 days. This is the amount typically supplied by a normal furrow or flood irrigation to cracking clay soils. Therefore, furrow or flood irrigation should be planned every 12 to 14 days in the absence of rain. An overhead irrigation system that applies 1.5 gross in. (1.2 to 1.28 net in. at 80 to 85% efficiency) at each application (assuming no runoff) should be scheduled to irrigate about every 5 to 6 days. This simple approach results in a successful strategy for irrigating soybeans in most situations and ensures that clay soils are irrigated before severe cracking occurs.

Crusting soils with a low capacity for water infiltration and shallow soils that have a relatively low total water-holding capacity will experience runoff if large amounts of water are applied. In this situation, less water must be applied at each irrigation, but irrigation should be more frequent. On a silt loam site at Stoneville, MS, runoff of irrigation water applied through an overhead system occurred when the application rate exceeded 0.75 in./acre/event (Heatherly et al., 1992). Thus, frequency of irrigation on this site was greater than on sites discussed above.

Experience in Arkansas (Tacker et al., 1994) indicates that inadequate irrigation and/or improper timing of irrigations are the major reasons for lower-than-expected soybean yield responses from irrigation. Their experience also indicates that a water-balance approach has the most potential for properly irrigating soybeans. They use two irrigation scheduling methods that are based on soil moisture accounting procedures. The Arkansas Checkbook Method uses a daily water use chart and a computation table for updating soil moisture content. The University of Arkansas Surface Irrigation Scheduler operates basically the same way, but uses a computer program to perform the computations. The computer program requires the emergence date, the soil moisture deficit at planting, and a predetermined allowable soil moisture deficit of 2, 3, or 4 in. The daily information required to use either of these methods is maximum air temperature, rainfall, and irrigation amounts. For information about either of these scheduling approaches, contact the Arkansas Cooperative Extension Service.

Starting and continuing irrigation

Numerous studies in the southeastern U.S. have investigated yield response of monocropped determinate soybean varieties to both full-season irrigation (water applied as needed during both the vegetative and reproductive phases of development) vs. irrigation during reproductive development only (water applied as needed from R1 to R6). The results of these studies have been summarized by Reicosky and Heatherly (1990). The conclusions from these many studies are as follows: (1) irrigation before R1 (beginning bloom) produced no appreciable yield advantage above that realized from irrigation applied only during reproductive development, and (2) irrigation efficiency, defined here as the increase in seed yield per acre per inch of water applied, was usually higher for the reproductive phase irrigation. Thus, irrigation of monoculture soybeans prior to R1 appears to be of little benefit, even though atmospheric demand for water increases through R1. In some years, significant drought during vegetative development may justify irrigation prior to bloom to ensure adequate growth. Most soil types, assuming periodic rainfall, can supply the water necessary to meet atmospheric demands and support adequate growth during the vegetative phase. Exceptions to this are those soils that have a shallow rooting depth (Griffin et al., 1985) or low available water-holding capacity, or doublecropped soybeans that are planted in dry soil. Calculations from 6 years of irrigation studies conducted at Stoneville, MS from 1980 through 1985 (Table 9.6) verified that water

Table 9.6 Gross Water Received [in./day; rainfall
(includes runoff) + irrigation] during the Vegetative (Veg.)
and Reproductive (Reprod.) Periods of Irrigated MG V, VI,
and VII Soybeans at Stoneville, MS (1980 through 1985)

MG[a]	Planting date	Growth period	
		Veg.	Reprod.
V	May 1–15	0.17	0.32
	May 26–June 4	0.16	0.27
VI	May 1–15	0.16	0.31
	May 26–June 4	0.18	0.32
VII	May 1–15	0.16	0.35
	May 26–June 25	0.15	0.31

[a] More than one variety in each MG.

All plantings in 40-in. rows, all irrigation applied by furrow method, all irrigations started at or near R1 and ended at near R6.

Source: Heatherly, L. G. in *Proc. Delta Irrig. Workshop*, Greenwood,
MS, Feb. 28, Mississippi Cooperative Extension Service, Starkville,
113–121, 1986. With permission.

use by determinate soybeans prior to R1 was significantly less than water use from R1 to
R6; thus, irrigation is needed much more and should provide greater benefit when applied
during the reproductive period, or the period of greatest use. The predicted time for this
initiation can be calculated from the equations in Table 9.5.

Delaying initiation of irrigation until the full pod stage (R4) or the beginning of seed
development (R5) in years when rainfall is limited during early reproductive stages results
in seed yields that are lower than those realized from irrigation started at or about R1
(Reicosky and Heatherly, 1990). The number of pods and seeds can be increased if irrigation occurs during early reproductive development, but only weight of seeds can be
increased if irrigation is delayed until later stages. Where drought stress is severe but
alleviated by irrigation during early reproductive development, the biggest percentage
yield increase comes from the increase in number of seeds. If irrigation is applied only
after pods are set and seeds are filling, increase in weight of individual seeds is the major
contributor to increased yields. Numerous research reports support the conclusion that
the major effect of drought stress on seed yield is a reduced number of seeds (Reicosky
and Heatherly, 1990).

Irrigation that is started during early reproductive development must be continued
into the seedfill stage (Griffin et al., 1985; Reicosky and Heatherly, 1990; Heatherly and
Spurlock, 1993; see Table 9.7) so that soil moisture is readily available through the full
seed stage. This ensures the realization of the yield potential that is established by starting
with the early irrigation. Stress that occurs during seedfill results in smaller seeds, but
will not reduce the total number of seeds below the number produced by plants that are
irrigated during all stages of reproductive development (Reicosky and Heatherly, 1990).
Thus, the number of seeds that are set is maintained during drought stress that occurs
after seed formation (except in extremely severe drought conditions), but maximum
weight of individual seeds is not realized if drought occurs during seedfill. Irrigation
during the full reproductive period is required to maximize both number of seeds (established by early alleviation of drought stress) and weight of seeds (maximized by later
irrigations).

Based on the preponderance of data, and in the absence of rainfall, irrigation during
the entire reproductive phase is the most desirable to maximize yield and net returns
(Heatherly and Spurlock, 1993). This is because:

Table 9.7 Seed Yield (bu/acre) and Net Returns (NETRET —
$/acre) of A 5980 and Leflore Soybeans as Affected by
Reproductive Stage[a] at Which Furrow Irrigation (ITRT) Was
Terminated at Stoneville, MS (from 1987 through 1990)

A 5980			Leflore		
ITRT	Yield	NETRET	ITRT	Yield	NETRET
1987					
			NI	16.4d	−16d
			R2.2	18.9d	−43e
			R2.5	27.6c	4cd
			R2.8	29.2c	11c
			R4.0	34.9b	40b
			R5.5	44.0a	86a
			R5.9	43.2a	80a
1988					
NI	39.3b	193b	NI	32.9c	144b
R1.0	54.8a	263a	R1.0	39.6b	151ab
R3.0	50.0a	225ab	R2.8	41.7ab	163ab
R5.0	53.2a	247a	R4.5	41.5ab	159ab
R5.7	54.4a	253a	R5.6	45.9a	189a
1989					
NI	39.8a	113a	NI	26.5a	38a
R5.1	39.9a	72a	R3.5	28.1a	4a
R6.0	41.2a	78a	R5.6	31.9a	25a
1990					
NI	19.0c	−3b	NI	15.9e	−22d
R1.0	25.0b	−11b	R2.3	24.2d	−17d
R2.7	26.1b	−7b	R2.8	24.0d	−21d
R4.4	26.8b	−5b	R3.5	32.2c	23c
R5.1	28.9b	4b	R4.5	41.3b	72b
R5.6	41.4a	72a	R5.2	49.5a	115a
R5.9	44.3a	85a	R6.0	47.1a	99ab

Values in individual columns within years followed by the same letter
are not significantly different at a probability level of 0.05.

[a] Reproductive stage at last irrigation; R1 = beginning bloom; R2 = full
bloom; R3 = beginning pod; R4 = full pod; R5 = beginning seed; R6 =
full seed. Fractional portion of number indicates when a treatment
was terminated relative to the time between the preceding and following whole number stages; i.e., R5.5 is halfway between R5 and
R6. NI = nonirrigated.

Source: Heatherly, L. G. and Spurlock, S. R., *Agron. J.* 85, 1103–1108,
1993. With permission.

1. Reproductive development occurs during the later portion of the growing season
   when rainfall amounts are below the potential evapotranspiration, and soil water
   supply has been significantly diminished because of water used by early season
   plant growth;
2. Leaf area is at or near maximum levels and potential evapotranspiration will be
   maximum from this leaf area; and
3. Components of seed yield (pods, seeds/pod, and weight of seeds) are being determined during this time.

Drought stress during any portion of this phase limits yield by limiting the contribution of one or more yield components. For maximum yield in dry years, irrigation must be started near or at the beginning of flowering. In the absence of rainfall, irrigation must be continued in a consistent and timely manner so that the soil stays relatively moist until seeds are near full size (Heatherly and Spurlock, 1993).

There may be cases where only a limited amount of irrigation water is available, and it is not enough for full reproductive-phase irrigation. It can be allocated for use during early reproductive development to establish a maximum number of seeds, or to the latter stages of reproductive development to maximize weight of seeds. However, neither of these practices produces the maximum yield that may be required to maximize net returns unless adequate rainfall is received during the times of no irrigation (Heatherly, 1983; Heatherly and Spurlock, 1993). The use of limited irrigation early in the reproductive phase can be advantageous if rains are received during the latter stages of reproductive development. Late-occurring rain will have the greatest effect if relatively large numbers of seed are set as a result of irrigation during early reproductive development. However, the probability of late-occurring rain in the Midsouth is low in late summer (August and September). The use of limited irrigation during the seedfill period can be advantageous for ensuring maximum weight of seeds that were set as a result of favorable soil moisture conditions during early reproductive development. The probability of receiving rain during the early reproductive period is higher than the probability of receiving rain during the seedfill period. Thus, in cases of limited irrigation water, irrigation during the seedfill period appears to provide the greatest probability for maximizing yield. This appears less risky than using it earlier and depending on late-season rain to enlarge seeds that were set as a result of irrigation during early reproductive development. This premise assumes that a reasonably high number of seeds were set in the absence of irrigation during early reproductive development.

Ending irrigation

Irrigation that exceeds that amount necessary to maximize profit or net returns is a waste of a valuable resource. Drought still may occur late in the season, but the ability of the soybean plant to use added water for additional increases in seed dry matter is limited at some point by the physiological processes of the maturing plant system.

Studies were conducted from 1987 through 1990 at Stoneville, MS. Leflore (MG VI, determinate) was grown in all years, and A 5980 (MG V, determinate) was grown from 1988 through 1990. Row spacing was 40 in. and planting dates were May 6, 1987; May 25, 1988; May 8, 1989; and May 2, 1990. Irrigation treatments consisted of (1) no irrigation or (2) furrow irrigation started at or near stage R1 (beginning bloom) with planned termination at successive intervals through R6 (full seed). Actual plant stage at the last irrigation is shown in Table 9.7. Further details of this study are presented by Heatherly and Spurlock (1993).

The 1987 and 1990 seasons were hot and dry during soybean reproductive development. Yields and net returns were greatest when furrow irrigation was started during the bloom stage (R1 to R3) and terminated at about mid-seedfill (R5.5) (Table 9.7). Thus, an effective surface irrigation at stage 5.5 supplied enough soil moisture to finish filling seeds since irrigation later in the season did not increase yield or net returns from either variety. Termination of irrigation at an earlier stage resulted in lower yields and net returns. When irrigation is supplied by overhead systems that may apply less water per event than is applied by surface methods, the last irrigation should be later since an overhead irrigation applied at stage 5.2 to 5.5 may not provide enough water to finish filling seeds. Irrigation terminated during early bloom (R1 to R2), full bloom (R2 to R3), podset (R3 to R4), and full pod/beginning seedfill (R4 to R5) stages resulted in negative or only slightly positive net returns, even though yields were increased significantly above nonirrigated yields by all of these early-terminated treatments. In 1988 (moderately wet), both yield and net returns were increased similarly by all irrigation treatments, while in 1989 (wet), irrigation had no effect.

These results point out the importance of continuing irrigation that is started during the early reproductive development of soybeans well into the seedfill period (surface application methods) or through the seedfill period (overhead application methods). Irrigation that is stopped too early results in less than maximum economic returns. In years with little or only moderate drought stress, irrigation may result in reduced economic return. These results should be transferrable to early-maturing and/or indeterminate soybean varieties.

Conclusions about timing

The key to irrigation management for soybeans in the midsouthern U.S. is to initiate irrigation near or at beginning bloom and continue it in a consistent and timely manner to provide adequate soil moisture until seeds are fully developed (touching in pods). This will ensure maximum podset, maximum number of seeds, and maximum seed size. This practice is based on the assumption that the soil profile is fully wet at planting, which is usually the case in the midsouthern U.S. If soil moisture is lacking at soybean planting time due to severe drought prior to planting or due to doublecropping, it is beneficial to irrigate at planting to get a uniform stand and before beginning bloom to develop plants of sufficient vegetative stature to provide an adequate canopy.

Optimum irrigation management attempts to maximize the number of seeds and seed size or weight. The important thing is to start irrigation soon enough to ensure setting the greatest number of seeds, since the majority of yield increases resulting from irrigation are derived from increases in this component. Starting irrigation at or during flowering will ensure this, and continuing irrigation through seedfill will ensure the greatest weight of each seed. Highest yields from irrigation are achieved by maximizing both components. Starting surface irrigation too soon (no yield response realized) will be less economically harmful ($4 to 6 cost/acre; Spurlock, 1995) than will stopping irrigation too early ($40 to 60 less net returns/acre; Heatherly and Spurlock, 1993).

Too few data are available to develop prediction equations for reproductive stages of MG IV varieties grown in the midsouthern U.S. Limited data show that mid to late MG IV varieties that are planted from mid-April to mid-May will start blooming 5 to 7 weeks after planting, which is 2 to 3 weeks earlier than for determinate MG V varieties. Starting irrigation at beginning bloom of the early-maturing varieties may not be necessary if soil moisture is still plentiful. If soil moisture is limiting, the response of early-maturing varieties to irrigation should be similar to that of later-maturing varieties. If MG V varieties are planted at about the same time as and in close proximity to the early-maturing MG IV varieties, they could be used to schedule the beginning of irrigation for the MG IV varieties until further data are available. Monitoring soil moisture status in earlier-than-normal plantings will be the most effective way of ensuring that irrigation is applied as early as needed.

Irrigation and management practices

Variety selection

The majority of the research conducted with irrigation of soybeans has incorporated the use of more than one variety. Varieties often have been from more than one maturity group; therefore, performance of varieties can often be more accurately defined as performance of individual maturity groups as represented by selected varieties from those groups. Whenever varieties from the same or different maturity groups have been evaluated with and without irrigation, differences generally occur. Lodging is sometimes a factor (Reicosky and Heatherly, 1990), and can result in reduced response to irrigation. This is especially true when overhead irrigation is used and irrigation is applied before beginning bloom of determinate varieties in the midsouthern U.S. (Boquet, 1989). In reality, plant height (will it lodge when irrigation is applied?) and rate or time of reproductive development in relation to drought stress that may be alleviated by irrigation probably are more important than variety per se. With proper irrigation, drought stress in any variety can be alleviated regardless of the rate of development of that variety; therefore, any variety should be able to realize its maximum yield potential with irrigation.

Data in Table 9.8 indicate a difference in irrigation management that is related to maturity group is the number of irrigations that may be required. Early-planted, early-maturing varieties sometimes require less irrigation than later-maturing varieties planted at conventional times (May to June). The irrigation period (R1 to R6) of the conventional plantings of later-maturing varieties is closely aligned with the period experiencing the greatest moisture deficit. A safer statement on this subject is that potentially greater irrigation will be required by later-maturing varieties because of the later calendar dates of their reproductive phase. For example, a MG V variety planted on May 15 will be at stage R1 on July 7 and at R6 on Sept. 7, for a potential irrigation period of 62 days (Table 9.4). A May 15–planted MG VI variety will be at R1 on July 17 and at R6 on Sept. 20, or a potential irrigation period of 65 days. The difference in the potential irrigation period between the MG V and MG VI varieties is only 3 days; however, the potential irrigation period for the MG VI variety starts 10 days after that of the MG V variety, which subjects its reproductive development to greater risk of drought stress and subsequently a potentially greater need for irrigation. This can affect net return differences between varieties of the two maturity groups if yields are equal. This difference can be even more dramatic when comparing reproductive development times of MG V and VI varieties with those of MG IV varieties (see Table 9.2 in Chapter 8).

Row spacing and plant population

Theoretically, soybean plant arrangement should be designed to give a complete canopy as quickly as possible. Narrow-row (20 in. or less) culture of soybeans is an effective way of doing this. As a result of the quicker and more complete canopy closure resulting from a narrow-row system, soybean water use is greater in narrow rows (Heatherly, 1984; Reicosky and Heatherly, 1990). Evidence indicates that irrigated soybeans grown in narrow rows produce about 2.5 to 5.0 bu/acre more yield than irrigated soybeans grown in 36- to 40-in. rows. The economic impact of this small yield difference is unknown. When both irrigation and row spacing are considered, proper irrigation is more important; i.e., much greater yield responses can be achieved with irrigation of any row spacing than can be achieved by changing row spacing in the absence of adequate water. In fact, the higher water use by soybeans in narrow rows can compound drought problems in a season of severe drought. Plant population appears to be even less critical than row width when irrigation is used.

Planting date

From 1980 through 1997, various experiments that have involved irrigation of soybeans planted in April, May, and June have been conducted at Stoneville, MS. Years, varieties,

Table 9.8 Irrigation and Planting Date Effects on Seed Yield
(bu/acre) and Number of Irrigations (No.) for Soybeans
Grown on Sharkey Clay at Stoneville, MS

Year	Variety (MG)	Planting date	Seed yield[a]			No.
			I	NI	I – NI	
1980	Bedford (V)	May 12	40.6	14.7	25.9	7
		June 3	46.8	17.2	29.6	5
	Bragg (VII)	May 12	52.4	19.8	32.6	7
		June 3	44.3	22.6	21.7	5
1981	Bedford (V)	May 13	41.3	14.6	26.7	3
		June 4	35.3	15.6	19.7	2
	Braxton (VII)	May 13	48.7	15.3	33.4	4
		June 4	43.7	25.2	18.5	3
1982	Bedford (V)	May 12	33.4	14.5	18.9	3
		May 28	24.8	13.1	11.7	3
	Braxton	May 12	40.4	15.0	25.4	4
		May 28	34.9	17.8	17.1	3
1984	Braxton	May 14	53.1	20.9	32.2	5
		June 25	46.3	23.5	22.8	4
1985	Braxton	May 2	44.0	27.7	16.3	6
		June 24	28.2	24.6	3.6	3
1986	Braxton	May 15	40.0	1.6	38.4	7
		June 24	21.2	3.9	17.3	4
1986	Leflore (VI)	May 6	53.5	—	—	7
		June 16	40.7	—	—	5
1987	Leflore	May 5	43.1	—	—	7
		May 28	37.4	—	—	7
1992	RA 452 (IV)	Apr. 15	62.2	42.3	19.9	2
		May 27	45.2	32.4	12.8	2
	A 5979 (V)	Apr. 15	64.2	52.9	11.3	2
		May 27	43.7	33.2	10.5	2
1994	RA 452	Apr. 21	50.0	39.4	10.6	4
		May 13	48.3	32.1	16.2	4
	A 5979	Apr. 21	51.2	38.6	11.6	4
		May 13	50.1	33.7	16.4	4
1995	DP 3478 (IV)	Apr. 18	66.1	43.2	22.9	3
		May 9	53.9	30.3	23.6	3
	A 5979	April 18	57.2	25.9	31.3	4
		May 9	57.9	20.9	37.0	4
1996	DP 3478	April 30	57.1	32.3	24.8	4
		May 15	52.3	29.0	23.3	5
	Hutcheson (V)	April 30	62.5	45.2	17.3	5
		May 15	61.2	45.2	16.0	5
1997[b]	DP 3478	April 9	62.6	30.0	32.6	4
		May 12	61.8	30.4	31.4	6
	Hutcheson	April 9	53.9	36.0	17.9	5
		May 12	63.1	33.3	19.8	7

[a] NI = nonirrigated; I = irrigated; I – NI = irrigated minus nonirrigated yield.

[b] 1997 irrigation scheduled more frequently.

Sources: Heatherly and Elmore, 1986; Heatherly, 1988; 1996; unpublished; Heatherly et al., 1990.

planting dates, and yields from nonirrigated and irrigated treatments are presented in Table 9.8.

The yield data lead to several general but unmistakable conclusions. First, irrigation of soybean varieties planted in April or early May almost always resulted in greater yields than did irrigation of the same varieties planted later. Prior to 1992, when the earliest plantings were in early to mid-May and the late plantings were in late May to late June, more irrigations of the early plantings were required to achieve these higher yields. From 1992 to 1996, when the early plantings were in April and the late plantings were in early to late May, an equal number of irrigations was applied to soybeans in each planting. The earlier trend of more irrigations being required for the May plantings recurred in 1997. Second, in the absence of irrigation, planting dates ranging from early May to late June (1980 through 1987) had little effect on soybean yield. This trend changed when the earliest planting was in April compared with early May (1992 through 1997). These first two conclusions lead to a third. For those acres that are to be irrigated, plant at the earliest acceptable time for an individual variety and location to provide opportunity for the maximum seed yield with the least irrigation input. For nonirrigated acres, April planting is especially important.

Irrigation for emergence

Research results about this are nonexistent. However, a few generally accepted truths can be presented here to address common questions. One dilemma sometimes faced is whether to water before or after planting in dry soil to effect emergence. Crusting soils should be watered before planting to prevent the crusting and subsequent emergence problems that occur if irrigation follows planting. These soils dry quickly following preplant irrigation and planting can proceed with little delay. Shrink-swell soils (clays) can be watered before or after planting, but the advantages of watering after planting outnumber the disadvantages or the advantages of watering before planting. The advantages of watering after planting are as follows:

1. Planting is not delayed by the wait for the soil to dry that occurs if watering precedes planting on these soils with slow internal drainage;
2. Rows planted in dry soil can be preplant cultivated for furrow formation to effect rapid downslope movement and drainage of postplant irrigation water;
3. Preemergence herbicides will be activated;
4. Seeds can be planted shallowly to effect quick emergence; and
5. Irrigation water is not wasted as a result of watering unplanted acres that may receive rain before planting can occur.

The early planting option offered by after-planting irrigation is valuable as discussed earlier. The disadvantage of after-planting irrigation is the effect that rain following irrigation and subsequent prolonged wet soil, along with low air temperatures, can have on germinating seeds and emerging seedlings. This option will require seed treatment protection against soil-borne pathogens that may reduce germination and emergence in cool, wet soils in the spring.

Weed control

Irrigation results in the development of the greatest plant height and main stem and branch leaf area, and the longest maintenance of canopy structure. Therefore, an assumption for

Table 9.9 Average Seed Yield (bu/acre) and Net Returns (NETRET —
$/acre) from Irrigated and Nonirrigated A 5980 Soybeans Grown
under Various Weed Control Systems (WTRT) at Stoneville, MS

	Irrigated		Nonirrigated	
WTRT[a]	Yield	NETRET	Yield	NETRET
1988–1989				
1	45.5b	168b	20.5c	43c
2	51.0a	191a	31.5ab	101a
3	51.6a	183a	30.9b	81b
4	50.2a	179ab	34.5a	114a
5	49.8a	182a	30.0b	85b
1987–1990				
1	38.5a	69a	20.4a	8b
2	34.0b	70a	20.0a	35a
3	27.6c	32b	16.6b	16b

Values in individual columns within year groups followed by the same letter are
not significantly different at a probability level of 0.05.

[a] 1988–1989: 1 = preplant foliar-applied (PFA) Roundup; 2 = PFA Roundup +
preemergent (PRE) Sencor/Lexone; 3 = PFA Roundup + postemergent (POST)
broadleaf and grass herbicides; 4 = PFA Roundup + PRE + POST herbicides;
5 = planted no-till with same weed control as 4. 1987–1990: 1 = PFA Roundup
to kill weeds at planting + PRE broadleaf and grass herbicides + POST-directed
herbicides; 2 = PFA Roundup to kill weeds at planting; 3 = PFA paraquat to
desiccate weeds at planting.

Sources: Heatherly et al., 1993, 1994.

any weed control strategy in irrigated soybeans is that this optimum canopy development
and maintenance will facilitate or enhance weed control. In other words, it is possible that
irrigation can serve effectively as an additional weed control tool through its enhancing
effect on the soybean canopy. It follows, then, that weed control strategies for irrigated
soybeans should be no more complex, and may be even less complex, than those used for
nonirrigated production. They still should be based on a knowledge of field history, row
spacing and expected crop canopy development, proper weed species identification, proper
herbicide selection, proper timing of herbicide applications, and economics. In order to
validate some of these assumptions, two experiments were conducted at Stoneville, MS.

Stale seedbed plantings of A 5980 soybeans were made in different fields in 1988
through 1989 and 1987 through 1990 in 40-in. rows where pre- and postemergent herbicides and postemergent cultivation were used for weed control. Treatment descriptions
are given in Table 9.9. Further details are given by Heatherly et al. (1993, 1994). In the
1988–1989 experiments where cocklebur and pigweed were present, yields and net returns
from soybeans in the irrigated treatment were not different among treatments that received
pre- and/or postemergent herbicides in addition to preplant foliar-applied (PFA) Roundup
and postemergent cultivation. In the nonirrigated treatment, net returns were highest
where preemergent Sencor/Lexone alone or in combination with postemergent herbicides
was applied in addition to PFA Roundup and postemergent cultivation. PFA Roundup
plus only postemergent herbicides did not provide adequate weed control to achieve
maximum net returns when irrigation was not used.

In the 1987 through 1990 experiment where cocklebur and pigweed were not present,
use of pre- and postemergent herbicides in addition to PFA Roundup resulted in higher
yields from the irrigated treatment, but no greater net returns than when PFA Roundup
was the only herbicide used and postemergent cultivation was done. In the nonirrigated
treatment where yields were lower, use of pre- and postemergent herbicides in addition

to PFA Roundup resulted in lower net returns than when PFA Roundup was the only herbicide used and cultivation was done.

The following conclusions are drawn from these limited data. In stale seedbed plantings where problem weeds occur, either pre- or postemergent herbicides provide sufficient weed control for greatest yield and net returns in irrigated soybeans, but only preemergent herbicides are required in nonirrigated plantings. This difference in weed control requirements between the nonirrigated and irrigated plantings is probably a result of the complete canopy formed by only the irrigated crop in the wide rows. In stale seedbed plantings where problem weeds do not occur, PFA herbicide plus postemergent cultivation results in lower yield but maximum net returns in irrigated plantings, and equal yield but higher net returns in nonirrigated plantings. Thus, a weed control program beyond PFA herbicide and postemergent cultivation in wide-row stale seedbed soybean systems should be determined based on whether or not irrigation will be used during the season and on what weed species are present. Obviously, the use of postemergent cultivation in wide-row soybean culture is an important component of weed control in both irrigated and nonirrigated plantings.

Insects and soybean cyst nematode

The following information is known about soil water status and its effect on soybean looper feeding on soybean leaves.

1. Insect-susceptible soybean plants grown under less than optimum soil water conditions have an adverse effect on soybean looper development (Lambert and Heatherly, 1991; 1995). The longer plants were grown under soil water deficit conditions, the greater the effect. Conversely, insect-susceptible soybean plants that were grown under well-watered conditions appeared to not limit soybean looper development.
2. Defoliation by soybean looper of susceptible soybean plants that were irrigated can be over 50% greater than defoliation of nonirrigated plants.
3. Percentage yield reduction resulting from uncontrolled soybean looper infestations in susceptible varieties was greater for irrigated than for nonirrigated plants.
4. Soybean plants growing under water-deficit conditions may allow a delay in initiation of soybean looper control measures, especially since yield potential and resulting profit potential are low. Conversely, soybean plants growing in a well-watered soil should be monitored closely for timely insect control, since leaf defoliation and subsequent yield and economic losses can be greater.

The following information is known about soil water status and its effect on soybean cyst nematode (SCN).

1. Optimally wet silt loam soil will promote and sustain increased populations of SCN, while dry soil will not (Heatherly et al., 1982).
2. SCN infection of susceptible soybean varieties reduces seed yields, and irrigation will not overcome the total effect of this SCN-induced stress (Young and Heatherly, 1988; Heatherly and Young, 1991; Heatherly et al., 1992).
3. Irrigation may slightly increase yield of susceptible soybean varieties grown in SCN-infested fields, but yields will be well below yields from irrigated, noninfested fields or from irrigated, resistant varieties grown in infested fields. Thus, irrigation efficiency will be low and irrigation probably will be unprofitable. Therefore, irrigation does not appear to be a viable SCN management strategy for infested fields. Conventional management strategies using crop rotation and resistant varieties in both irrigated and nonirrigated fields are more viable alternatives.

Flood irrigation

Surveys indicate that the majority of irrigation applied to soybeans in the lower Mississippi River Delta uses a surface (furrow or flood) vs. overhead method, presumably because of relatively level topography and the higher costs associated with overhead (center pivot, lateral move) irrigation (Spurlock, 1995). A significant portion of this surface irrigation is done by the flood method. Questions often arise about the management of flood irrigation applied to soybeans, and about using furrow irrigation data to draw conclusions about flood irrigation. The following information is presented to answer some basic questions about management of flood irrigation, variety performance under flood irrigation, and flood vs. furrow irrigation yield responses.

Land in the Mississippi River alluvial plain is nearly flat, and a significant area has been graded to facilitate surface drainage and furrow irrigation. However, the practice of grading is expensive. A less costly alternative, contour flood irrigation, uses levees installed at the same elevation around or through a relatively flat but ungraded field. This method requires little or no grading of relatively flat fields. Straight levee irrigation uses levees that run perpendicular or parallel to the slope of a field to confine water to defined areas in fields that have been graded to slope in only one direction. This method requires moderate grading to ensure that the entire field slopes in the same direction. With either method of flood irrigation, the levees separate field areas of different elevation, and irrigation water flows from the highest to lowest point in the area enclosed by the levees. This results in soil in the lower elevations of the enclosed areas being submerged for some period of time. Water is released from areas enclosed by levees when the entire area within a levee system is wetted.

Whatever levee method is used to enclose the area to be irrigated completely, the result is an inundation of a field with water amounts that result in standing water on some portion of the enclosed area. The flow rate of the water source and the size of the paddy or enclosed area determine the time required to complete the flood. During the time of flooding, an increasingly larger area is covered with water until the entire area within the levees or borders is finally inundated.

Levees that are constructed parallel to the major field grade facilitate what is known as border irrigation, which is really a hybrid of furrow and flood irrigation. Gated pipe (rigid or rollout) or flume pads are used to deliver irrigation water to the area between the parallel levees. This method does not involve the complete enclosure of a field area. It is used to confine irrigation water to a relatively small width of a field at any one time to ensure that the entire area between the parallel levees is watered completely from top to bottom before water is applied to an adjacent border area. This method is used most efficiently in fields that have a slope in only one direction; i.e., it works best on fields that have zero side slope. This method ensures the relatively rapid watering of the area receiving water, and should result in the avoidance of complications associated with anaerobic conditions induced by flood irrigation.

Studies conducted by Nathanson et al. (1984) and Troedson et al. (1989) and the summary by Lawn and Byth (1989) provide details of the processes that occur in soybeans during and immediately following an initial acclimation phase that occurs when soil is saturated with water. During and following soil saturation or flooding (the period of flood stress), plants develop leaf symptoms that are characteristic of N-deficiency stress. The development of these symptoms coincides with a decline in soil nitrate and total available soil N levels. Roots in the saturated zone die, which apparently contributes to the expression of N-deficiency symptoms in the plants (Nathanson et al., 1984). Soil O_2 levels fall to zero and remain there until removal of flooded conditions. Recovery to normal O_2 levels can take from 4 days (2-h flood) to 14 days (72-h flood) (D.A. Pennington, 1990, personal communication). Damage from excessive flooding has been attributed to this reduction in soil O_2 and concurrent accumulation of microbial toxins (Jackson, 1983). Nodule function is also reduced (Troedson et al., 1981). Bennett and Albrecht (1984), however, reported that the nitrogenase activities of soybean nodules returned to levels that were similar to those of a well-watered control after removal from a flooded environment. Both Nathanson et al. (1984) and Troedson et al. (1989) found that after the initial acclimation to a saturated soil, soybean nodule weight was increased significantly over that in conventionally irrigated soybeans, and shoot growth rate was also greater. This was with seedlings and resulted from repositioning of the effective roots. Roots that were formed prior to flooding largely died. New roots formed at the saturated zone/oxygen interface and development resumed.

Numerous studies have been conducted in the midsouthern U.S. to investigate the response of soybeans to flood irrigation. Results from some of this work are presented below to provide support for conclusions about flood irrigation management.

1982–1984 Crowley, LA study

Ransom (MG VII) variety was planted in late May and used to determine the effect that length of flood duration had on yield of drill-seeded (10-in. row spacing) soybeans following rice on a Crowley silt loam soil (Griffin et al., 1988). Plots were flooded at the V6 (six-leaf) vegetative stage, and at the R2 (full bloom) and R2 + R5 (seedfill stage) stages. Flood water was applied to a depth of 3 in. and allowed to remain on the area for 0 (nonflooded control), 1, 2, 4, and 8 days before surface drainage. Conclusions derived from the results (Table 9.10) of this research were (1) soybeans exposed to longer than 2 days of standing water were more tolerant during the vegetative than the reproductive period and (2) attempts should be made when flooding soybeans after flowering to remove water within 2 days to avoid significant yield reductions.

Table 9.10 Influence of Plant Development Stage and Length of Flooding (days) on 3-year (1982–1984) Average Yield (bu/acre) of Ransom Soybeans Grown at Crowley, LA

Time of flooding	Flood length	Yield
None	0	39.8
V6 vegetative	1	44.7
	2	43.8
	4	43.5
	8	40.3
R2 Bloom	1	40.1
	2	40.1
	4	35.5
	8	27.5
R2 + R5 seedfill	1	43.9
	2	42.5
	4	34.2
	8	22.6
$LSD_{0.05}$		2.7

Source: Griffin, J. L. et al., Louisiana Agric. Exp. Sta. Bull. 795, 1988.

Table 9.11   Flood Duration (days) Effect on Seed Yield
(bu/acre) of Soybeans Grown on Two Soils and
Flooded at V4 (four-leaf stage) and R2 (full bloom)
Stages at Keiser and Stuttgart, AR (1987)

| Days | Crowley soil | | Sharkey soil | |
flooded	V4	R2	V4	R2
2	48.0 a	41.6 ab	37.0 a	33.9 a
4	46.5 a	47.4 a	30.5 b	25.1 b
7	47.0 a	35.2 b	26.4 c	13.3 c
14	38.4 b	26.9 c	13.8 d	5.3 d

Values within a column followed by the same letter are not
significantly different.

Source: Scott, H. D. et al., *Agron. J.* 81, 631–636, 1989. With
permission.

Table 9.12   Seed Yield (bu/acre) of Centennial (Cent.) and Sharkey
Soybeans Grown under Various Flood Irrigation Treatments (ITRT)
on Sharkey Clay at Stoneville, MS (1986 through 1988)

ITRT[a]	1986		1987		1988	
	Cent.	Sharkey	Cent.	Sharkey	Cent.	Sharkey
NI	4.6c	= 5.6c	9.2c	= 10.5c	26.6c	< 33.9c
1	36.5a	< 39.5ab	32.8a	< 41.1a	29.0b	< 36.7b
2	32.9ab	< 39.2ab	33.1a	< 41.3a	28.0b	< 37.7b
3	30.8b	< 40.8ab	27.2b	< 34.5b	22.5d	< 32.2c
4	38.0a	< 43.9a	31.6ab	< 40.0ab	28.8b	< 39.8a
5	33.8ab	< 37.2b	33.6a	< 39.0ab	31.5a	< 41.1a

Values in individual columns followed by the same letter are not significantly different at a probability level of 0.05; differences between varieties
indicated by = or < (significantly less than).

[a] NI = nonirrigated; 1 = 24 h to full flood; 2 = 24 h to full flood plus 24-h
moving flood; 3 = 24 h to full flood plus 48-h moving flood; 4 = 24-h
standing flood; 5 = flood and drain within 2 h — no standing water.

Source: Heatherly, L. G. and Pringle, H. C., III, 1991, *Agron. J.* 83, 231–236.
With permission.

1987 Arkansas study

Eight determinate soybean varieties were grown in 32- or 38-in. rows at Keiser and
Stuttgart, AR to determine the influence of flood irrigation on yield of soybean grown on
Crowley silt loam and Sharkey clay soils (Scott et al., 1989). Plots were flooded to a depth
of 1+ in. for 2, 4, 7, and 14 days at the V4 (four-leaf) and R2 stages. Results (Table 9.11)
showed that prolonged flooding significantly decreased yield. The damaging effect of
prolonged flooding was more severe at the full bloom stage, and more severe for soybeans
grown on the Sharkey clay.

1986–1988 Stoneville, MS study

Centennial and Sharkey (MG VI) varieties were grown to determine length of flood irri-
gation duration that could be tolerated without adversely affecting yield (Table 9.12).
Experiments were planted on June 3, 1986, May 11, 1987, and May 16, 1988 in 40-in. rows.
A description of the moving water flood treatments is given in Table 9.12 (Heatherly and

Table 9.13   Height (in.) at Maturity and Seed Yield (bu/acre)
of Soybean Varieties Grown on Sharkey Clay and Flood
Irrigated at Stoneville, MS (1990 through 1992)

Variety (MG)	Plant height			Seed yield		
	1990	1991	1992	1990	1991	1992
A5980 (V)	28a	—	—	45b	—	—
DP 105 (V)	25a	28a	—	48ab	52a	—
Forrest (V)	21b	20c	23a	42b	40b	32b
P 9592 (V)	26a	27a	25a	54a	51a	39a
DP 415 (V)	21b	21bc	23a	47ab	49a	33b
Hartz 5164 (V)	21b	—	—	42b	—	—
Hutcheson (V)	—	22bc	—	—	43b	—
A5979 (V)	—	23b	23a	—	53a	34b
Centennial (VI)	29bc	31bc	34bc	42b	36c	35b
Young (VI)	31b	32b	36ab	46b	39bc	44a
Sharkey (VI)	37a	38a	38a	43b	40b	38b
A6785 (VI)	30bc	29c	31cd	52a	44a	37b
Tracy M (VI)	33b	—	—	43b	—	—
P 9641 (VI)	26c	25d	28d	54a	47a	44a
Hartz 6686 (VI)	—	30bc	30d	—	37bc	37b

Values in columns within each MG group followed by the same letter
are not significantly different at a probability level of 0.05.

Pringle, 1991). A 1984 flood study using Centennial was abandoned because the treatments involved rapid inundation to achieve a standing-water flood, and this resulted in death of plants or zero podset and yield.

In 1986 and 1987, Sharkey and Centennial yields from the nonirrigated treatment were not different, while in 1988, the nonirrigated yield of Sharkey was greater than that of Centennial (Table 9.12). In all irrigated treatments, yields of Sharkey were greater than those of Centennial. Thus, Centennial appeared to be more sensitive to flood irrigation and resulting oxygen deprivation than did Sharkey.

The clear-cut trend shown by data in Table 9.12 is that flood irrigation duration of 3 days (treatment 3) resulted in less than maximum yield increase from irrigation, while that of 2 days or less ensured the greatest yield increase. Thus, for highest yield response from flood irrigation, it should be managed so that all field areas have the process started and finished within 2 days. Longer flood irrigation periods will result in a yield increase compared with no irrigation, but the increase will be less due to soil oxygen deprivation.

1990–1992 Stoneville, MS study

To further explore the performance of varieties under flood irrigation, MG V and VI varieties (Table 9.13) were planted on May 8, 1990, May 16, 1991, and May 27, 1992. Flood irrigation was applied as described for treatment 2 in Table 9.12.

Ranges in height and yield in all years (Table 9.13) indicate that growth and yield of soybean varieties can differ when flood irrigation is used, but this difference is not necessarily because of flood irrigation since a nonirrigated check was not used for comparison. Within MG V, Forrest was always among the shortest and lowest-yielding varieties. DP 105 and Pioneer Brand 9592 were always among the tallest and highest-yielding varieties. Within MG VI, Centennial and Hartz 6686 were always among the shortest and lowest-yielding varieties. Sharkey was always the tallest variety, but Pioneer Brand 9641 was

Table 9.14 Average Yield (bu/acre) of Nonirrigated (NI) and Furrow (FU) and Flood (FL) Irrigated MG IV and V Soybeans Grown Following Rice, and Number of Irrigations (No.) at Stoneville, MS (1993–1996)

MG	NI	FU	FL	No.
May 21, 1993 planting date				
MG IV	37.4	42.9	44.4	4
MG V	40.9	50.0	51.9	4
April 21, 1994 planting date				
MG IV	38.6	50.0	50.0	4
MG V	46.1	59.8	58.6	4
April 18, 1995 planting date				
MG IV	38.0	57.5	57.9	3
MG V	35.6	54.0	53.8	4
April 30, 1996 planting date				
MG IV	40.1	58.6	45.7	4,3[a]
MG V	57.0	58.9	56.0	5,4

[a] Four and five furrow irrigations for MG IV and V varieties, respectively; three and four flood irrigations for MG IV and V varieties, respectively.

Average of four varieties in each MG.

always among the highest yielders, while Sharkey was not. These results reemphasize the importance of variety selection for fields that are to be flood irrigated.

1993–1996 Stoneville, MS study

Maturity group IV and V soybean varieties were planted in 20-inch rows on May 21, 1993, April 21, 1994, April 18, 1995, and April 30, 1996 following rice in a 1:1 rice:soybean rotation study at Stoneville, MS (Table 9.14). Nonirrigated, furrow-irrigated, and flood-irrigated environments were used. The flood-irrigated portion was watered at the same time as the furrow-irrigated portion except in 1996, and the flood period was 24 to 30 h at each watering. Average yields of MG IV varieties from the furrow and flood environments were similar to each other except in 1996, whereas average yields of MG V varieties from the two irrigated environments were similar to each other every year. These data indicate that properly timed and managed flood irrigation resulted in yields of soybeans that were comparable to those resulting from proper furrow irrigation.

Quality of harvested seeds

Drought and high temperatures during seed development of soybeans can have a significant effect on germination of harvested seeds. In years of severe drought (large seed yield increases from irrigation), irrigation can result in significantly increased germination of harvested seed of MG V and VI soybean varieties (Heatherly, 1993). In fact, irrigation can result in germination being increased from unacceptable (<80%) to acceptable levels, even though daytime temperatures may be relatively high (up to 93°). It appears that the effect on germination of high daytime temperatures that occur along with drought during seedfill can be overcome by irrigation in these varieties. Further research is needed to clarify the relationship between irrigation and germination of harvested seeds of indeterminate, early-maturing varieties. Information about the effects of irrigation on germination of soybean seeds developed during moderate or nonstress years is not available. Thus, soybeans grown for planting seed in the midsouthern U.S. should be irrigated during the full reproductive period in dry years to ensure highest yield of seeds with acceptable germination.

In 1993 and 1994, the United Soybean Board sponsored work at Stoneville, MS that was designed to determine the effect of Benlate fungicide on germination of harvested seeds from nonirrigated and irrigated soybean varieties (Bowers et al., 1995). One application of Benlate (0.5 lb/acre) applied near or at the beginning of seedfill (R5) had no significant effect on germination of harvested seeds of MG IV and V soybean varieties that were planted on April 29 and June 3, 1993 and April 21 and May 13, 1994. In 1993, germination percentage ranged from 82 to 98%, and in 1994 all seed samples germinated greater than 89%. Irrigation and planting date had little effect on germination of harvested seeds in either year. In earlier work at Stoneville, Heatherly and Sciumbato (1986) applied Benlate at stages R3 and R5 of MG VI soybean varieties planted on May 15, 1981 and May 9, 1983. These applications resulted in small, uneconomical yield increases in furrow- and overhead-irrigated environments, and no yield increases in nonirrigated environments. Germination of harvested seeds was not measured.

Acknowledgments

I appreciate the contributions of James Thomas and Phil Tacker, Extension Irrigation Specialists for Mississippi and Arkansas, respectively.

References

Anonymous. 1995. Mississippi Agriculture, 1970–1994, Agric. Econ. Dept., Mississippi State University, Starkville.

Anonymous. 1996. Mississippi Weekly Weather Crop Report, May 7, 1996, Miss. Agric. Stat. Serv., Jackson.

Anonymous. 1997. Mississippi Weekly Weather Crop Report, May 5, 1997, Miss. Agric. Stat. Serv., Jackson.

Bennett, J. M., and S. L. Albrecht. 1984. Drought and flooding effects on N_2 fixation, water relations, and diffusive resistance of soybean, *Agron. J.* 76:735–740.

Boquet, D. J. 1989. Sprinkler irrigation effects on determinate soybean yield and lodging on a clay soil, *Agron. J.* 81:793–797.

Bowers, G. R., L. O. Ashlock, J. L. Rabb, and L. G. Heatherly. 1995. Improving Efficiencies of Early Planted, Early Maturing Soybean Cropping Systems in the Southcentral U.S., Final Prog. Rept. to United Soybean Board, St. Louis, MO.

Boykin, D. L., R. R. Carle, C. D. Ranney, and R. Shanklin. 1995. Weather data summary for 1964–1993, Stoneville, MS, *MAFES Tech. Bull.*, 201.

Griffin, J. L., R. W. Taylor, R. J. Habetz, and R. P. Regan. 1985. Response of solid-seeded soybeans to flood irrigation. I. Application timing, *Agron. J.* 77:551–554.

Griffin, J. L., R. J. Habetz, and R. P. Regan. 1988. Flood Irrigation of Soybeans in Southwest Louisiana, Louisiana Agric. Exp. Sta. Bull. 795.

Heatherly, L. G. 1983. Response of soybean cultivars to irrigation of a clay soil, *Agron. J.* 75:859–864.

Heatherly, L. G. 1984. Soybean Response to Irrigation of Mississippi River Delta Soils, U.S. Department of Agriculture, Agric. Res. Serv., ARS-18, U.S. Government Printing Office, Washington, D.C., 49 pp.

Heatherly, L. G. 1986. Water use by soybeans grown on clay soil, in *Proc. Delta Irrig. Workshop*, Greenwood, MS, Feb. 28, Mississippi Cooperative Extension Service, Starkville, MS, 113–121.

Heatherly, L. G. 1988. Planting date, row spacing, and irrigation effects on soybean grown on clay soil, *Agron. J.* 80:227–231.

Heatherly, L. G. 1993. Drought stress and irrigation effects on germination of harvested soybean seed, *Crop Sci.* 33:777–781.

Heatherly, L. G. 1996. Yield and germination of harvested seed from irrigated and nonirrigated early and late planted MG IV and V soybean, *Crop Sci.* 36:1000–1006.

Heatherly, L. G. and C. D. Elmore. 1986. Irrigation and planting date effects on soybeans grown on clay soil, *Agron. J.* 78:576–580.

Heatherly, L. G. and H. C. Pringle, III. 1991. Soybean cultivars' response to flood irrigation of clay soil, *Agron. J.* 83:231–236.

Heatherly, L. G. and G. L. Sciumbato. 1986. Effect of benomyl fungicide and irrigation on soybean seed yield and yield components, *Crop Sci.* 26:352–355.

Heatherly, L. G., and S. R. Spurlock. 1993. Timing of furrow irrigation termination for determinate soybean on clay soil, *Agron. J.* 85:1103–1108.

Heatherly, L. G. and L. D. Young. 1991. Soybean and soybean cyst nematode response to soil water content in loam and clay soils, *Crop Sci.* 31:191–196.

Heatherly, L. G., L. D. Young, J. M. Epps, and E. E. Hartwig. 1982. Effect of upper-profile soil water potential on numbers of cysts of *Heterodera glycines* on soybeans, *Crop Sci.* 22:833–835.

Heatherly, L. G., C. D. Elmore, and R. A. Wesley. 1990. Weed control and soybean response to preplant tillage and planting time, *Soil Till. Res.* 17:199–210.

Heatherly, L. G., H. C. Pringle, III, G. L. Sciumbato, L. D. Young, M. W. Ebelhar, R. A. Wesley, and G. R. Tupper. 1992. Irrigation of soybean cultivars susceptible and resistant to soybean cyst nematode, *Crop Sci.* 32:802–806.

Heatherly, L. G., R. A. Wesley, C. D. Elmore, and S. R. Spurlock. 1993. Net returns from stale seedbed plantings of soybean (*Glycine max*) on clay soil, *Weed Technol.* 7:972–980.

Heatherly, L. G., C. D. Elmore, and S. R. Spurlock. 1994. Effect of irrigation and weed control treatment on yield and net return from soybean (*Glycine max*), *Weed Technol.* 8:69–76.

Hodges, H. F. and L. G. Heatherly. 1983. Principles of water management for soybean production. Bull. 919, Mississippi State University, Starkville.

Jackson, M. B. 1983. Plant responses to waterlogging of the soil, *Aspects Appl. Biol.* 4:99–116.

Lambert, L. and L. G. Heatherly. 1991. Soil water potential: effects on soybean looper feeding on soybean leaves, *Crop Sci.* 31:1625–1628.

Lambert, L., and L. G. Heatherly. 1995. Influence of irrigation and susceptibility of selected soybean genotypes to soybean looper, *Crop Sci.* 35:1657–1660.

Lawn, R. J. and D. E. Byth. 1989. Saturated soil culture — a technology to expand the adaptation of soybean, in *Proc. World Soybean Res. Conf. IV*, Buenos Aires, Argentina, 5–9 March, 576–581.

Nathanson, K., R. J. Lawn, P. L. M. de Jabrun, and D. E. Byth. 1984. Growth, nodulation and nitrogen accumulation by soybean in saturated soil culture, *Field Crops Res.* 8:73–92.

Reicosky, D. A. and L. G. Heatherly. 1990. Soybean, in B. A. Stewart and D. A. Nielsen, Eds., *Irrigation of Agricultural Crops*, Agronomy Monograph 30:639–674. American Society of Agronomy; Madison, WI.

Scott, H. D., J. DeAngulo, M. B. Daniels, and L. S. Wood. 1989. Flood duration effects on soybean growth and yield, *Agron. J.* 81:631–636.

Spurlock, S. R. 1995. 1995 Planning Budgets: Soybeans, Mississippi State University, Dept. of Agric. Econ. Rept. 60.

Tacker, P. L., E. D. Vories, and L. O. Ashlock. 1994. Drainage and irrigation, in L.O. Ashlock, Ed., *Technology for Optimum Production of Soybeans*, Publ. AG411-12-94, University of Arkansas Coop. Ext. Serv., Fayetteville, AR.

Troedson, R. J., R. J. Lawn, and D. E. Byth. 1981. Growth and nodulation of soybeans in high water table culture, in A. H. Gibson and W. E. Newton, Ed., *Current Perspectives in Nitrogen Fixation*, Australian Academy of Science, Canberra, A.C.T., Australia, 464.

Troedson, R. J., R. J. Lawn, D. E. Byth, and G. L. Wilson. 1989. Response of field-grown soybean to saturated soil culture. I. Patterns of biomass and nitrogen accumulation, *Field Crops Res.* 21:171–187.

Wesley, R. A., L. G. Heatherly, C. D. Elmore, and S. R. Spurlock. 1994. Net returns from eight irrigated cropping systems on clay soil, *J. Prod. Agric.* 7:109–115.

Wesley, R. A., L. G. Heatherly, C. D. Elmore, and S. R. Spurlock. 1995. Net returns from eight non-irrigated cropping systems on clay soil, *J. Prod. Agric.* 8:514–520.

Young, L. D. and L. G. Heatherly. 1988. Soybean cyst nematode effect on soybean grown at controlled soil water potentials, *Crop Sci.* 28:543–545.

chapter ten

Doublecropping wheat and soybeans

Richard A. Wesley

Contents

Introduction

Doublecropping winter wheat and soybeans is an accepted practice. In 1994, approximately 1.5 million acres of winter wheat were harvested for grain in the midsouthern states (Arkansas, Louisiana, Mississippi, and Tennessee). After wheat harvest, the majority of this acreage was then planted to doublecrop soybeans and this accounted for approximately 19% of the total soybean acreage in this area (USDA National Agricultural Statistic Services, 1994). Doublecropping will continue to be an important cropping practice in areas where the length of the growing season is adequate and soil moisture from rainfall and/or irrigation is available to assure prompt seedling emergence and adequate plant development of the summer crop.

The potential advantages of doublecropping are (1) increased cash flow that results from better utilization of climate, land, and other resources; (2) reduced soil and water losses by having the soil covered with a plant canopy most of the year; and (3) more intensive land use and utilization of machinery, labor, and capital investments.

It is well known that doublecropping requires careful and timely management for success, and this has been one of the major factors that has limited its use. Doublecropped soybeans that are not managed properly usually yield less than monocropped soybeans, and since wheat has had relatively low cash value, the two crops in a wheat–soybean doublecrop system can be less profitable than monocropped soybean alone. However, with irrigation, yields of doublecropped soybeans may be comparable to the yield of monocropped soybeans. Thus, profits from the wheat crop become a bonus that provides a strong incentive to grow wheat as a winter crop in the doublecrop system.

Several factors should be considered relative to the economic feasibility of double-cropping. First, doublecropping spreads fixed costs of farm machinery, particularly com-bines, and thus leads to some improvement in total farm income. A second factor, and perhaps the most important feature of doublecropping, is increased cash flow. Sales of wheat in June provide some money to be put into the production process for other crops grown during the remainder of the growing season. Because money does not have to be borrowed, interest on operating capital is reduced, thereby reducing production costs and increasing farm income.

Most farmers in the midsouthern U.S. generally decide to plant wheat based on the expected price of wheat the following summer and the government programs in force at that time. The decision to plant soybeans after wheat in a doublecropping pattern is influenced by both agronomic and economic factors. Agronomic factors include the harvest date for wheat, which dictates time of soybean planting and soil moisture availability for timely planting, germination, and emergence of the soybean seed. However, price is probably the dominant factor in the decision to doublecrop soybeans. The higher the soybean price, the greater the tendency to produce soybeans behind wheat.

Wheat production practices

General

In the midsouthern U.S. winter wheat (almost exclusively soft red winter wheat) is grown primarily under nonirrigated conditions. This is possible because of the abundance of rainfall received during the winter and early spring. However, wet spring conditions sometimes contribute to severe disease problems and rank growth. Therefore, most wheat varieties grown are usually awnless since the awned characteristic is associated with higher incidence of disease. Most varieties grown today have been selected based on yield level, disease tolerance, and short stature to reduce lodging caused by rank growth (Banks, 1990). Generally, if wheat is planted during the recommended time frame, the yield is about the same whether monocropped or doublecropped following a soybean crop. How-ever, nitrogen requirements and costs may be reduced due to the addition of nitrogen by the soybean crop.

Seedbed preparation

In general, tillage prior to wheat planting involves only a disk harrow to remove weeds and a portion of the residue of the previous crop and to prepare a seedbed. On sandy soils, one pass with the disk harrow followed by a drag, spike-tooth harrow, or a field cultivator will prepare an adequate seedbed. However, for soils with a high clay content, two or more disking and harrowing operations may be needed. When wheat is to be planted following a summer crop of corn or sorghum, three diskings are often required to incorporate the residue prior to smoothing with a field cultivator. In contrast, some producers use no-till grain drills to plant wheat directly into the residue of the previous crop. Regardless of the type of seedbed preparation, no other tillage operation is required until after wheat harvest.

Planting

In the midsouthern U.S., winter wheat is normally planted during October and November with a standard or no-till grain drill in rows spaced 6 to 10 in. apart. Some producers broadcast the seed and lightly disk the soil to cover. However, this planting practice often results in erratic emergence due to variable seeding depths and poor distribution of seed. Erratic stands usually increase weed problems that often must be controlled in the spring. Conversely, a good uniform stand of wheat normally provides adequate shading and adequate weed control. Wheat seeding rates range from 90 to 120 lb/acre when drill seeded and up to 120 to 150 lb/acre when broadcast seeded or air seeded.

Fertilization

The alluvial soils of the Mississippi Delta in the midsouthern states contain adequate levels of organic matter and the essential nutrients (Brady, 1974). Generally, soil test recommendations indicate only an addition of nitrogen. A preplant application of 20 to 30 lb/acre of nitrogen is recommended if wheat follows a summer grass crop (corn, sorghum, etc.). Conversely, no preplant nitrogen is recommended if wheat follows a legume crop such as soybeans (Baskin, 1978). For grain production, an application of 90 to 120 lb/acre of nitrogen is recommended from mid-February to mid-March, depending on soil texture. On soils with poor internal drainage such as clays, split applications of nitrogen are more advantageous.

For many soils of the region, it is necessary to apply lime periodically. The recommended soil pH for wheat production ranges from 5.6 to 6.2. Generally, at soil pH levels below 5.6, aluminum toxicity to wheat occurs and greatly reduces root growth and development. A standard soil test should be made annually to determine the soil pH and the amount of lime recommended. Good soil fertility practices not only enhance grain yields but also aid in weed control. Wheat that grows vigorously is much more competitive with winter annual weeds than nutrient-deficient plants.

Weed control

Weed control in winter wheat in the midsouthern U.S. is usually minimal. Most weed control efforts involve the application of postemergence herbicides only if an infestation warrants control. The most common broadleaf and grass weeds in wheat are mayweed, vetch, wild garlic, henbit, wild mustards, chickweed, buttercup, and curly dock (French, 1986; Elmore et al., 1995). However, the most troublesome weeds are wild garlic, wild mustards, curly dock, Italian ryegrass, and cutleaf eveningprimrose.

Most of the broadleaf weeds can be controlled with a single application of 2,4-D after the wheat is fully tillered and before joints are formed in the stem. Occasionally, Banvel, MCPA, or Harmony Extra are used to control troublesome weeds such as wild garlic and wild radish (Ashburn, 1983). Application of these herbicides should be made after wheat reaches the two-leaf stage but before the flag leaf is visible. The most serious perennial broadleaf weed problem is curly dock (Banks, 1990). Moderate to heavy infestations of this weed can be very damaging to wheat production because of competition and harvesting interference. This weed seems to be more prevalent in minimum and no-till production systems since there is no disruption of the plant's life cycle.

Italian ryegrass is the most serious grass problem in winter wheat in the region and infests up to 15% of the planted acreage (Banks, 1990). Moderate infestations of this weed can seriously reduce wheat yields and result in significant dockage due to contamination by its seed in wheat (Hinkle, 1975). However, it can be adequately controlled with preplant incorporated, preemergence, or postemergence applications of Hoelon (Robinson and Banks, 1983). Postemergence applications should be made before the ryegrass develops beyond the five-leaf stage. Biennial rotations of a wheat–soybean doublecrop system with either corn or sorghum the following year significantly reduces ryegrass infestation levels (Banks, 1990; Wesley et al., 1994b). Other annual grasses such as cheat, little barley, and downy brome also occur but are of lesser importance.

Harvest

Wheat is mature and ready for harvest in late May or early June. Harvest is accomplished by direct combining of the standing wheat when grain moisture is approximately 14%. Higher than acceptable moisture is often attributed to green weed parts in the harvested grain caused when harvest is delayed, or due to poor wheat stands that allow weeds to become very large as the wheat matures. Perennial vines, such as redvine and trumpet creeper, sometimes become sufficiently established at harvest and cause mechanical harvest problems as well as elevated moisture contents. Numerous other weeds such as common ragweed, common lambsquarters, and occasionally johnsongrass and horseweed often interfere with harvest. However, Italian ryegrass and wild garlic contribute to the most dockage because they contaminate the harvested grain and are classified as foreign material that must be removed by cleaning. Wild garlic has little effect on wheat yield but its presence in the harvested grain is easily detected and greatly reduces the value of the wheat since the bulblets are difficult to remove by mechanical cleaning.

Straw management

Wheat straw management is an important part of a successful wheat–soybean doublecrop system. A 60 bu/acre wheat yield may produce from 3 to 4 tons of straw per acre. If not managed properly, it can present major problems during the production of the following soybean crop.

If soybeans are to be planted no-till, it is recommended that wheat be combined at a height of 6 to 12 in. above ground level. This height assures that adequate weed leaf surface area is left for herbicide contact. Stubble heights greater than 12 in. tend to decrease soybean yields which is caused primarily from etiolation of soybean seedlings and less effective weed control. The etiolated soybean seedlings tend to lodge more in the tall stubble, and weed infestations are more severe because of ineffective herbicide application. When wheat is cut at heights less than 6 in. above ground level, the mass of straw and chaff deposited on the surface interferes with planting and reduces soybean stands. All straw going through the combine should be chopped and evenly spread to provide ground cover, to reduce soil and water losses in areas subject to soil erosion, and to facilitate soybean planting.

Field burning of wheat straw, although criticized by some as environmentally unacceptable, is still allowed and practiced on a sizable acreage, especially clay soils (Boquet and Walker, 1984; Sanford, 1982). Research has shown that burning the straw does not adversely affect the yield of soybeans compared with planting directly into wheat stubble (Rogers et al., 1971, Wesley et al., 1994b). In fact, the general trend shows better stands and increased yields of soybean where wheat straw was burned on fine-textured soils (Boquet and Walker, 1984; Elmore et al., 1995; Sanford, 1982). Burning wheat stubble destroys numerous weeds and weed seed on the soil surface, reduces seedling diseases, and makes planting and cultivating of soybean easier.

Soybean production practices

General

High yields of doublecropped soybeans require the successful management of several agronomic and cultural factors. The time lapse between harvesting wheat and planting soybeans is critical. Yields decline rapidly when soybean planting dates are delayed, especially at this late time in the planting season. Factors that affect soybean performance include planting equipment, row spacing, seedbed preparation, weed control, and supplemental irrigation.

Planting equipment

Selecting the proper planting equipment is a prerequisite to the successful production of soybeans in a doublecropped system. The planting unit selected, whether planter or drill, should cut through the remaining trash and residue, place the seed at the desired depth, and firmly press the soil around the seed for good seed-to-soil contact necessary for prompt and uniform emergence. Seeding rates should be increased 10% to compensate for less than desirable soil moisture conditions. If soybeans are drilled, seeding rates should be increased an additional 5% to compensate for less precise seed placement.

No-till planters are generally equipped with coulters, disk openers, downpressure springs, gauge wheels, and press wheels. To ensure good seed-to-soil contact, the planting unit should be set up so that the coulters cut through the residue to a depth of 1 in. below the desired seeding depth. Damp straw and residue are hard to cut even with the best coulters and disk openers. Uncut straw may "hairpin" into the seed furrow and prevent good seed-to-soil contact. Thus, planting should be delayed until the dew has dried. Generally, seed are placed 1 to 2 in. deep into moist soil; however, deeper depths are required if soil moisture is lacking. Disk openers and down-pressure springs assist in proper seed placements, whereas gauge wheels are necessary to maintain uniform planting depths in fields that contain different soil types. Press wheels are furrow-closing devices that ensure that the seed dropped at the predetermined and optimum depth has good contact with the surrounding soil.

Row spacing

Generally, soybeans are planted in a broadcast pattern or in row spacings that range from 10 to 40 in. apart. Numerous research studies have shown that soybeans planted in broadcast patterns or in row spacings of 20 in. or less (narrow row) produce higher yields than soybeans planted in row spacings of 30 in. or more (Hinkle, 1975; Crabtree and Rupp,

1980; Ewen et al., 1981; Boquet and Walker, 1984). In a Louisiana study, yields of double-cropped soybeans planted broadcast and in 10 and 20 in. spacings were similar to each other, but 5.5 bu/acre higher (27.8 vs. 22.3) than those planted in row spacings of 40 in. (Boquet and Walker, 1984). Soybeans grown in narrow-row patterns are generally more competitive and shade the soil surface earlier, which aids in weed control. When a complete canopy is formed, less than 1% of the incoming radiation reaches the soil surface, and thus reduces water loss through evaporation and shades emerging weeds to a non-competitive basis (Bradley, 1995).

Seedbed Preparation

Soybeans may be planted in either a conventionally prepared or minimum-tilled seedbed, or no-tilled directly into the wheat residue with no seedbed preparation. For soybean planted in a conventional seedbed, yields are often higher and weed control is usually better (Jeffers et al., 1973). However, this system requires more machinery, labor, and energy inputs than the minimum and no-till systems. The minimum and no-till systems conserve soil (McGregor et al., 1975; Langdale et al., 1979; McDowell and McGregor, 1980), soil moisture (Jeffers et al., 1973), and fossil fuels (Council of Agricultural Science and Technology (CAST), 1977), and require less machinery and labor input (Jeffers et al., 1973). When compared with conventional production systems, weed control can be a greater problem in some fields planted no-till because preplant-incorporated herbicides cannot be used. Wheat straw intercepts pre- and postemergence herbicides, thereby decreasing their efficiency. Mechanical cultivation for weed control is also more complicated (Sanford et al., 1973; Kapusta, 1979). Burning of the wheat straw gives better early-season weed control and makes cultivation for weed control much easier (Sanford, 1982).

Conventional

In fields where johnsongrass, red vine, and other perennial vines exist, conventional land preparation is recommended where erosion is not a serious threat. Also, certain soils and weather conditions may dictate that a seedbed be mechanically prepared for soybean planting. This is especially true on clayey soils when combines create deep ruts during wheat harvest due to wet soil. Generally, these conditions require two disk passes and a single pass with a field cultivator to reestablish a suitable seedbed for soybeans.

In prepared seedbeds, weed problems can be reduced through tillage; however, this approach to soybean production requires extra fuel and time, and increases evaporative losses of soil moisture needed for timely germination and emergence. In separate studies in Louisiana, Boquet and Walker (1984) reported yields from doublecropped soybeans grown on Sharkey clay and Mhoon silty clay soils averaged 7.5 bu/acre higher (31.5 vs. 24.0) when planted in a tilled vs. no-till seedbed. However, Hutchinson et al. (1988) concluded that yields of doublecropped soybeans grown in tilled and no-till environments on a Gigger silt loam were similar and averaged 30.6 bu/acre.

Generally, the wheat straw and residue are incorporated into the soil by disking twice. This increases costs, and valuable time is lost that is critical to the normal growth and maturity of the soybean crop since soil moisture will not be adequate following this tillage for timely planting, germination, and emergence of soybean. Yields decline rapidly if soybean are planted later than June 20 (Kluse et al., 1976; Heatherly, 1984b). Disking for seedbed preparation also exposes the land to soil erosion, increases evaporative loss of soil water, and, on soils susceptible to crusting, can result in less water infiltration into the soil later.

No-till

In doublecrop systems, soybean planted no-till after wheat harvest is a widely used practice in both standing and burned stubble (Sanford, 1982). No-till planting provides the least delay in establishing a second crop, thereby increasing chances for success in doublecropping. Soybean may be successfully planted in standing stubble if the wheat straw is chopped and evenly spread over the field. No-till planting of soybean should be seriously considered in rolling hill areas where soil erosion occurs.

Soybeans may also be planted no-till after burning wheat straw and residue. When the straw is dry, it burns quickly and leaves the field almost completely bare. In areas not subject to high rainfall levels and large soil erosion losses, there are few data to document specific advantages of wheat residue as a mulch as opposed to its known disadvantages in doublecropping soybean following wheat. Generally, the removal of the wheat straw by burning does not significantly reduce soybean yields. In a wheat–soybean doublecrop study on a clay soil with and without supplemental irrigation, yields of soybeans grown in a burned environment under both irrigation regimes were significantly higher than yields produced by soybeans grown in standing stubble (Wesley and Cooke, 1988). In another study on a Brooksville silty clay, a doublecrop production system in which the wheat straw was burned and followed by no-till seeding of soybean and mechanical cultivation gave the greatest average soybean yield of the numerous systems studied (Sanford, 1982).

Weed control

Successful doublecropping in the midsouthern U.S. is frequently associated with management's ability to control weeds in soybean. Many weed control options are available and there is general agreement that a system that utilizes standard weed control practices can usually be found for the weed complex present. This weed complex is often dictated by stubble management, tillage system, and seeding method selected.

On a clay soil, the degree of weed cover in standing and burned wheat stubble was determined to be similar (Elmore et al., 1995). In this study, certain weed species were inhibited by the mulch cover provided by no-till planting in standing wheat stubble. However, other weed species were uninhibited and able to compensate. Weed species were different and altered over time by tillage systems. The burned stubble seedbed had greater populations of nodding spurge, barnyard grass, pricky sida, and fall panicum. This tends to indicate the wheat straw was affecting some weeds. These data also indicate there is a complex interaction among weed species in burned and standing wheat stubble environments. Some weed species may be suppressed by standing wheat stubble that may allow other surviving weed species to flourish with less competition.

Irrigation

Supplemental irrigation is probably the most important factor in successfully doublecropping soybean after wheat. When soil and weather conditions dictate a conventional seedbed be prepared by disking, soil moisture deficits are created through evaporative losses of moisture during the tillage operations. However, the option to irrigate assures adequate soil moisture for timely planting and emergence and provides immediate activation of preemergent herbicides. When soybeans are planted no-till into standing or burned wheat stubble, irrigation also prevents delays in emergence associated with soil moisture deficits that are often present after wheat harvest.

Table 10.1 Yields of Corn, Sorghum, Soybeans, and Wheat Grown in Monocrop, Doublecrop, and Rotational Cropping Systems over an 8-year Period on Tunica Clay near Stoneville, MS

| Crop year | Crop | Grain yield by cropping system and irrigation level,[a] bu/acre | | | | | | | |
| | | SB | | WSB | | CWSB | | SGWSB | |
		I	NI	I	NI	I	NI	I	NI
1984	Wheat			48.0	47.5				
	Soybean	38.2	12.6	37.5	8.9				
	Corn					107.0	46.0		
	Sorghum							118.0	84.0
1985	Wheat			38.0	35.3	33.3	29.5	28.1	25.3
	Soybean	37.8	23.5	37.7	21.1	43.4	29.1	42.4	31.1
1986	Wheat			42.6	28.7				
	Soybean	30.5	1.6	31.0	0.8				
	Corn					137.0	89.2		
	Sorghum							64.0	62.7
1987	Wheat			42.7	37.6	66.7	53.8	60.3	52.0
	Soybean	45.6	20.8	20.9	2.0	37.2	7.5	35.9	7.5
1988	Wheat			63.8	45.0				
	Soybean	48.9	0.0[b]	45.3	0.0[b]				
	Corn					100.0	0.0[b]		
	Sorghum							117.7	0.0[b]
1989	Wheat			49.5	57.4	60.6	58.3	53.2	71.1
	Soybean	47.0	49.8	22.7	22.4	26.5	16.9	24.4	21.3
1990	Wheat			40.8	26.7				
	Soybean	42.3	22.9	37.7	13.9				
	Corn					156.2	25.7		
	Sorghum							113.6	86.1
1991	Wheat			25.5	27.7	34.7	30.8	35.0	32.0
	Sobyean	38.7	40.8	27.8	17.0	45.8	26.2	46.5	28.4
Average	Wheat			43.9	38.2	48.8	43.1	44.2	45.1
	Soybean	41.1	21.5	32.6	10.8	38.2	19.9	37.3	22.1
	Corn					125.1	40.2		
	Sorghum							103.3	58.2

[a] SB = soybean monocrop, WSB = wheat–soybean doublecrop, CWSB = corn–wheat–soybean rotation, and SGWSB = sorghum–wheat–soybean rotation; I = irrigated and NI = nonirrigated.

[b] Crops not planted due to lack of moisture for germination.

Irrigation has been shown to increase soybean yields consistently and significantly in years where moisture deficits occur during the reproductive stages of development (Heatherly, 1984a, b). The yield response to irrigation on clayey soils is often more than the response obtained on coarser-textured soils. The response to irrigation is more closely related to the stage of development of the plant when moisture deficits occur rather than differences in variety or irrigation methods. Moisture deficits generally occur in mid to late summer when extended periods of high temperature are combined with relatively low and/or poorly distributed rainfall.

Properly timed irrigation of doublecropped soybeans often produces yields similar to those produced by irrigated monocropped soybeans. Data from an 8-year study on Tunica clay near Stoneville, MS (Table 10.1) indicate yields of soybeans in irrigated monocrop and wheat–soybean doublecrop systems were similar in 5 of the 8 years (Wesley et al., 1994a,b). Yields of irrigated soybeans in rotational doublecrop systems (wheat–soybean rotated with corn or sorghum) were greater than from irrigated monocropped soybeans in 1985 and 1991. Soybeans in all doublecrop systems were planted later than monocropped soybeans all years. In 1987, 1989, and 1991, yields of irrigated soybeans in the wheat–soybean doublecrop system were lower and this was attributed to the 7 to 8 week later planting date these years.

In contrast, the production of doublecropped soybeans without supplemental irrigation is risky on clay soil. In the same study, the yield of nonirrigated soybeans after wheat ranged from 0 to 17.0 bu/acre in 6 of the 8 years, and averaged only 10.8 bu/acre over the 8 years, whereas yield of irrigated soybeans after wheat ranged from 20.9 to 45.3 bu/acre and averaged 32.6 bu/acre over the 8-year study.

Another irrigated study conducted on a Gigger silt loam in Louisiana indicated irrigation significantly increased soybean yields 4 out of 5 years (Hutchinson et al., 1988). The largest yield increase with irrigation, 20.3 bu/acre, occurred in an extremely dry year. Over the 5-year study, irrigation of soybeans grown under burned and nonburned conditions increased the average yield by 13.1 and 9.2 bu/acre, respectively.

Economics of doublecropping

Several research studies have been conducted to determine the profit potential of various wheat–soybean doublecrop systems. Five doublecrop systems that involved various straw and tillage management practices on a nonirrigated Brooksville silty clay were evaluated and compared with a monocrop soybean system (Sanford, 1982). Data indicate wheat yields from all doublecrop systems were similar and averaged 50.0 bu/acre, whereas soybean yields ranged from 15.0 to 20.4 bu/acre. Monocropped soybean yields averaged 28.7 bu/acre. Gross income to wheat greatly exceeded the income from soybeans in all doublecrop treatments. Net returns to the doublecrop treatments ranged from $61 to 134/acre, whereas net returns to monocropped soybeans averaged $73/acre.

In the Mississippi River Delta of Arkansas, there are several million acres of soils with low organic matter (<0.8%), and they are prone to having severe cyst nematode problems. In this region, most of the nonirrigated acreage is either planted to cotton, continuous soybeans, or wheat–soybean doublecrop systems. A 4-year study was conducted on a Loring–Calloway–Henry (Alfisol) silt loam to determine the yields and net returns from continuous soybeans and wheat–soybean doublecrop systems in nonirrigated environments (Keisling et al., 1995). The continuous soybean and wheat–soybean doublecrop systems were grown both conventionally and no-till. In addition, the doublecrop system had residue management treatments, where the stubble and residue were either burned or left standing. "Forrest" soybeans were planted in each system between June 1 and June 15 each year. Yields of soybeans from all monocrop and doublecrop systems were similar and ranged from 24.2 to 30.2 bu/acre. Net returns from the wheat–soybean doublecrop system, regardless of tillage practice and stubble management, were the largest and ranged from $117 to 153/acre. Net returns to the continuous no-till and conventional soybean systems averaged $48 and 65/acre, respectively. Because soybean yields from these systems were similar, the higher net returns to each doublecrop system were attributed to income generated by the wheat crop.

Net returns above specified costs (Table 10.2) were determined from the yields presented in Table 10.1. Net returns to wheat in the continuous wheat–soybean and rotational doublecrop systems were highly positive all years except 1991. Over the study period, the average net returns to wheat in all systems ranged from $74.30 to 99.93/acre. Net returns to soybeans in the irrigated wheat–soybean doublecrop system were positive 5 of the 8 years, whereas in the nonirrigated counterpart net returns to soybean were positive only in 1989 and 1991.

Table 10.2   Net Returns above Total Specified Costs for Corn, Sorghum, Soybean, and Wheat from Monocrop, Doublecrop, and Rotational Cropping Systems over an 8-year Period on Tunica Clay near Stoneville, MS

| Crop year | Crop | Net returns by cropping system and irrigation level,[a] $/acre | | | | | | | |
| | | SB | | WSB | | CWSB | | SGWSB | |
		I	NI	I	NI	I	NI	I	NI
1984	Wheat			86.35	87.62				
	Soybean	46.38	−34.83	53.33	−55.30				
	Corn					140.75	6.80		
	Sorghum							56.65	41.58
1985	Wheat			96.08	84.47	59.92	43.74	23.29	11.25
	Soybean	3.52	5.65	28.92	−9.74	61.17	32.85	56.13	48.16
1986	Wheat			141.94	72.16				
	Soybean	−58.08	−87.47	−5.85	−82.38				
	Corn					89.95	42.37		
	Sorghum							−43.99	9.91
1987	Wheat			140.74	116.87	222.90	162.53	192.99	154.15
	Soybean	69.94	3.62	−54.10	−99.95	30.99	−76.02	23.60	−76.02
1988	Wheat			183.34	106.45				
	Soybean	161.32	−14.77[b]	146.39	0.0[b]				
	Corn					77.33	−151.56[b]		
	Sorghum							174.08	−6.73[b]
1989	Wheat			96.87	127.13	126.70	117.89	98.36	166.91
	Soybean	88.65	176.98	−59.92	21.67	−22.34	−7.87	−34.39	15.36
1990	Wheat			67.77	11.66				
	Soybean	71.80	31.07	64.08	−6.90				
	Corn					193.48	−98.56		
	Sorghum							93.39	77.72
1991	Wheat			−17.70	−11.96	−9.80	−19.97	−9.02	−16.84
	Soybean	39.28	117.57	19.57	20.55	119.26	71.50	131.46	83.69
Average (by crop)									
	Wheat			99.42	74.30	99.93	76.05	76.41	78.87
	Soybean	52.85	24.73	24.05	−26.51	47.27	5.12	44.20	17.80
	Corn					125.38	−50.24		
	Sorghum							70.03	30.62
Average (all years)[c]		52.85	24.73	123.47	47.79	136.29	15.47	95.32	63.65

[a] SB = soybean monocrop, WSB = wheat–soybean doublecrop, CWSB = corn–wheat–soybean rotation, and SGWSB = sorghum–wheat–soybean rotation; I = irrigated and NI = nonirrigated.

[b] Crops not planted due to lack of moisture for germination.

[c] Represents overall net returns above total specified costs for each cropping system.

Over the study period, net returns to irrigated doublecropped soybeans were slightly positive ($24.05), whereas net returns to nonirrigated doublecropped soybeans were negative ($–26.51) due to an average yield of 10.8 bu/acre over the study period. Net returns to irrigated soybeans in the rotational doublecrop systems were positive and higher than returns in the continuous wheat–soybean doublecrop system. Net returns to corn in the corn–wheat–soybean rotation were highly positive ($125.38) when grown with supplemental irrigation, but highly negative when grown without irrigation ($–50.24). Net returns to sorghum in both the irrigated and nonirrigated environments were highly positive.

Over the study period, net returns to all irrigated doublecrop systems were greater than returns to irrigated monocropped soybeans. For the nonirrigated systems, net returns

Table 10.3   Estimated Costs and Returns per acre, Wheat before Soybean,
Clayey Soil, Delta Area, MS, 1995

Item	Unit	Price, $	Quantity	Amount, $	Your farm, $
Income					
Wheat	bu	4.00	37.0	148.00	
Direct expenses					
Custom labor	acre	3.50	1.00	3.50	
Fertilizer	acre	18.44	1.00	18.44	
Haul seed	acre	5.18	1.00	5.18	
Seed	acre	9.90	1.00	9.90	
Operator labor	h	5.63	0.67	3.71	
Unallocated labor	h	5.63	0.57	3.20	
Diesel fuel	gal	0.67	3.17	2.12	
Repair & maintenance	acre	14.13	1.00	14.13	
Interest on operating capital	acre	2.41	1.00	2.41	
Total direct expense				62.59	
Total fixed expense				26.72	
Total specified expenses				89.31	
Returns above specified expenses				58.69	

to the continuous wheat–soybean doublecrop system and the sorghum–wheat–soybean rotational system were greater than those from monocropped soybeans. Net returns to the nonirrigated corn–wheat–soybean system were the lowest, which was attributed to the extremely low average yield of corn (40.2 bu/acre) on this clayey soil. Long-term studies indicate net returns to a continuous wheat–soybean doublecrop system were among the highest of the eight cropping systems evaluated in both irrigated and nonirrigated environments (Wesley et al., 1994a; 1995). However, the major portion of these returns were attributed to the wheat crop.

The accompanying tables show the estimated costs and returns for the conventional production of wheat (Table 10.3) and soybean (Table 10.4) grown in 1995 on clayey soil in the Delta Area of Mississippi (Soybean Budgets Committee, 1994). These costs and returns have been modified slightly but represent a close approximation of the actual production costs.

Table 10.3 shows direct expenses per acre associated with wheat production totals $62.59. This includes charges for labor, fertilizer, seed, fuel, oil, repairs on machinery and equipment, interest, and hauling expenses. Fixed expenses associated with machinery and equipment used to produce wheat were calculated to be $26.72 per acre. Total specified expenses for wheat production (direct plus fixed expenses) are $89.31. Returns per acre above specified expenses ($58.69) represent the difference in projected income ($148.00) and total specified expenses ($89.31). As long as this figure is positive, it is economical to produce wheat before soybean in a doublecrop enterprise.

Table 10.4 is an approximation of the costs and returns per acre associated with the production of nonirrigated soybeans in a wheat–soybean doublecrop system on clay soil. Direct, fixed, and total specified expenses are estimated as $85.53, 23.68, and 109.20/acre, respectively. In this enterprise, net returns — the difference in projected income ($114.00) and total specified expenses ($109.20) — are $4.80/acre. Under these production conditions, nonirrigated soybeans grown on a clay soil following wheat are only marginally profitable. For farm enterprises with supplemental irrigation, soybean yields approximate 35.0 bu/acre. Estimated income and total specified expenses would approximate $210 and $150/acre, respectively. Thus, the estimated net returns become highly positive

Table 10.4 Estimated Costs and Returns per acre, Soybean after Wheat, Clayey Soil, 20-in. 12-row Equipment, Delta Area, MS, 1995[a]

Item	Unit	Price, $	Quantity	Amount, $	Your farm, $
Income					
Soybeans	bu	6.00	19.0	114.00	
Direct expenses					
Custom labor	acre	1.13	1.00	1.13	
Herbicide	acre	41.73	1.00	41.73	
Insecticide	acre	3.32	1.00	3.32	
Haul seed	acre	3.04	1.00	3.04	
Seed	acre	9.20	1.00	9.20	
Operator labor	h	5.63	0.90	5.06	
Unallocated labor	h	5.63	0.77	4.36	
Diesel fuel	gal	0.67	3.55	2.38	
Repair & maintenance	acre	12.93	1.00	12.93	
Interest on operating capital	acre	2.38	1.00	2.38	
Total direct expense				85.53	
Total fixed expense				23.68	
Total specified expenses				109.20	
Returns above specified expenses				4.80	

[a] Estimated yield from irrigated soybean after wheat would approximate 35.0 bu/acre; estimated income, total specified expenses, and returns above specified expenses would approximate $210, 150, and 60/acre, respectively.

($60.00/acre) and thereby justify the production of irrigated soybean on a clay soil following wheat.

In order to justify the production of wheat and soybean in a continuous wheat–soybean doublecrop system economically, a producer must first determine the cost of production, breakeven price, and profit potential for each crop. These figures can be determined by utilizing the columns provided in Tables 10.3 and 10.4. For individual farm operations these estimates should be adjusted to reflect differences in production practices for specific regions and farm situations. Once producers have determined their actual cost of production, the breakeven price and profit potential can be determined based on the expected price and yield level. If the profit potential of the doublecrop system is highly positive, then serious consideration should be given to a wheat–soybean doublecrop enterprise.

Production guidelines

The following guidelines are provided for wheat–soybean doublecrop systems in the midsouthern U.S.

1. Prepare a seedbed for wheat. Drill wheat at a rate of 90 to 120 lb/acre. Provide adequate drainage of all field areas.
2. Make a weed assessment after wheat emergence. If ryegrass is a problem, make fall applications of Hoelon before ryegrass reaches the five-leaf stage. Use a postemergence application of Harmony Extra to control wild garlic, curly dock, and winter annual broadleaf weeds, if needed.
3. Apply 90 to 120 lb/acre nitrogen from mid-February to mid-March. On soils with poor internal drainage, split applications of nitrogen are common.

4. Harvest wheat in a timely manner. Cut wheat 6 to 12 in. above the ground level if soybeans are to be planted no-till in standing wheat stubble. Chop and evenly spread all residue. If soybeans are to be planted in burned stubble, wait until dew has dried so that a uniform and complete burn is possible.
5. Plant soybeans in a narrow-row pattern as soon as possible after wheat harvest. Plant no-till in burned or standing stubble so that the soybean crop can be established with the least time delay. Prepare a conventional seedbed if johnsongrass or perennial vines warrant, or if adverse weather conditions caused severe rutting during wheat harvest. Increase seeding rate 10 to 15%.
6. Weed control in doublecropped soybean is mandatory. Eliminate all emerged weeds with a burndown herbicide at planting. Apply pre- and postemergence grass and broadleaf herbicides as dictated by the weed complex present. If cultivation is utilized, band applications of herbicides are adequate; otherwise, broadcast applications are recommended.
7. If supplemental irrigation is available, use it for timely planting, germination, emergence, and growth of soybeans in doublecrop systems. Timely irrigation of doublecropped soybeans often produces yields similar to irrigated monocropped soybeans.
8. Production of doublecropped soybeans after wheat on clayey soils without irrigation is not recommended. Extremely low yields are normal due to the severe and lengthy periods of drought many years.

References

Ashburn, E. L. 1983. Wild Garlic Control in Wheat, Univ. Tenn. Agric. Ext. Serv. Publ. 1012, 7 pp.

Banks, P. A. 1990. Weed control in wheat in the southeast, in W.W. Donald, Ed., *Systems of Weed Control in North America*, Monograph Series, Weed Science Society of America, 182–190.

Baskin, C. C. 1978. Small Grain Production, MS Coop. Ext. Serv., Mississippi State University, Inf. Sheet 961, 2 pp.

Boquet, D. J. and D. M. Walker. 1984. Wheat–Soybean Doublecropping: Stubble Management, Tillage, Row Spacing, and Irrigation, La. Agric. Exp. Stn. Bull. 760.

Bradley, J. 1995. No-till doublecrop soybeans, a natural, *Mid-South Farmer*, May, 13.

Brady, N. C. 1974. *The Nature and Properties of Soil*, 8th ed., Macmillan, New York, 639 pp.

Council for Agricultural Science and Technology (CAST). 1977. Energy Use in Agriculture: Now and in the Future, Rep. No. 68, CAST, Ames, IA.

Crabtree, R. J. and R. N. Rupp. 1980. Double and mono cropped wheat *Triticum aestivum* and soybeans *Glycine max* under different tillage and row spacings, *Agron. J.* 72:445–448.

Elmore, C. D., L. G. Heatherly, and R. A. Wesley. 1995. Weed control in no-till doublecrop soybean following winter wheat on a clay soil, *Weed Technol.* 9:306–315.

Ewen, L. S., E. M. Smith, and D. B. Egli. 1981. Doublecropped soybean planting variables, *Trans. ASAE* 25:43–44.

French, C. M., Ed. 1986. *Research Report, South*, Weed Science Society, 215 pp.

Heatherly, L. G. 1984a. Soybean Response to Irrigation of Mississippi River Delta Soils. USDA-ARS-18, U.S. Government Printing Office, Washington, D.C.

Heatherly, L. G. 1984b. A summary of USDA soybean irrigation research at Stoneville, Mississippi — 1979–1983, in *Proc. Delta Irrigation Workshop*, Greenville, MS, 6–7 Mar., Mississippi Cooperative Extension Service, Mississippi State, 13–33.

Hinkle, D. A. 1975. Use of No Tillage in Doublecropping Wheat with Soybeans or Grain Sorghum, Arkansas Agric. Exp. Stat. Rep. Ser. No. 223, 11 pp.

Hutchinson, R. L., T. R. Sharpe, and T. P. Talbot. 1988. Soybean-Wheat Doublecropping Systems for the Loessial Terrace Soils of Northeast Louisiana, Spec. Bull. 88-1, August, Tupelo, MS.

Jeffers, D. L., G. B. Triplett, and J. E. Beyerlein. 1973. Management is the key to success — doublecropped soybeans, *Ohio Rep. Res. Dev.* 58:67–69.

Kapusta, G. 1979. Seedbed tillage and herbicide influence on soybean (*Glycine max*) weed control and yield, *Weed Sci.* 27:520–526.

Keisling, T. C., C. R. Dillon, L. R. Oliver, J. M. Faulkner, and A. G. Flynn. 1995. Profitability of seven nonirrigated soybean cropping rotations on a shallow silt loam soil, in *Proc. 1995 Southern Conservation Tillage Conference for Sustainable Agriculture*, Spec. Bull. 88-7. June, Jackson, MS.

Kluse, C. E., J. G. Shannon, and L. H. Duclos. 1976. Growth, Yield, and Date of Planting Studies with Irrigated Soybean Varieties in Southeast Missouri, Missouri Agric. Exp. Stn. Bull., 1014.

Langdale, G. W., A. P. Barnett, R. A. Leonard, and W. G. Flemings. 1979. Reduction of soil erosion by the no-till system in the Southern Piedmont, *Trans. ASAE* 22:82–92.

McDowell, L. L. and K. C. McGregor. 1980. Nitrogen and phosphorus losses in runoff from no-till soybeans, *Trans. ASAE* 23:643–648.

McGregor, K. C., J. D. Greer, and G. E. Gurley. 1975. Erosion control with no-till cropping practices, *Trans. ASAE* 18:918–920.

Robinson, E. L. and P. A. Banks. 1983. The effectiveness of Diclofop for Control of Italian Ryegrass (*Lolium multiflorum*) in Winter Wheat (*Triticum aestivum*), Univ. Ga. Agric. Exp. Stn. Res. Rep. 428, 6 pp.

Rogers, H. T., D. L. Thurlow, and G. A. Buchanan. 1971. Soybean Production — Recent Research Findings. Cropping Systems and Other Cultural Practices, Alabama Agric. Exp. Stat. Bull. 413:31–38.

Sanford, J. O. 1982. Straw and tillage management practices in soybean-wheat doublecropping, *Agron. J.* 74:1032–1035.

Sanford, J. O., D. L. Myhre, and N. C. Merwine. 1973. Doublecropping systems involving no-tillage and conventional tillage, *Agron. J.* 65:78–982.

Soybean Budgets Committee. 1994. Soybean Planning Budgets, 1995, Mississippi Agric. For. Exp. Stn. Ag. Econ. Report 66.

USDA National Agricultural Statistics Service. 1994. Crop Production Summary, U.S. Government Printing Office, Washington, D.C.

Wesley, R. A. and F. T. Cooke. 1988. Wheat soybean doublecrop systems on clay soil in the Mississippi valley area, *J. Prod. Agric.* 1:166–171.

Wesley, R. A., L. G. Heatherly, C. D. Elmore, and S. R. Spurlock. 1994a. Net returns from eight irrigated cropping systems on clay, *J. Prod. Agric.* 7:109–115.

Wesley, R. A., L. G. Heatherly, and C. D. Elmore. 1994b. Effects of Crop Rotation and Irrigation on Soybean and Wheat Doublecropping on Clay Soil: An Economic Analysis, USDA-ARS-119, U.S. Government Printing Office, Washington, D.C.

Wesley, R. A., L. G. Heatherly, C. D. Elmore, and S. R. Spurlock. 1995. Net returns from eight non-irrigated cropping systems on clay soil, *J. Prod. Agric.* 8:514–520.

chapter eleven

Crop rotation systems for soybeans

Richard A. Wesley

Contents

Introduction

Crop rotation refers to the growing of different crops in sequence and has been shown to increase crop yields. The cause of the higher yields is related to increased soil fertility, to improved soil physical properties, to improved weed control, or to reduced incidences of diseases, nematodes, and insect pests.

Fahad et al. (1982) attributed the increase in crop yields to the enhanced water infiltration rate of the soil caused by the rotation. Baird and Bernard (1984) and Young et al. (1986) claim crop rotations tend to control plant parasitic nematode populations, whereas Boquet et al. (1986) suggest that the reduction in disease incidence is a vital factor. Recent research in Louisiana indicated that soybeans grown after either sorghum or summer fallow consistently outyielded soybeans following soybeans. This favorable response was attributed to the reduction of soybean cyst nematode populations by rotation (Dabney et al., 1988). In corn–wheat–soybean and sorghum–wheat–soybean rotation sequences, crop yields were enhanced and johnsongrass was effectively controlled during the soybean sequence (Litsinger and Moody, 1976).

Organic matter content increased when a sod crop was used in different crop rotations (Spurgeon and Grissom, 1965). However, no other changes were reported for the rotations studied. In a 4-year rice–cotton rotation, rotations with rice did not result in an increase in subsequent seed cotton yield (Snipes et al., 1990). The appreciable amount of rice straw at the end of the growing season also failed to increase the soil organic matter content at either the 6-in. or 12-in. depth.

Doublecropping wheat and soybeans can be an effective way of increasing net farm income (Sanford, 1982; Sanford et al., 1986), especially with irrigation (Wesley and Cooke, 1988). Yield of doublecropped soybeans, however, is usually reduced below that of soybeans grown alone (Crabtree and Rupp, 1980; Sanford, 1982; Sanford et al., 1986; Dabney et al., 1988; Wesley and Cooke, 1988), probably because of the later planting and the considerable amount of moisture removed from the soil by the maturing wheat crop (Heatherly, 1988; Heatherly et al., 1990). However, doublecropping offers a number of potential advantages over monocropping. Among these are more extensive use of fixed resources, reduced soil erosion, improved cash flow, and increased net returns (Hairston et al., 1984; Sanford et al., 1986).

A feasible alternative to continuous doublecropping is rotational doublecropping, where a summer grain crop such as corn or grain sorghum is grown in rotation with the conventional wheat–soybean doublecrop sequence. In this system, wheat follows the summer grain crop, and winter fallow follows the soybean crop. This rotational system produces three cash crops every 2 years.

Biennial rotations of two summer crops is an attempt to improve the yield of one or both crops. Biennial rotations of soybeans and corn have produced significant increases in the yield of both soybeans (Crookston and Kurle, 1989; Peterson and Varvel, 1989a; Meese et al., 1991) and corn (Crookston et al., 1988; Crookston and Kurle, 1989; Meese et al., 1991). Similar results were obtained with biennial rotations of grain sorghum and soybeans (Clegg, 1982; Dabney et al., 1988; Peterson and Varvel, 1989a). The effects of corn and grain sorghum on the yield of the following crop are generally equal (Peterson and Varvel, 1989a). However, in some instances under nonirrigated conditions, the effect of soybeans on the succeeding corn grain yields has been varied over a period of years (Edwards et al., 1988; Peterson and Varvel, 1989b).

Cropping systems studies

General

Nonirrigated and irrigated field studies were conducted from 1984 through 1991 on a Tunica clay near Stoneville, MS to determine crop yields and to compare net returns from the selected cropping systems (Wesley et al., 1994a, b; 1996). Each study included eight cropping systems (treatments) composed of three monocrops, two biennial rotations, one

Table 11.1 Cropping System and Crop Production Sequences for Corn, Soybean, Sorghum, and Wheat Grown on Tunica Clay near Stoneville, MS (1984 through 1991)

		Crop year			
		1984, 1986, 1988, 1990		1985, 1987, 1989, 1991	
Cropping	Crop	Summer	Winter	Summer	Winter
1	Corn–corn	Corn	—	Corn	—
2	Soybean–soybean	Soybean	—	Soybean	—
3	Sorghum–sorghum	Sorghum	—	Sorghum	—
4	Corn–soybean	Corn	—	Soybean	—
5	Sorghum–soybean	Sorghum	—	Soybean	—
6	Wheat–soybean	Soybean	Wheat	Soybean	Wheat
7	Corn–wheat–soybean	Corn	Wheat	Soybean	—
8	Sorghum–wheat–soybean	Sorghum	Wheat	Soybean	—

Sources: Wesley et al., 1994; 1996.

doublecrop, and two rotational doublecrop systems. Monocrop systems were corn, soybeans, and sorghum. Biennial rotations were corn–soybeans and sorghum–soybeans. The doublecrop system included soft red winter wheat and soybeans, whereas the rotational doublecrop systems included summer crops of corn and sorghum one year, followed by a wheat–soybean doublecrop sequence the next year. Crop production sequences for the eight treatments or cropping systems are presented in Table 11.1. Land area available for these studies was limited; thus, it was not possible to evaluate all components of a rotation each year.

Corn and sorghum in the monocrop, biennial rotation and rotational doublecrop systems were planted on undisturbed beds that had been formed the previous fall. Corn and sorghum were planted in 40-in.-wide rows with a conventional planter with double-disk openers. Soybeans were planted in 40-in. rows in a flat, undisturbed seedbed with the same planter; however, doublecrop soybeans were planted no-till after burning the wheat straw. Summer crops were mechanically cultivated for weed control as needed during the early part of each growing season. Wheat was planted in prepared seedbeds (disked and harrowed) with a conventional grain drill with double-disk openers spaced 8 in. apart.

All crops in the nonirrigated study were totally dependent on rainfall. An overhead lateral-move sprinkler irrigation system was used to irrigate all crops in the irrigated study. Each crop was irrigated only during its reproductive period. Irrigation of corn began at tassel emergence and ended at near dent stage. Irrigation of soybeans started at beginning bloom and ended at full-seed stage, whereas irrigation of sorghum began at boot stage and ended at hard-dough stage. Water deficits that occurred during the reproductive stage of each crop were monitored by tensiometers positioned 12 in. beneath the planted row. The amount of water applied and the number of irrigations each crop received were recorded each year of the study.

Economic analysis

Incomes and expenses on a per-acre basis were estimated annually for each cycle of each cropping system. Application rates for all the variable inputs were those recommended and used for crop production in these studies. Crop prices used were the seasonal average prices as reported by the Mississippi Agricultural Statistics Service (1984–1991). The estimated net returns did not include charges for land, management, and general farm

Table 11.2   Average Crop Yield (bu/acre) and Net Returns (NETRET —
$/acre) from Eight Nonirrigated and Irrigated Cropping Systems
on Tunica Clay near Stoneville, MS (1984 through 1991)

Cropping system[a]	Crop	Nonirrigated Yield	NETRET[b]	Irrigated Yield	NETRET[b]
1	Corn	60.8	−9e	116.8	79c
2	Soybean	21.5	25cd	41.1	53d
3	Sorghum	79.9	60ab	93.8	19e
4	Corn	53.8	42bc	126.8	110ab
	Soybean	40.3		49.8	
5	Sorghum	59.8	76a	105.7	84c
	Soybean	44.0		51.6	
6	Wheat	38.2	48abc	43.9	123a
	Soybean	10.8		32.6	
7	Corn	40.3	16de	125.1	136a
	Wheat	43.1		48.8	
	Soybean	19.9		38.2	
8	Sorghum	58.2	64ab	103.3	95bc
	Wheat	45.1		44.2	
	Soybean	22.0		37.3	

[a] Cropping systems are as follows: 1 = monocrop corn; 2 = monocrop soybean; 3 = monocrop sorghum; 4 = biennial rotation of corn–soybean; 5 = biennial rotation of sorghum–soybean; 6 = continuous wheat–soybean doublecrop; 7 = biennial rotation of corn and wheat–soybean doublecrop; 8 = biennial rotation of sorghum and wheat–soybean doublecrop.

[b] Values in individual columns followed by the same letter are not significantly different at a probability level of 0.05.

Source: Wesley et al., 1994, 1996.

overhead. Net returns for the wheat–soybean doublecrop system, and for the wheat–soybean sequence in the rotational doublecrop systems represent the sum of net returns for both the wheat and soybean crops produced in each respective system.

Yields

In the nonirrigated study, yield of monocrop soybeans averaged 21.5 bu/acre over the 8-year study (Table 11.2). Yield of soybeans grown in the corn–soybean and sorghum–soybean rotations during the odd years of the study averaged 40.3 and 44.0 bu/acre, respectively, whereas yield of monocrop soybeans these years averaged 33.7 bu/acre. These higher yields are attributed to the benefits derived from rotations. Also, yields of soybeans from the corn–wheat–soybean and the sorghum–wheat–soybean systems during the odd years averaged 19.9 and 22.0 bu/acre, respectively, whereas yields of soybeans in the continuous wheat–soybean doublecrop system averaged 15.6 bu/acre these years and only 10.8 bu/acre over the 8-year study. Soybeans in these doublecrop systems were planted no-till as soon as feasible following wheat harvest and burning of wheat straw and residue. However, soybeans in these systems were always planted later than soybeans in the monocrop and biennial rotations.

In the irrigated study, yields of all crops were higher than those produced in the nonirrigated study. As in the nonirrigated study, soybean yields from the corn–soybean and sorghum–soybean rotation grown during the odd years were the highest and averaged 49.8 and 51.6 bu/acre, respectively. Yield of monocrop soybeans averaged 42.3 bu/acre during the odd years. Soybean yields from the monocrop soybean system and the

Table 11.3   Average Net Returns above Specified
Expenses for Eight Nonirrigated Cropping Systems on
Tunica Clay near Stoneville, MS (1984 through 1991)

Cropping system[a]	Net returns, $/acre		
	Even years[b]	Odd years[c]	All years
1	−39	20	−9
2	−26	76	25
3	30	90	60
4	−6	91	42
5	37	116	76
6	33	62	48
7	−50	81	16
8	31	97	64

[a] Cropping systems are as follows: 1 = monocrop corn; 2 = monocrop soybean; 3 = monocrop sorghum; 4 = biennial rotation of corn–soybean; 5 = biennial rotation of sorghum–soybean; 6 = continuous wheat–soybean doublecrop; 7 = biennial rotation of corn and wheat–soybean doublecrop; 8 = biennial rotation of sorghum and wheat–soybean doublecrop.

[b] Average net returns for crop years 1984, 1986, 1988, and 1990, when corn and sorghum were the component crop in the rotations.

[c] Average net returns for crop years 1985, 1987, 1989, 1991, when soybeans were the component crop in the rotations.

corn–wheat–soybean and the sorghum–wheat–soybean systems during the odd years were nearly identical and averaged 42.3, 38.2, and 37.3 bu/acre, respectively. As in the nonirrigated study, soybean yields from the continuous wheat–soybean doublecrop system were the lowest, but averaged 32.6 bu/acre.

Net returns

In the nonirrigated study, net returns from monocrop corn, soybeans, and sorghum averaged −$9, $25, and $60/acre, respectively (Table 11.2). Average net returns of $76/acre from the 2-year rotation of sorghum–soybean were greater than the $42/acre from the corn–soybean rotation. Average net returns from the sorghum–wheat–soybean system were $64/acre while the average from the corn–wheat–soybean system was $16/acre. The wheat–soybean doublecrop system provided average net returns of $48/acre, but average soybean yield in this system was only 10.8 bu/acre. Thus, grain sorghum was determined to be the more desirable component crop for rotation with soybean and the wheat–soybean doublecrop sequence under nonirrigated conditions in this study.

During the even years of the study when grain sorghum was grown in cropping systems 3, 5, and 8 and corn was grown in cropping systems 1, 4, and 7, the respective average net returns for the grain sorghum component ($30, $37, and $31) were among the highest and greater than respective returns (−$39, −$6, and −$50) for the comparable corn component (Table 11.3). During the odd years of the study, net returns from soybean rotated with corn ($91) and with grain sorghum ($116) were higher than net returns from monocrop soybeans ($76). Average net returns from the wheat–soybean doublecrop sequences rotated with corn ($81) and grain sorghum ($97) were higher than net returns from the continuous doublecrop sequence ($62). These established trends favor rotated crop sequences over continuous monocrop and wheat–soybean doublecrop culture.

Table 11.4 Average Net Returns above Specified
Expenses for Eight Irrigated Cropping Systems on
Tunica Clay near Stoneville, MS (1984 through 1991)

Cropping system[1]	Net returns, $/acre		
	Even years[b]	Odd years[c]	All years
1	97	61	79
2	55	51	53
3	22	16	19
4	136	84	110
5	74	94	84
6	184	62	123
7	126	146	136
8	70	120	95

[a] Cropping systems are as follows: 1 = monocrop corn; 2 = monocrop soybean; 3 = monocrop sorghum; 4 = biennial rotation of corn–soybean; 5 = biennial rotation of sorghum–soybean; 6 = continuous wheat–soybean doublecrop; 7 = biennial rotation of corn and wheat–soybean doublecrop; 8 = biennial rotation of sorghum and wheat–soybean doublecrop.

[b] Average net returns for crop years 1984, 1986, 1988, and 1990, when corn and sorghum were the component crop in the rotations.

[c] Average net returns for crop years 1985, 1987, 1989, and 1991, when soybeans were the component crop in the rotations.

In the irrigated study, average net returns from all cropping systems except monocrop sorghum were higher than those in the nonirrigated study (Table 11.2). Average net returns from the corn–wheat–soybean ($136/acre), wheat–soybean ($123/acre), corn–soybean ($110/acre), sorghum–wheat–soybean ($95/acre), and sorghum–soybean ($84/acre) systems in the irrigated study were greater than average returns from the monocrop systems of the component crops. Net returns to all cropping systems that included corn were greater than comparable cropping systems with sorghum as the component crop. Thus, under irrigated conditions, cropping systems that included corn provided maximum profit potential on this clayey soil.

During the even years of the irrigated study, net returns from the irrigated corn component of cropping systems 1, 4, and 7, respectively, averaged $97, $136, and $126/acre (Table 11.4), and greatly exceeded the respective net returns from the irrigated grain sorghum component of cropping systems 3, 5, and 8 ($22, $74, and $70). As in the nonirrigated study, net returns to soybeans grown in the odd years in rotated sequences with corn ($84) and with grain sorghum ($94) were higher than returns to monocrop soybean ($51). Average net returns from the wheat–soybean sequence rotated with corn ($146) and with grain sorghum ($120) were higher than from the continuous doublecrop sequence ($62). These trends also favor rotated sequences.

Summary

All rotational systems with irrigation provided greater net returns than corresponding monocrop systems. Without irrigation, biennial rotations resulted in improved soybean yields, but net returns were not always significantly improved. These results indicate that a doublecrop wheat–soybean system should be used only with irrigation, and that rotation of soybeans with sorghum produced the highest net returns in nonirrigated environments, whereas rotations with corn produced the highest net returns when irrigation was used.

Table 11.5 Influence of a Rice–Soybean Rotation on Yield of Soybean Grown on Sharkey Clay at the Delta Branch Experiment Station, Stoneville, MS (1983–1990)

Crop rotation system	Soybean yield, bu/acre							
	1983	1984	1985	1986	1987	1988	1989	1990
1 Rice: 1 Soybean		26.8		15.4		37.3		25.4
2 Rice: 1 Soybean			39.1			40.1		
3 Rice: 1 Soybean				16.7				33.1
1 Rice: 2 Soybean		24.8	35.4		11.2	32.7		20.8
2 Rice: 2 Soybean			40.5	12.1			27.0	29.4
Continuous soybean	14.0	19.1	29.4	8.1	4.3	30.1	24.4	17.7
LSD (0.05) = 9.0[a]								
LSD (0.05)[b]	NS	NS	8.5	4.7	NS	4.6	NS	9.8

[a] For comparison of any two means across years of continuous soybean.

[b] For comparison of any two means within a single year.

Source: Kurtz, M.E. et al., MAFES Bull. 994, 1993. With permission.

Table 11.6 Influence of a Rice–Soybean Rotation on Yield of Rice Grown on Sharkey Clay at the Delta Branch Experiment Station, Stoneville, MS (1983–1990)

Crop rotation system	Rice yield, bu/acre							
	1983	1984	1985	1986	1987	1988	1989	1990
1 Rice: 1 Soybean	126.4		160.2		141.2		124.6	
2 Rice: 1 Soybean	139.5	148.1		123.1	125.9		126.6	95.2
3 Rice: 1 Soybean	137.8	146.1	139.0		142.2	118.0	116.6	
1 Rice: 2 Soybean	134.0			139.6			134.6	
2 Rice: 2 Soybean	141.5	143.4			155.5	138.4		
Continuous Rice	138.1	150.9	139.8	93.5	98.8	92.1	113.0	89.9
LSD (0.05) = 22.9[a]								
LSD (0.05)[b]	NS	NS	19.2	12.5	29.7	39.4	NS	NS

[a] For comparison of any two means across years of continuous rice.

[b] For comparison of any two means within a single year.

Source: Kurtz, M. E. et al., MAFES Bull. 994, 1993. With permission.

Rice–soybean rotation studies

1983–1990 study

A long-term rice–soybean rotation study was conducted on Sharkey clay near Stoneville, MS (Kurtz et al., 1993). The study included continuous rice, continuous soybeans, and five rotational sequences of rice and soybeans that were being utilized by producers in the region. Specifically, the treatments evaluated were (1) continuous rice, (2) continuous soybeans, (3) 1 year rice:1 year soybeans, (4) 2 years rice:1 year soybeans, (5) 3 years rice:1 year soybeans, (6) 1 year rice:2 years soybeans, and (7) 2 years rice:2 years soybeans. Rice was drill-seeded in mid-April to early May each year at a seeding rate of 90 lb/acre. Soybean (Centennial variety, MG VI) were planted in 40 in. rows in mid-May to early June each year at a seeding rate of 50 lb/acre. Soybean were not irrigated in any year.

Yields

Soybean and rice yields are presented in Tables 11.5 and 11.6, respectively. During the 8-year experiment, soybeans occurred 17 times in the various rotations following rice

Table 11.7 Net Returns above Specified Costs from an 8-year Rice–Soybean Rotation on Sharkey Clay at the Delta Branch Experiment Station, Stoneville, MS (1983–1990)

	Net returns above specified costs,[a] $/acre		
Crop rotation system	8-year avg	Last 4-year avg	Difference
1 Rice: 1 Soybean	66.75 a	74.15 ab	+7.40
2 Rice: 1 Soybean	63.75 a	38.69 cd	−25.06
3 Rice: 1 Soybean	58.25 a	51.96 bc	−6.29
1 Rice: 2 Soybean	56.87 b	37.16 cd	−19.71
2 Rice: 2 Soybean	83.50 a	93.11 a	+9.61
Continuous soybean	7.88 c	12.55 d	+4.67
Continuous rice	0.15 c	−52.59 e	−52.44

[a] Means followed by the same letter do not significantly differ (Duncan's MRT, $P = 0.05$).

Source: Kurtz, M. E. et al., MAFES Bull. 994, 1993. With permission.

(Table 11.5). Eight of these occurrences resulted in significant soybean yield increases over the continuous soybean system. When grown in the 1:1 system, soybean yields increased two out of the four times this rotation occurred (7.3 bu/acre in 1986 and 7.2 bu/acre in 1988). In the 2:1 (9.8 bu/acre average increase) and 3:1 (12.0 bu/acre average increase) systems, soybean yields when in rotation with rice were always significantly higher than continuous soybeans. No yield increase was measured for soybeans in the 1:2 rotation. Soybean yield increases occurred two out of four times in the 2:2 system (11.1 bu/acre in 1985 and 11.7 bu/acre in 1990).

Rice yields usually benefited from rotations with soybeans. After year one, rice occurred 18 times in various soybean rotations. Rotational rice yields were greater than yields from the continuous rice system in eight of those occurrences (Table 11.6). After year one, where rice was grown in all plots except the continuous soybeans, rice yields increased in the 1:1 rotation two out of three times this rotation occurred (20.4 bu/acre in 1985 and 42.4 bu/acre in 1987), one out of five times in the 2:1 system (29.6 bu/acre in 1986), one out of five times in the 3:1 system (43.4 bu/acre in 1987), one out of two times in the 1:2 rotation (46.1 bu/acre in 1986), and two out of three times in the 2:2 system (56.7 bu/acre in 1987 and 46.3 bu/acre in 1988). In the continuous rice system, yields decreased after the third year and never yielded as high as in the first year (Table 11.6).

Costs and Returns

The average net returns above specified costs for the 8-year experiment indicate all rotations out-performed continuous rice and continuous soybeans (Table 11.7). During the last 4 years of the experiment, all rotational sequences and continuous soybeans provided higher net returns than continuous rice; however, only the 2:2 (rice:soybean) and 1:1 rotations and continuous soybeans resulted in increased net returns compared to the 8-year average. The 2:2, 1:1, and 3:1 rotations provided significantly higher net returns than continuous soybeans. These rotational systems resulted in the highest net returns of $93, $74, and $52/acre/year, respectively, above specified costs during the last 4 years. Net returns from the 2:2 system were significantly greater than those from the 3:1 system. These values do not include management fees, land costs, or general farm overhead.

Table 11.8 presents 8-year average yields, gross values, specified production costs, and net returns of continuous soybeans and soybeans produced in the 1:1 and 2:1 rotations with rice. Average yields of soybeans from the rotations exceeded those from the continuous soybean system by 9.3 bu/acre (27.7 vs. 18.4 bu/acre). Average net returns for soybeans

Table 11.8   Yield, Net Returns, and Costs for Continuous Soybeans Compared with Those from Soybeans Following 1 or 2 years of Rice, Delta Branch Experiment Station, Stoneville, MS (1983–1990)

Cropping system	Average[a] yield, bu/acre	Gross[b] value, $acre	Average production cost, $/acre	Net return, $/acre
Continuous soybeans	18.4	112.79	104.50	8.29
Soybeans following 1 or 2 years of rice	27.7	169.80	104.50	65.30

[a] Average yield represents the 8-year average for the specified crop sequence.
[b] Gross value determined as the product of yield and a seasonal average price of $6.13/bu.
Source: Kurtz, M. E. et al., MAFES Bull. 994, 1993. With permission.

following 1 or 2 years of rice were $65.30/acre, whereas average net returns to continuous soybeans were only $8.29/acre.

Summary
When soybeans were grown behind 1 or 2 years of rice, the average soybean yield increased 9.3 bu/acre compared with continuous soybeans (Table 11.8), resulting in $57.01/acre increased net returns. The rotations that resulted in the highest returns above specified costs during the last 4 years were the 2:2 and 1:1 rotations, with a 3:1 (rice:soybean) rotation being equal to the 1:1 rotation. Each of these rotational systems returned a higher dollar value than either continuous soybeans or continuous rice. These data clearly indicate that rice makes a valuable rotational crop with soybeans and the returns far exceed those for either crop grown in continuous monoculture.

1994–1995 study

Another rice–soybean rotation study was conducted at Stoneville, MS to determine the effect of irrigation on the yield of MG IV and V soybeans grown continuously and in a 1:1 rice:soybean rotation (Heatherly, 1995, personal communication). Soybeans in each production system were either grown in a nonirrigated or a furrow irrigated environment. All production inputs within a year were identical for all varieties and environments.

Yields
Soybeans in this study were planted on April 21, 1994 and April 18, 1995. The rotation had been in place for 4 years before the yield data were collected. In the nonirrigated environment, yields of MG IV and V soybeans following rice averaged 5.9 bu/acre higher than yields of soybean after soybean in 1994, and 5.1 bu/acre higher in 1995 (Table 11.9). In the irrigated environment, average yields of MG IV and V soybeans following rice and after soybean were nearly identical each year.

Furrow irrigation increased yields of continuous soybeans and soybean following rice both years. In 1994, irrigation increased the average yield of soybean 19.2 bu/acre in the continuous soybean system and 12.3 bu/acre in the rice:soybean rotation (Table 11.9). Furrow irrigation produced similar results in 1995 with increases of 26.0 and 19.3 bu/acre, respectively. In the irrigated environment, the average yield of soybeans in the continuous soybean system and in the rice rotation were nearly identical each year.

Summary
In a nonirrigated environment, the production of soybeans in a rice–soybean rotation produced yields that averaged 5 to 6 bu/acre more than those produced in a continuous

Table 11.9   Yield (bu/acre) of Nonirrigated and Furrow Irrigated
Soybeans Following Rice (RICE) and Continuous Soybeans (SOY)
on Sharkey Clay at Stoneville, MS (1994 and 1995)

Variety (MG)	Nonirrigated		Irrigated	
	RICE	SOY	RICE	SOY
April 21, 1994 planting date				
RA 452 (IV)	41.9	35.3	53.6	57.3
HBK 49 (IV)	38.0	33.9	52.7	55.5
DP 3499 (IV)	36.8	29.9	47.3	50.1
P 9592 (V)	43.6	39.5	56.2	55.4
A 5979 (V)	50.7	43.3	63.9	59.5
DP 3589 (V)	44.1	37.7	55.0	56.8
Average	42.5	36.6	54.8	55.8
April 18, 1995 planting date				
RA 452(IV)	37.5	32.6	57.1	59.6
DP 3478 (IV)	40.4	38.2	62.5	66.1
P 9501 (IV)	40.9	32.0	60.2	62.2
P 9592(V)	30.2	27.6	46.2	46.2
DP 3589(V)	32.3	26.1	52.2	52.2
Hutcheson(V)	39.6	33.9	58.2	60.1
Average	36.8	31.7	56.1	57.7

Source: Heatherly et al., 1995.

soybean system. Furrow irrigation of continuous soybeans and soybeans following rice increased the average yield of MG IV and V soybeans 19.2 bu/acre. However, with furrow irrigation the yield of continuous soybeans and soybeans following rice were nearly identical. Thus, the use of irrigation alleviated drought stress and masked the beneficial effects provided by the rotation.

Short-term crop rotation systems

General

In Louisiana, monocropped soybean is the most popular cropping system and has been continually grown on some land since the mid-1960s. During the late 1970s, Louisiana soybean farmers noted that soybean yields were declining with time, whereas soybean farmers in Arkansas were reporting yield increases where soybeans were rotated with grain sorghum (Beatty and Eldridge, 1980). Field studies were initiated in 1982 in Tangipahoa Parish, LA to identify short-term rotations that would increase soybean yields above those from monocropped soybeans and to determine the cause of the yield decline in monocropped soybeans (Dabney et al., 1988).

The experiment included eight 2-year crop rotations on a Providence silt loam that had been planted to soybeans during each of the previous 15 years. Only three of the rotations had soybeans planted on the same plots every year. These were (1) continuous soybeans (S–S); (2) continuous soybeans with a winter ryegrass cover crop (S,R–S,R); and (3) doublecropped soybeans and wheat (S,W–S,W). The other five rotations had a soybean crop only in alternate years. These were (4) soybeans alternated with wheat interseeded with alfalfa (S–W,A); (5) soybeans alternated with grain sorghum (S–GS); (6) soybeans alternated with grain sorghum, with each crop followed by a ryegrass winter cover crop (S,R–GS,R); (7) soybeans alternated with fallow (S–F); and (8) soybeans alternated with fallow with a ryegrass cover crop seeded each winter (S,R–F,R). In this study, fallow refers

Table 11.10 Mean Yields of Centennial and Davis Soybeans Grown
in Several Crop Rotation Systems on a Providence Silt Loam
in Southeast Louisiana (1983 to 1986)

Crop rotation system[a]	Crop sequence	Centennial, bu/acre	Davis, bu/acre
1	S–S	30.2	21.3
2	S,R–S,R	32.4	23.0
3	S–W	22.2	15.4
4	S–W,A	31.6	20.5
5	S–GS	35.0	26.8
6	S,R–GS,R	34.0	28.4
7	S–F	34.2	26.0
8	S,R–F,R	34.4	27.8

[a] Cropping systems are as follows: 1 = monocrop soybeans; 2 = monocrop soybean with a winter ryegrass cover crop; 3 = doublecrop soybeans and wheat. Crop rotation systems 4 to 8 had a soybean crop only in alternate years as follows: 4 = soybeans alternated with wheat interseeded with alfalfa; 5 = soybeans alternated with grain sorghum; 6 = soybeans alternated with grain sorghum with each crop followed by a ryegrass winter cover crop; 7 = soybeans alternated with fallow; and 8 = soybeans alternated with fallow with a ryegrass cover crop seeded each winter.

Source: Dabney, S. M. et al., *Agron. J.*, 80, 197–204, 1988. With permission.

to volunteer vegetation that occupied the land in the absence of tillage. It should be noted that rotation 3, the wheat–soybean doublecrop, produced two cash crops each year, whereas rotations 1, 2, 4, 5, and 6 produced one cash crop each year. Rotations 7 and 8 produced only one cash crop every 2 years. Centennial and Davis soybeans (MG VI varieties) were used in each of the rotations. Nematode populations were determined three times during each growing season: at planting, midseason, and harvest. Insect populations were monitored on a weekly basis from June to October of 1982 to 1985.

Yields

Over the 4 years, yields of Centennial and Davis soybeans were increased by rotation (Table 11.10). Centennial yielded significantly more than Davis all years of the study. Soybeans rotated with grain sorghum or alternated with fallow produced the greatest increase in yield above that produced by continuous soybeans. Doublecropped soybeans yielded significantly less than full-season soybeans all years except 1985. This exception is attributed to the relatively early planting date of doublecropped soybeans in 1985. There were no positive effects attributed to the presence of ryegrass. However, yields were reduced when rotations involved wheat or alfalfa.

Soybean cyst nematode numbers

Soybean cyst nematode (SCN) populations were relatively low for both soybean varieties in 1982. However, at that time SCN populations under the Davis variety were significantly higher than under the Centennial variety. Over the study period, a tremendous increase in SCN occurred in continuous soybean culture, especially under the Davis variety. A steady but less dramatic increase occurred under continuous Centennial culture. Continuous production of a specific soybean variety year after year, even a resistant variety, is an important factor in the development of SCN populations. Data in Table 11.11 indicate continuous soybean culture resulted in higher SCN populations than those resulting when soybeans were rotated with either grain sorghum or fallow.

Table 11.11 Number of Soybean Cyst Nematodes per Liter at Harvest
of Centennial and Davis Soybeans Grown in Several Crop Rotation
Systems on a Providence Silt Loam in Southeast Louisiana (1984 to 1986)

Crop rotation system[a]	Crop sequence	Centennial, no. L^{-1}	Davis, no. L^{-1}
1	S–S	768	2427
2	S,R–S,R	389	3950
3	S–W	69	154
4	S–W,A	392	732
5	S–GS	155	922
6	S,R–GS,R	207	854
7	S–F	0	364
8	S,R–F,R	69	233

[a] Cropping systems are as follows: 1 = monocrop soybeans; 2 = monocrop soybean with a winter ryegrass cover crop; 3 = doublecrop soybeans and wheat.
Crop rotation systems 4 to 8 had a soybean crop only in alternate years as
follows: 4 = soybeans alternated with wheat interseeded with alfalfa; 5 = soybeans alternated with grain sorghum; 6 = soybeans alternated with grain sorghum with each crop followed by a ryegrass winter cover crop; 7 = soybeans
alternated with fallow; and 8 = soybeans alternated with fallow with a ryegrass
cover crop seeded each winter.

Source: Dabney, S. M. et al., *Agron. J.,* 80, 197–204, 1988. With permission.

Insect populations

Insect pressure at the test site was low. Economic thresholds for any species were reached
only twice in 1982, and not at all in 1983, 1984, and 1985. Treatment with insecticides in
1982 rapidly reduced populations below threshold levels. Thus, it is unlikely that insect
damage limited yields in any rotation except possibly the doublecrop rotation.

Summary

Soybeans in the rotated systems responded positively to crop rotations. Centennial produced higher yields each year than Davis; however, Davis showed a greater response to
rotation. Doublecropped soybeans yielded less than full season soybeans all years except
1985. Ryegrass in the rotations failed to affect yields, SCN, or insect populations. High
SCN populations in continuous soybean culture tended to reduce soybean yields, whereas
rotations reduced SCN populations and enhanced soybean yields. Continuous production
of a single variety, even a resistant variety, on the same land area over time is not
recommended.

Conclusions

Biennial rotations of corn–soybean and sorghum–soybean produced soybean yields that
were significantly higher than yields from monocropped soybeans in both nonirrigated
and irrigated environments on Tunica clay. These high yields are attributed to the timely
planting of soybeans each year and the positive benefits derived from rotations. In nonirrigated environments, net returns to cropping systems that included sorghum as a
component crop were higher than net returns from cropping systems that included corn.
Thus, sorghum is the more desirable rotation crop with soybeans in nonirrigated cropping
systems on clayey soils. Continuous wheat–soybean doublecrop systems produced
extremely low soybean yields without irrigation, and thus are not recommended for

nonirrigated clay soils. In the irrigated environment, net returns to all cropping systems with corn as the component crop were higher than returns to cropping systems with sorghum as the component crop. Therefore, cropping systems that include corn provide the maximum profit potential in irrigated environments. Soybean–sorghum rotations reduced soybean cyst nematode populations and thereby enhanced soybean yields on a Providence silt loam.

Rotations of soybeans and rice on a Sharkey clay increased the yield of each crop above that produced by continuous monocropping of each crop. This was true when Centennial soybeans were grown following 1 or 2 years of rice. In these rotations, nonirrigated soybean yields averaged 9.3 bu/acre greater than yields from continuous monocropped soybeans; net returns averaged $57.01/acre higher.

In a rice–soybean rotation study on Sharkey clay, yields of nonirrigated MG IV and V soybeans rotated with rice averaged 5 to 6 bu/acre higher than the yield of continuous soybeans. Furrow irrigation significantly increased the yields of soybeans in the continuous soybean culture and in the rice rotation. However, with irrigation the yield of soybeans from the monocrop soybean treatments and from the rice–soybean rotation were nearly identical. Thus, furrow irrigation alleviated drought stress and masked the beneficial effects of rotation.

References

Baird, S. M. and E. C. Bernard. 1984. Nematode populations and community dynamics in soybean-wheat cropping and tillage regimes, *J. Nematol.* 16:379–386.

Beatty, K. D. and I. L. Eldridge. 1980. 1979 Results from crop rotation study, Keiser, *Ark. Farm Res.* 29:6.

Boquet, D. J., A. B. Coco, and D. E. Summers. 1986. Cropping systems for higher yields, *LA. Agric.* 30:4–5,7.

Crabtree, R. J. and R. N. Rupp. 1980. Double and monocropped wheat and soybeans under different tillage and row spacings, *Agron. J.* 72:445–448.

Clegg, M. D. 1982. Effect of soybean on yield and nitrogen response of subsequent crops in eastern Nebraska, *Field Crops Res.* 5:233–239.

Crookston, R. K. and J. E. Kurle. 1989. Corn residue effect on the yield of corn and soybean grown in rotation, *Agron. J.* 81:229–232.

Crookston, R. K., J. E. Kurle, and W. E. Lueschen. 1988. Relative ability of soybean, fallow, and triacontanol to alleviate yield reduction associated with growing corn continuously, *Crop Sci.* 28:145–147.

Dabney, S. M., E. C. McGawley, D. J. Boethel, and D. A. Berger. 1988. Short-term crop rotation systems for soybean production, *Agron. J.* 80:197–204.

Edwards, J. H., D. L. Thurlow, and J. T. Eason. 1988. Influence of tillage and crop rotation on yields of corn, soybean, and wheat, *Agron. J.* 80:76–90.

Fahad, A. A., L. N. Mielke, A. D. Flowerday, and D. Swartzendruber. 1982. Soil physical properties as affected by soybean and other cropping sequences, *Soil Sci.* 46:377–381.

Hairston, J. E., J. O. Sanford, J. C. Hayes, and L. L. Reinschmiedt. 1984. Crop yield, soil erosion, and net returns from five tillage systems in the Mississippi Blackland Prairie, *J. Soil Water Conserv.* 39:391–395.

Heatherly, L. G. 1988. Planting date, row spacing, and irrigation effects on soybean grown on clay soil, *Agron. J.* 80:227–231.

Heatherly, L. G., C. D. Elmore, and R. A. Wesley. 1990. Weed control and soybean response to preplant tillage and planting time, *Soil Till. Res.* 17:199–210.

Kurtz, M. E., C. E. Snipes, J. E. Street, and F. T. Cooke, Jr. 1993. Soybean Yield Increases in Mississippi Due to Rotations with Rice, MAFES Bull. 994.

Litsinger, J. A. and K. Moody. 1976. Integrated pest management in multiple cropping systems, in R.I. Papendick et al., Ed., Multiple Cropping, ASA Spec. Publ. 27 ASA, CSSA, and SSSA, Madison, WI, 293–316.

Meese, B. G., P. R. Carter, E. S. Oplinger, and J. W. Pendleton. 1991. Corn/soybean rotation effect as influenced by tillage, nitrogen, and hybrid/cultivar, *J. Prod. Agric.*4:74–80.

Mississippi Agricultural Statistics Service. 1984–1991. Mississippi Agricultural Statistics Suppl. 24, Mississippi Department of Agriculture and Commerce, Jackson, MS.

Peterson, T. A. and G. E. Varvel. 1989a. Crop yield as affected by rotation and nitrogen rate. I, Soybean, *Agron. J.* 81:727–731.

Peterson, T. A. and G. E. Varvel. 1989b. Crop yield as affected by rotation and nitrogen rate. III, Corn, *Agron. J.* 81:735–738.

Sanford, J. O. 1982. Straw and tillage management practices in soybean-wheat doublecropping, *Agron. J.* 74:1032–1035.

Sanford, J. O., B. R. Eddleman, S. R. Spurlock, and J. E. Hairston. 1986. Evaluating ten cropping alternatives for the midsouth, *Agron. J.* 78:875–880.

Snipes, C. E., M. E. Kurtz, and J. E. Street. 1990. Productivity of cotton following rice, *J. Prod. Agric.* 3:209–211.

Spurgeon, W. I. and P. H. Grissom. 1965. Influence of Cropping Systems on Soil Properties and Crop Productions, MAFES Bull. 710.

Wesley, R. A. and F. T. Cooke. 1988. Wheat–soybean doublecrop systems on clay soil in the Mississippi Valley area, *J. Prod. Agric.* 1:166–171.

Wesley, R. A., L. G. Heatherly, C. D. Elmore, and S. R. Spurlock. 1994a. Net returns from eight irrigated cropping systems on clay soil, *J. Prod. Agric.* 7:109–115.

Wesley, R. A., L. G. Heatherly, C. E. Elmore, and S. R. Spurlock. 1994b. Effects of Crop Rotation and Irrigation on Soybean and Wheat Doublecropping on Clay Soil: An Economic Analysis, USDA-ARS, ARS-119, U.S. Government Printing Office, Washington, D.C.

Wesley, R. A., L. G. Heatherly, C. D. Elmore, and S. R. Spurlock. 1996. Net returns from eight non-irrigated cropping systems on clay soil, *J. Prod. Agric.* 8:514–520.

Young, L. D., E. E. Hartwig, S. C. Anand, and D. Widick. 1986. Response of soybeans and soybean cyst nematodes to cropping sequences, *Plant Dis.* 70:787–791.

chapter twelve

Weed management

Krishna N. Reddy, Larry G. Heatherly, and Alan Blaine

Contents

Introduction

Weeds are able to infest and thrive on cropland despite extensive control efforts. They rapidly adapt to a variety of soils, climatic conditions, and cultural practices. The density and distribution of weed species may vary depending on crop production practices. Even a slight reduction in weed control efforts may result in reinfested fields despite years of carefully planned weed control programs. Weeds continually pose a serious threat to profitable crop production, and the extent of monetary loss depends on the degree of weed infestation. Nationwide surveys estimate that over 316 million bu of soybeans were lost annually to weeds during 1975 to 1979 in the U.S. This represents a loss of from 13 to 27% of the total annual production. During the same period, over 54 million bu of soybeans were lost annually to weeds in the Delta states (Arkansas, Louisiana, Mississippi), which amounts to 16% of the total annual production (Chandler et al., 1984).

In 1996, soybeans were planted on 64.2 million acres in the U.S., with a total production of 2382 million bu (U.S. Department of Agriculture, 1997). New varieties, improved weed, insect, and disease control; and better management practices have made substantial contributions to increases in soybean yield. As weeds are a major deterrent to soybean production, they must be controlled to protect profits.

Soybean yield losses due to weed interference must be minimized and the cost of weed control must be lowered. Development of safe, effective, and relatively inexpensive herbicides coupled with advances in application technology during the past 40 years have provided many successful weed control options in soybean production. Herbicides provide cost-effective, timely weed control and have helped producers to be highly productive and remain economically viable. Use of herbicides to kill weeds selectively in crops is an integral part of any modern weed management system.

Soybean is grown as monoculture (continuously on the same land), as doublecrop following winter wheat, or in rotation with other crops (cotton, corn, sorghum). Crop production practices exert selection pressure on weed communities and create niches that favor or disfavor various species (Buhler, 1995). Effective control of weeds requires the use of specific weed management strategies. The current weed management strategies used in soybean production with emphasis on herbicide-use patterns and emerging trends in weed control are briefly summarized.

Weeds

A weed is defined as a plant that is objectionable or interferes with the activities or welfare of humans. In simple terms, a weed is a plant growing where it is not desired. Some crop plants may be considered weeds. For example, a corn plant in a soybean field is a weed. Some plants are generally recognized and accepted as weeds because of their competitive, persistent, and pernicious characteristics. For example, pigweeds, velvetleaf, foxtails, and barnyardgrass are considered undesirable and immediately recognized as weeds. Weeds encompass all types of undesirable plants such as broadleaf plants, grasses, sedges, shrubs, vines, and aquatic as well as parasitic flowering plants. Weeds may have different names in different parts of the country or world. For example, prickly sida is also known as teaweed, spiny sida, indian mallow, and false-mallow. Weeds are given a scientific name under the plant classification system to allow them to be identified anywhere in the world.

Weeds may be classified based on habitat, life cycle, and morphology (Ross and Lembi, 1985; Ashton and Monaco, 1991). Based on habitat, weeds are classified as (1) terrestrial, or plants that live on land, and (2) aquatic, or plants that live in or on water (e.g., hydrilla, duckweed, water hyacinth, etc.). Based on life cycle, weeds are classified into three groups: annuals, biennials, and perennials. Annuals are plants that complete their life cycle in 1 year. Most common weeds are annuals (summer annuals and winter annuals). Summer annuals germinate in the spring or summer, grow in the summer, and die in the fall (e.g., barnyardgrass, common cocklebur, crabgrasses, pigweeds, foxtails). Winter annuals germinate in the fall or winter and die in the spring or summer (e.g., chickweed, dandelion, downy brome, henbit, shepherdspurse). Biennials are plants that live 1 to 2 years (e.g., wild carrot, common mullein). Perennials are plants that live more than 2 years (e.g., johnsongrass, quackgrass, purple nutsedge). Based on morphology, weeds are classified as monocots (grasses) and dicots (broadleaf weeds). Holm et al. (1977) estimated that out of 250,000 plant species, about 250 species have become known as weeds of the world.

The distribution and density of weed species vary depending on soil properties, weather (such as rainfall and temperature patterns), and cultural practices. Warm temperatures, relatively high rainfall, and the fertile soils of the midsouthern U.S. provide an environment suitable for a wide array of weed species. Successful weeds in soybean are usually those with the same life cycle and growth habits as soybean. Weed populations will vary depending on tillage systems. Under no-tillage systems, weed populations usually increase in number and diversity unless adequate control is substituted for tillage (Buhler, 1995). The most commonly found weeds in soybean fields are listed in Table 12.1.

Weeds reduce crop yields and lower crop quality by contaminating the harvested produce with trash and weed seeds or by interfering with farm equipment operations. Weeds harbor crop pests such as insects, plant pathogens, and nematodes. Weeds also reduce efficient land use and lower human and machinery efficiency. Some weeds produce and release inhibiting chemicals (allelochemicals) that adversely affect crop plants. Weeds compete with crops for nutrients, water, light, and space. Crop losses from weed competition depend on composition of the weed species and their spatial distribution since weeds compete with each other as well as the crop. It has been estimated that U.S. farmers spend $3.6 billion annually for chemical weed control and $2.6 billion for cultural and other methods of weed control. The total cost of weeds in the U.S. could approach $15 to 20 billion/year (Ashton and Monaco, 1991)

Chemical weed control

The ultimate objective of the farmer is to control weeds adequately to maximize crop yields. Generally, some combinations of chemical, mechanical, and cultural methods are used to achieve effective weed control. However, chemicals (herbicides) that kill or suppress the growth of plants are more commonly used for weed control than are the other two methods. Development of safe and effective herbicides remains one of the most significant technological advances in the history of agriculture. It is now difficult to imagine crop production without the use of herbicides.

Herbicides now lead all other pesticide groups in terms of total usage and total sales in the U.S. Overwhelming success of herbicides as tools of weed control was due to several reasons. Herbicides control weeds beyond the reach of a cultivator. Weeds within the crop row as well as in reduced-tillage production systems can be managed with herbicides. Control of weeds that directly compete with crop plants in the row is especially important to minimize crop yield losses. No-tillage crop production is now feasible because of herbicides. Herbicide use has resulted in energy-efficient and economical weed control. Farmers have adopted herbicides for weed control because they increase weed control efficacy, are cost-effective, offer production flexibility, and reduce time and labor requirements for weed management.

Herbicides are grouped in several ways for easy understanding. They are classified by chemical families (e.g., phenoxy acids, sulfonylureas, imidazolinones, dinitroanilines), by mode of action (e.g., photosynthetic inhibitors, acetolactate synthase inhibitors), or by time of application (e.g., preplant, preplant incorporated, preemergence, postemergence).

In the U.S. herbicides are sold only after the U.S. Environmental Protection Agency has reviewed the labeling and registered the product uses. Producers should always read and follow the label before using any herbicide. Check herbicide labels for any "restrictive use." Proper calibration of herbicide applications is essential for safe, economical, and

Table 12.1   Common Weeds in Soybeans in the Midsouthern U.S.[a]

Common name	Scientific name	Life cycle
Annual bluegrass	*Poa annua* L.	Annual
Barnyardgrass	*Echinochloa crus-galli* (L) Beauv.	Summer annual
Bermudagrass	*Cynodon dactylon* (L.) pers.	Perennial
Bittercress	*Cardamine* spp.	Annual or biennial
Broadleaf signalgrass	*Brachiaria platyphylla* (Griseb.) Nash	Summer annual
Browntop millet	*Brachiaria ramosa* (L.) Stapf	Annual
Buttercups	*Ranunculus* spp.	Winter annual
Carpetweed	*Mollugo verticillata* L.	Summer annual
Carolina foxtail	*Alopecurus carolinianus* Walt.	Winter annual
Common chickweed	*Stellaria media* (L.) Vill.	Annual or winter annual
Common cocklebur	*Xanthium strumarium* L.	Summer annual
Common lambsquarters	*Chenopodium album* L.	Summer annual
Common purslane	*Portulaca oleracea* L.	Annual
Common ragweed	*Ambrosia artemisiifolia* L.	Summer annual
Crabgrass	*Digitaria* spp.	Summer annual
Cressleaf groundsel	*Senecio glabellus* Poir.	Annual
Cupgrass	*Eriochloa* spp.	Annual
Curly dock	*Rumex crispus* L.	Perennial
Dandelion	*Taraxacum officinale* Weber in Wiggers	Winter annual or perennial
Eclipta	*Eclipta prostrata* L.	Annual
Eveningprimroses	*Oenothera* spp.	Winter annual or biennial
Fall panicum	*Panicum dichotomiflorum* Michx.	Summer annual
Foxtails	*Setaria* Spp.	Summer annual
Geraniums	*Geranium* spp.	Winter annual or biennial
Goosegrass	*Eleusine indica* (L.) Gaertn.	Annual
Hemp sesbania	*Sesbania exaltata* (Raf.) Rydb. ex A.W. Hill	Annual
Henbit	*Lamium amplexicaule* L.	Winter annual or biennial
Honeyvine milkweed	*Ampelamus albidus* (Nutt.) Britt.	Perennial
Horseweed	*Conyza canadensis* (L.) Cronq.	Summer or winter annual
Italian ryegrass	*Lolium multiflorum* Lam.	Winter annual or biennial
Jimsonweed	*Datura stramonium* L.	Annual
Johnsongrass	*Sorghum halepense* (L.) pers.	Perennial
Maypop passionflower	*Passiflora incarnata* L.	Perennial
Mayweed chamomile	*Anthemis cotula* L.	Winter annual
Morningglories	*Ipomoea* spp.	Annual, some perennial
Mustards	*Brassica* spp	Winter or summer annual
Nutsedges	*Cyperus* spp.	Annual, some perennial
Pigweeds	*Amaranthus* spp.	Summer annual
Plains coreopsis	*Coreopsis tinctoria* Nutt.	Annual
Prickly lettuce	*Lactuca serriola* L.	Winter annual or biennial
Prickly sida	*Sida spinosa* L.	Annual
Quackgrass	*Agropyron repens* (L.) Beauv.	Perennial
Red rice	*Oryza sativa* L.	Summer annual
Red sprangletop	*Leptochloa filiformis* (Lam.) Beauv.	Summer annual
Redvine	*Brunnichia ovata* (Walt.) Shinners	Perennial
Shepherdspurse	*Capsella bursa-pastoris* (L.) Medik	Winter annual
Sicklepod	*Senna obtusifolia* (L.) Irwin and Barneby	Annual
Smartweeds	*Polygonum* spp.	Annual, some perennial
Speedwell	*Veronica* spp.	Annual or winter annual
Spurges	*Euphorbia* spp.	Annual, some perennial
Spurred anoda	*Anoda cristata* (L.) Schlecht	Annual
Swinecress	*Coronopus didymus* (L.) Sm.	Annual or biennial

Table 12.1 (continued)	Common Weeds in Soybeans in the Midsouthern U.S.[a]	
Common name	Scientific name	Life cycle
Tall goldenrod	*Solidago altissima* L.	Perennial
Trumpetcreeper	*Campsis radicans* (L.) Seem. ex Bureau	Perennial
Velvetleaf	*Abutilon theophrasti* Medik.	Annual
Viginia pepperweed	*Lepidium virginicum* L.	Annual or winter annual

[a] The distribution and population density of the weed species vary significantly from field to field. Overall impact of weeds on soybean yield depends on type of weed species and density.

effective weed control. For up-to-date information on herbicide use, producers should refer to a product label and the weed control guidelines that have been developed based on research by several Land Grant universities.

There are about 30 herbicides used in soybeans, and many of them are limited to specific situations. Herbicides are applied in six different ways: (1) preplant foliar (burn-down), (2) preplant incorporated, (3) preemergence, (4) postemergence, (5) directed-poste-mergence, and (6) spot treatments. The most commonly used herbicides in soybean under each application method are listed in Table 12.2. More-detailed information on herbicides, use rates, weeds controlled, and time of application is available in *Herbicide Handbook* (Weed Science Society of America, 1994), *Crop Protection Reference* (Anonymous, 1998), weed control guidelines put out every year by the respective state cooperative extension services (e.g., *Weed Control Guidelines for Mississippi,* a yearly publication, MAFES and MCES, 1998), and manufacturers' labels.

Mode of action of herbicides

The mode of action of a herbicide can be described as the physiological and biochemical interaction of the herbicide with a plant. Herbicide-treated plants may take from 1 h to 1 week or longer to develop visible injury symptoms (Devine et al., 1993). The most rapid visible symptoms of phytotoxicity are desiccation (e.g., paraquat) and epinasty (e.g., 2,4-D). The mode of action of a herbicide is a complicated and multifaceted process. It begins with herbicide contact with plant surfaces and ends with cellular death. Soil-applied herbicides enter the plant at the root or emerging shoot, while foliar-applied herbicides enter at the shoot or leaf. The process by which a herbicide passes from one system into another is usually referred to as herbicide absorption (uptake). After herbicides are taken into the plant, they are translocated to the site of action (target site) within the plant. In some instances, the site of action may be one or two cell layers from the point of entry (e.g., Paraquat). Some herbicides in the plant can move a relatively short distance (cell to cell), while others move long distances (root to shoot, shoot to root) or undergo metabolic conversion (either activation or deactivation). Herbicide movement is a function of the chemical nature of the herbicide, the plant species, plant stress, and the environment. The site at which a herbicide acts on a plant can be at the organ (root or leaf), tissue (mesophyll or meristem), cellular, or organelle (mitochondria or chloroplast) levels (Duke, 1990; 1992; Devine et al., 1993). Herbicide action at the target site is followed by a series of toxic consequences that result in partial or complete death of the plant.

Soybean herbicides exhibit several different modes of action. Numerous herbicides inhibit photosynthesis — triazines (e.g., Sencor/Lexone), bipyridiliums (e.g., Gramaxone), benzothiadzoles (e.g., Basagran). Glyphosate (Roundup), sulfonylureas (e.g., Classic), imi-dazolinones (e.g., Scepter), triazolopyrimidines (e.g., Broadstrike), and glufosinate (Liberty) inhibit amino acid synthesis. Several herbicides from diphenyl ether (e.g., Blazer) and aryltriazolinone (e.g., Authority) chemical classes interfere with porphyrin synthesis

Table 12.2   Labeled Soybean Herbicides[a]

Herbicide	Trade name	Remarks
Preplant foliar (burndown)		
2,4-D	2,4-D (several)	Apply 7 to 30 days before planting, separate equipment or extra care in cleaning of spray equipment required
Glyphosate	Roundup Ultra	Apply before soybeans emerge
Paraquat	Gramoxone Extra	Apply before soybeans emerge
Thifensulfuron + tribenuron	Harmony Extra	Apply 45 days before planting
Preplant incorporated		
Alachlor + trifluralin	Freedom	—
Clomazone	Command	Ground equipment; drift
Clomazone + trifluralin	Commence	Ground equipment; volatile; drift
Dimethenamid	Frontier	Poor control of large-seeded broadleafs
Flumetsulam + metolachlor	Broadstrike SF + Dual	Ground equipment
Flumetsulam + trifluralin	Broadstrike + Treflan	Ground equipment
Imazaquin	Scepter	—
Imazaquin + dimethenamid	Detail	—
Imazaquin + pendimethalin	Squadron	—
Imazaquin + trifluralin	Tri-Scept	—
Metribuzin + chlorimuron	Canopy	Rate depends on soil texture; sensitive varieties
Pendimethalin	Prowl	Volatile, incorporation preferred
Sulfentrazone + chlorimuron	Authority Broadleaf	Sensitive varieties
Trifluralin	Trifluralin (several)	Volatile, incorporation preferred
Trifluralin + metribuzin	Salute	Sensitive varieties
Preemergence		
Alachlor	Alachlor (several)	Rainfall within 10 days of application
Clomazone	Command	Ground equipment; drift
Dimethenamid	Frontier	Poor control of large-seeded broadleafs
Flumetsulam + metolachlor	Broadstrike SF + Dual	Ground equipment
Imazaquin	Scepter	—
Imazaquin + dimethenamid	Detail	Can be applied up to 3rd trifoliate stage
Imazaquin + pendimethalin	Squadron	—
Metolachlor	Dual	Rainfall for optimum control
Metolachlor + metribuzin	Turbo	Sensitive varieties
Metribuzin	Lexone/Sencor	Sensitive varieties; soil texture
Metribuzin + chlorimuron	Canopy	Rate depends on soil texture; sensitive varieties
Sulfentrazone + chlorimuron	Authority Broadleaf	Sensitive varieties
Postemergence		—
Acifluorfen	Blazer	—
Acifluorfen + bentazon	Storm	—
Acifluorfen + bentazon + clethodim	Conclude B + G	—
Bentazon	Basagran	—
Chlorimuron	Classic	—
Clethodim	Select	—
Fomesafen	Reflex	—
Fluazifop-P	Fusilade DX	—
Fluazifop-P + fenoxyprop	Fusion	—
Fluazifop-P + fomesafen	Typhoon	—
Glyphosate	Roundup Ultra	Roundup Ready varieties; avoid drift

Table 12.2 *(continued)*	Labeled Soybean Herbicides[a]	
Herbicide	Trade name	Remarks
Imazaquin	Scepter	Do not tank-mix with POST grass herbicides
Imazaquin + acifluorfen	Scepter OT	—
Imazethapyr	Pursuit	Avoid within 85 days of soybean harvest
Lactofen	Cobra	Avoid within 90 days of soybean harvest
Sethoxydim	Poast Plus	Avoid within 90 days of soybean harvest; do not tank-mix with Basagran
Quazilofop-P	Assure II	Petroleum-base crop oil adjuvant
Directed sprays		
2,4-DB + linuron	2,4-DB (several) + Linuron (several)	Lower one third of soybean stem; low pressures
2,4-DB + metribuzin	2,4-DB (several) + Lexone/Sencor	Lower one third of soybean stem; low pressures
Paraquat	Gramoxone Extra	Lower one-third of soybean stem; low pressures
Spot treatment		—
Clethodim	Select	Wet foliage thoroughly
Fluazifop-P	Fusilade DX	Wet foliage thoroughly
Quizalofop-P	Assure	Wet foliage thoroughly
Sethoxydim	Poast	Wet foliage thoroughly
Glyphosate	Roundup Ultra	Wet foliage thoroughly; non-Roundup Ready soybeans are killed

[a] For more-detailed information on herbicides, use rates, weeds controlled, time of application, and rotational crop restrictions, the reader may refer to herbicide labels, *Herbicide Handbook* (WSSA, 1994), or *Weed Control Guidelines for Mississippi*, a yearly publication (MAFES and MCES, 1998).

and cause accumulation of protoporphyrin IX. Protoporphyrin IX is phytotoxic to plants because in the presence of light and molecular oxygen it induces the formation of highly reactive singlet oxygen that breaks down cell membranes. Lipid synthesis is the target site for aryloxyphenoxy propionate (e.g., Fusilade), cyclohexanedione (e.g., Poast), and chloroacetamide (e.g., Lasso) herbicides. Dinitroanilines (e.g., Prowl) inhibit cell division. Pyridazinones (e.g., Zorial) and isoxazolidinones (e.g., Command) inhibit carotinoid synthesis. The exact molecular site of action of some herbicides is still unknown; for example, chloroacetamides (Lasso, Dual), besides their doubtful effects on lipid synthesis, have been recognized as inhibitors of cell division and cell growth (Devine et al., 1993). The mode of action and visual symptoms of some soybean herbicides are summarized in Table 12.3.

The knowledge of the mode of action of a herbicide is essential to make appropriate weed control management decisions. Crop and weed species exhibit differential sensitivity to different herbicides. For example, bentazon (Basagran) kills some annual broadleaf weeds and fluazifop-P (Fusilade) kills some annual grasses, but neither kills soybean. Herbicides with different modes of action can be rotated in soybean weed control programs to minimize selection pressure that may result in development of weed resistance to the herbicides.

Timing of herbicide application

Application timings are defined with respect to the growth stage of the weed or crop. Herbicides can be applied either to the soil or to the foliage of weeds. Some herbicides are only effective when applied to the foliage, while others are effective when applied to

Table 12.3   Mode of Action of Different Soybean Herbicides

Physiological site	Herbicide family	Examples	Visible symptoms
Amino acid synthesis	Glyphosate	Glyphosate	Chlorosis, necrosis, regrowth with malformed leaves; development of multiple shoots; foliage sometimes reddish purple in certain species; plant death within 4 to 20 days
	Sulfonylureas	Chlorimuron	Stunted growth, chlorosis and reddening, death after several days; plant death within 7 to 21 days
	Imidazolinones	Imazaquin	Inhibition of growth followed by purple discoloration, chlorosis, and necrosis; plant death in 1 to 2 weeks or more
	Triazolopyrimidine	Flumetsulam	Sensitive species are killed before emergence; emerged species exhibit stunting, interveinal chlorosis, veinal discoloration (purpling), and necrosis within 1 to 3 weeks
	Glufosinate	Glufosinate	Chlorosis and wilting within 3 to 5 days, followed by necrosis in 1 to 2 weeks
Photosynthesis	Triazines	Metribuzin	Seedlings emerge through soil and become chlorotic and necrotic, followed by death
	Bipyridiliums	Paraquat	Rapid wilting and desiccation of foliage within a few hours; complete necrosis in 1 to 3 days
	Benzothiadiazole	Bentazon	Chlorosis in 3–5 days followed by desiccation and necrosis; foliar bronzing may occur on tolerant crops such as soybean
Tetrapyrrole synthesis	Diphenyl ethers	Acifluorfen, Fomesafen	Plants become chlorotic and necrotic within 1–2 days; sublethal rates may produce bronzing on young leaves
	Aryltriazolinones	Sulfentrazone	Plants emerging from treated soils turn necrotic and die shortly after exposure to light; post-treated plants exhibit rapid desiccation and necrosis
Lipid synthesis	Aryloxyphenoxy propionates	Fluazifop-P, Quazilofop-P	Cessation of growth, chlorosis, meristematic necrosis, collapse of foliage; death within 1 to 3 weeks
	Cyclohexanediones	Sethoxydim	Cessation of growth; leaf chlorosis and necrosis within 1–3 weeks; leaf sheaths become brown and mushy; older leaves often turn purple, orange, or red before becoming necrotic
	Chloroacetamides	Alachlor, Metolachlor	Susceptible weeds fail to emerge; inhibits seedling growth, especially root growth; monocots if they emerge, appear twisted and malformed with leaves unable to unroll normally; broadleaf seedlings have cupped and crinkled leaves

	Table 12.3 (continued)	Mode of Action of Different Soybean Herbicides	
Physiological site	Herbicide family	Examples	Visible symptoms
Carotenoid synthesis	Pyridazinones	Norflurazon	Interveinal whitening of leaf and stem becomes apparent as seedlings emerge; growth may continue, but seedlings lack chlorophyll and soon become necrotic
	Isoxazolidinones	Clomazone	Seedlings emerge from soil bleached white followed by necrosis
Cell division	Dinitroanilines	Trifluralin, Pendimethalin	Susceptible weeds fail to emerge; inhibits seedling growth; roots appear stubby with tips becoming thickened and swollen
Growth hormone (auxin-type)	Phenoxys	2,4–D	Epinastic bending and twisting of stems and petioles, stem swelling (particularly at nodes) and elongation, and leaf cupping and curling; chlorosis at the growing points, growth inhibition, wilting, and necrosis

Source: Extracted from *Herbicide Handbook* (WSSA, 1994).

both soil and foliage. For example, glyphosate (Roundup) is active only as a postemergence (POST), whereas imazaquin (Scepter) can be applied preplant incorporated (PPI), preemergence (PRE), or postemergence. No single herbicide has the capability to control all weeds in all situations without causing injury to the crop. Knowledge of weed populations and careful timing of applications when weeds can be controlled determine the success of a weed control program. In soybean monoculture, it is advisable to avoid using the same herbicide year after year to avoid a shift in weed spectrum and/or a buildup of herbicide-resistant weed populations. Rotating crops and herbicides with different modes of action should be part of every soybean weed control program.

Preplant foliar (burndown) herbicides

In a conventional soybean production system, several tillage operations are performed prior to planting. These tillage operations usually control most winter weeds. However, rainfall patterns often prevent tillage operations during much of the winter and early spring, especially on poorly drained shrink-swell clay soils. High soil moisture and relatively warm temperatures (average daily air temperatures of 54 to 64°F) stimulate germination and encourage robust growth of weeds. These weeds must be controlled by herbicides. Glyphosate (Roundup), paraquat (Gramoxone), glufosinate (Liberty), and 2,4-D are the most common burndown herbicides used to control winter weeds before planting. Burndown herbicides are used in single or split applications as needed. Burndown herbicides such as glyphosate (Roundup) or paraquat (Gramoxone) can be tank-mixed with several PRE herbicides and applied at planting if desired. To avoid excessive weed growth in the early spring months and to keep the seedbed free of weeds at planting, a split application of burndown herbicides often proves beneficial. A frequently used program is to apply a nonselective foliar herbicide (e.g., Roundup) early to kill winter weeds and follow by a second application of paraquat (Gramoxone) tank-mixed with a PRE herbicide at soybean planting. In fields that follow wheat or a cover crop, burndown applications of glyphosate (Roundup) or paraquat (Gramoxone) may be made just before or at soybean planting. Although many options exist, the number of applications, the burndown materials used, and the rates are all dependent on weeds present, weed size, and the subsequent crops to be grown.

Preplant foliar-applied herbicides are most commonly used in reduced-tillage systems. One popular form of a reduced tillage system is the stale seedbed planting system (Heatherly and Elmore, 1983). In a stale seedbed system, waiting until planting to kill winter vegetation may result in large weeds that are difficult to control, require higher herbicide rates, not to mention the difficulties that may result at planting due to a buildup of residue (Bruff and Shaw, 1992a; Shaw, 1996). Allowing weeds to persist until planting depletes soil moisture and often results in poor soybean stand establishment (Hydrick and Shaw, 1995; Shaw, 1996). However, in a wet spring, vegetation will prevent fields from drying out, possibly contributing to delayed planting. Application of nonselective foliar herbicides such as glyphosate (Roundup) and paraquat (Gramoxone) early enough to control winter weeds before they become difficult to control, followed by a tank-mix of a nonselective foliar herbicide with PRE herbicides at planting, usually provides effective weed control. Soil-applied herbicides such as metribuzin + chlorimuron (Canopy), pendimethalin + imazaquin (Squadron), dimethenamid + imazaquin (Detail), metribuzin (Lexone/Sencor), or imazaquin (Scepter) have foliar activity and could be considered as additional options in a burndown program. Research has shown that in a stale seedbed system, the use of herbicides such as Sencor/Lexone or Canopy, which have POST activity, resulted in increased soybean yields when tank-mixed with paraquat (Gramoxone) compared to yields obtained with paraquat (Gramoxone) alone (Hydrick and Shaw, 1995). Tank-mix combinations of paraquat (Gramoxone) with herbicides such as Scepter or Sencor/Lexone which have POST activity in stale seedbed systems have resulted in yields equivalent to those of conventional tillage (Bruff and Shaw, 1992b).

Preplant incorporated and preemergence herbicides

Preplant incorporated herbicides are usually applied within several weeks of planting or right prior to planting and should be incorporated 1 to 3 in. following application. Incorporation is essential to ensure effective weed control. Soil incorporation reduces volatilization, decreases photodecomposition, and provides uniform distribution of the herbicide in the seed germination zone. Incorporation can be done with a disk or field cultivator. Sometimes light rainfall soon after application can aid in incorporation of a herbicide; however, mechanical incorporation is desirable since rainfall patterns are not dependable. Rainfall, however, is extremely beneficial for incorporating PRE herbicides. In fallow or reduced-tillage soybean production, PPI herbicides are not used. Preplant foliar and PRE herbicides are applied instead to control existing weeds and to provide residual weed control.

PRE herbicides are applied at or following soybean planting but before soybean emergence. These herbicides are applied either broadcast or in a band centered over the crop row. Rainfall and soil texture will affect the performance of PRE herbicides. Rainfall following a PRE herbicide application ensures the effectiveness of soil-applied herbicides by leaching the herbicide into the weed seed germinating zone. If there is no rainfall within 10 days after application of a PRE herbicide but enough moisture for the weed seeds to germinate and emerge, then weeds may escape the preemergence herbicide treatment. Although rainfall is essential for herbicide activation, heavy rainfall can cause poor weed control. Heavy rains may leach a herbicide below the seed-germinating zone and also move herbicides off-site in surface runoff. The amount and time of rainfall after a herbicide application is critical for optimum incorporation and optimum weed control. Scouting following prolonged periods of no rainfall or heavy rainfall after PRE applications should be conducted to determine the need for follow-up POST treatments.

In reduced-tillage systems, PRE herbicides are often tank-mixed with nonselective burndown herbicides and applied at planting. The level and consistency of weed control with PRE herbicides in reduced tillage systems are influenced by timely rain that is necessary to move the herbicides into the soil from plant residues since neither incorporation nor cultivation is feasible. During the last 10 years, the level and consistency of weed control in reduced-tillage systems have improved with the commercialization of herbicides such as chlorimuron + metribuzin (Canopy) and imazaquin (Scepter) that have both PRE and POST activity, do not require incorporation, and provide residual control.

No single herbicide can control all weeds. Some herbicides are active on grasses and some on broadleaf weeds. Every herbicide has a weakness. Since PPI and PRE herbicide applications are prophylactic in nature, care should be taken in selecting a particular herbicide and use rate. Weeds that were present in the previous crop season provide a useful guide in herbicide selection. In fact, this may be the only method for deciding on a herbicide to use based on its spectrum of control.

Herbicides applied to soils must remain in an active and available form to achieve best weed control. The amounts of sand, silt, clay, and organic matter determine the efficacy of soil-applied herbicides. As the soil particle size decreases, the number of particles and soil surface area increase significantly. Thus, silty and clayey soils have more surface area than sandy soils. Organic matter formed by decay of plant material is the most active component of a soil due to its larger surface area and greater capacity to adsorb other materials.

Herbicide availability to plants and its persistence in the soil are influenced by adsorption, movement, and degradation processes. Adsorption is a process by which a herbicide associates with the surface of a soil particle. Adsorption is the most important process affecting herbicide availability to plants and its persistence in the soil. Although almost all herbicides are adsorbed to some extent, the amount of herbicide adsorbed to the soil depends upon the amount of clay, organic matter, soil moisture content, and ionization properties of the herbicide. A herbicide adsorbed by the soil is not readily available for plant uptake, lateral or vertical movement (leaching), or chemical and microbial degradation processes. Herbicide must be desorbed from soil particles for bioavailability. Effective weed control depends on the amount of herbicide that is present in the soil solution.

Generally, as the clay and organic matter content of the soil increase, there is greater adsorption of herbicides to the soil. For example, chlorimuron (Classic) adsorption increased as organic matter levels increased in several soils under different tillage systems (Reddy et al., 1995b). Thus, the rates of some herbicides must be based on these soil properties. On some coarse-textured soils with low organic matter levels, certain herbicides leach so readily that they may fail to adequately control weeds. Manufacturers recommend use rates of some herbicides based on soil texture and organic matter content of the soil. For example, labeled use rates for pendimethalin (Prowl) increase from 0.5 to 0.75 lb ai/acre in coarse soils to 1 to 1.5 lb ai/acre in fine-textured soils. Authority Broadleaf, Canopy, Detail, Dual, and Sencor/Lexone are common PRE herbicides that require rate adjustments for soil texture. PRE herbicide labels should be checked for rate changes based on soil properties.

Soil-applied herbicide losses can occur via surface runoff, leaching, or a combination of both. Many agricultural fields are flat or have small slopes that may aid in minimizing runoff losses of soil-applied herbicides. The downward movement of herbicides in a soil profile is common because of the percolating action of rainfall or irrigation water. The amount and depth of herbicide leaching depends on the extent of adsorption to the soil and the total amount of rainfall or irrigation. Soil-applied herbicides are subjected to degradation by chemical or biological processes. A wide array of soil microorganisms are primarily responsible for the biological degradation of herbicides in the soil. Overall, the degradation of herbicides is influenced by the chemical structure of a herbicide, the degree of soil adsorption, the composition and populations of microbes, and the microbial environment (soil moisture, temperature, aeration, pH, organic matter). The bioavailability (extent of weed control) and persistence (carryover in soil) of herbicides are a function of complex interactions among herbicides, soils, and plants.

Postemergence herbicides

PPI and PRE herbicide applications often fail to provide adequate season-long weed control for various reasons. Weed control failure could be related to misapplication, weather, or the tremendous weed pressure often observed in the midsouthern U.S. The time and amount of rainfall received before and after herbicide application are critical as they affect the ability of weeds to absorb the herbicide and herbicide movement in the soil profile. Herbicides that have short residual activity fail to control weeds emerging late in the season. Weeds often escape herbicide activity because of the location of weed seed in the soil. The control of susceptible weed species by PRE herbicides may create an environment for other tolerant weed species to become more competitive. Not controlling escapes may reduce costs temporarily, but it will perpetuate weed problems by permitting a shift of weed populations from susceptible to tolerant species and allow for competition among the crop and weeds. POST applications are often necessary to control escaped weeds to prevent them from competing and producing seed. Scouting fields for density and distribution of weed species is essential to plan POST herbicide strategies.

The activity of a POST herbicide is a function of complex interactions among the herbicide (herbicide chemistry), weeds (weed species, density, growth stage, plant stress), adjuvants, and environmental conditions (wind velocity, temperature, relative humidity, rainfall). These factors will affect uniform deposition of herbicide spray on plant foliage, retention of spray droplets on leaf surfaces, and absorption of herbicide into shoot tissue. POST applications made when air temperatures are 85 to 90°F, relative humidity is above 60%, weeds are small and actively growing, soil moisture is high, and no rainfall is received for 4 to 6 h after application should provide satisfactory control of weeds (MAFES and MCES, 1998).

Most POST herbicides require adjuvants to maximize their efficacy. Adjuvants are either included by the manufacturer in the formulated product (e.g., Roundup Ultra) or added to the spray tank by the applicator. These adjuvants may function as activators, spreaders, stickers, antifoamers, wetting agents, compatibility agents, suspension agents, penetrating agents, or buffering agents. Thousands of adjuvants with different physicochemical properties are being marketed in the U.S. The most commonly used adjuvants are formulated as blends of petroleum oils (e.g., Agri-Dex, Dash), crop oils (e.g., Soy Spreader, Cotton Oil Plus), conventional hydrocarbons (e.g., X-77, Induce), or organosilicones (e.g., Kinetic HV, Silwet L-77). The enhancement of POST herbicide activity by adjuvants has been attributed to reduced surface tension, improved coverage (leaf wetting and spreading), induced stomatal entry, and increased cuticle penetration. The addition of adjuvants to the spray solution where advised should be done in accordance with the manufacturer's label to enhance the effectiveness of POST herbicides.

Differential response of certain plant species to adjuvants has been reported. Organosilicone-based adjuvants have increased the activity of bentazon (Basagran) and imazaquin (Scepter) on sicklepod and hemp sesbania compared with a nonionic adjuvant; however, both adjuvants had a similar effect on smooth pigweed (Reddy et al., 1995a; Reddy and Locke, 1996). Adjuvants help enhance herbicide spread, retention of spray on the leaf, and subsequent uptake into the leaf. The appropriate selection of adjuvants greatly increases the effectiveness of a herbicide. Rain can reduce the efficacy of POST herbicides. Rainfall at 1 or 24 h after herbicide application washed off a considerable amount of residues of lactofen (Cobra), bentazon (Basagran), and imazaquin (Scepter) from plant foliage (Reddy et al., 1994; 1995a; Reddy and Locke, 1996).

Some POST herbicides require a minimum rain-free period after application for better activity. The rain-free periods range from 1 h (e.g., Poast, Assure, Fusilade, Select) to 4 h (e.g., Reflex, Classic) to 6 h (e.g., Blazer). Tank-mixing of POST herbicides often aid in increasing the spectrum of weeds controlled and reduce application costs. Manufacturers market package mixtures for some herbicides (e.g., Storm, Fusion, Typhoon). In some instances, producers prefer to tank mix broadleaf and grass herbicides for a "one-pass" application. Care should be taken, since antagonistic (less weed control than expected) interactions can result with such tank mixtures. For example, when Basagran is tank mixed with Poast, the Poast rates need to be increased by 50% to offset antagonism.

Often, POST herbicide label rates vary depending on weed species and size. For example, the Blazer label rate for a four-leaf palmer amaranth is 0.5 pt/acre and for a four-leaf entireleaf morningglory is 1.5 pt/acre. Higher rates are usually required for satisfactory control of bigger weeds. For example, the Poast rate for barnyardgrass control is 0.75 pt/acre at the four-leaf, 1.0 pt/acre at the eight-leaf, and 1.5 pt/acre at the 12-leaf stage. In addition to rate corrections, complete coverage of plant foliage with herbicide sprays is essential for effective weed control. This is extremely critical for contact herbicides. For example, Blazer is active only through contact action; therefore, the target weeds must be thoroughly covered with the spray solution. If cultivation is intended, do not cultivate within 1 week before or after a POST herbicide application as this may reduce weed control.

Preemergence vs. postemergence herbicides

PRE herbicide applications are very much like insurance, which ensure adequate weed control from planting. However, this is an expense with no guarantees. For example, Canopy at 10 oz/acre costs $23.90 and Dual at 1 qt/acre costs $14.72. The cost of herbicides reported herein reflect the current market prices; however, these prices may vary in the future depending on market conditions. Inadequate or no rainfall can result in poor germination and emergence of soybean, and late-season stresses can result in low yields and income even though a high cost was incurred for PRE weed control. POST applications provide backup for the lack of control of weeds by PRE applications, or for the control of late emerging weeds. POST applications should be used for weed control when an economic threshold is reached, but their effectiveness depends on timely application with regard to size and age of weeds and on climatic conditions (rainfall, wind, temperature, light, and relative humidity). A combination of preventative treatments (e.g., PPI or PRE herbicides followed by POST herbicides) may be more effective and economical in fields where weed infestations are known to be potentially high, and yield potential of soybean is high.

POST herbicides are considered more environmentally friendly than PRE herbicides because most have little or no soil residual activity and they are applied only when needed. POST applications also facilitate the use of scouting, economic thresholds, and allow reduced rate applications for smaller weeds. In general, research has shown that a period of 4 to 6 weeks without weed competition at the beginning of the growing season is required for maximum soybean yields under most environmental conditions (Stoller et al., 1987; Jordan et al., 1987). Any weeds emerging after this initial weed-free period will not compete effectively with soybean and will have a limited impact on yield (Stoller et al., 1987; Jordan et al., 1987).

Adopting a total POST weed control program can be successful, but it has certain constraints. Frequent rainfall during the initial 2 to 6 weeks of the growing season can prevent the timely application of POST herbicides using ground equipment and contribute to frequent germination/emergence of weeds. This failure to achieve timely weed control will have adverse yield consequences, and shows the need for a well-balanced weed control program where economical.

Banding vs. broadcast applications

PRE application of herbicides in a band over the crop row coupled with timely cultivation of row middles can eliminate early weed competition in rows that are 30 in. or greater in width. Band applications of herbicides coupled with cultivation of wide-row middles also reduces herbicide use rates compared with broadcast applications. On the other hand, broadcast applications are required in narrow-row or drilled soybean plantings since cultivation is not practical in narrow-row middles, and precision of banding over narrow rows may be inadequate when not done at planting.

POST-directed sprays are used to prevent herbicide application to the upper, more sensitive portions of soybean plants that may exhibit phytotoxicity symptoms when some POST herbicides are applied. This application method also ensures that the herbicide is deposited on the weeds that are usually underneath the crop canopy. It may be worthwhile to invest in equipment that will allow POST-directed spraying to exploit these advantages. Recently, Ginn and co-workers (1998) developed a sprayer that could be used to direct POST herbicides to undercanopy areas of crops planted in rows less than 20 in. The sprayer can be attached to a 40- to 100-hp row-crop tractor having a PTO and standard Category II three-point hitch attachment. Use of a sprayer such as this can ensure the effective application of POST herbicides to the target weeds under the canopy, and reduce the herbicide drift compared with applications made above the canopy.

Nonchemical weed control methods

In nonchemical methods of weed control, all production practices are directed toward creating the most favorable environment for the crop and the least favorable environment for weeds. This often results in the creation of conditions just as suitable for certain weed species as for the crop (Buchanan, 1992). Management decisions such as seedbed preparation, variety selection, time of planting, use of clean seed, soil fertility, planting pattern, and soil moisture can all be managed so that soybean is favored over weeds.

Control of weeds on row ends and turn rows to prevent weed seed production is as important as control of weeds in the crop. Crop rotation permits the use of a wider range of alternative herbicides. In addition, reduced-tillage systems play a role in reducing replenishment of weed seeds in the germination zone. Pareja and Staniforth (1985) reported that 85% of all weed seeds were in the 0 to 2 in. soil depth in a reduced-tillage system compared with only 25% in a conventional tillage system. This may be due to redistribution of weed seeds in the germinating zone under conventional tillage systems. Changes in seed depth in the soil and species differences in emergence from various soil depths may contribute to shifts in weed spectrum under different tillage systems. For example, giant foxtail seed placed on the soil surface had higher (86%) establishment compared with seeds planted at 2.4 in. (50%) depth. In contrast, velvetleaf seed planted at 2.4 in. had higher (100%) establishment compared with seeds placed on the soil surface (18%) (Buhler, 1995). Koskinen and McWhorter (1986) reported increased perennial and biennial weeds with no-tillage systems. Use of cover crops should be considered as part of the cultural approach to weed control. Aside from reducing soil erosion and improving soil fertility and crop performance, some cover crops exhibit allelopathic effects on weeds. Winter cover crops offer the potential to overcome weed problems in the winter and spring which are otherwise unmanageable.

Cultivation

Cultivation after soybean emergence can complement chemical weed control. Cultivation controls escapes in row middles and extends the longevity of weed suppression. However, available moisture is the major determining factor as to what weed pressure will be like after cultivation. Weeds directly in the crop row are generally inaccessible to mechanical cultivation and must be controlled by herbicides. Cultivation kills weeds by uprooting plants, cutting shoots from roots, and burying of plant shoots (Ashton and Monaco, 1991). These actions lead to weed desiccation and depletion of plant food reserves. The best results from cultivation are obtained when weeds are small, since larger weeds are difficult to bury and have sufficient roots to escape total separation from soil. Seedlings of perennial weeds are easily controlled by cultivation, but older plants can survive the disturbance. Cultivation of dry soil promotes desiccation, while wet soil allows weeds to transplant. Rainfall soon after cultivation often results in incomplete desiccation and allows regrowth to occur from the reproductive organs of perennial weeds. In addition to weed control, cultivation will help break surface crusts, increase soil aeration, and increase rainfall infiltration (Ashton and Monaco, 1991). Reduced-tillage systems often limit the use of cultivation, because plant residues on the soil surface can interfere with cultivation; however, several cultivators have been developed to work in plant residues.

Results of several studies have shown that a PRE herbicide followed by a POST herbicide in combination with a well-planned cultivation program provide season-long control of most weeds (Shaw and Coats, 1988; Poston et al., 1992; Miller and Griffin, 1994; Newsom and Shaw, 1996). However, weed management systems using cultivation alone are unsatisfactory for controlling weeds in soybean, especially in irrigated culture. Cultivation may also reposition weed seeds near the soil surface, which may stimulate their germination and subsequent emergence (Newsom and Shaw, 1996).

The main purpose of cultivation of row middles is to control weeds. The use of cultivation depends on weed problems (population and density), weed control budget, and row spacing. Cultivation is often an expensive operation, especially if herbicides are broadcast. Estimated 1997 costs for cultivation of 30-in.-wide rows in the Delta area of Mississippi are $5.84/acre for an early cultivation and $3.91/acre for a late cultivation, for a total of $9.75/acre for two cultivations (Anonymous, 1997). Results of recent research show that a broadcast herbicide application to include the cultivated row middles vs. a band application over only the row (excluding cultivated row middles) results in inflated expense with no increase in yield or income, and thus a lower net return. Therefore, the use of both cultivation and herbicides to control weeds in the row middles is not economical. If herbicides are applied broadcast, then rows could be narrowed to ensure quicker canopy closure.

Row spacing

Narrow rows help reduce weed interference in soybeans by providing a faster and greater degree of canopy closure. When herbicides suppress early weed growth, less weed biomass is produced in narrow rows than in wide rows. In narrow rows, soybeans shade the ground earlier and enhance competitiveness of soybeans with weeds that germinate late. However, in narrow (≤20 in.) rows, all weed control will be relegated to herbicides since cultivation is not an option. Broadcast applications of POST herbicides over-the-top may fail to control weeds underneath the canopy due to spray interception by the crop canopy and will require a directed spray to ensure effective control of weeds. Quicker canopy closure is a potential benefit, but this is not always in time to offset weed management, especially on clay soils. Canopy closure by many varieties grown on clay soil is often as late as 60 days after planting because of the slow growth on these soils. On the other hand, the only purpose for planting in wide rows is to allow cultivation of row middles to control weeds; however, herbicides should be banded for any cost benefit to accrue. Some producers are more comfortable with the option of cultivation to control weeds even with broadcast herbicide applications. If so, they should consider the cost savings of banded herbicide application.

Three well-timed cultivations in wide rows (40 in.) will cost $13.66/acre ($5.84 + $3.91 + $3.91). This is almost identical to the cost of one broadcast application of Storm (material and application cost), which is equivalent to only one application of the POST (Storm controls mostly broadleaf weeds) measures that may be used in narrow rows. In fact, it is reasonable to assume that weed control costs for narrow-row soybean production will be more expensive since all weed control is by herbicides and all applications must be broadcast. Therefore, narrow-row production systems, to be economically viable, must produce a yield advantage over wide rows. Currently, at the Southern Weed Science Research Unit, USDA-ARS, Stoneville, MS the senior author (KNR) is conducting studies on the economics of weed control in narrow vs. wide rows with PRE only, POST only, and PRE + POST herbicide applications.

Planting date

Altering planting dates may also alter crop and weed emergence patterns. Changes in crop planting dates may shift the competitive advantage to the crop, thereby reducing weed interference (Oliver, 1979; Heatherly et al., 1990; Klingman and Oliver, 1994). Soybean was more competitive with entireleaf morningglory and sicklepod when planted during early May compared with early June (Klingman and Oliver, 1994). Soybean yield loss due to interference of both weeds increased as planting date was delayed from early May (10%) to early June (20%). Overall, weed control has been found to be similar regardless of planting date from May to June (Heatherly et al., 1990).

The trend is toward early (April) planting of early-maturing MG IV and V soybean varieties in the midsouthern U.S. When these varieties are planted in April, summer broadleaf weeds such as morningglory, prickly sida, and cocklebur may not be emerged. Thus, planting at this time can be done in a stale seedbed and existing weeds can be controlled using low rates of burndown herbicides. Also, soybeans may emerge before weeds even without a PRE herbicide application. This allows use of POST herbicides for weed control on an as-needed basis. In essence, planting in April may lessen the weed control aspect of soybean production by avoiding some of the competitive weeds that emerge in early May to early June under normal soybean systems.

Crop rotation

Crop rotation, along with associated herbicide options, is an excellent way of reducing weed populations and a potential weed shift in soybean. The successful management of some weeds leads to infestation of other difficult-to-manage weeds. This replacement of one problem species or group of species by another is viewed as a weed shift.

Repeated use of the same herbicide or herbicides with the same mode of action can result in a shift in weed spectrum and create an environment (selection pressure) to develop resistance to the herbicides. Rotation of crops allows use of different herbicides and reduces the potential for resistance to develop. Rotation with summer crops such as corn and cotton or a winter crop such as wheat with associated herbicides can provide more effective weed control systems. Certain weeds are easier to control in some crops than in others, and a single approach to weed control is not as successful as multiple approaches. In the absence of crop rotation, it is better to alternate soybean herbicides having different modes of action to avoid or delay potential resistance problems.

Soybean variety

Soybean varieties differ in their ability to compete with weeds. For example, the short variety 'Tracy-M' had the least effect on sicklepod, whereas the tall variety 'Biloxi' produced a more competitive crop canopy and had a greater effect on sicklepod (Shilling et al., 1995). Although there were some differences in competitiveness between 'Centennial' and 'Forest' soybeans with five weed species (common cocklebur, johnsongrass, palmer amaranth, sicklepod, tall morningglory), yields of both varieties were similar (Monks and Oliver, 1988). Variety selection and row spacing can play a significant role in weed suppression although weed control costs will be similar regardless of soybean variety. Yield potential continues to be the best criterion for selecting a soybean variety. Row spacing can be narrowed to compensate for shortness of growth, and thus provide sufficient competitiveness with weeds. Since no variety will totally outcompete weeds, it is most appropriate to select the highest-yielding variety to match planting date, soil texture, and time-of-maturity needs.

Herbicide usage pattern

The soybean crop was the second largest user of herbicides (68 million lb) following corn (224 million lb) in 1992 (U.S. Department of Agriculture, 1996). Total herbicide use in the U.S. on major crops increased from about 55 million lb in 1964 to 465 million lb in 1982, which accounted for over 75% of total pesticide use. Herbicide use then declined to 388 million lb in 1992. The decline in herbicide use coincided with a decline of about 25 million planted acres coupled with reduced herbicide use per acre. Lower herbicide rates in the 1980s were due to the introduction of low-use rate herbicides such as those of the sulfonylurea and imidazolinone families. Results of the survey conducted on 1993 cropping practices for soybean in selected states are summarized in Table 12.4.

Pest scouting is a common practice in soybean and was conducted on 69% of the surveyed soybean acres. Most of the scouting was done by farmers, followed by dealers, crop consultants, and extension workers. Economic thresholds were used by 59% of farmers to decide about herbicide applications in soybean. In major soybean-producing states, about 67% of the acres were under crop rotation for weed control. In the eight states surveyed, about one third of the acres in each state were treated with PRE only, POST only, or both. Less than 5% of the acres were treated with band applications, a majority of the acres were treated as broadcast, and about 6% of the acres were treated with both band and broadcast applications. Banding vs. broadcasting herbicides saves material costs; however, banding often requires mechanical cultivation to control weeds in the row middles, and this can be done only to wide-row soybeans (>20 in). In 1993, 38% of soybean acres in the surveyed states were cultivated an average of 1.5 times for weed control. Spot treatment with herbicides, which is an effective way to control isolated heavy infestations of weeds, was used on about 3% of soybean acres. Herbicides were rotated on 55% of soybean acres.

Results of surveys conducted in the 14 states during the 1995 crop year indicated that an average of 97% of the 51.8 million acres planted to soybean were treated with herbicides. The major herbicides used on soybean during 1995 are shown in Table 12.5. Imazethapyr (Pursuit) was applied to 44% of the acreage, with pendimethalin (Prowl) and glyphosate

Table 12.4 Herbicide Decision Making, Application Timing, and Application Method in Major Soybean-Producing States in 1993[a]

Item	Unit	Soybeans
Scouting for weeds, insects, or diseases	% of acres	69.2
Source of scouting[b]		
Farmer	% of acres	64.9
Extension	% of acres	1.2
Dealer	% of acres	11.6
Consultant	% of acres	2.7
Processor (fieldman)	% of acres	0.1
Other	% of acres	0.1
Economic thresholds used with scouting information for herbicides		
Used	% of acres	59.4
Not used	% of acres	9.7
Crop rotation to control weeds[c]	% of acres	66.6
Herbicide application timing[d]		
Preemergence only		
Area treated	%	30.6
Rate per acre	lb	1.3
Postemergence only		
Area treated	%	31.8
Rate per acre	lb	0.6
Pre- and postemergence		
Area treated	%	37.6
Rate per acre	lb	1.3
Herbicide application method[d]		
Banding only, area treated[e]	%	4.5
Broadcast only, area treated[e]	%	89.2
Banding and broadcast	%	6.4
Row cultivation to control weeds		
Area cultivated[f]	%	37.9
Times cultivated	number	1.5
Weed spot treatments[f]	%	3.0
Rotating pesticides to slow resistance to herbicides[d]	%	54.8

[a] Includes eight soybean producing states: AR, IL, IN, IA, MN, MO, NE, and OH.

[b] The sum within scouting sources exceeds the percent of acres scouted because some acres were scouted by more than one source.

[c] Excludes fields where the same crop was planted in the preceding year.

[d] Percentages are of treated acres.

[e] Banding includes in-furrow and banded in/over row. Broadcast includes ground and aerial broadcast, chemigation, and directed spray.

[f] Percentages are of planted acres.

(Roundup) applied to 26 and 20%, respectively. In terms of amount of active ingredient, pendimethalin (Prowl) was used extensively (12.9 million lb), followed by trifluralin (Treflan), metolachlor (Dual), and glyphosate (Roundup) (U.S. Department of Agriculture, 1996).

Herbicide injury

Soybean varieties differ in tolerance to herbicides. Results of Barrentine et al. (1982) have indicated that 'Tracy-M' and 'Centennial' are more tolerant to metribuzin than 'Tracy'. Chlorimuron (Classic) and imazaquin (Scepter) cause injury to soybeans under some

Table 12.5 Herbicides Used on Soybeans in 14 Major
Soybean-Producing States during 1995.[a]

Herbicides	Total applied, 1000 lb	Area applied,[b] %
2,4-D	2418	10
2,4-DB	40	1
Acifluorfen	1487	12
Alachlor	3930	4
Bentazon	4272	12
Chlorimuron-ethyl	154	16
Clethodim	262	5
Clomazone	1223	4
Dimethenamid	483	1
Ethalfluralin	262	1
Fenoxaprop-ethyl	395	6
Fluazifop-*P*-butyl	363	10
Flumetsulam	65	2
Fomesafen	630	4
Glyphosate	6318	20
Imazaquin	726	15
Imazethapyr	1332	44
Lactofen	202	5
Linuron	348	2
Metolachlor	6995	7
Metribuzin	1397	11
Paraquat	667	2
Pendimethalin	12930	26
Quizalofop-ethyl	160	6
Sethoxydim	690	7
Thifensulfuron	20	12
Trifluralin	8329	20

[a] Area planted in 1995 for the 14 major states was 51.8 million acres. States included are AR, GA, IL, IN, IA, KY, LA, MN, MS, MO, NE, NC, OH, TN.

[b] Percentages of area applied exceeds 100%, as more than one herbicide was applied on same acre.

conditions. Newsom and Shaw (1992a, b; 1995) evaluated 20 determinate soybean varieties commonly grown in Mississippi for tolerance to chlorimuron (Classic) and imazaquin (Scepter). Most of the varieties were sensitive to these herbicides, but the degree of sensitivity was related to growing conditions. Overall, soybean varieties tolerate herbicides very well; however, some varieties were more sensitive to chlorimuron (Classic) in combination with excessive moisture. Under optimum moisture conditions, chlorimuron (Classic) reduced the yield of nine varieties, but imazaquin (Scepter) did not affect yield. In an evaluation of 32 soybean varieties from five maturity groups for tolerance to dimethenamid (Frontier) and metolachlor (Dual), tolerance was variable. 'Bryan' was the most sensitive variety to both herbicides, while 'Brim' was the most tolerant of all varieties evaluated (Osborne et al., 1995). Research has shown that soybean varieties exhibit a range of tolerance to soil-applied sulfentrazone (Authority). The varieties 'Centennial' and 'Stonewall' are tolerant, and 'Hutcheson' and 'Asgrow 6785' are susceptible to sulfentrazone (Walker, 1994; Zhaohu et al., 1997). Soybean producers should be aware of the differential tolerance of soybean varieties to herbicides. This information should be requested from the suppliers, or should be listed on the label.

Herbicides are the most effective and commonly used tool in weed control programs of modern agriculture. They must be used in accordance with label directions to ensure safety to crops, animals, and the environment. Diagnosing herbicide injury is difficult, and in many cases ascribing symptoms to any one cause is impossible. Symptoms are usually most clearly expressed early, and as time passes they may either disappear or develop into severe necrosis or death. A survey concerned with causes of herbicide injury and poor weed control revealed that approximately 80% of the complaints could be attributed to improper application or misuse (Monaco et al., 1986). The most common causes of herbicide injury are improper calibration of spray equipment, herbicide drift, improper tank-mixes, contamination, herbicide residues in soils, varietal sensitivity, and adverse weather conditions (Monaco et al., 1986). Symptoms of herbicidal injury are related to mechanism of action (Table 12.3). In general, herbicides having a similar mechanism of action produce similar injury symptoms in plants. Often injury symptoms that appear right after application may not be a problem in the end. For example, AC 263,222 (Cadre) and chlorimuron (Classic) applied POST injured soybean from 12 to 32% at 3 days after treatment, but had no measured effect on soybean yields (Newsom and Shaw, 1994).

When soybean is grown in rotation with other crops, persistent herbicides used in previous crops may cause serious injury to soybeans. In general, herbicide persistence varies with soils (texture, organic carbon, pH) and climate (temperature, rainfall), and consequently rotational crop restrictions vary accordingly. For example, to avoid unacceptable injury to soybeans, do not plant within 60 days after a oxyfluorfen (Goal) application, 6 months after a fluometuron (Cotoran) application, and 10 months after a pyrithiobac (Staple) application. Consult all labels for rotational restrictions.

Preharvest desiccation

Preliminary consideration has been given to and current research efforts are addressing the use of a preharvest desiccant in early soybean production system plantings. This consideration has been deemed advantageous since early-maturing varieties planted early have an earlier open canopy. Ramifications of this early maturity have been addressed already.

Until research produces sufficient results to make recommendations about the use of a preharvest desiccant, the following thoughts are presented.

1. A preharvest desiccant will not be needed if:

    a. Weeds present at maturity emerged late in the growing season and their small size will not interfere with harvest;

    b. The species present are small-statured annual grasses, small-statured and small stemmed broadleaf weeds such as spurges, and/or small-stemmed perennial vines whose presence will not reduce harvest efficiency;

    c. The weeds present have not produced mature seeds that will contaminate the grain;

    d. The desiccant cannot be applied sufficiently ahead of harvest so as to ensure that the weeds are dry at harvest (this may be the case in the high-temperature, low-humidity conditions of late August and early September when the time between maturity of soybean or 95% mature pod color and harvest may be as little as 5 to 7 days);

    e. The weed vegetation that is cut by the combine during soybean harvest is returned to the field with the soybean residue so that no foreign matter enters the grain sample; and

    f. The row spacing and variety selection were compatible (i.e., the row spacing was sufficiently narrow to allow the selected variety to form a canopy) and an effective herbicide program was in place.

2. It is economical to control only those weed infestations that will result in foreign matter in the harvested grain in an amount sufficient to cause dockage.
3. Effective control of early-season weeds will result in fewer significant weed infestations late in the season when the soybean canopy opens at maturity. Late-emerging weeds will be much less important at harvest than those that are present but underdeveloped beneath the soybean canopy during the growing season. A few weeds have the capacity to emerge and reach a problematic size in the short time between soybean leaf senescence and harvest, but this will be highly dependent on the growing season.

Weed resistance

Producers often apply the same herbicides at the same rate every year regardless of actual weed infestations. Sometimes the right herbicide was applied, but it failed to control weeds. There are several possible reasons for this failure of herbicides to control weeds. Rarely does any herbicide kill all target weeds. Herbicides fail to control weeds because some weeds may have developed resistance. Each time a herbicide is used, it selectively kills the most susceptible weeds. Some weeds cannot be killed by the herbicide. Weeds that are not killed may pass along to their offspring the trait that allowed them to survive. When one herbicide is used repeatedly in the same field, the surviving weed population becomes more resistant to the herbicide than the original population. Common cocklebur has developed resistance in some fields to imazaquin (Scepter) and imazethapyr (Pursuit), goosegrass to pendimethalin (Prowl) and trifluralin, and johnsongrass to fluazifop (Fusilade), quizalofop (Assure), fenoxaprop, pendimethalin, and trifluralin. Rotating herbicides helps to reduce the development of herbicide-resistant weed populations and it should be used wherever possible.

Herbicide-resistant soybeans

Herbicides control a portion of the weed species present in the crop, thus necessitating the use of other herbicides or management strategies to get totally effective weed control. Traditionally, herbicides have been largely tailored to be used with crops rather than the crops being bred to tolerate the herbicide. During the past decade, advances in biotechnology coupled with plant breeding have led to the development of herbicide-resistant crops. A herbicide-resistant crop is made to tolerate a specific herbicide. Roundup Ready soybeans (resistant to glyphosate) have been recently commercialized (Padgette et al., 1996). Soybeans resistant to glufosinate (Liberty) and sulfonylureas are under development (Dyer et al., 1993; Dekker and Duke, 1995; Duke, 1996).

Glyphosate (Roundup) is a nonselective herbicide that kills most annual and perennial grasses and broadleaf weeds. Recent advances in plant biotechnology have made it possible to insert a gene into soybeans to provide crop tolerance specifically to the herbicide Roundup (Padgette et al., 1996). Roundup Ready soybeans will provide producers the flexibility to control a broad spectrum of weeds in the crop with minimal concern for crop damage. Roundup is considered an environmentally safe herbicide. Roundup inhibits an important enzyme found only in plants, bacteria, and fungi, but not in animals (low mammalian toxicity). Furthermore, after years of use, no weed resistance to Roundup has developed (Dekker and Duke, 1995). Thus, Roundup Ready soybeans offer a tremendous advantage in the management of weed resistance. Use of Roundup in Roundup Ready soybeans offers an excellent option for the control of johnsongrass, common cocklebur, and goosegrass, which are resistant to other herbicides, and for problem weeds such as sicklepod and red rice.

There are other advantages for using Roundup for weed control in Roundup Ready soybean varieties.

1. Roundup can be used to control both broadleaf and grass weeds. Thus, there is no concern about whether a broadleaf or grass herbicide should be applied first.
2. Control of weeds of the same species that differ in size can be attained by simply increasing the rate of Roundup. This means that the time of application is of less concern than for conventional herbicides that have very stringent weed size limitations.
3. Cost of weed control should be less.
4. Tillage and PRE herbicides are commonly used as prophylactic treatments to pre-empt weed problems. Effective control of many weed species in herbicide-resistant crops could encourage adoption of reduced-tillage production practices, thus reducing soil erosion and herbicide movement to ground water (Dyer et al., 1993; Duke, 1996).

Soybean farmers employ a wide range of weed management practices and strategies. It may be difficult to convince them to purchase a herbicide-resistant crop unless there is a clearly demonstrable advantage or benefit, such as a reduced total price for weed management, an improved spectrum of hard-to-control weeds in that crop, or an increase in net returns resulting from use of the herbicide-resistant strategy. Use of any herbicide technology that is paid for with the seed purchase must be viewed from such an economic standpoint. Data in Table 12.6 point out the importance of this. This information is based on using Hutcheson, a public variety, DP 3588, a private variety, and DK 5961, a Roundup Ready variety. The systems shown assume planting in narrow rows in a stale seedbed and applying a preplant foliar herbicide before and/or at planting. They are also based on the assumption that all weed management will be by herbicides that are broadcast. The costs (1997 prices) shown for the herbicides assume no rebates, no cost break for buying in bulk, no special offers, and using full rates of labeled herbicides. The costs are based on those expected for an irrigated system of soybean production where grasses are expected to be prevalent. The inputs and costs shown assume that weed pressure and subsequent management will be intense.

Table 12.6 data show several important points to consider in comparing conventional and Roundup Ready weed management systems. First, seed of Roundup Ready varieties costs more. Second, expected herbicide costs in the conventional systems are higher. The magnitude of this higher cost will depend on the use of PRE + POST vs. POST-only systems, the number of applications made, and the choice of herbicides. Third, less total herbicide is applied in Roundup Ready systems. Fourth, these hypothetical data show that if Roundup Ready varieties yield as much as conventional varieties, net returns should be slightly higher compared with those from all of the conventional weed management systems.

There are some potential risks associated with the use of Roundup Ready soybeans.

1. Roundup Ready soybeans are in the early stages of commercialization. Long-term effects of growing Roundup Ready soybeans year after year on the same land on weed density and population shifts is not clear.
2. Potential for weeds to develop resistance to Roundup due to continuous use of Roundup Ready soybeans cannot be completely ruled out.
3. Can Roundup Ready soybeans completely eliminate the need for PRE herbicide applications? Fields with heavy weed populations may still require PRE herbicides for good soybean germination and stand establishment.

Table 12.6 Comparisons of Costs of Weed Control in Hutcheson (conventional public) and DP 3588 (conventional private) Soybean Varieties with Roundup-Based Weed Control in Roundup Ready Soybeans

Variety	Weed control program	Inputs	Rate	Cost ($/acre)
Hutcheson	PRE + POST	Seed[a]	45 lb/acre	8.62
		PRE: Canopy	10 oz/acre	23.90
		POST: Poast Plus[b]	1.5 pt/acre	9.92
		POST: 2,4-DB + Lorox	15 oz + 1.0 lb/acre	14.15
		Total		56.59
	POST	Seed[a]	45 lb/acre	8.62
		POST: Storm[c]	1.5 pt/acre	14.41
		POST: Poast Plus[b]	1.5 pt/acre	9.92
		POST: 2,4-DB + Lorox	15 oz + 1.0 lb/acre	14.15
		Total		47.10
DP 3588	PRE + POST	Seed[a]	45 lb/acre	14.85
		PRE: Canopy	10 oz/acre	23.90
		POST: Poast Plus[b]	1.5 pt/acre	9.92
		POST: 2,4-DB + Lorox	15 oz + 1.0 lb/acre	14.15
		Total		62.82
	POST	Seed[a]	45 lb/acre	14.85
		POST: Storm[c]	1.5 pt/acre	14.41
		POST: Poast Plus[b]	1.5 pt/acre	9.92
		POST: 2,4-DB + Lorox	15 oz + 1.0 lb/acre	14.15
		Total		53.33
DK 5961 RR	POST	Seed[a]	45 lb/acre	23.36
		POST: Roundup	24 oz/acre	8.25
		POST: Roundup	24 oz/acre	8.25
		Total		39.86

[a] Based on 45 lb seed/acre of Hutcheson at $11.50/60 lb bag, DP 3588 at $16.50/50 lb bag, and DK 5961 RR at $25.95/50 lb bag.

[b] Includes herbicide at $7.54/acre + crop oil at $2.38/acre (1 qt/acre).

[c] Includes herbicide at $13.22/acre + crop oil at $1.19/acre (1 qt/acre).

4. Complete dependence on POST applications of Roundup for timely weed control also involves risk. In years with heavy rainfall (e.g., 1997), wet soil can prevent the use of ground equipment for timely POST applications. Although aerial application under these conditions is an option, a potential to damage nontarget crops by Roundup drift exists.

Currently, several scientists at universities, the U.S. Department of Agriculture, and industries are conducting studies on various aspects of Roundup Ready soybean production systems that may provide the needed information on long-term implications of using Roundup Ready soybeans in years to come.

Disclaimer

Mention of a trademark or proprietary product does not constitute a guarantee or warranty of the product by the U.S. Department of Agriculture or Mississippi State University and does not imply their approval to the exclusion of other products that may also be suitable.

References

Anonymous. 1997. Soybeans 1997 Planting Budgets, Agric. Econ. Rep. 78, Mississippi State University, Mississippi State.

Anonymous. 1998. *Crop Protection Reference*, 14th ed., C and P Press, New York.

Ashton, F. M. and T. J. Monaco. 1991. *Weed Science: Principles and Practices*, 3rd ed., John Wiley and Sons, New York, 466.

Barrentine, W. L., E. E. Hartwig, C. J. Edwards, Jr., and T. C. Kilen. 1982. Tolerance of three soybean (*Glycine max*) cultivars to metribuzin, *Weed Sci.* 30:344–348.

Bruff, S. A. and D. R. Shaw. 1992a. Early season herbicide applications for weed control in stale seedbed soybean (*Glycine max*), *Weed Technol.* 6:36–44.

Bruff, S. A. and D. R. Shaw. 1992b. Tank-mix combinations for weed control in stale seedbed soybean (*Glycine max*), *Weed Technol.* 6:45–51.

Buchanan, G. A. 1992. Trends in weed control methods, in C. G. McWhorter and J. R. Abernathy, Eds., *Weeds of Cotton: Characterization and Control*, Reference Book Series No. 2, The Cotton Foundation, Memphis, TN, 47–72.

Buhler, D. D. 1995. Influence of tillage systems on weed population dynamics and management in corn and soybean in the central U.S.A, *Crop Sci.* 35:1247–1258.

Chandler, J. M., A. S. Hamill, and A. G. Thomas. 1984. *Crop Losses Due to Weeds in Canada and the United States*, Weed Science Society of America, Champaign, IL, 22.

Dekker, J. and S. O. Duke. 1995. Herbicide-resistant field crops, *Adv. Agron.* 54:69–116.

Devine, M. D., S. O. Duke, and C. Fedtke. 1993. *Physiology of Herbicide Action*, PTR Prentice-Hall, Englewood Cliffs, NJ, 441.

Duke, S. O. 1990. Overview of herbicide mechanisms of action, *Environ. Health Perspect.* 87:263–271.

Duke, S. O. 1992. Modes of action of herbicides used in cotton, in C. G. McWhorter and J. R. Abernathy, Eds., *Weeds of Cotton: Characterization and Control*, Reference Book Series No. 2, The Cotton Foundation, Memphis, TN, 403–437.

Duke, S. O. 1996. Herbicide-resistant crops — background and perspectives, in S. O. Duke, Ed., *Herbicide-Resistant Crops: Agricultural, Environmental, Economic, Regulatory, and Technical Aspects*, CRC Press, and Lewis Publishers, Boca Raton, Florida.

Dyer, W. E., F. D. Hess, J. S. Holt, and S. O. Duke. 1993. Potential benefits and risks of herbicide-resistant crops produced by biotechnology, *Hort. Rev.* 15:367–408.

Ginn, L. H., L. G. Heatherly, E. R. Adams, and R. A. Wesley. 1998. A sprayer for application of herbicides under a crop canopy, *J. Prod. Agric.* 11:196–199.

Heatherly, L. G. and C. D. Elmore. 1983. Response of soybeans (*Glycine max*) to planting in untilled, weedy seedbed on clay soil, *Weed Sci.* 31:93–99.

Heatherly, L. G., C. D. Elmore, and R. A. Wesley. 1990. Weed control and soybean response to preplant tillage and planting time, *Soil Tillage Res.* 17:199–210.

Holm, L. G., D. L. Plucknett, J. V. Pancho, and J. P. Herberger. 1977. *The World's Worst Weeds: Distribution and Biology*, University Press of Hawaii, Honolulu, 609.

Hydrick, D. E. and D. R. Shaw. 1995. Non-selective and selective herbicide combinations in stale seedbed soybean (*Glycine max*), *Weed Technol.* 9:158–165.

Jordan, T. N., H. D. Coble, and L. M. Wax. 1987. Weed control, in J. R. Wilcox, Ed., *Soybeans: Improvement, Production, and Uses*, Agronomy Series 16, American Society of Agronomy, Madison, WI, 429–460.

Klingman, T. E. and L. R. Oliver. 1994. Influence of cotton (*Gossypium hirsutum*) and soybean (*Glycine max*) planting date on weed interference, *Weed Sci.* 42:61–65.

Koskinen, W. C. and C. G. McWhorter. 1986. Weed control in conservation tillage, *J. Soil Water Conserv.* 41:365–370.

MAFES and MCES, 1998. Weed Control Guidelines for Mississippi, Mississippi Agricultural and Forestry Experiment Station and Mississippi Cooperative Extension Service, Mississippi State University, Mississippi State, 167.

Miller, D. K. and J. L. Griffin. 1994. Comparison of herbicide programs and cultivation for sicklepod (*Cassia obtusifolia*) control in soybean (*Glycine max*), *Weed Technol.* 8:77–82.

Monaco, T. J., A. R. Bonanno, and J. J. Baron. 1986. Herbicide injury: diagnosis, causes, prevention, and remedial action, in N. D. Camper, Ed., *Research Methods in Weed Science*, 3rd ed., Southern Weed Science Society, Champaign, IL, 399–428.

Monks, D. W. and L. R. Oliver. 1988. Interactions between soybean (*Glycine max*) cultivars and selected weeds, *Weed Sci.* 36:770–774.

Newsom, L. J. and D. R. Shaw. 1992a. Soybean (*Glycine max*) cultivar tolerance to chlorimuron and imazaquin with varying hydroponic solution pH, *Weed Technol.* 6:382–388.

Newsom, L. J. and D. R. Shaw. 1992b. Soybean (*Glycine max*) response to chlorimuron and imazaquin as influenced by soil moisture, *Weed Technol.* 6:389–395.

Newsom, L. J. and D. R. Shaw. 1994. Improving soybean (*Glycine max*) tolerance to AC 263,222 by limiting terminal exposure, *Weed Sci.* 42:608–613.

Newsom, L. J. and D. R. Shaw. 1995. Soybean (*Glycine max*) response to AC 263,222 and chlorimuron as influenced by soil moisture, *Weed Technol.* 9:553–560.

Newsom, L. J. and D. R. Shaw. 1996. Cultivation enhances weed control in soybean (*Glycine max*) with AC 263,222, *Weed Technol.* 10:502–507.

Oliver, L. R. 1979. Influence of soybean (*Glycine max*) planting date on velvetleaf (*Abutilon theophrasti*) competition, *Weed Sci.* 27:183–188.

Osborne, B. T., D. R. Shaw, and R. L. Ratliff. 1995. Response of selected soybean (*Glycine max*) cultivars to dimethenamid and metolachlor in hydroponic conditions, *Weed Technol.* 9:178–181.

Padgette, S. R., D. B. Re, G. F. Barry, D. E. Eichholtz, X. Delannay, R. L. Fuchs, G. M. Kishore, and R. T. Fraley. 1996. New weed control opportunities: development of soybeans with a Roundup Ready™ gene, in S. O. Duke, Ed., *Herbicide-Resistant Crops: Agricultural, Environmental, Economic, Regulatory, and Technical Aspects*, CRC Press and Lewis Publishers, Boca Raton, FL, 53–84.

Pareja, M. R. and D. W. Staniforth. 1985. Soil-seed microsite characteristics in relation to seed germination, *Weed Sci.* 33:190–193.

Poston, D. H., E. C. Murdock, and J. E. Toler. 1992. Cost-efficient weed control in soybean (*Glycine max*) with cultivation and banded herbicide applications, *Weed Technol.* 6:990–995.

Reddy, K. N. and M. A. Locke. 1996. Imazaquin spray retention, foliar washoff and runoff losses under simulated rainfall, *Pest. Sci.* 48:179–187.

Reddy, K. N., M. A. Locke, and C. T. Bryson. 1994. Foliar washoff and runoff losses of lactofen, norflurazon, and fluometuron under simulated rainfall, *J. Agric. Food Chem.* 42:2338–2343.

Reddy, K. N., M. A. Locke, and K. D. Howard. 1995a. Bentazon spray retention, activity, and foliar washoff in weed species, *Weed Technol.* 9:773–778.

Reddy, K. N., R. M. Zablotowicz, and M. A. Locke. 1995b. Chlorimuron adsorption, desorption, and degradation in soils from conventional tillage and no-tillage systems, *J. Environ. Qual.* 24:760–767.

Ross, M. A. and C. A. Lembi. 1985. *Applied Weed Science*, Burgess Publishing Company, Minneapolis, MN, 340.

Shaw, D. R. 1996. Development of stale seedbed weed control programs for southern row crops, *Weed Sci.* 44:413–416.

Shaw, D. R. and G. E. Coats. 1988. Herbicides and cultivation for sicklepod, *Cassia obtusifolia*, control in soybeans, *Glycine max*, *Weed Technol.* 2:187–190.

Shilling, D. G., B. J. Brecke, C. Hiebsch, and G. MacDonald. 1995. Effect of soybean (*Glycine max*) cultivar, tillage, and rye (*Secale cereale*) mulch on sicklepod (*Senna obtusifolia*), *Weed Technol.* 9:339–342.

Stoller, E. W., S. K. Harrison, L. M. Wax, E. E. Regnier, and E. D. Nafziger. 1987. Weed interference in soybeans (*Glycine max*), *Rev. Weed Sci.* 3:155–181.

U.S. Department of Agriculture. 1996. Agricultural Chemical Usage in Field Crops and AREI/Production Inputs, dated 3/25/96, Division of Economic Research Service, National Agricultural Statistics Service, U.S. Department of Agriculture. Washington, D.C. Website: http://www.usda.gov/nass/.

U.S. Department of Agriculture. 1997. *Crop Production 1996 Summary*, National Agricultural Statistics Service, U.S. Department of Agriculture, Washington, D.C., January 1997, pp. A1 to B21.

Walker, R. H. 1994. F-6285 applied postemergence in soybean, *Proc. South. Weed Sci. Soc.* 47:64.

Weed Science Society of America. 1994. *Herbicide Handbook*, 7th ed., Weed Science Society of America, Champaign, IL, 352.

Zhaohu, L., R. H. Walker, and G. R. Wehtje. 1997. Laboratory studies for predicting response of soybean cultivars to sulfentrazone, *Proc. South. Weed Sci. Soc.* 50:177.

chapter thirteen

Seed quality, production, and treatment

Bennie C. Keith and James C. Delouche

Contents

0-8493-2301-0/99/$0.00+$.50
© 1999 by CRC Press LLC

Introduction

Soybean is a seed-propagated crop. The genetic traits and physiological qualities of the seeds establish the production potential of the crop under the varying soil and climatic conditions it is cultivated. Most soybean producers appreciate that the emergence of an adequate stand is the essential first step in a successful production program. Many producers, however, are not fully aware of the attributes of seed quality that affect the percent, rate, and uniformity of emergence and early seedling growth, all important aspects of stand establishment. Moreover, recent and continuing changes in soybean varieties and production practices have greatly increased demands on seed performance. These changes require both seed users and seed producers to be more knowledgeable about the qualities of seeds that affect performance under varying soil and climatic conditions. The most significant of these changes are:

1. Increased seed prices for transgenic varieties e.g., varieties with herbicide resistance, and a corresponding increase in farmer expectations regarding performance;
2. Early planting of early-maturing varieties (Group IVs and Vs in the southern U.S.), which increases the probability of cool and wet seedbed conditions unfavorable for emergence; and
3. Adoption of so-called conservation production systems such as no-till, minimum-till, and stale seedbed in which an array of microenvironmental stresses can interact to adversely affect germination and emergence.

Most soybean growers resort occasionally, some frequently, to "seed saving" supply arrangements. Others would like to share in the added-value of seed production as contract seed producers for breeding and production companies or independent producer–suppliers. In either case the soybean farmer needs a good understanding of production practices and factors that determine or influence the quality of seeds produced. Specifically, they should be well informed about the importance of land selection, preharvest climatic factors, harvest procedures, and postharvest practices for the production of high-quality seeds.

The first part of this chapter presents information important to all soybean growers as users of seeds, while the second part contains information that will be of most interest and help to those who are involved or want to become involved in seed production.

Seed and related legislation

The production, marketing, and use of seeds are regulated or affected by the provisions, requirements, or effects of seed laws and related legislation at the state and federal levels. Both seed producers and seed users need to keep informed about their rights, obligations, and opportunities under the several seed and related laws.

Seed laws and regulations

The importance of the quality of seeds planted for economical crop production programs and the difficulties or inability of farmers to assess seed quality adequately by visual inspection were first recognized in seed legislation enacted in Europe in the late 19th century. At present, both seed *production* and *marketing* are regulated in most European countries. In the U.S., on the other hand, seed legislation enacted at both the state and federal levels during the first half of this century and periodically amended regulates only the *marketing* of seeds. The legislation is based on the general philosophical right of customers (e.g., farmers) to truthful information regarding the characteristics and utility of products (e.g., seeds) offered in the market, and protection against any harmful aspects or properties. These rights are fulfilled in the truth-in-labeling provisions of the seed laws, which require the vendors of seeds to label them with the characteristics and status of significant and designated seed properties and quality attributes, and prohibit or restrict the marketing of seeds with properties that are or can be harmful.

Under our federal–state system, each state has a seed law that regulates intrastate commerce in seeds and the marketing of seeds shipped into the state from other states. The Federal Seed Act regulates interstate commerce in seeds and the importation of seeds from other countries (U.S. Department of Agriculture, Federal Seeds Act, undated). The various state seed laws are quite uniform in terms of general requirements, but differ in specifications such as those relating to restricted and prohibited weed seeds (Mississippi Department of Agriculture and Commerce, Mississippi Pure Seed Law, undated). While the core of all state seed laws is the truth-in-labeling requirements, there are some minimum requirements relating to purity and germination. The state seed laws and the Federal Seed Act require that seeds offered in the market must be truthfully labeled with the following information:

1. Name and address of the individual or company (vendor) labeling the seeds;
2. Kind and variety (kind only in some laws);
3. Lot number or other approved identification;
4. Net weight;
5. Percentages of pure seed, other crop/variety seed, weed seed, and inert material;
6. Percent germination (and hard or dormant seeds as applicable);
7. Month and year of the germination test on which the label information is based;
8. Name and number per pound of restricted noxious weed seeds;
9. Name of any chemical used to treat the seeds with specified cautionary statement, specifically, "Do not use for food, feed, or oil purposes or for processing," and the skull and crossbones poison symbol as appropriate for the seed-treatment chemical utilized.

Seed certification

Seed certification is an officially sanctioned, well-established, and effective system for multiplication and production of seeds. It principally assures (certifies) varietal identity and purity, but also establishes minimum standards for germination, physical purity, freedom from other crop and weed seeds, and, in some cases, freedom from specified seed-borne diseases. The certification system encompasses all aspects of seed production from source of the seeds planted for production (multiplication) through packaging and labeling. The process involves:

1. Authentication of the source and varietal identity of the seeds used for multiplication (pedigree records);
2. Determination of the eligibility of producers and suitability of land and facilities for production of certified seeds in accord with established criteria;
3. Field, equipment, and facility inspections to determine compliance with regulations and standards; and
4. Seed sampling and testing to assess quality and establish that the seeds meet minimum quality standards.

Certification is organized on a state basis. One agency in each participating state is designated, legally sanctioned, and granted monopolistic rights to organize and offer seed-certification services in accordance with purposes and principles set forth in the sanctioning legislation and/or charter. The agency authorized to conduct certification services in Mississippi is the Mississippi Seed Improvement Association. Seed certification in the U.S. is voluntary or optional in contrast to compulsory certification of seeds of the main crops in the European Union and many other countries. Seed producers are free to join the certification agency as member producers or not to join in accordance with their inclination.

Many seed companies rely on in-house systems or programs to assure and control the quality of seeds they produce and market, and use certification services only in special cases or situations, or not at all. States in the U.S. with certification agencies and Canada are organized as the Association of Official Seed Certifying Agencies (AOSCA). Membership in AOSCA was recently opened to agencies in countries outside North America. The main purposes of AOSCA are to establish and promulgate minimum genetic and quality standards for certified seeds, assist state certifying agencies with promotional and educational programs for certified seeds, and monitor the adherence of the individual state associations to accepted procedures and standards (AOSCA, 1994).

Certified seeds are labeled with distinctly colored and inscribed labels or tags: *white* for foundation-class seeds used by seed producers to produce registered- or certified-class seeds; *purple* for registered-class seeds intended primarily for seed production; and *blue* for certified-class seeds marketed for crop production. The certified labels contain *all* the labeling information specified under the state seed law as well as that specifically related to certification, and the seeds are subject to marketplace inspections, verification testing, and other requirements in the same manner as noncertified seeds.

Plant variety protection and patents

The Plant Variety Protection Act (PVPA) was enacted in 1970. Its objective is, "To encourage the development of novel cultivars (i.e., *varieties*) of sexually and asexually reproduced plants and to make them available to the public, providing protection to those who breed, develop, or discover them, and thereby promoting progress in agriculture in the public interest." The PVPA was amended in 1994 to strengthen protection, eliminate loopholes, and conform to the 1991 convention of the Union for the Protection of Varieties (U.S. Department of Agriculture, Plant Variety Protection Act and Regulations, 1994).

The PVPA grants exclusive rights for multiplication, production, and marketing of seeds of a variety to its breeder or developer, an individual, or a company. The rights have to be applied for in the manner prescribed in the regulations under the PVPA and the criteria for protection have to be met fully and unequivocally. There is a Title V option under the PVPA which, if elected, requires that the protected seeds be sold as a class of certified seeds; i.e., certification is compulsory. The period of rights is 18 years for varieties protected under the 1970 law and 20 years for those protected under the 1994 law. The owner of the rights can license or assign the rights to others on mutually agreeable terms. The most important exception to these rights under both the 1970 and 1994 PVPAs is the so-called farmer's privilege or exemption that permits farmers to reproduce (or save) enough seeds of a protected variety acquired in a legal manner for planting the crop the next season on their own farm.

Congressional authorization of the U.S. Patent Office to issue patents for specific genes within a variety or for the variety itself added another type of protection more restrictive than that granted under the PVPA. Farmers, especially those who save or sell seeds, need to understand fully their rights and the restrictions related to seeds of protected and patented varieties. These have been summarized by Spears (1997).

- *Unprotected Variety.* There is no restriction on the saving and selling of seeds of an unprotected variety other than those under the seed law. It should be noted that most unprotected varieties are old varieties.
- *PVPA Protection — 1970, Varieties Protected before April 5, 1995.* A farmer may *save enough seeds for planting* on his or her own farm. Seeds not needed or used for on-farm planting *can be sold without permission* but not more than the amount saved for planting. Farmer-saved seeds of a variety protected under the *Title V* option must be sold as a class of certified seeds.
- *PVPA Protection — 1994, Varieties Protected on or after April, 5, 1995.* A farmer may *save enough seeds for planting* on his or her own farm. Seeds not needed or used for on-farm planting *cannot be sold without permission* from the owner of the variety. If permission is granted, seeds of a variety protected under the *Title V* option *must be sold as a class of certified seeds*.
- *Patented Varieties and Genes.* A farmer *cannot save seeds* of a patented variety or a variety with a patented gene for planting on his own farm or elsewhere and *cannot sell* any of the production as seeds for planting.

In view of the several levels and combinations of protection that can be attached to varieties in the marketplace, farmers should contact the county cooperative extension office, state certification agency, or department of agriculture when there is any doubt about the level and type of protection attached to varieties they have or want.

Seed quality and performance

Seed quality in soybeans encompasses many attributes or aspects most of which apply both to individual seeds and seed populations or seed lots. Seed companies and farmers are interested in seed lot quality since lots or populations are marketed, purchased, and planted. The quality of a seed lot is determined by standardized and special tests that evaluate the level or status of different quality attributes. The results of the different tests are expressed as a percentage with 100% or 0% designating the highest quality level depending on whether the attribute is positive, e.g., pure seed, or negative, e.g., other crop/variety seeds, or as a rate of occurrence, e.g., number-per-pound of restricted weed seeds.

Five quality attributes that influence significant aspects of seed and/or crop performance are discussed in the following sections. These attributes are varietal or genetic purity, crop and physical purity, germination, seed vigor, and seed health. Seed laws at present require labeling only for the first three attributes listed.

Varietal or genetic purity

Varietal purity problems are of two general sorts. First, the identity of a seed variety selected for planting is incorrect and, thus, not the one desired. Second, seeds of the desired variety are contaminated with those of other varieties with very different characteristics. Both problems can severely affect production.

Seed certification and rigorous internal quality assurance procedures have greatly reduced the frequency of varietal identity and purity problems in seeds produced and marketed by commercial (and professional) seed producers. They are, however, all too frequent in seeds acquired from noncommercial sources such as another farmer or a "trucker," and in seeds saved from production for planting the next crop. Most state seed laws require truthful labeling of seed lots offered for sale in the market for both crop *kind* and *variety*, and labeling for variety is commonly practiced by the seed trade even in those states where it is not required. In view of the paramount influence of variety on production, farmers should purchase seeds in commercial market channels from professional and reliable companies and take special precautions to maintain varietal identity when saving seeds from production for planting the next season.

Crop and physical purity

The crop and physical purity of a seed lot refers to how much of the material in a lot is of the desired variety i.e., the "pure seed," and how much is inert material, such as fragments of seeds, other plant parts, soil particles, seeds of other crops, and weed seeds. High crop and physical purity of a seed lot facilitates precision planting, prevents the introduction of objectionable weeds and other crops, and is a good indicator of high values for other quality attributes. For example, a high level of "inert" seed fragments, say 3% or more, is usually indicative of a high incidence of mechanically damaged, intact seeds that are very susceptible to attack by microorganisms in the seedbed that cause seed rot. Soybean seeds are relatively easy to clean, but seeds of some crops and weeds such as corn, cowpea, okra, purple moonflower, and ballonvine shown in Figure 13.1 frequently occur as contaminants because they are similar in size and shape to soybean seed and, thus, difficult to impossible to remove.

Labeling requirements for crop and physical purity under state seed laws and the Federal Seed Act are percent inert material; percent other crop (species and variety) seeds; percent weed seeds; and the name and rate of occurrence (number/lb or oz) of seeds of weedy species classified as noxious. Seeds of some very troublesome and hard to eradicate weed species are legally declared as "prohibited" and seed lots that contain such seeds cannot be marketed. Seeds from reputable dealers are usually well cleaned, but the farmer should always carefully read the information on the seed label to detect above-normal percentages of any of the above-listed and other contaminants.

Germination

The term *germination* has quite different legal and biological meanings. Plant physiologists and biochemists define germination in terms of their interest in growth and differentiation

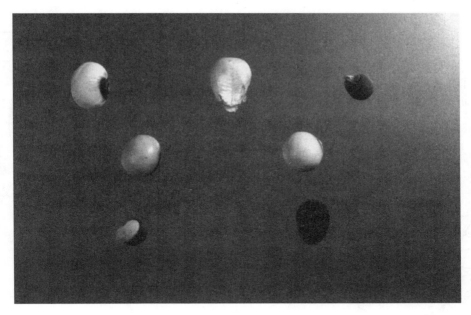

Figure 13.1   Other crop and weed seed contaminants that are very difficult to separate from soybean seed. Center row, seed of two varieties of soybean for comparison; top row from left, cowpea, corn, okra; bottom row from left, balloonvine, purple moonflower.

as "emergence of the embryonic root (radicle) through the seed covering," i.e., resumption of active growth of the embryo. Seed legislation, the Association of Official Seed Analysts (AOSA), commercial practice, and seed technologists, on the other hand, define germination practically and functionally in terms of performance or capability as "the emergence and development from the seed embryo of those essential structures which, for the kinds of seed in question, are indicative of the ability to produce a normal plant under favorable conditions" (AOSA, 1993). A seedling that meets this definition is termed a "normal seedling," which is further legally defined as, "seedlings possessing the essential structures that are indicative of their ability to produce plants under favorable conditions." An individual seed is germinable if it produces a normal seedling or nongerminable if it decays or produces an abnormal seedling. Normal and abnormal seedlings of soybean are illustrated in Figure 13.2. The percentage of normal seedlings produced by a seed lot is termed the *percent germination*. State seed laws and the Federal Seed Act require that seed lots offered for sale in the market be labeled for the percent germination and the month and year (termed the *test date*) in which a germination test was conducted to obtain the information for labeling. Under the Mississippi Pure Seed Law the germination percentage stated on the label is valid for a period not to exceed 9 months exclusive of the month of the test date, whereas under the Federal Seed Act the period of validity in interstate shipments of seeds is only 5 months.

Ability to germinate is the essential seed quality. Seed lots that germinate less than 60% are declared as legally worthless for planting and illegal for sale in Mississippi, regardless of the level of varietal and crop purity. The minimum germination for certified seeds is 80%. Most seed companies strive to produce seeds that germinate substantially higher than the minimum standard. Farmers who have to plant seed lots of lower-than-desired germination due to nonavailability of lots of higher germination, or who elect to plant such lots for "economy" or convenience, generally increase planting rate substantially to compensate for the low germination. This practice usually produces satisfactory

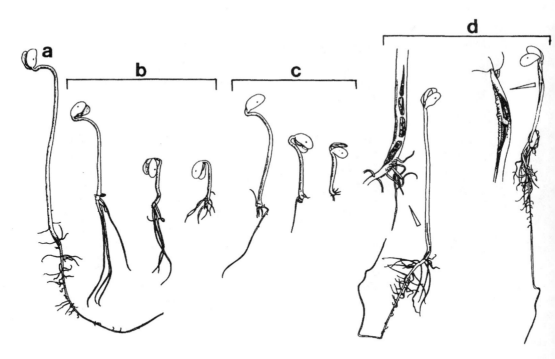

Figure 13.2 Soybean seedlings from germination test. a, normal seedling; b, normal seedlings without primary root but with adequate secondary roots; c, abnormal seedlings with poor root development; d, abnormal seedlings with deep lesions in hypocotyl. (From Association of Official Seed Analysts, *Seedling Evaluation Handbook,* Contrib. 35, Omaha, NE, 1992.)

results when germination is at least 70%, seed vigor is not also very poor (see below), and the seeds are planted in a well-prepared seedbed when the soil temperature is warm but not hot, such as from mid-May to about mid-June. The risks and potential losses or costs, however, are much greater when the seedbed conditions are not favorable for germination and emergence. Cool and wet soil conditions frequently occur in April plantings of early-maturing (e.g., Group IV and V) varieties, while hot and dry conditions are frequently encountered in plantings delayed into late June.

Considering that the establishment of an adequate stand of healthy, rapidly developing seedlings is the first milestone in successful crop production, it pays to plant seeds with a high stand-establishing capability. One indicator of high emergence potential is germination of 90% or higher. Most seed companies, as mentioned above, strive to produce seeds of 90% or higher germination. Some companies, however, take a very conservative approach and practice so-called standard labeling. All seed lots that germinate 80% or higher are labeled as 80% even though some lots germinate even higher than 90%. The consequences of standard germination labeling are elimination of any basis for discrimination among seed lots for germination and wasted seed from excessive seeding rates for high germination lots. Farmers confronted with standard 80% germination labeling should ask the vendor if the seed lot has been inspected and, if it has been, to see a copy of the official inspection report which will show the actual germination percentage for the lot.

Seed lots should be tested for germination at least twice before marketing or planting. Seed companies test the seeds soon after harvest while they are still in the hands of contract producers to determine if they meet the minimum germination standard which is 80%, or more frequently 85% in recent times. The seeds are tested again in March–April before marketing to establish that germination has been maintained and to comply with the

Table 13.1 Comparison of Laboratory
Germination and Field Emergence Percentages
of Samples from 94 Soybean Seed Lots

Field Emergence (%)	Laboratory germination (%)			
	90–94	85–89	80–84	Total
	No. of samples			
90 +	5	0	0	5
80–89	9	12	0	21
70–79	10	14	6	30
60–69	2	8	4	14
50–59	3	6	6	15
40–49	0	3	2	5
40–	0	4	0	4
Total	29	47	18	94

Emergence tests made at Mississippi State, MS in 1969.
Source: Delouche, J. C. *Proc. 3rd Soybean Res. Conf.,* 3, 56–72, 1973. With permission.

5-month limitation on test date under the Federal Seed Act. Farmers saving seeds for planting should also have them tested at least twice: soon after harvest to determine if the seeds are suitable for saving as planting seeds and a month or so before planting to determine whether they are still suitable for planting.

Seed vigor

The percent germination of a seed lot is determined by a germination test made under near optimum laboratory conditions and is, in every sense, a measure of the maximum potential of the seed lot for producing normal seedlings. Crop seeds, however, are planted in the field where conditions for germination and emergence vary widely and rapidly from relatively favorable to very adverse. Thus, field emergence is frequently lower than the laboratory germination percentage. Commonly observed differences between the germinative performance of seeds in the laboratory and field emergence led to development of the concept of *germination energy* in the early years of this century (Heydecker, 1972; Copeland and McDonald, 1995). This concept recognized differences in "vitality" or "energy" among the germinable seeds determined in laboratory tests that had a great influence on emergence and stand establishment. Thus, it was recognized very early in the development and implementation of seed-testing procedures that the standard germination test was deficient as a measure of the potential field performance of seeds. This deficiency is illustrated rather dramatically in the results of field emergence tests for soybean seeds in Table 13.1. Soybean seed lots with essentially the same germination percentage performed quite differently in terms of field emergence under the same conditions. Note especially that field emergence of the 47 lots that germinated between 85 and 89% ranged from less than 40% (four lots) to 80% or higher (12 lots).

By the end of the 1950s, the term *germination energy* was replaced by *seed vigor*, and the cold test for corn seed had been developed and gained wide acceptance for assessing the potential field performance, or vigor, of corn seeds planted in cold, wet seedbeds. Subsequent research and development refined the concept of seed vigor, extended it to seeds of most major crops, and produced an official definition and several vigor tests. The AOSA adopted a definition of seed vigor in 1981 (McDonald, 1980), and published a *Seed Vigor Testing Handbook* in 1983 (AOSA, 1983), which is revised periodically. The AOSA defines seed vigor as

Seed vigor comprises those seed properties which determine the
potential for rapid, uniform emergence, and development of normal
seedlings under a wide range of field conditions.

The *Seed Vigor Testing Handbook* describes in detail "suggested" vigor tests for many
kinds of crop seeds. Only one vigor test, however, has been advanced from "suggested"
to "recommended" status: the *accelerated aging test* for soybean seeds. This test evaluates
the germinative capacity retained by seeds subjected to high temperature (41°C or 106°F)
and humidity (100%) stresses for 72 h. Field trials over many years under a variety of
conditions throughout the soybean-growing area have confirmed the high correlation
between accelerated aging test results and field emergence.

Increasingly, seed vigor is recognized as one of the most important quality attributes
affecting field emergence. Reduced seeding rates for high-priced seed of transgenic varieties, early and late plantings, and conservation tillage practices require good and consistent performance of seeds under a variety of field conditions. The accelerated aging test
is the preferred method for evaluating the vigor of soybean seeds. Farmers should request
available information on seed vigor from the supplier for the seed lots they select for
planting. Those who save seeds for planting should have them tested for vigor as well as
for purity and germination.

Seed health

Seed health has two aspects: the healthiness or freedom from seed-borne diseases of seeds
to be planted, and *maintenance* of seed/seedling health *after planting* (Agrawal and Sinclair,
1987). The healthiness of seeds per se is an important but largely ignored attribute of seed
quality. Maintenance of seed health after planting involves treatment of seeds with chemicals or biologicals. Seed treatment is a routine practice for cotton, corn, rice, wheat, and
sorghum while its use for soybean is limited but increasing. It is discussed in a later section.

Sinclair (1982) lists 30 seed-borne fungi in soybean but only a few, such as *Fusarium*
and the *Phomopsis/Diaporthe* complex, consistently reduce germination and field emergence
(Roy et al., 1994). Several bacteria and viruses are also seed-borne and transmitted. Seeds
infected with the soybean mosaic virus are often mottled and reduced in germinability.

Pod and stem blight caused by the *Phomopsis/Diaporthe* complex is perhaps the most
common and important seed-borne disease affecting seed quality in soybean. Infected
seeds are usually shriveled, misshapen, cracked, and chalky in texture and color. Many
of these are removed from the seed lot during conditioning (cleaning) with the air-screen
cleaner, spiral separator, and gravity table. Some infected seeds, however, are symptomless
and inseparable from the healthy seeds. These constitute the problem (Jeffers and Reichard,
1982). The laboratory germination of *Phomopsis*-infected seeds can be severely reduced
due to the rapid spread of seed/seedling rot under the warm and wet conditions of the
test. Chemical seed treatment can be applied to prevent or reduce the spread of seed rot
before retesting when a good estimate of germination percentage is needed despite the
severity of infection. A rather common observation of seed analysts is that the manifes-
tation of *Phomopsis* infection in germination tests diminishes as time in storage increases.
Thus, a germination test made in November soon after harvest will usually be more severely
affected than a test made in March after 4 to 5 months of storage. This suggests that the
fungus dies during storage under certain conditions. Purple seed stain (*Cercospora kikuchii*)
is a rather common seed-borne disease, but there is no agreement on its effects on germination and emergence. It is not generally recognized as a significant seed health problem.

Seed health testing is very limited in the U.S. compared with Europe, and very few
laboratories offer such tests. There are several likely reasons for the lack of attention to

seed health and seed health testing in the U.S. The wide variety of seed production environments and large quantities of seeds available permit considerable selectivity of locations for production and seed lots for marketing. Efficient and rigorous seed cleaning eliminates a large portion of infected seeds that are low in density, shriveled, and mis-shapen. Additionally, seed treatment, planter-box, and/or in-furrow applications of chemicals are used for most kinds of seed.

Seed health is at present receiving increasing attention in the U.S. This is due to increasingly rigorous phytosanitary requirements in other countries for imported seeds, rising seed prices, decreasing planting rates, and increasing litigation related to seed-borne diseases. There is also a growing awareness among crop producers that maximization of production in terms of yield and/or profits requires elimination of as many limiting factors as feasible, including poor health of planted seeds.

Other seed attributes affecting quality

Several physical properties and conditions of soybean seed affect quality primarily through their influence on the quality attributes discussed above. Mechanical damage, for example, influences both purity and germination. The physical properties of seed size and density and the condition of seed damage are discussed below because they are matters of interest and frequent concern to both seed producers and growers.

Seed size

Seed size in soybean is an inherited trait strongly affected by environment. It is usually included as a varietal characteristic expressed as seed number per pound. Within a variety, however, the mean seed size and range can vary considerably among years and production locations. This variability is a frequent source of concern for many soybean growers who question the quality of seed lots with seeds substantially smaller than "normal." Reduced seed size for a variety is associated with poor or deficient growing conditions such as drought during seed development (Aguiar, 1974; Heatherly, 1993). When the onset of stresses such as a hot, dry period during the later stages of seed fill is rather sudden and severe, many of the seeds are shriveled and flat as well as small. These are easily separated from well-filled seeds during cleaning and conditioning. In many cases, however, chronic, less severe deficiencies produce a population of smaller-size, well-rounded, smooth, good-color seeds which cannot be efficiently size-separated. The latter type of seed-size variability is well illustrated in the seed-size data shown in Table 13.2 for Bragg, Lee 68, and Dare varieties produced in South Carolina, Mississippi, and Texas in 1972 (Aguiar, 1974). The range in seed size and the mean seed size within each variety varied considerably among the three production locations (environments). In most instances, however, only the seeds in the smallest and largest size classes were lower in germinability than those in the intermediate size classes. Based on these responses, the many other quality evaluations, and results reported by others, Aguiar contended there was no basis for discrimination among seed lots of the same variety that are smaller than "normal" in mean seed size.

The relationship between seed size and quality in soybean was comprehensively researched in the mid-1970s by Wetzel (1978). He used the Lee variety and two isolines of Lee, one smaller and one larger in mean seed size, to determine the effects of seed size within and across lines on emergence and stand establishment. The data in Table 13.3 are from these studies. Based on the results of his studies and those from the earlier work of Possamai (1976), Wetzel reached the following conclusions:

1. Within lines (varieties), germination and emergence generally decreased as seed size decreased below the mean size. Seeds in the very largest size class tended to

Table 13.2 Seed Size Distribution (S) by Weight and Percent Germination (G) by Size Class for Nine Lots of Three Soybean Varieties Produced in Different Locations in 1972

Variety	Lot no.	Origin (state)	Factor	Seed diameter (64th-in.), %								Mean diameter 64th-in.
				11	12	13	14	15	16	17	18	
Bragg	2	SC	S	0	5	21	43	24	6	1	0	14.1
			G	—	88bc	90ab	94a	93a	87bc	84c		
	4	TX	S	0	0	0	2	6	32	46	14	16.8
			G				87c	98a	93ab	96a	89bc	
	5	LA	S	0	0	5	21	31	29	12	2	15.3
			G			74b	80ab	86a	86a	84ab	82ab	
Dare	8	SC	S	0	1	8	39	39	12	1	0	14.6
			G		73d	82ab	80bc	87a	86a	75cd		
	10	TX	S	0	1	2	12	37	38	9	1	15.4
					84d	89bcd	87cd	93ab	94ab	96a	96a	
	12	MS	S	0	2	7	24	34	27	5	1	14.9
			G		60bc	71a	72a	70a	68ab	52c	54c	
Lee 68	13	SC	S	1	5	25	52	15	2	0	0	13.8
				71b	92a	93a	95a	94a	93a			
	16	TX	S	0	0	1	3	14	42	33	7	16.3
			G			86ab	90a	84abc	82bc	76c	76c	
	18	MS	S	0	1	3	20	43	29	4	0	15.1
			G		55c	84b	96a	94a	96ab	94a		

Factor means in rows not followed by the same letter differ at the 0.01 level of probability (DMRT).

Source: Aguiar, P. A. A., Ph.D. dissertation, Mississippi State University, Mississippi State, 1974.

also be lower in germination than those in the other size classes down to the mean, apparently as a result of their greater vulnerability to mechanical damage during harvest and handling.

2. There were no consistent differences in germination and emergence among seeds of the same size from different lines even though the smaller seeds of the large-seeded line were the same size as the large seeds of the small-seeded line.

3. The smallest, mostly immature, very low germination seeds were mostly eliminated during combining and seed cleaning. The smaller seeds of lower germination and emergence within a line that remained after normal cleaning comprised such a small percentage of the seeds in most lots that their removal did not significantly improve germination and emergence.

It should be noted that Edwards and Hartwig (1971) found that soybean lines with small mean seed size generally emerged better than those from lines with large mean seed size, especially in clay soils. The apparent reason for this response is that soybean seed are epigeal; i.e., the cotyledons or seed leaves emerge, and less force is required to "pull" the cotyledons of small seeds with small cross-sectional area through dense, often crusted, soil than the large cotyledons of large seeds. Seedlings of epigeal seeds often fail because the cotyledons remain stuck in the soil crust.

Seed density

Aspirators and gravity tables have been effectively used for many years to upgrade the quality of corn, cotton, and wheat seed lots by removing the low-density seeds (i.e., low

Table 13.3   Seed Size Distribution by Weight and Percent
Germination for Each Size Class for Three Seed
Size Isolines of Lee Soybean

Seed diameter (64th-inch)	Seed Size Isoline, %					
	Large line		Medium line		Small line	
	Seeds	Germ.	Seeds	Germ.	Seeds	Germ.
21	1	92a	—		—	
20	2	95a	—		—	
19	17	93a	—		—	
18	27	90a	1	96a	—	
17	23	91a	8	97a	1	95a
16	16	88ab	28	96a	4	97a
15	8	80b	33	93a	18	95a
14	4	54c	20	90ab	87	97a
13	1	40cd	6	86b	26	96a
12	1	36de	2	63c	10	93a
11	1	7e	1	53d	2	90
10	—		1	9e	1	61b
9	—		—		1	—
8	—		—		—	
Mean size	17.2		15.1		13.6	
Unsized		90		92		92

Means in columns not followed by the same letter differ at the 0.01 level of probability (DMRT).

Source: Wetzel, C., Ph.D. dissertation, Mississippi State University, Mississippi State, 1978.

weight per unit volume) from the lot. Low-density seeds are usually immature, diseased, badly weathered, rotten, and low in germination and vigor. Soybean seed conditioning, however, seldom included density separations until the early 1980s. At present, most seed companies routinely density-grade the soybean seed lots they handle. Density grading with removal of the lightest 5 to 10% of the seeds, usually with visible evidence of weathering or mechanical abuse, can increase the germination and vigor of the lot a few points and greatly improves its appearance.

Selected physical properties and germination of seeds from gravity table separations of high-, medium-, and low-germination seed lots in Assman's (1983) studies of seed density and quality relationships in soybean are shown in Table 13.4. The physical appearance was better and germination higher for seeds in the two highest-density fractions for each of the three lots compared with those in the two lowest-density fractions. These results, however, should not and do not suggest that low-germination, low-vigor seed lots can be transformed into good-quality seed lots through rigorous density grading. Experience has shown that while very substantial improvements in germination and appearance can be effected in badly weathered seed lots by removal of 15 to 20% of the lowest-density seeds, those retained usually have marginal or lower-than-acceptable vigor.

Mechanical damage

Seed lots with a considerable percentage of visibly damaged seeds *after cleaning*, say, more than 5%, should not be used for planting when there is an alternative supply of sound seeds of good germination. Those with 3 to 5% visibly damaged seeds should be treated with an appropriate fungicide or planted with a hopper box treatment.

Table 13.4 Effects of Gravity Table Separation on the Physical Properties, Quality, and Performance of High-, Medium-, and Low-Quality Soybean Seeds (average of four lots each quality class)

Attribute	Unit	Seed quality class	Original lot	Gravity table separate Heaviest 1	2	3	Lightest 4
Seeds	%	High	100.0	28	27	29	16
		Medium	100.0	30	28	27	15
		Low	100.0	31	27	28	14
Volume weight	lb/bu	High	58.0	58	58	58	57.6
		Medium	56.2	57.0	56.8	56.4	53.9
		Low	56.3	57.0	56.6	56.1	53.9
Seed weight	g/100 seeds	High	14.0	14.9	14.3	13.9	13.2
		Medium	13.4	14.1	13.6	13.1	12.5
		Low	13.6	14.6	13.8	13.1	12.4
Weather damage	%	High	12 b	8 a	9 a	12 b	19 c
		Medium	19 b	12 a	14 a	20 b	31 c
		Low	21 c	13 a	15 b	22 b	40 d
Mechanical damage	%	High	9 b	5 a	8 b	11 c	17 d
		Medium	15 c	8 a	12 b	19 d	35 f
		Low	17 b	9 a	15 b	26 c	41 d
Germination	%	High	94 b	97 a	97 a	94 b	90 c
		Medium	81 c	88 a	85 b	78 d	68 f
		Low	72 b	82 a	82 a	71 b	59 c
Field emergence	%	High	86 b	90 a	89 a	86 b	77 c
		Medium	69 c	78 a	73 b	62 d	42 e
		Low	53 c	68 a	65 b	45 d	20 e

Attribute means in rows not followed by the same letter differ significantly at the 5% level of probability (SNK).
Source: Assman, E. J., Ph.D. dissertation, Mississippi State University, Mississippi State, 1983.

Seed testing and quality evaluation

Standard and special analyses

There are frequent references to seed testing and admonitions to have seeds tested in the foregoing discussions of seed quality and performance. Tests and/or procedures are available for establishing the status and level of all of the attributes and aspects of seed quality. Most of these are rather technical and need to be carried out in properly equipped laboratories by experienced analysts. Some important information, however, can be obtained by simple tests or procedures described later that can be carried out by a farmer or local seed company and grain elevator personnel.

Seed-testing services are available in most states from public laboratories of the department of agriculture, agricultural experiment station, or cooperative extension service, and from commercial laboratories. The Mississippi State Seed Testing and Regulatory Laboratory located on the campus of Mississippi State University is a unit of the Mississippi Department of Agriculture and Commerce, Jackson, MS. The laboratory conducts official tests of seed samples taken by inspectors from seed lots offered for sale in the marketplace in enforcement of provisions of the state seed law, and from seed lots eligible for certification to determine if they meet minimum seed standards. It also offers testing services on a fee basis to seed companies, farmers, and other users of seeds such as contractors. Under the terms of the seed law, bonafide farmers can get a free "standard analysis" for *one* sample of each crop kind per year. A standard analysis consists of the purity test, noxious weed seed examination, and germination test. Other seed-quality evaluations and determinations available for soybeans from the state and commercial seed testing laboratories on a fee basis are seed moisture content, vigor (accelerated aging), rapid viability (tetrazolium), and special variety verifications including herbicide resistance. The county cooperative extension office has information on sampling seeds and the addresses of the state and commercial laboratories that offer testing services.

Some rather simple procedures are available to the farmer–seed grower and contract production supervisor for obtaining important information related to seed quality. Several of these are discussed in the sections that follow.

Visual examination

Visual examination of seeds hand-threshed in the field before harvest, during harvesting, and in bulk storage can reveal important seed defects and quality problems. These include seed damage; shriveled and misshapen seeds caused by drought; discoloration of seeds associated with weathering and field deterioration; chalky, grainy seeds symptomatic of pod and stem blight; different hilum colors indicative of variety contamination; and weed seeds that are difficult to separate. Very importantly, the visual examination can also reveal the absence of these indicators of problems. Recently harvested seeds that are well filled and rounded, nondamaged, and bright are nearly always high in germination and vigor.

Seed moisture content

Knowledge of the moisture content of seeds is important for scheduling harvest, adjusting the combine, and proper aeration and drying. While some experienced farmers and seed production supervisors can estimate seed moisture quite closely by the "bite" test, use of a moisture meter is recommended. Most farmers do not have an electronic seed moisture meter, but can access one at a nearby seed-processing plant or grain elevator. The test is rapid, simple, and accurate if a good sample is taken.

Monitoring mechanical damage

Many seed companies strongly advocate strict monitoring of seed damage during harvest and handling to their contract seed growers. The *bleach test method* for monitoring and determining seed damage in soybean recommended by the Asgrow Seed Co. is simple and effective.

Bleach test method

The test can be used in the field or in the laboratory as a quick method to determine the number of fractured seed in a batch of soybeans.

Procedure:

1. Take a sample from the combine hopper or grain wagon and count out several (at least two) batches of 50 or 100 seeds into a plastic container;
2. Cover the seeds with bleach water made by mixing 1 oz of laundry bleach with 1 qt of water for about 30 s;
3. Pour off the bleach water, count the visibly swollen seeds, and convert to a percent swollen or damaged seeds.

Action: The goal is 98% or better intact seeds, thus, if 3 or more out of 100 seeds are swollen, the combine needs to be adjusted.

Rapid assessment of germination and vigor

A standard germination test takes a minimum of 7 days. Frequently, however, decisions have to be made within a few days of harvest on the disposition and handling of seed lots that require accurate information on their germinability. The visual examination discussed above provides good information, especially to an experienced production supervisor, but a rapid and powerful method for assessment of germination potential is available and widely used. The tetrazolium test can be completed in one day and provides accurate information not only on germination, but also vigor and the likely causes of seed quality problems (Delouche et al., 1962; AOSA, 1970). The test is based on a biochemical reaction that reduces the chemical tetrazolium from a soluble, colorless form to an insoluble, red pigment in living tissue. While the test procedures are simple and the materials inexpensive, training and experience are required for interpretation of the test results. The test is mentioned here not to suggest that farmers and seed production supervisors should master the test, but rather to remind both that this rapid and powerful test is available in most seed-testing laboratories including the quality control laboratories of most seed companies.

Seed production

Farmers obtain seeds for planting their soybean crops from the market, on-farm produced and saved stocks, so-called bin seeds, and occasionally from a neighbor or trucker. Seeds in the market are produced by seed companies, usually as proprietary or licensed varieties, and are subject to labeling requirements and minimum quality standards under state and federal seed laws. Bin seeds, on the other hand, are sometimes produced and saved by farmers with special care and attention that includes testing, but all too often they are simply taken from the grain bin. While the proportion of the soybean crop planted with bin seeds has decreased sharply in the past decade, it is still above 30% in Mississippi. The portion of the soybean crop planted with bin seed will further decline with increased use of patented transgenic varieties for which seeds cannot be saved for planting.

Seed companies have the expertise and experience to produce good- to high-quality seeds and to access additional and very specialized expertise when needed. Farmers who want to save seeds for planting or to become contract seed producers for seed companies must also develop a good understanding of the factors that influence seed quality and the "fine-tuning" and precautions in production practices required to produce good-quality seeds and minimize problems. The remainder of this section, therefore, presents background information on seed development and maturation and descriptions of certification and "in-house" systems for assurance of quality during seed multiplication and production that will be helpful to independent and contract seed producers and other farmers who save seeds for planting.

Seed development and maturation

The structural and physiological delicacy of the soybean seed relative to those of other major crops such as corn, wheat, and sorghum contributes in a major way to many of the germination and vigor problems encountered by both professional and farmer seed producers (Delouche, 1974). Some knowledge of seed development and morphology, therefore, is essential to appreciate and understand the complexity of factors involved in loss of germinability and vigor and possible means of minimizing these losses. The structure and components of the soybean seed and their development into a seedling are illustrated in Figures 13.3 and 13.4.

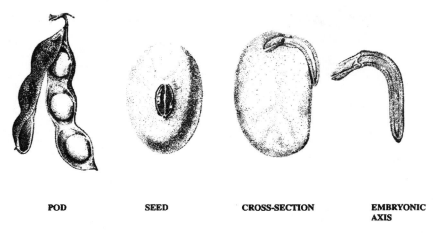

POD **SEED** **CROSS-SECTION** **EMBRYONIC AXIS**

Figure 13.3 Structure and components of soybean pod and seed. (Drawings courtesy of Seed Science Center, University Extension, Iowa State University.)

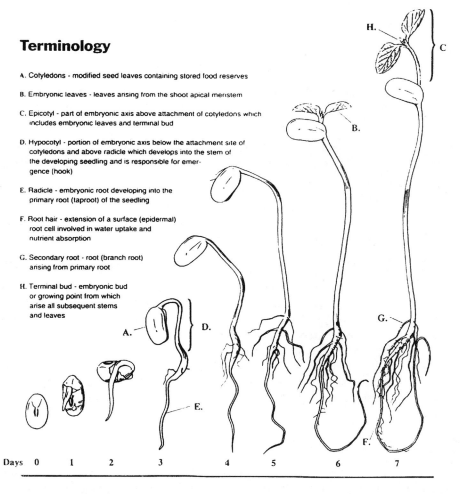

Terminology

A. Cotyledons - modified seed leaves containing stored food reserves

B. Embryonic leaves - leaves arising from the shoot apical meristem

C. Epicotyl - part of embryonic axis above attachment of cotyledons which includes embryonic leaves and terminal bud

D. Hypocotyl - portion of embryonic axis below the attachment site of cotyledons and above radicle which develops into the stem of the developing seedling and is responsible for emergence (hook)

E. Radicle - embryonic root developing into the primary root (taproot) of the seedling

F. Root hair - extension of a surface (epidermal) root cell involved in water uptake and nutrient absorption

G. Secondary root - root (branch root) arising from primary root

H. Terminal bud - embryonic bud or growing point from which arise all subsequent stems and leaves

Days 0   1   2   3   4   5   6   7

Figure 13.4 Soybean germination and seedling development during 7-day sand test. (Drawings courtesy of Seed Science Center, University Extension, Iowa State University.)

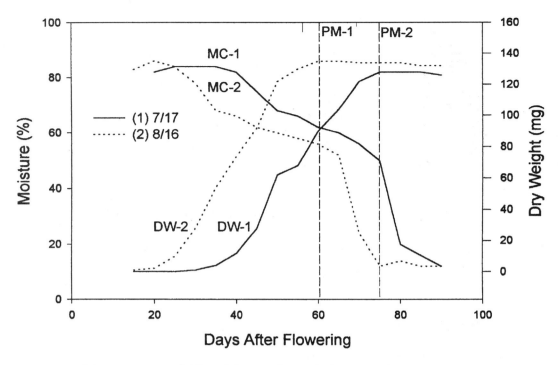

Figure 13.5 Moisture content (MC) and dry weight (DW) changes in developing Lee soybean seed. PM is physiological maturity; 1 and 2 designations refer to seed set from July 17 and August 16 flowers, respectively. (After Andrews, 1966.)

Development of soybean seed takes place in the ovary of the flower which matures into the pod. The egg cell and polar nuclei in an ovule are fertilized (double fertilization) with two sperm nuclei from the pollen. The two cotyledons and growing points of the embryo are fully differentiated within 2 weeks of fertilization. Growth and dry matter accumulation (seed fill) continue rather slowly through 20 to 30 days at which time moisture content of the developing seed decreases to less than 80%. Thereafter, dry matter accumulation progresses rapidly until maximum dry weight, an indicator of physiological maturity, is attained at 50 to 70 days after flowering depending on the maturity group of the variety and the temperature. A physiologically mature seed is well filled, rounded, yellow or cream in color (rather than greenish), no longer functionally connected to the pod, and has a moisture content of 50 to 55%. Physiological maturity is followed by a period of rapid dry-down to 12 to 15% moisture content, the harvest maturity stage, and even lower if harvest is delayed and drying conditions are good (TeKrony et al., 1987). The time course of dry weight and moisture changes in developing seeds of the determinate Lee variety, now obsolete, are shown in Figure 13.5 (Andrews, 1966). Note the more rapid development of seed set from the later flowers. This attribute contributes to a relatively high degree of uniformity in pod maturity within and among plants and facilitates harvest.

Production practices and management

Successful soybean seed production is based on good crop management. Selection of fields well suited for seed production is a first step in good management. The seed field should be productive and uniform, accessible to irrigation, free from weeds that are especially troublesome in soybeans, and have an acceptable cropping history. Cultural practices, including fertilization, irrigation, and weed and pest control, are basically the same for seed and grain production. Supplemental irrigation is especially important to prevent late-season drought stress that can reduce yield, mean seed size, appearance, and germination (Heatherly, 1993). There are, however, some additional critically important requirements for production of high-quality seeds, mostly but not entirely in the realm of management (Delouche, 1973). First, the varietal identity and purity of the seeds must be maintained. Second, contamination with inseparable weed and other crop seeds must be avoided. Third, losses in germinability and vigor of the seeds must be minimized.

Maintaining varietal purity

Seed source

The most important factor in maintenance of varietal purity is source and authenticity of the seeds planted. The classes of seeds intended for production of certified seeds i.e., foundation, registered, and certified in some cases, are produced in a limited-generation, multiplication system designed to maintain and assure varietal purity (AOSCA, 1994). For commercial production of noncertified seeds, seed classes essentially equivalent to those in the certification system are produced and authenticated within the company in-house quality assurance system. The source and authenticity of seeds used by farmer contract–seed producers are controlled and assured by the company contracting the production. Farmers producing seeds for their own use, however, have to make sure that the seeds planted are the desired variety and are high in purity. Certified class seeds or noncertified seeds from a reputable company are usually satisfactory for on-farm seed production. If all the steps important in on-farm seed production are taken, there is no reason seeds from a good source cannot be multiplied for several years without substantial reduction in varietal purity.

Cropping history and isolation

To prevent contamination of seed fields from volunteer plants, soybeans intended for seed use should not be planted in fields in which the previous crop was another variety of soybeans. When an intervening "winter" crop such as wheat is produced, however, the risk of contamination from volunteers is negligible. Although soybean is a self-pollinated species, the seed field should be separated from any other variety by a distance adequate to prevent mechanical mixture during harvest and postharvest handling.

Field and equipment/facility inspections

Both certified and contract, noncertified seed production are subject to field and equipment/facility inspections to determine compliance with established procedures, regulations, and standards. These inspections are performed by certification inspectors in the case of certified seeds, and by production and quality control supervisors or monitors from contracting seed companies for both certified and noncertified seed production. New seed producers usually undergo a precontract inspection to determine if their fields and equipment/facilities meet criteria for producers. Several field inspections or visits may be made during the production cycle, but the final inspection is always made just before harvest when varietal characteristics, weeds, and potential quality problems can best be observed.

The field inspections establish varietal authenticity and determine compliance with standards for other varieties, specified weeds, and isolation distance from another variety. Contract production supervisors also provide advice on plant protection measures and the scheduling of harvest. The seed producer is expected to clean thoroughly combine harvesters, grain wagons, handling devices, and bulk storage bins before harvest to prevent varietal contamination. The thoroughness of the cleanup is frequently verified by the certification inspector or seed company production supervisor.

Farmers who intend to save seeds for planting the next season must also inspect the designated seed field several times to make sure the variety is authentic, that off-type and other variety plants are within acceptable limits and/or "rogued," that weed control is satisfactory, and that the crop is healthy. The last walk-through should occur just before harvest. Equipment and facilities should be thoroughly cleaned before harvest, and the identity of soybeans intended for seed use should be maintained rigorously in well-marked separate storage before and after cleaning.

Minimizing contamination with other seeds

Careful selection of the production field has been emphasized. The field should be free of troublesome weeds and soybean and other crop volunteers. These features combined with careful harvest and handling should prevent any serious contamination with seeds of other varieties, other crops, and hard-to-separate weeds. When these measures are neglected or inadequately applied and the walk-through inspections reveal serious contamination problems, it is usually best, technically and economically, to abandon the field for seed use rather than attempt to salvage it.

Maintaining germination and vigor

The physiological qualities of seeds i.e., germinability, vigor, and storability, are mainly affected by the preharvest field environment, mechanical abuse during harvesting and handling, and the conditions of bulk and packaged seed storage.

Preharvest field environment

The germination and vigor of soybean seeds are strongly influenced and frequently established by climatic conditions during the late seed fill and postmaturation, preharvest period. Disruption or premature termination of seed fill by desiccation from drought or early freeze can result in a substantial proportion of smallish, shriveled, immature, and misshapen seeds with low germination and vigor. Drought effects, of course, can and should be prevented by timely irrigation.

Frequent and/or prolonged precipitation with associated fogs, heavy dews, and warm temperatures during the postmaturation, preharvest period are the most common and important causes of poor physiological quality of soybean seeds produced in the southern U.S. and other warm, humid areas (Delouche, 1974). The processes involved are termed "weathering" or field deterioration. After seed moisture has decreased to below 25%, adverse climatic conditions of the sorts described result in alternate wetting and drying of the seeds in the pod and reductions in germinability and vigor that can be very severe. Since any substantial rainfall after the seeds reach harvest maturity will delay harvest, the exposure to weathering and severity of deterioration increases as the frequency of rainfall increases. While weathering conditions are most common from about mid-September through the fall harvest season, they also occur and are *most damaging* during the late August and early September harvest period for early-maturing varieties when the temperature is very warm.

The effects of weathering associated with delayed harvest on germination of the Hill (Group IV) and Bragg (Group VII) varieties are shown in Table 13.5. Although these data were gathered more than 25 years ago, they still illustrate very well the problem of field

Table 13.5 Effect of Weathering on Moisture Content (M.C.) and Germination of Seed of the Hill and Bragg Soybean Varieties

Date of harvest[a]	Hill, %		Bragg, %	
	M.C.	Germination	M.C.	Germination
9/15	26	96		
9/22	13	97		
9/29[b]	17	90		
10/6[b]	20	78		
10/13	11	76	26	98
10/20[b]	19	71	18	98
10/27	12	53	13	93
11/3[b]	14	37	14	92
11/10			14	92
11/17[b]			20	89
11/24[b]			13	86
12/1			15	87
12/8			11	84
12/15[b]			14	84

[a] Seed hand harvested and threshed, then cleaned with hand screens and aspriator before germination test.

[b] One or more rains during preceding week.

Source: After Potts, H. C., *Proc. 8th Soybean Seed Res. Conf.,* 8, 33–42, 1978.

deterioration. After reaching harvest maturity with a very high germination of 97% about September 20, moisture content of seeds of Hill left in the field fluctuated from 11 to 20% in association with rainfall through November 3. Germination dropped below 80% by October 6 and then to 37% by November 3. Seeds of Bragg reached harvest maturity about 1 month later on October 20 and germination was 98%. The moisture content of Bragg seeds also fluctuated from 11 to 20% when harvest was delayed until December 5. Germination, however, decreased to only 84% during the period of more than 40 days, even though the frequency and quantity of rain was about the same as it was for Hill. The quite different responses of seeds of early-maturing Hill and late-maturing Bragg show clearly the pronounced effect of temperature during weathering on the rate and severity of seed deterioration. The mean field temperature during the main weathering period for Hill, approximately September 20 to October 13, was about 8° warmer than during the weathering period for Bragg. This "temperature effect" is the main reason it is especially difficult to produce good-quality seeds of early-maturing varieties in the Gulf Coast states from Texas to Florida. There is abundant anecdotal evidence that a single rainy period of 2 to 3 days in the heat of late August or early September when the seeds have dried to below 25% moisture content can reduce germination to well below the generally accepted marketing minimum of 80%. Seed companies recognize the risks in producing seeds of early-maturing varieties in the lower portion of the southern U.S., and generally arrange for production in the most northern regions of the zone of adaptation.

While the consequences of weathering are most severe for early-maturing varieties, field deterioration is a chronic problem for all maturity groups in the southern U.S. and other warm, humid areas. Since very little can be done about the weather, research and development efforts to reduce field deterioration have focused on genetic resistance to weathering, application of fungicides to control field fungi that accelerate deterioration processes, and management practices that minimize the time seeds are exposed to weathering (Delouche et al., 1995).

Table 13.6 Responses of the Dare Variety and D-1 Hardseed Line
of Soybeans to Postmaturation Field Weathering in 1973 and 1974

Harvest date	Rain	Dare, %		D-1, %			
		SMC[a]	Germ.	SMC	Germ.	HS	TVS
			1973				
Oct. 8		21.0	86	16.0	90	8	98
Oct. 14		12.0	89	12.0	60	35	95
Oct. 21	*[b]	15.0	90	11.0	61	31	92
Oct. 29	*	16.0	93	11.0	57	40	97
Nov. 5	*	32.0	72	18.0	39	47	88
Nov. 12	*	17.0	51	10.0	30	62	92
			1974				
Oct. 4		15.5	94	10.0	82	15	97
Oct. 12		9.3	92	9.0	50	46	96
Oct. 19	*	9.7	90	8.2	45	46	91
Oct. 26		9.0	88	7.4	35	59	94
Nov. 2		13.1	86	9.1	44	49	93
Nov. 9		10.4	72	8.0	44	48	92
Nov. 16	*	15.8	63	9.2	42	50	92
Nov. 23	*	12.0	50	8.5	46	42	90
Nov. 30	*	14.0	40	10.8	45	43	88
Dec. 7		14.4	32	12.0	46	38	84

[a] SMC = seed moisture content; Germ. = germination; HS = hard seeds;
   TVS = total viable seeds (Germ. + HS).

[b] Indicates rain of more than 5 mm during the preceding week.

Source: After Potts, H. C. et al., *Crop Sci.*, 18, 221–224, 1978.

The genetic resistance-to-weathering approach has mainly involved use of the hard seed or water-impermeable seed coat (ISC) character to reduce "rewetting" and increases in the moisture content of seeds after they first dry to below about 14% moisture. The excellent protection against field deterioration provided by the ISC character was first demonstrated by Potts and co-workers in 1978 (Table 13.6). This demonstration and subsequent work by Mississippi State University and U.S.DA workers (Potts, 1978, 1985; Hartwig and Potts, 1987; Keith, 1991; Heatherly and Kenty, 1995) attracted the attention and efforts of many other researchers in warm, humid, and rainy locations such as Florida (Calero et al., 1981), Puerto Rico (Minor and Paschal, 1982), West Africa (Kueneman, 1983), and Indonesia (Nugraha et al., 1991). Many hard seed lines with good resistance to field deterioration have been developed, but most are experimental and none is yet in full commercial production (Keith, 1991). The main problem connected with the use of the hard seed character to develop weathering-resistant varieties is delayed and nonuniform germination and emergence. Seed populations from hard seed lines range widely in seed coat permeability to water from very permeable, which permits rapid germination and emergence, to nearly impermeable which can delay germination and emergence for several days to several weeks. Effective and economical methods for eliminating or greatly reducing water impermeability of the seeds before planting (e.g., mechanical scarification, heat treatments) must be developed before the weathering resistance in hard seed lines can be commercially exploited.

Several management practices are used by seed companies to reduce field deterioration of soybean seeds they produce. These include strong emphasis on harvesting *as soon as seed moisture content drops below 15%*, limiting seed production contracts to farmers with a favorable acreage/combine harvester ratio to permit timely and rapid harvest, and producing seeds in the more northern portions of the zone of adaptability to the degree that is feasible (Delouche, 1974). Harvest-aid herbicides and desiccants have also been used to "speed up" dry-down and permit earlier harvest but with very limited effectiveness (Ratnayake and Shaw, 1991). Preharvest desiccants do not speed up the drying of mature seeds.

Control of pod and stem blight with properly timed foliar sprays of benomyl, a systemic fungicide, can reduce the loss in germinability often caused by this late-season disease (Ellis and Sinclair, 1976; McGee and Brandt, 1979; Miller and Roy, 1982; Jeffers and Reichard, 1982). Various rating systems for timing of foliar sprays have been developed. Seed companies that advocate foliar sprays usually rely on one of the systems to advise their contract seed producers on whether or not a foliar treatment is needed and when to spray. Extension plant pathologists are the best source of information on spray schedules for farmers. It should be noted that the beneficial effects of late-season foliar applications of fungicides are limited to improvements in seed quality, which are usually modest and sometimes negligible. Yield is not increased.

Harvesting and handling

Harvesting is probably the most critical phase in the overall soybean seed production operation (Delouche, 1974; TeKrony et al., 1987). Weathering or field deterioration of seeds is likely to be associated with any delay in harvest beyond the harvest maturity stage when seed moisture content first drops below 15%. Delays caused by inclement weather are the worst since there is rain and considerable wetting and drying of the seeds. Delays resulting from insufficient equipment for timely harvest, however, can be equally bad since any delay in harvest increases the likelihood of weathering and field deterioration.

The combine is one of the most frequent sources of varietal contamination. It must be inspected, thoroughly cleaned, and reinspected before harvest. Many producers have found that high-pressure water is very effective in cleaning and preparing combines for inspection and harvest.

Mechanical abuse during the harvesting operation is the most important cause of physical injury to soybean seed. Very severe injuries produce fragments or "splits," which can be separated from the good seeds during cleaning. However, this results in excessive losses in both salable product and seed production premium or price which is usually based, at least in part, on the percentage of intact or whole seeds. Seeds with less-severe injuries such as a fractured cotyledon or chipped seed coat may be intact but incapable of germination, succumb to rot in the seedbed, or produce abnormal, weak seedlings susceptible to seedling disease. The incidence and severity of mechanical damage are largely determined by the moisture content of the seeds at the time of harvest and the setting and operation of the combine harvester.

The susceptibility of soybean seeds to mechanical abuse increases sharply as moisture content decreases below about 13% (Table 13.7). With heavy dew and early morning fog and good midday drying conditions that are common in September and October, seed moisture content can vary a critical 2 to 3% from early morning to midafternoon. This may represent a moisture content change from 14 to 11% during a single harvest day. Thus, combine settings made in the early morning for good threshing with minimum seed loss and damage can cause excessive seed loss and damage in midafternoon when seed moisture content has dropped to its low level. Harvest equipment and procedures are essentially the same for grain and seeds. Seed harvest, however, requires closer management and greater caution to minimize mechanical damage, maintain good germination, and prevent contamination. Some important "rules of thumb" or guidelines for harvesting soybean seeds are summarized below.

Table 13.7A Effect of Height of Drop and
Seed Moisture Content on Percentage
Germination in Soybeans

Seed moisture content	Height of drop (ft.), % germination			
	0	5	10	20
6	98	88	80	67
8	98	88	78	70
10	98	90	82	73
12	98	97	94	87
14	98	97	97	97
16	98	98	96	96
18	98	96	97	96

Table 13.7B Effect of Successive 10-ft Drops
on Percentage Germination of Soybeans
at Various Seed Moisture Contents

Seed moisture content	No. of 10 ft drops, % germination			
	0	1	2	4
6	98	80	73	62
8	98	78	65	53
10	98	82	73	62
12	98	94	88	81
14	98	97	96	98
16	98	98	98	96
18	98	97	97	97

1. Weed-free, uniform stands facilitate adjustments of the combine to minimize damage.
2. Thoroughly clean combine, grain wagon, and conveyors before harvest to remove residual seeds and other contaminants.
3. Commence harvesting when seed moisture content first drops below about 15% and complete harvest as rapidly as possible.
4. Adjust clearance and cylinder speed just to the point where complete threshing is achieved and operate the combine at a uniform ground speed. One seed company advises its contract producers to "open" and "slow" adjustments fully and then close and speed them just to the point of satisfactory threshing.
5. Visually inspect threshed seeds periodically to determine the incidence and severity of damage, and readjust the combine during the day as drying conditions improve. A simple and rapid procedure for determining the incidence of damage is described in an earlier section.
6. Combine harvesters with rotary threshing mechanism(s) are considered to be less damaging to soybean seeds than those with the conventional rasp-bar or spike-tooth cylinder (TeKrony et al., 1987).

Soybean seeds must also be handled carefully after harvest to prevent damage, especially when the seed moisture content is 12% or lower. Loading bulk bins with inclined augers or screw conveyors should be avoided to the extent possible. Belt-bucket elevators for lifting and belt or drag-flight conveyors for horizontal conveying are preferred. Cushion boxes or "slow-downs" should be used in long spout runs to prevent seeds from

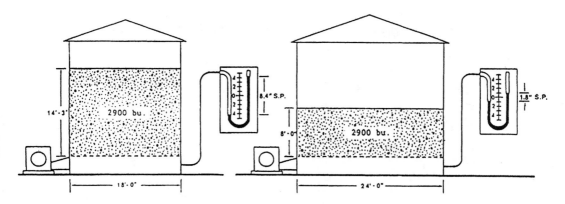

Figure 13.6 Typical round, metal aeration/drying and bulk storage bin used for soybean seed and grain. Note the effect of seed depth (14.25 vs. 8.0 ft) on resistance to airflow at 4 cfm/bu rate as measured in inches of static pressure (S.P.), water gauge. Both bins contain 2900 bu of soybean.

attaining excessive velocity. Discharge of seeds from spouts should not be against the bin wall but into a so-called bean ladder, spiral let-down, or other cushioning device to lower the seeds gently into the bin. When an auger has to be used, it should be in good mechanical condition, uniformly fed, and operated at near full capacity.

Postharvest operations and practices

Aeration, drying, and bulk storage

Soybean grain harvest traditionally begins at about 13 to 14% seed moisture content. Harvesting at moisture contents lower than 12% increases mechanical damage and may result in increased field losses due to shattering. Harvest of soybeans for seed production should commence at seed moisture contents a few points higher i.e., 14 to 15%, to minimize weathering and/or avoid low-moisture-content seeds that are very susceptible to damage. Seeds harvested at a moisture content higher than 13%, however, need to be aerated or dried.

Harvested soybeans are typically loaded into on-farm metal bins for bulk storage until they are removed for conditioning and packaging or for sale as grain if it is decided they are unfit or not needed for seed. The period in bulk storage ranges from a few weeks to 2 to 3 months for contract seed producers, and generally from harvest to planting for farmer-saved seeds.

The bulk storage bins used for seeds should be equipped for aeration and, at least, emergency drying as illustrated in Figure 13.6. Seeds loaded into bulk storage at a moisture content less than 14% can usually be satisfactorily conditioned for safe storage by aeration. Aeration removes field "heat," equalizes temperature within the seed mass which prevents moisture migration, and cools the seeds as ambient temperature drops with the advance of fall. It can also reduce seed moisture content 1 to 1.5% when operated during good drying days at airflow rates of 1 to 1.5 cubic feet per minute (cfm) per bushel (bu) rather than the lower 0.1 to 1.0 cfm/bu normally used for aeration. Aeration is best controlled with a humidistat set to cut off the fan whenever relative humidity rises above 60 to 65%, which is in equilibrium with a seed moisture content of about 12.5%. The equilibrium moisture content of soybean seeds at different percentages of relative humidity and temperature are represented graphically in Figure 13.7. It can be seen from the chart that aeration during periods of high humidity, say, above 70%, can increase seed moisture to

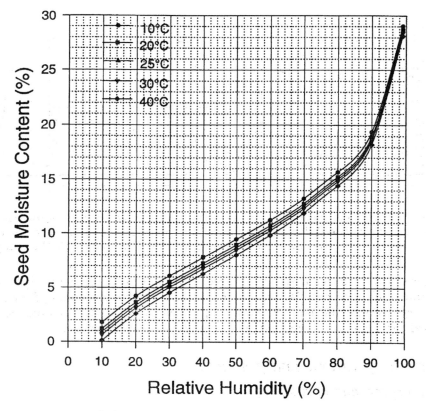

Figure 13.7 Equilibrium moisture content of soybean seed at different levels of relative humidity and temperature.

above 13%, an unsafe level, when continued for several days. The aeration system should be turned off after "field heat" has been removed from the seed mass and the moisture content is stabilized at 13% moisture content or less. This takes a few days to a week after loading, depending on weather conditions. Thereafter, the system should be run periodically on cool, dry days to reduce and equalize the seed mass temperature to the ambient level.

Seeds harvested at a moisture content above 14% need to be dried to 13% or less to maintain good germination and vigor in bulk storage. Batch drying in a metal bin with an elevated perforated floor and air plenum beneath (see Figure 13.4) using natural (unheated) air at flow rates of 2 to 3 cfm/bu is generally satisfactory for drying seeds with moisture content *less than 16%*. Seed depth should not be above about 6 ft since airflow rates of the level needed require a lot of horsepower when the greater seed depths typical of aeration are used.

Heated air is needed to dry seeds bulked at 16% or higher moisture contents. A burner or heater is required. Airflow rate should be 3 to 5 cfm/bu, and air temperature should not exceed 100°F. Like aeration, drying can be best and most safely controlled by a humidistat set to maintain drying air relative humidity between 40 and 60%. Drying air humidity lower than about 40% can cause cracking or "checking" of the seed coat as a result of too rapid drying. The seeds should be dried to a moisture content between 12 and 13%. Heated-air-drying of 14 to 16% moisture seeds also produces excellent results but is more costly than natural or unheated air-drying. After drying the seeds should be periodically aerated as previously discussed.

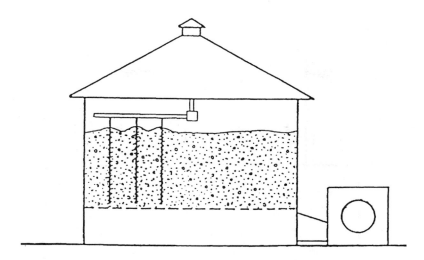

Figure 13.8 Bulk storage bin fitted with stirring devices. Stirring minimizes stratification of seed moisture and temperature during aeration, drying, and storage.

Stirring devices (Figure 13.8) are employed in grain aeration and drying to speed moisture loss, improve efficiency, and produce a seed mass relatively uniform in temperature and moisture content. There is no general agreement on the suitability of the stirring devices for seeds. Some companies encourage the use of stirrers, while others fearful of excessive damage recommend that they not be used. The general experience seems to be that properly operated and maintained stirrers cause little if any damage to seeds with minimal weathering and harvest damage.

Conditioning and packaging

Combine-run soybean seeds usually contain substantial amounts of foreign material, such as fragments of pods, stems, leaves, split seeds, weed fruits and seeds, and even some soil particles or "peds." These need to be removed by processing or conditioning to raise purity and reduce contaminants to acceptable levels, upgrade germination and vigor to the extent feasible, improve flowability and plantability of the seeds, and prepare them for storage. Seeds produced by contract seed growers move from on-farm bulk storage to the company bulk storage for conditioning (cleaning and upgrading) and packaging. Conditioning and packaging are accomplished in sequential operations with one to several cleaners and separators and a weighing–packaging system. A typical conditioning plant layout is shown in Figure 13.9.

Seed handling in the conditioning line is by belt-bucket elevator, gravity flow, and horizontal conveyors with surge bins above each machine to establish a uniform flow of seeds through the line. Basic cleaning is accomplished with an air and screen machine in which the seeds are aspirated one or two times to remove lightweight and irregularly shaped material, and sieved two to four times to remove material smaller and larger than the good seeds. Basic cleaning is sufficient for conditioning the seeds of some lots to the desired quality standards for purity and appearance, and they are moved directly to the packaging system. More frequently, however, additional cleaning and separation operations are required to achieve quality standards.

The spiral separator is usually the second machine in the processing line. It removes nonround material such as a seed halves or splits, low-quality "bean-shaped" seeds, shriveled seeds, flat-sided weed seeds, and soil particles. The final machine is the gravity

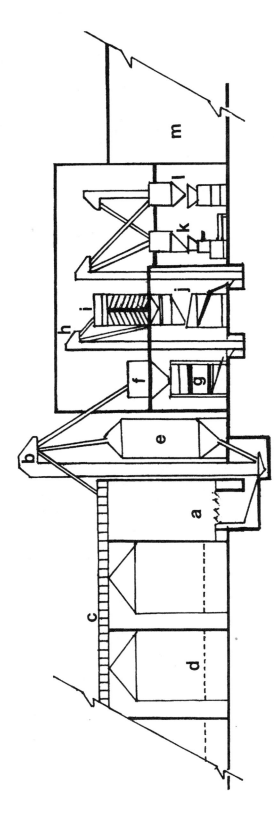

Figure 13.9 Side elevation (unscaled) of typical soybean seed bulk storage, conditioning, and flat storage facility. a, receiving area and dump pit; b, main receiving belt and bucket elevator; c, drag flight or belt conveyor for loading bulk storage bins (unloading belt conveyor for conveying seeds from bin to main elevator is not shown); d, aeration/drying bulk storage bin; e, holding bin; f, surge bin over air and screen cleaner; g, air and screen cleaner for basic cleaning; h, bucket elevator for in-line transfer of seed; i, spiral separator for separation of "nonround" contaminants; j, surge bin and gravity table separator for density separations; k, surge bin and automatic weigher–bagger for packaging seeds; l, surge bin and treater for treating seeds; m, flat or packaged seed storeroom. In most soybean-conditioning plants density separation (gravity table) and treatment are options which can be and are bypassed for many of the seed lots handled.

table or separator which removes low-density seeds and material such as badly weathered, stink bug damaged, mechanically damaged intact seeds, and heavy rounded soil particles. The cleaners and separators, of course, do not remove all of the material described, but if properly operated they will remove most of it and produce a very attractive and plantable seed product. It should be noted again that seeds of crops such as cowpea, okra, and corn (the small rounds from the tip of the ear) that volunteer in soybean fields and weeds such as balloonvine and purple moonflower are very difficult or impossible to separate from soybean seeds. These difficult-to-separate seeds were shown along with soybean seeds in Figure 13.1.

Commercial soybean seeds are commonly packaged in multiwall paper bags pre-printed with the company logo, kind of seed, and other information. Packaging is accomplished with a weighing–bagging system that weighs the desired quantity of seeds and dumps or "blows" it into the bag, which is closed by sewing or, most commonly, by the pressure of the seeds against a valve opening. A label or tag containing the information required by the seed law is affixed to each package of seeds by sewing, adhesive, or preprinting on the bag. The packaged seeds are moved to a so-called flat or packaged seed storehouse until sold or shipped to other distribution locations. Some seed companies treat some lots of seeds with fungicides and other chemicals before packaging, usually on the request or order from the customer. This is discussed in a later section.

Farmer-saved seeds are often taken directly from the storage bin to the planter. This is satisfactory only when the combine-run seeds are exceptionally clean. In most cases the seeds should be cleaned. This can be accomplished with a small on-farm-type air and screen cleaner or, if one is not available, by taking the seeds to a custom seed conditioner (cleaner). Custom-conditioned seeds can be handled entirely in large bulk bags or packaged in unprinted (brown) paper bags.

Storage

Soybean seeds are in storage from the time they attain physiological maturity in the field until they germinate in the seedbed. During this long period encompassing many operations, the seeds are subject to the effects of temperature and seed moisture content, the two main physiological and environmental factors that control deterioration processes in seeds and determine longevity. The discussions in previous sections on the importance of early harvest, careful handling, aeration in bulk storage, and prompt and adequate drying when needed have dealt primarily with quality maintenance during the various segments of the storage period. Most losses in quality, as emphasized in previous discussions, occur in the field before harvest as a consequence of weathering, during harvest, and during the time the seeds are in bulk storage. If the quality of seeds at the time of packaging is good, maintenance of quality during the packaged seed segment of the overall storage period, 1 to 6 months, is usually not difficult except perhaps in warmer and more humid regions, such as Florida, south Texas, and Louisiana.

Soybean seeds, like other kinds of seeds, are hygroscopic. They exchange moisture with the atmosphere and increase or decrease in moisture content until they come into equilibrium with the ambient relative humidity (see Figure 13.5). During storage, therefore, seed moisture content is either a function of relative humidity or relative humidity is a function of seed moisture content as in a mass of seeds in nonaerated bulk storage or in a moisture-vapor-retardant package. The relative humidity–seed moisture content equilibrium is established gradually over several days. In humid areas, seed moisture content can increase to 14% or higher over time, which can substantially reduce germination and vigor if the temperature is warm, say, 75°F or higher.

Although soybean seeds are inherently short-lived as compared with seeds of rice, corn, cotton, and other major crops, germination and vigor of good-quality seeds can be maintained in properly attended bulk and packaged seed storage in the generally cool conditions that prevail from harvest to planting if moisture content is maintained below about 13.5%. Seeds that are not planted, however, should not be "carried over" in the southern U.S. *except when kept in a conditioned storeroom* maintained at or below 70°F and 60% relative humidity. They rapidly and drastically decrease in germination and vigor in a bulk bin or nonconditioned warehouse.

Seed treatment

Seeds of major field crops except soybean are routinely treated with at least one fungicide for protection against microorganisms in the soil and control of some seed-borne diseases. There are several reasons soybean seeds have not been routinely treated. First, soybean seeds, until rather recently, were planted in late May or June when soil temperature is typically warm (above 80°F) but not too hot, and soil moisture is more likely to be at a favorable level than earlier or later. Seed treatments, especially of the seed protectant class, are most beneficial when conditions are *unfavorable* for germination and emergence, e.g., cool soil temperature combined with either excessive or deficient moisture, *rather than favorable*. Second, soybean seeds that are treated with a slurry or film coating of materials but not planted in the current season are difficult to "carry over" in normal storage until the next season, a common practice with treated corn, sorghum, cotton, rice, and wheat seeds, and conditioned storage is very expensive . Since treated seeds *can only be used for planting* and disposal of discarded treated seeds is costly and must be handled as hazardous waste, seed companies have viewed the risks of treatment as too great for routine practice.

Seed treatments for soybeans have generally been recommended in Mississippi and other states only under special conditions. These are (1) the field has a history of severe seedling disease; (2) planting in April or early May; (3) seed quality is marginal; (4) soybeans follow wheat; and (5) the seeds are infected with pod and stem blight or stem canker (Sciumbiato, 1997). The case for seed treatment, however, is strengthening not only because early planting has become a well-established practice, but also because other practices, such as conservation tillage systems that are gaining in popularity, often produce rather unfavorable conditions for germination, emergence, and stand establishment. Additionally, farmers are responding to high seed prices, especially for the new transgenic varieties, by reducing seeding rate and expecting near fail-safe seeds from suppliers. Seed treatment is a powerful option available to seed companies and farmers to meet these changes in production practices, varieties, and expectations.

Benefits of soybean seed treatment

Soybean seeds are treated with fungicides with the expectation of beneficial results. The value of the benefits to a soybean farmer provides the reason or justification for using treated seeds and paying the extra cost. Many reasons for or benefits from treating soybean seeds are claimed (Brunoehler, 1995; Gustafson, Inc., undated; *Soybean Digest*, 1995). These are summarized below.

1. Stand establishment is improved for soybeans planted early or into reduced-tillage conditions.
2. The risk of stand failure and need to replant is minimized.
3. Seedlings from treated seed are more vigorous than those from nontreated seeds and produce an earlier, more complete, and uniform canopy.

4. A substantial body of research in the midwestern U.S. has demonstrated the value of seed treatment, especially for farmers interested in reducing production costs and maximizing returns through early planting, reduced seeding rate, and use of narrow-row, no-till, and reduced-till production systems (Gustafson, Inc., 1994; Oplinger et al., 1996)).
5. Germination test results of seeds infected with stem canker, pod and stem blight, and other diseases are improved.

Diseases controlled and treatment chemicals used

The main seed and seedling diseases controlled by seed treatment are *Rhizoctonia* root rot, *Phytophthora* root rot, *Pythium* seedling disease, and pod and stem blight (*Phomopsis* spp.). Since the disease organisms live in the soil or on crop residue (the pod and stem and stem canker fungi are also seed borne), the main fungicides at present used are of the protectant class. Since seed treatment recommendations change, the extension service should be consulted for the latest recommendations.

Rhizoctonia *root rot*

The development of *Rhizoctonia* root rot is favored by cool, wet conditions followed by hot and dry conditions. Infected seedlings wilt and may die when the stem is girdled by rot at the soil line. The pattern of disease in the field ranges from single infected plants to groups of infected plants in a row or circular areas where there were early-season wet spots. *Seed Treatment*: Vitavax — Thiram; Rival (combination proprietary formulation by Gustafson, Inc.); Rival — Apron XL.

Phytophthora *root rot*

This disease is most severe in cold, wet, clayey soils with buried crop residue. Infected plants turn yellow, wilt, and die at all stages of development. The disease is active throughout the growing season. *Seed Treatment*: Apron XL; and Rival — Apron XL.

Pythium *seedling rot*

Seedling rot can be severe in cold, wet, clayey soils with buried residue. Infected seeds may rot or develop into weak seedlings that can "damp-off" at any stage up to the four- to six-leaf stage. *Seed Treatment*: Apron XL; Rival — Apron XL; Captan or Thiram.

Pod and stem blight

Pod and stem blight is a late-season disease. It is seed borne but the major source of inoculum is from crop residues. Distinguishing symptoms are small black fruiting structures arranged in linear rows on the stems that appear when the plant is maturing, poorly developed pods, and various degrees of seed shriveling, wrinkling, and seed coat cracking. The seeds may be partially covered with a white "mold." The seeds are treated to control the seed-borne and/or soilborne infections. *Seed Treatment*: Rival, Rival — Apron XL; Vitavax — Thiram, Thiram-Thiabendazole.

Commercially, seed treatments are applied as a final step in conditioning, just before packaging. Modern seed treatment formulations and procedures coat the seeds uniformly with a nondusting, non-rub-off film of the active ingredient(s), binders, and other materials. Since unsold, treated seeds are a major risk, the trend among seed companies is treatment on request. Few if any seeds are treated without a firm order. Some seed companies offer custom-treating services to farmers for farm-saved seeds. On-farm slurry-type treaters are also available for farmers who want to treat their own seeds.

Treated seeds flow more slowly than those not treated. Planters, therefore, should be set accordingly or readjusted as necessary to plant the desired rate.

Farmer-applied (hopper-box) treatments

Hopper-box application of seed treatment materials is an alternative to dealer-applied treatment and farmer-applied slurry formulations. The liquid or powder treatment material is mixed with the seeds in the planter box. Captan, Vitavax — Captan, Vitavax, Thiram, Apron XL and other formulations are available with and without molybdenum, a trace element that enhances nitrogen fixation.

At present, farmer-applied, hopper-box treatment accounts for more than 75% of the soybean acreage that is planted with treated seeds in the southern U.S. Beltwide, it is estimated that hopper-box treatments are applied to about 12 to 14 million acres (Rushing, 1997). There is, however, a strong trend, especially in the Midwest, toward dealer application which requires less-active material (more environmentally friendly), usually provides superior protection because of more uniform application and better coverage, and is more convenient for the farmer as compared with the hopper-box method.

Summary of quality assurance measures in seed production

The information and discussions in this chapter have emphasized the maintenance and assurance of quality of soybean planting seeds. The aims of all seed quality assurance programs are (1) maintenance of varietal purity; (2) prevention and/or elimination of other crop and weed seed contaminants; (3) minimization of mechanical injury; and (4) maintenance of germination and vigor at the highest levels economically feasible. Achievement of these aims in large measure is necessary to meet the expectations of farmer-customers and to assure continuation of the company reputation. Farmers who save seeds for planting need similar assurances.

The main points for assurance of high-quality soybean seeds are summarized below.

- Select productive, well-drained, uniform fields for seed production with provisions for irrigation and without serious weed or potential volunteer plant problems.
- Plant varietally authentic and pure seeds of high germination and vigor to obtain a uniform stand of healthy, rapidly developing plants. Seed treatment should be seriously considered.
- Manage crop in a manner similar to that for high-yield soybean grain production except as noted below.
- Walk through seed field several times with final visit just before harvest to determine any problem areas that should be excluded from the harvest and to identify contamination problems that need to be corrected by roguing or that may render the field unfit for seed production.
- Harvest crop as soon as seeds first dry down below 15% moisture and as rapidly as possible with a clean combine. Adjust combine to minimize both seed losses and seed damage and operate at a uniform ground speed. Monitor seed condition during harvest and subsequent handling and readjust combine and handling equipment as needed to keep damage below maximum level.
- Handle seeds carefully into and out of bulk storage. Augers should be avoided.
- Aerate seeds in bulk storage to remove field heat, even out and reduce temperature with the advance of cooler fall days, and reduce moisture content of 13 to 14% moisture seeds to 13% or less. Seeds from 14 to 16% moisture should be promptly dried with unheated air at airflow rates of 2 to 3 cfm/bu to 13% or less. Seeds with 16% moisture or higher require heated air-drying.

- Thoroughly clean seeds to include density grading if available. Companies should package seeds in attractive bags with required label and store in a clean storeroom through distribution and marketing. Farmer-saved seeds should also be thoroughly cleaned and stored in bulk or bags in a clean, dry place until planting.
- Farmers practicing conservation tillage, planting in April or in June behind wheat, or with fields that have a history of seed rot and seedling disease should purchase treated seeds or, at least, use a hopper-box treatment.
- Seed companies should make available treated seeds on request or order of farmers.

References

Agrawal, V. K. and J. B. Sinclair. 1987. *Principles of Seed Pathology*, Vols. I and II, CRC Press, Boca Raton, FL.

Aguiar, P. A. A. 1974. Some Relationships between Seed Diameter and Quality in Soybean [*Glycine max* (L.) Merrill], Ph.D. dissertation, Mississippi State University, Mississippi State.

Andrews, C. H. 1966. Some Aspects of Pod and Seed Development in Lee Soybean, Ph.D. dissertation, Mississippi State University, Mississippi State.

Assman, E. J. 1983. Seed Density and Quality Relationships in Gravity Graded Soybean [*Glycine max* (L.) Merrill], Ph.D. dissertation, Mississippi State University, Mississippi State.

Association of Official Seed Analysts (AOSA). 1970. *Tetrazolium Testing Handbook for Agricultural Seeds*, Contrib. 29, AOSA, Omaha, NE.

Association of Official Seed Analysts (AOSA). 1983. *Seed Vigor Testing Handbook*, AOSA Handbook on Seed Testing, Contrib. No. 32, 88 pp., with amendments. AOSA, Lincoln, NE.

Association of Official Seed Analysts (AOSA), 1992. *Seedling Evaluation Handbook*, Contrib. 35, AOSA, Omaha, NE, 101 pp.

Association of Official Seed Analysts (AOSA). 1993. *Rules for Testing Seeds*, AOSA, Omaha, NE. 123 pp.

Association of Official Seed Certifying Agencies. 1994. (a) Operational Procedures, (b) Genetic and Crop Standards, AOSCA, Mississippi State.

Brunoehler, R. 1995. Seed treatments put the squeeze on bean disease, *Farmer Ind. News*, March, 1995.

Calero, E., S. H. West, and K. Hinson. 1981. Water absorption by soybean seed and associated causal factors, *Crop Sci.* 21:926–933.

Copeland, L. O. and M. B. McDonald, Jr. 1995. *Principles of Seed Science and Technology*, 3rd ed., Chapman & Hall, New York.

Delouche, J. C. 1973. Seed vigor in soybeans, *Proc. 3rd Soybean Seed Res. Conf.* 3:56–72, American Seed Trade Association, Washington, D.C.

Delouche, J. C. 1974. Maintaining soybean seed quality, in *Soybean, Production, Marketing and Use*, NFDC, TVA, TVA Bull. Y-69, Muscle Shoals, AL, 46–62.

Delouche, J. C. 1977. Soybean seed storage beyond one year, *Proc. 7th Soybean Seed. Res. Conf.* 7:60–73, American Seed Trade Association, Washington, D.C.

Delouche, J. C., T. W. Still, M. Raspet, and M. Lienhard. 1962. The Tetrazolium Test for Seed Viability, Miss. Agr. and For. Exp. Sta. Tech. Bull. 51.

Delouche, J. C., E. R. Cabrera, and B. C. Keith. 1995. Strategies for Improving Physiological Seed Quality, Miss. Agr. and For. Exp. Sta. Bull. 1029.

Edwards, C. J. and E. E. Hartwig. 1971. Effect of seed size upon the rate of emergence in soybean, *Agron. J.* 63:429–431.

Ellis, R. H. and J. B. Sinclair. 1976. Effect of benomyl field sprays on internally borne fungi, germination, and emergence of late-harvested soybean seeds, *Phytopathology* 66:680–682.

Gustafson, Inc. (undated). Top ten reasons for treating soybean seed with fungicides. Informational material, mimeo. 12 pp., with attachments, Gustafson, Inc., Dallas, TX.

Gustafson, Inc. 1994. Diseases rob soybean yields in conservation tillage, Soybean Backgrounder, Gustafson, Inc., Dallas, TX.

Hartwig, E. E. and H. C. Potts. 1987. Development and evaluation of impermeable seed coats for preserving soybean seed quality, *Crop Sci.* 27:506–508.

Heatherly, L. G. 1993. Drought stress and irrigation effects on germination of harvested soybean seed, *Crop Sci.* 33:777–781.

Heatherly, L. G. and M. M. Kenty. 1995. Irrigation during seedfill and germinability of soybean with impermeable seed coat character, *Crop Sci.* 35: 205–208.

Heydecker, W. 1972. Vigour, in E. H. Roberts, Ed., *Viability of Seeds*, Syracuse University Press, Syracuse, NY., Chap. 8, 208–252.

Jeffers, D. L. and D. L. Reichard. 1982. Seed-borne fungi, quality, and yield of soybean treated with benomyl fungicide by various application methods, *Agron. J.* 74:886–890.

Keith, B. C., Jr. 1991. Evaluation of Hardseeded Genotypes of Soybean for Agronomic Traits, Hard Seed Content and Resistance to Seed Deterioration, Ph.D. dissertation. Mississippi State University, Mississippi State.

Kueneman, E. A. 1983. Genetic control of seed longevity in soybeans, *Crop Sci.* 23:5–8.

McDonald, M. B., Jr. 1980. Vigor test subcommittee report, *Assoc. Off. Seed Anal. News Lett.* 54:37–40.

McGee, D. C. and C. L. Brandt. 1979. Effect of foliar application of benomyl on infection of soybean seeds by *Phomopsis* in relation to time of inoculation, *Plant Dis. Rep.* 63:675–677.

Miller, W. A. and K. W. Roy. 1982. Effect of benomyl on the colonization of soybean leaves, pods and seeds by fungi, *Plant Dis. Rep.* 66:918–920.

Minor, H. C. and E. H. Paschal. 1982. Variation in storability of soybeans under simulated tropical conditions, *Seed Sci. Technol.* 10:131–139.

Mississippi Department of Agriculture and Commerce. (Undated). Mississippi Pure Seed Law and Regulations, MSDAC Publication 1462, Jackson, MS.

Nugraha, U., Soejadi, and Sumarno. 1991. Association of seed coat color and permeability in soybean with resistance to weathering stress and adverse storage conditions, *Indones. J. Crop Sci.* 6:1–10.

Oplinger, E. S., J. M. Gaska, and C. R. Grau. 1996. Fungicide Seed Treatments for Soybeans, Agronomy Advice, Field Crops 27.425, Cooperative Extension Programs, Agronomy Department, University of Wisconsin, Madison.

Possamai, E. 1976. Some Influences of Seed Size on Performance of Soybean [*Glycine max* (L.) Merrill] Seed, M.S. thesis, Mississippi State University, Mississippi State.

Potts, H. C. 1978. Hardseeded soybeans, *Proc. 8th Soybean Seed Res. Conf.* 8:33–42. American Seed Trade Association, Washington, D.C.

Potts, H. C. 1985. Seed coat permeability and deterioration studies in soybean, *Proc. 15th Soybean Seed Res. Conf.* 15:71–77, American Seed Trade Association, Washington, D.C.

Potts, H. C., J. J. Duangpatra, W. G. Hairston, and J. C. Delouche. 1978. Some influences of hard-seededness on soybean seed quality, *Crop Sci.* 18:221–224.

Ratnayake, S. and D. R. Shaw. 1992. Effects of harvest-aid herbicides on soybean (*Glycine max*) seed yield and quality, *Weed Technol.* 6:339–344.

Roy, K. W., B. C. Keith, and C. H. Andrews. 1994. Resistance of hardseeded soybean lines to seed infection by *Phomopsis*, other fungi and soybean mosaic virus, *Can. J. Plant Pathol.* 16:122–128.

Rushing, K. W. 1997. Private communication. Gustafson, Inc., Dallas, TX, Feb. 7.

Sciumbiato, G. 1997. Private communication to B. Keith. Delta Research and Extension Center, Mississippi State University, Stoneville, Jan. 25.

Sinclair, J. B. 1982. *Compendium of Soybean Diseases*, The American Phytopathological Society, St. Paul, MN.

Soybean Digest. 1995. Safeguard seedlings, *Soybean Dig.* March.

Spears, J. F. 1997. Farmer-Saved Seed: Understanding Federal Laws and State Seed Regulations, North Carolina Cooperative Extension Service, Information Sheet AG-555, North Carolina State University, Raleigh.

TeKrony, D. M., D. B. Egli, and G. M. White. 1987. Seed production and technology, in J. R. Wilcox, Ed., *Soybeans: Improvement, Production and Uses*, 2nd ed. American Society of Agronomy, Madison, WI. Chap. 8, 285–354.

United States Department of Agriculture. Undated. Federal Seed Act Regulations, Part 201–202, Agricultural Marketing Service, U.S.D.A., Hyattsville, MD.

United States Department of Agriculture. 1994. Plant Variety Protection Act and Regulations, Plant Variety Protection Office, Hyattsville, MD.

Wetzel, C. 1978. Some Effects of Seed Size on Performance of Soybean [*Glycine max* (L.) Merrill], Ph.D. dissertation, Mississippi State University, Mississippi State.

chapter fourteen

Soybean disease management

Glenn R. Bowers and John S. Russin

Contents

0-8493-2301-0/99/$0.00+$.50
© 1999 by CRC Press LLC

Introduction

Soybean diseases may be classified as infectious or noninfectious. Infectious diseases are caused by agents that can be transmitted from an infected to a healthy plant and cause disease when conditions are favorable for infection. Fungi, bacteria, viruses, and nematodes cause infectious diseases. Various unfavorable environmental and nutritional conditions cause noninfectious diseases.

More than 100 pathogens are known to affect soybean; about 35 are important economically. This chapter discusses 15 of the most common fungal diseases found in the midsouthern U.S. Diseases caused by bacteria and viruses occur in this region but are not recognized as causing significant economic problems in soybean production. Nematodes are discussed in another chapter.

One or more diseases can generally be found in fields wherever soybean is grown. All parts of the soybean plant are susceptible to a number of pathogens that reduce quality and quantity of seed yield. The extent of losses depends upon the pathogen, the state of plant development and health when infection occurs, the severity of the disease on individual plants, and the number of plants affected. In many diseases, a latent period elapses between infection and symptom expression. This latent period can range up to 2 months.

A pathogen may be very destructive one season and difficult or impossible to find the next season. Many pathogens can initiate an epidemic only under rather specific environmental conditions. Thus, the extent and severity of soybean diseases also depend on the degree of compatibility between the host and the pathogen and the influence of the environment on this association.

During the growing season in the midsouthern U.S., days and nights are warm, humidity is high, and dew periods are long. These are environmental conditions most fungal pathogens require for infection and rapid colonization of the host plant. Collectively, diseases are a major constraint to profitable soybean production in the midsouthern U.S. By some estimates, during the period 1984 to 1994 diseases reduced yields an average of 5% (Wrather et al., 1995). Losses can be much higher in individual fields. However, disease can be effectively dealt with if one remembers that the goal of disease management is not to eliminate disease but to maximize profit and contribute to a sustainable production system. This chapter describes and discusses fungal diseases of soybean and established disease management principles to provide the reader with knowledge needed to help assure profit from soybean production in the midsouthern U.S.

Root and lower stem diseases

Charcoal rot

Symptoms and losses

Charcoal rot is caused by the fungus *Macrophomina phaseolina* (Tassi) Goid. This pathogen has both a broad host range and wide geographic distribution. In addition to soybean, known hosts include numerous other cultivated crop species. The disease is prevalent throughout the world but is most severe in regions between 35° north and 35° south latitude.

Symptoms of charcoal rot are evident in soybean plants of all ages. Infected seed can be asymptomatic or show black spots of variable size. Black microsclerotia may be visible in seed coat cracks or completely cover the seed. Infected seeds may germinate, but seedlings that develop from infected seeds usually die within a few days. Seed-borne inoculum invades cotyledon and embryo tissues and produces microsclerotia within 3 to

5 days. Diseased seedlings typically show hypocotyl lesions that are reddish brown initially but subsequently turn gray to black. Seedling disease is greatest when soil temperature is warm and soil moisture is low. Surviving seedlings may show little or no external symptoms and thus serve as latent sources of inoculum.

Plants infected after the seedling stage generally show no aboveground symptoms until after midseason. Diseased plants initially show nonspecific symptoms such as reduced leaf size, stem height, etc., which indicate loss of vigor. McGee (1992) described occasional superficial lesions that extend from the soil line. Wilting, which is an indicator of root dysfunction, also may be evident. Beginning at flowering, a light gray discoloration develops on the epidermal and subepidermal tissues of both tap and secondary roots and lower stems. Black microsclerotia are produced in these tissues and in pith as well. They can be so abundant that diseased tissue appears grayish black (hence, the name charcoal rot) or sufficiently sparse so that they appear as random black specks. These can be particularly noticeable at stem nodes. The fungus also can cause reddish brown discoloration in vascular tissues of roots and stems.

Results from controlled studies showed that the pathogen can reduce plant height, root volume, and root weight by more than 50%. These deleterious effects on roots are most evident during the pod formation and filling stages when demand is high for water and nutrient absorption. Because diseased plants have smaller root systems, resulting seeds tend to be fewer and lighter. Diseased plants also will mature several weeks earlier, which further contributes to yield loss. Accurate figures for yield loss are difficult to determine, but Wrather et al. (1995) estimated average yield losses in 16 southern states at nearly 1% annually for 1988 through 1994. This translates into a yearly loss of over 7 million bu during this period. Prior to 1988, charcoal rot was not recognized as a serious problem across the south.

Disease cycle

The charcoal rot fungus survives in soil and soybean debris as microsclerotia, the primary source of inoculum. These are black, spherical to oblong, and typically measure 0.002 to 0.008 in. in diameter. Microsclerotia produced in host tissues are released into soil as plant tissues decay. Populations of microsclerotia in soil can vary based on cropping history. These populations are closely correlated with the number of years that a given field had been cropped to soybean or corn. Redistribution of microsclerotia in soil occurs with normal tillage practices. Microsclerotia germinate on the surfaces of or in close proximity to roots throughout the growing season. Wyllie (1976) reported that 100% of soybean plants can be infected by *M. phaseolina* within 3 to 4 weeks after planting. Hyphae contact host roots and penetrate by means of mechanical pressure combined with digesting or dissolving the root cells by enzyme activity. Invasion of the root cortex is followed by colonization of vascular tissues. The fungus then grows throughout the root and stem tissue, and eventually produces characteristic microsclerotia later in the growing season. Late in the growing season is when the disease is most evident; however, most root damage and yield reduction have already occurred. Affected plants mature prematurely, normal leaf abscission is not initiated, foliage may appear chlorotic and senescent, and pods generally fail to fill completely.

Environmental conditions

Numerous environmental factors influence production and germination of microsclerotia, root colonization, and symptom expression. Microsclerotia are produced abundantly under favorable conditions, but production can increase in response to various stresses. Microsclerotia production was greater in very acidic (pH 4.3) and alkaline soils (pH 8.0) than in more neutral soils, in low (40°F) and high (104°F) temperature soils than at more moderate temperatures, at low soil moistures, and at low to moderate soil bulk densities. Production of microsclerotia was less at deeper soil depths and at lower oxygen concentration. Microsclerotia germination is affected by root exudates, soil moisture and oxygen levels, organic soil amendments, and inorganic N. Severity of charcoal rot generally correlates positively with soil temperature and inversely with soil moisture. Sinclair and Backman (1989) reported that disease severity was greatest from 82 to 95°F.

Several studies examined effects of herbicides on *M. phaseolina*. Colonization of soybean roots by this fungus was increased by 2,4-DB and chloramben, but decreased by trifluralin and alachlor. Saprophytic colonization of soybean stems by *M. phaseolina* decreased following application of desiccant herbicides. Recovery of microsclerotia from a sandy loam soil was reduced by several herbicides, including alachlor. Growth of this pathogen was reduced by ametryn at high concentrations. Atrazine, metolachlor, and alachlor stimulated microsclerotia production, whereas slight (≤10%) reductions in microsclerotia germination were induced by both metolachlor and alachlor (Russin et al., 1995).

Control

Charcoal rot can be managed by (1) rotation with a less susceptible crop and (2) minimizing plant stress by avoiding excessive seeding rates, fertilizing when necessary, and irrigating to keep soil moisture high during pod development periods.

Soybean varieties resistant to charcoal rot have not been identified. Corn, grain sorghum, and cotton also are hosts for the pathogen but generally support lower populations of microsclerotia in soil than does soybean. Wyllie (1988) reported that 15 microsclerotia per 0.04 oz of soil is an acceptable level for production of soybean. Rotation for 1 or more years with less-susceptible crops should be sufficient to reduce soil populations to this level. Once this level is achieved, rotating away from soybean for 1 year using any of the above crops should maintain soil microsclerotia populations in this acceptable range.

The remaining control recommendations generally focus on reducing stress and maintaining crop vigor. Avoiding excessive seeding rates will minimize competition stress (Bowen and Schapaugh, 1989), and proper fertilization will ensure vigorous, rapidly growing plants.

Host resistance generally is not effective for control of charcoal rot. However, Pearson et al. (1984) found that varieties that mature later in the season may escape some of the stress associated with high temperature and low moisture in soil.

Phytophthora rot

Symptoms and losses

Phytophthora rot is caused by the soilborne fungus *Phytophthora megasperma* Drechs. f. sp. *glycinea* (Hildeb.) Kuan and Erwin. It is a destructive disease of soybean, particularly in the north central states. The disease is present in many of the soybean-growing regions of the world, but apparently not in Central and South America or China (Schmitthenner, 1989). Soybean is the only economic host for the pathogen. Disease severity is greatest on fine-textured soils or those with relatively poor drainage. To date, 25 races of *P. megasperma* f. sp. *glycinea* have been described, and many isolates have been found that do not fit into any of these races. This is an important fact when disease resistance is considered.

The first detectable symptoms of *Phytophthora* rot are seed death as well as pre- and postemergence damping-off. These symptoms can occur in all susceptible varieties, and they generally are not distinguishable from those caused by other seedling pathogens such as *Pythium* and *Rhizoctonia*. *Phytophthora* rot symptoms in older seedlings are more distinctive. Susceptible varieties often show yellowing and wilting of leaves, and plants may be severely stunted. Lateral roots are almost completely destroyed and the main root shows dark discoloration, which can progress up the stem until the seedling is killed. If susceptible plants survive the seedling stage, root and stem rot may appear throughout the season. Brown discoloration may extend up the stem for several nodes before the plant is killed. Varieties that are tolerant to _Phytophthora_ rot may show early-season symptoms that are limited to lateral root rot and tap root discoloration. By mid to late season, these plants may appear stunted and less thrifty than healthy counterparts, and these symptoms resemble those of mild nitrogen deficiency. Root rot almost always is present in such plants. This hidden damage also may be accompanied by elongated stem lesions that resemble those caused by stem canker.

Disease losses are difficult to estimate accurately. Severe disease in the seedling stage can result in 100% mortality of susceptible plants (Schmitthenner, 1989). Sinclair and Backman (1989) reported that losses up to 40% can occur in fields where plants show mild stunting, chlorosis, and stem lesions. In a study designed to examine effects of the fungicide metalaxyl on _Phytophthora_ rot, Anderson and Buzzell (1982) found that plant loss due to _Phytophthora_ rot was 54% in control plots but only 10% in treated plots. This resulted in yields of 28.9 and 43.6 bu/acre, respectively, a difference of 14.7 bu/acre (nearly 34% yield loss).

Disease cycle

The pathogen can survive in soil for many years as oospores. Upon flooding of soil through excessive rainfall or irrigation, oospores produce sporangia and release motile zoospores, which move in water. Consequently, the disease is more prevalent on wet, poorly drained, fine-textured clay soils (Athow, 1985). Zoospores are attracted to and aggregate on seedling roots, then germinate and infect. Excessive soil moisture during this period may greatly worsen the disease. As the fungus ramifies through the host, oospores are produced in host tissues. Many more oospores are produced in susceptible or tolerant varieties than in resistant varieties (Athow, 1985). These are released into the soil as infected host tissue decomposes. Root infection on older plants can occur throughout the season if conditions are favorable. However, it is believed that most severe losses result from early-season infection.

Environmental conditions

Several environmental and cultural factors are reported to influence _Phytophthora_ rot. Kittle and Gray (1979) reported that a soil temperature of 60°F, which is 20 to 30° below optimum for soybean root growth, was most favorable for root infection. However, disease development is much more rapid when air temperatures are above 77°F.

Flooding of soil is essential for release and movement of fungal zoospores. Disease severity generally is greater under conditions of excessive soil moisture and less on well-drained soils. Results from Ohio indicated that disease was more severe when flooding occurred soon after planting, which suggests that damage may be less if fields are flooded later. Excessive soil moisture may be less important if tolerant or resistant varieties are used. Schmitthenner (1989) found that drainage of excess water significantly increased the yield of a susceptible variety but a less-susceptible tolerant variety was affected less. Consequently, tactics such as tiling fields to improve drainage and use of conventional rather than reduced tillage are successful for managing this disease in areas of the country where it is severe.

Control

Phytophthora root and stem rot can be managed by (1) planting race-resistant varieties; (2) using Apron® 50W, Apron® FS, or Apron XL™ LS as a seed or in-furrow treatment; and (3) optimizing cultural practices such as good drainage and conventional tillage. Use of resistant varieties is by far the best approach to management of *Phytophthora* rot. There are at least 25 different races of *P. megasperma.* f. sp. *glycinea.* Varieties resistant to prevalent races are widely available, but it is important to know the race prevalent in a field and to select varieties known to have resistance to that race. In addition, not all susceptible varieties are equally damaged by *Phytophthora* rot. Some can survive and yield fairly well even when infected (Schmitthenner, 1989). Such varieties are described as having tolerance, rate-limiting resistance, or age-related resistance. The exact mechanisms involved in this resistance are not known. Such varieties may be combined with other cultural practices to minimize disease-related losses.

Standard seed treatments are not effective against *P. megasperma* f. sp. *sojina.* Metalaxyl (Apron 50W and Apron FS) seed treatment provides some control depending on variety and disease severity. Disease control was most effective on a tolerant variety under conditions of moderate disease severity. Control was poor on a susceptible variety regardless of disease severity. Several reports indicated that in-furrow application of metalaxyl at planting provided good disease control. Low rates were effective in combination with tolerant varieties, but even high rates may not be effective with susceptible varieties.

Phytophthora rot generally is more severe under reduced or minimum tillage. This may result from several factors, including cooler soil temperatures and elevated soil moisture levels with reduced tillage. Conventional tillage may improve drainage, decrease soil bulk density, or mix inoculum so that it is not concentrated in a narrow band near the soil surface.

Sudden death syndrome

Symptoms and losses

Sudden death syndrome is caused by certain isolates of the soilborne fungus *Fusarium solani* (Mart.) Appel & Wollenw. emend. Snyd. & Hans. This disease was first observed in Arkansas in 1971 and occurred sporadically there until 1984. In 1984–1985, sudden death syndrome also occurred in six other states that border the Mississippi River, the southernmost of which was Mississippi (Rupe et al., 1993). In all of these states, the disease continues to cause periodic problems.

Symptoms of sudden death syndrome first appear as interveinal chlorotic spots at or near flowering. This usually occurs beginning at midseason but can continue through pod fill. Plants are apparently healthy prior to development of foliar symptoms. Spots can become necrotic or can coalesce and expand into chlorotic interveinal streaks which also can become necrotic (Sinclair and Backman, 1989). In severely affected foliage, only the leaf veins remain green. There is no bending or obvious wilting of foliage. Symptomatic leaflets often abscise, leaving only petioles attached to stems. Root systems of diseased plants are smaller, and vascular tissue shows light brown discoloration that can extend several nodes up the stem. Pith tissue, however, remains white. In infested fields, symptoms appear in somewhat circular or elongated patches, which can coalesce into large, irregular areas of diseased plants .

Yield losses can be sporadic, because sudden death syndrome is not a severe problem every year or in every part of an affected field. However, yield losses can be great in areas where the disease is severe. Sudden death syndrome affects yields by reducing both seed size and seed number. Reductions in seed size occur through reductions in leaf area that result from leaf disease and premature defoliation, whereas seed number is reduced by flower and pod abortion.

Disease cycle

Relatively little is known about the sudden death syndrome disease cycle. The pathogen penetrates root tissue early in plant growth. Roy et al. (1989a) reported development of leaf symptoms 3 to 4 weeks after planting in a greenhouse test, which indicates that root colonization occurred somewhat earlier than that. The lower portion of the tap root apparently is first to become discolored (Scott, 1988). The fungus also is able to colonize roots and stems of older plants (Rupe, 1989). In greenhouse tests, the fungus produced reddish brown external stem lesions at the soil line beginning 7 days after inoculation. These extended both up the stem and down into the roots from the point of inoculation. Foliar symptoms developed subsequently and were most severe 20 to 43 days after inoculation. There may be differences in levels of virulence among isolates of *F. solani*. Different fungal isolates produced symptoms with varying degrees of severity.

Environmental conditions

Severity of sudden death syndrome may be affected by soil moisture, soil fertility, and population densities of soybean cyst nematode. Cool, wet weather around the time of flowering has been associated with symptom development. Symptoms generally are less severe in dry years and more severe in irrigated areas of fields.

Soil factors associated with sudden death syndrome were increased levels of available soil phosphorus, soluble salts, organic matter, and exchangeable sodium, calcium, and magnesium (Rupe et al., 1993). This supports original observations that sudden death syndrome was commonly found in fields with high yield potential. A cropping system study showed that disease was more severe under no-till than conventional tillage, and under continuous soybean than in corn–soybean rotations (Von Qualen et al., 1989). However, Rupe (1988) indicated that crop rotation was not an effective control.

The role of soybean cyst nematode in sudden death syndrome is still unclear. In a field screening test, varieties susceptible to cyst nematode race 6 developed symptoms of sudden death syndrome 10 days earlier and had higher levels of disease than did varieties that were resistant to race 6. Varieties resistant to soybean cyst nematode were less affected by sudden death syndrome than were susceptible varieties. Dual inoculation with both *F. solani* and soybean cyst nematode caused foliar symptoms that were more severe than those caused by the fungus alone.The nematode apparently is not required for disease development, but can increase disease severity.

Control

Sudden death syndrome can be managed by (1) use of less-susceptible varieties and (2) control of soybean cyst nematode. Soybean varieties differ in susceptibility to sudden death syndrome (Hershman et al., 1990; Rupe et al., 1991). These differences were associated with the reactions of these varieties to soybean cyst nematode (Rupe et al., 1991). In a Kentucky study, early (May) plantings resulted in more severe disease than did later (mid-June to early July) plantings (Hershman et al., 1990). However, late planting is not recommended because of the reduced plant growth and lower yield potential of late-planted soybean compared with earlier plantings. Control of soybean cyst nematode normally is accomplished through planting resistant varieties. This practice should prove helpful to reduce but not eliminate the disease.

Southern blight

Symptoms and losses

Southern blight, also called *Sclerotium* blight, southern stem rot, and white mold, is caused by the soilborne fungus *Sclerotium rolfsii* Sacc. The pathogen has been reported to infect 500 plant species, including soybean, and is found but not restricted to the warmer regions of the world (Sinclair and Backman, 1989). It occurs primarily in the southern U.S.

Southern blight generally develops on isolated or small groups of plants scattered throughout the field. Symptoms can develop at all growth stages and include light brown lesions that quickly darken and girdle the hypocotyl or stem. Yellowing and wilting of foliage follows quickly, and necrotic leaves can cling to the dead stem. Foliage may show circular, tan to brown lesions with dark brown margins. These lesions may contain several concentric rings of discolored tissue. Plants infected as seedlings may develop twin stems. A white mat of fungus mycelium commonly is seen on bases of infected stems as well as on leaf debris and the soil surface. The fungus forms tan to brown spherical sclerotia about 0.04 in. in diameter on diseased plant debris as well as the soil surface. Although symptoms of southern blight can be striking, it is generally considered to be a minor disease. Up to 10% seedling loss has been reported, but reliable estimates of yield loss on soybean are lacking.

Disease cycle

The fungus survives as sclerotia in soil and as mycelia in colonized plant debris, and sclerotia viability can remain high for 8 to 10 months. Infection occurs at or below the soil surface, and the fungus can use decomposing plant material as a food base for attacking healthy plants. Although symptoms can appear throughout the season, disease severity generally is greatest during the reproductive stages of plant growth.

Environmental conditions

Epidemiology of the southern blight is poorly understood, and most available information is from studies of this pathogen on peanut. Southern blight is more severe in wet years. Cool temperatures seem to limit disease development. Survival of sclerotia is influenced by both soil moisture and temperature. Beute and Rodríguez-Kabana (1981) reported that sclerotia survival was not influenced by temperature in dry soils, but that survival in moist field soil was reduced at temperatures greater than 68°F. Mycelia died rapidly in moist field soil but survived at least 6 months in dry soil. In crops other than soybean, addition of calcium fertilizer or ammonium bicarbonate reduced losses to this disease. Although survival mechanisms are unclear, they may result from increased levels of calcium in tissues and from direct toxicity to the fungus (Punja et al., 1986). Root injury by the root knot nematode, *Meloidogyne arenaria*, as well as stem girdling by threecornered alfalfa hopper are reported to predispose plants to infection.

Control

Southern blight can be managed by (1) rotation with a nonhost crop and (2) deep plowing. Soybean varieties resistant to southern blight have not been identified. Several crops, including corn, grain sorghum, and pasture grasses, are not known to be hosts for the fungus. To reduce levels of inoculum significantly in soil, 2 to 4 years of the nonhost crop may be required. Plowing to depths of 3 to 4 in. or greater inhibited sclerotia germination and reduced disease in carrot (Punja et al., 1986). Resistant varieties have been identified and should be used if a field has a consistent history of Southern blight.

Red crown rot

Symptoms and losses

Red crown rot, also known as Cylindrocladium root rot, black root rot, and *Calonectria* root rot, is caused by the soilborne fungus *Calonectria ilicicola* Boedijn and Reitsma. Red crown rot is common to Louisiana, where it is one of the main soybean diseases (Berner et al., 1986). It has been reported to occur in several southern states, including Mississippi.

This fungus also causes a serious disease of peanut. Losses on soybean have been difficult to determine. Roy et al. (1989b) estimated 25 to 30% yield loss in a single Mississippi soybean field in which the disease was uniformly distributed. Berner et al. (1988) predicted 50% yield loss at 100% infection of a susceptible variety, although actual disease incidence levels generally are around 20 to 30%.

Infected soybean plants show no symptoms until after plants enter reproductive stages. Symptoms begin as small, interveinal chlorotic spots on the upper leaves. These can occur on individual plants or on small groups of plants in a field. Lesions enlarge, chlorotic tissue becomes necrotic, and desiccation and defoliation follow. This progression of foliar symptoms proceeds rapidly so that a diseased plant can go from apparently healthy foliage to defoliated in 1 to 2 weeks. Disease detection based on foliar symptoms alone then becomes difficult because surrounding healthy plants with green foliage obscure diseased plants. Yield loss presumably occurs due to incomplete pod filling, although controlled studies to document this on soybean are lacking. In one study, red crown rot severity was greater on a clay soil than on a silt loam soil.

Diagnostic red perithecia are visible on lower stems of diseased plants near the soil line. Although diseased plants generally show both foliar symptoms and perithecia, it is possible to have only one of these symptoms or to have infected root systems with no aboveground signs or symptoms.

Disease cycle

Many of the currently understood aspects of this disease cycle have been studied on peanut but not on soybean. Microsclerotia are the overwintering structures for *C. crotalariae*. They can survive several years in soil or on peanut debris and are spread in soil by movement of root fragments. Kuruppu and Russin (1995) showed that colonization of soybean roots initially was detected within 3 to 4 weeks after planting and increased throughout the season. When planting was delayed 3 or 6 weeks, however, root colonization initially was low and remained low until harvest. When combined with other studies (Russin et al., 1985; Berner et al., 1988) which also showed that delayed planting reduced disease severity, these results suggest a critical period for root colonization that must occur early in the growing season. Even after root infection occurs, microsclerotia apparently are not produced quickly in host root tissue. Production of microsclerotia in peanut roots was detected 8 to 12 weeks after inoculation, and they presumably are released into soil as infected root tissues deteriorate.

Environmental conditions

Root infection in peanut apparently is favored by moderate soil temperatures (78 to 86°F) and high soil moisture. Infection of peanut roots was not detected at 92°F when exposed to the pathogen, which suggests an upper thermal limit for this pathogen. Germination decreased when infested soil was previously incubated at 14 or 43°F for 4 weeks, but incubation at 79°F had no effect on germination (Griffin et al., 1978; Roth et al., 1979)

Control

Red crown rot can be managed by (1) delayed planting and (2) planting on coarse-textured soils. Soybean varieties resistant to red crown rot have not been identified, although some are less susceptible than others. Several studies showed that delayed planting can effectively reduce red crown rot disease (Russin et al., 1985; Berner et al., 1988; Kuruppu and Russin, 1995). However, delayed planting may not always be a viable option due to potential conflicts with wheat–soybean doublecropping and to normal yield constraints that operate if soybean is planted excessively late.

Seedling diseases

Symptoms and losses

Seedling diseases are caused by a number of different fungi and bacteria that are present wherever soybean is grown. These pathogens can be present in or carried on seed or present in soil or plant debris. They can act individually or together to produce symptoms that are relatively similar regardless of the pathogen involved. Damage may occur before seeds germinate or before or after seedlings emerge. Symptoms can be dark lesions or water-soaked or necrotic areas on roots or hypocotyls, necrosis of cotyledons or leaves, or seedling death, depending upon the specific pathogen involved. Diseased roots frequently are discolored and shriveled. Microscopic examination or laboratory culture techniques may be required to identify the pathogen.

Seedling death results in lower yields if plant populations fall below optimum. Stand reduction can result in inefficient use of fertilizers and pesticides, as well as slower soybean canopy development, which can encourage weed problems. Seedlings that survive may have reduced vigor and thus be less able to compete with weeds or produce adequate yield.

Environmental conditions

Seedling diseases generally are favored by early planting when soils are cool. Soil temperatures below 60°F reduce the rates of both seed germination and seedling growth, which result in longer periods during which pathogens can attack. Soil moisture also can be important in seedling disease development. It is generally accepted that excessive soil moisture can slow rates of seed germination, seedling emergence, and growth.

Control

Seedling diseases can be managed by (1) fungicide seed treatment (Apron 50W, Apron FS, Apron XL LS, Vitavax® CT, Vitavax — M, or Vitavax M DC), (2) planting high-quality seed, and (3) delaying planting until soil temperatures are at least 68°F. Fungicide seed treatments provide inexpensive insurance against seedling disease. Planting high-quality, disease-free seed will not only reduce losses due to poor stands but also will help prevent introduction of pathogens into fields where they are not a problem. Planting in warmer soils generally favors rapid seed germination and seedling growth, resulting in vigorous stands.

Stem, pod, and seed diseases

Stem canker

Symptoms and losses

In recent years, the most dramatic and destructive disease in the midsouthern U.S. has been stem canker caused by *Diaporthe phaseolorum* (Cke. & Ell.) Sacc. f. sp. *meridionalis* (Morgan-Jones). Symptoms first appear during early reproductive stages, with the development of a small, reddish brown superficial spot at the base of a branch or petiole on the lower part of the stem. In general, the spot is first observable in the leaf scar after the petiole has fallen from one of the first eight nodes. As the season progresses, the spot rapidly elongates into a slightly sunken lesion. This lesion or canker is dark brown to black and may girdle the stem. Leaf symptoms develop at this stage, with interveinal chlorosis and necrosis characteristic of diseases that restrict water movement in the plant. At the time the plant is killed, green stem tissue can be seen both above and below the canker. Leaves on dead plants wither but remain attached. The stem becomes brittle and is easily broken at the canker. Occasionally, a top dieback occurs, with characteristic shepherd's crook curling of the terminal bud.

Yield losses for naturally infected, susceptible varieties have been reported to be as high as 50 (Keeling, 1982) to 80% (Krausz and Fortnum, 1983). In some cases, production fields with a high disease incidence were not harvested.

Disease cycle

The pathogen can overwinter on diseased stems and in diseased seed (Backman et al., 1985; Athow, 1987). Physically moving diseased stems and diseased seed accounts for long distant spread of this disease. In a given field, stem canker can increase from negligible amounts in one year to epidemic levels the next. Typically, less than 5% of the seed from even the most severely infested fields in the midsouthern U.S. will contain the disease-causing organism. Even with low percentages of infected seed and associated reduced germination, severe cases of stem canker have developed with seed as the only apparent inoculum source. This is supported by observations indicating a clustering of diseased plants in these same fields, and may indicate possible secondary disease cycles originating from primary inoculum introduced on seed. The direction of disease development from a source of inoculum is influenced by field slope, row direction, and prevailing wind.

Fruiting structures of the fungus develop on crop debris in late winter, and spore release begins in late April and continues into June. It is these spores that are responsible for primary infections. Spore dispersal is by splashing and blowing rain. Spread from a point source of inoculum is limited to 6 to 7 ft and the direction of spread is related to wind direction (Damicone et al., 1990). Petioles, petiole bases, stems, and less frequently leaves are infected. Disease incidence and severity depend upon the plant growth stage at the time of exposure to inoculum. Maximum disease at R6 occurred when inoculum was applied at V3 about 20 to 21 days after planting. As the time of first exposure to inoculum is delayed from V3 to V10, there is progressively less disease. Even a very susceptible variety does not develop stem canker symptoms when exposed to inoculum after the vegetative stages. Stem canker spores have a low infection efficiency so a relatively large number of spores are required for infection to occur.

The fungus cannot be isolated from infected, symptomless plants until 3 weeks after inoculation. Thereafter, this pathogen routinely can be isolated from tissues without symptoms from inoculated susceptible and resistant varieties. In fact, initial infection levels of resistant and susceptible varieties are similar. The percentage of naturally infected plants from which the fungus can be isolated progressively increases from 20 to 100 days after planting. Events responsible for resistance apparently take place after infection occurs. However, for a susceptible variety, the incidence of infected plants early in the growing season is closely related to the final disease severity. Stem canker appears to have a long incubation period since symptoms generally are not visible until after R2 or R3, or about 50 to 80 days after infection. Foliar symptoms apparently are caused by a translocated toxin produced by the pathogen.

Researchers previously thought that stem canker only had a single cycle of infection, since fungal fruiting structures were not observed on infected plants during the growing season. The long incubation period of the stem canker was assumed to negate secondary spread. Recently, it was reported that fruiting structures of the fungus were produced in cankers during the growing season. Spores from these cankers were capable of infecting soybean plants in the greenhouse even at reproductive growth stages. This indicates inoculum produced in the field during the growing season is capable of causing infection late in the season. It is likely that plants inoculated by such late-season inoculum remain symptomless because of the long incubation period. However, these late season infections could add to the inoculum potential the following season.

The disease-causing fungus can be isolated from cotton seedlings. Soybean inoculated with such isolates showed typical stem canker symptoms, which suggests that infested cotton crop residue may also serve as a source of inoculum.

Environmental effects

The incidence and severity of stem canker has been observed to vary among years, even when the same susceptible variety was grown in the same field. This suggests that environmental conditions, particularly during the early vegetative stages, play a role in disease development. Maximum levels of infection were reported to occur when the air temperature was between 82 and 95°F, with lower levels of infection occurring at 50 to 72°F. Infection did not occur at 104°F. Fastest infection rates occurred between 72 and 86°F (Rupe and Sutton, 1994). Infection rates dropped sharply above and below this temperature range. Temperature may influence the distribution of the two forms of stem canker. Northern isolates were found to be nonpathogenic at temperatures over 86°F (Keeling, 1988).

Moisture plays an important role in the development of stem canker epidemics. Early-season rainfall during vegetative growth is critical for stem canker development. However, the exact moisture requirements are not clear. Damicone and co-workers (1990) stated that the cumulative rainfall and not the number of rainy days was related to disease incidence, while Subbarao and co-workers (1992) believed that the number of rainy days and not the total rainfall correlated with disease severity. Disease severity was higher on all varieties grown under sprinkler irrigation compared with nonirrigated plots, and appearance of symptoms was hastened by irrigation. Higher disease severity resulted from an increased rate of disease progress and extent and velocity of spread.

Moisture requirements for stem canker development are not rigid. Once spore deposition takes place, then a prolonged and continuous wetting event or many discontinuous wetting periods facilitate stem canker development. The length of the disease incubation period is influenced by the duration of the free moisture period. The longer water is available, the quicker disease symptoms appear. This relationship has epidemiological implications, since yield loss is due to plants dying prior to complete pod fill.

Conditions in the soil environment affect disease development. The addition of nitrogen, potassium, or phosphorus has reduced stem canker severity. Greater stem canker infection can be expected on soils with higher organic matter, pH, and soil-water content (Rhoton, 1989).

Other soybean pests can have either positive or negative impacts on stem canker development and effect. Stem canker lengths are longer on plants with girdles caused by threecornered alfalfa hoppers (Russin et al., 1986). These two organisms reduce yield more together than either alone. The more severe the defoliation by the soybean looper, the less severe the stem canker symptoms. Damage by stem canker is also less on plants damaged by the soybean cyst nematode (Russin et al., 1989b).

Control

Stem canker can be managed by (1) planting only resistant varieties, (2) using disease-free planting seed, (3) rotating soybean with other crops, and (4) plowing under crop residues. A program of residue destruction combined with the use of a resistant variety is the most effective method of controlling stem canker (Damicone et al., 1990).

Stem canker was prevalent in the midwestern U.S. during the late 1940s and early 1950s, but its impact diminished with a shift away from the use of highly susceptible varieties. A similar situation occurred in the midsouthern U.S. in the late 1970s and early

1980s. Most of the varieties grown at that time were very susceptible to stem canker (Weaver et al., 1984; Harville et al., 1986). Since that time, the development and use of stem canker resistant varieties has reduced the incidence of this disease.

A major problem facing breeders is the apparent association of stem canker susceptibility and soybean cyst nematode resistance. Therefore, breeders must be cautious when developing nematode-resistant varieties. A high incidence of stem canker could occur if susceptible varieties are grown again because the causal organism of this disease persists in the environment.

Seed harvested from fields having stem canker–infected plants should not be planted in noninfested fields. Treatment of seed with fungicides has been shown to reduce greatly but not eliminate stem canker (Crawford, 1984).

Foliar fungicides have been effective when applied during the early vegetative period (Backman et al., 1985), but disease control cannot be achieved on highly susceptible varieties. Varieties with intermediate susceptibility respond to fungicide treatment with yield increases commensurate with disease control, whereas no benefits were seen when resistant varieties were treated (Backman, 1984). For foliar and pod and stem diseases, fungicides are applied during the reproductive period. Attempting to use a fungicide for stem canker control would necessitate an additional application of fungicide that would only target this one disease. Using a foliar fungicide for stem canker would not be the most economical control strategy.

Benefits of crop rotation for control of stem canker have not been directly demonstrated in the midsouthern U.S. Since the fungus overwinters on crop debris, it is logical to assume that rotation with a nonhost crop would lead to decomposition of the debris and result in a reduction of primary inoculum. As previously mentioned, cotton can also be infected by *D. phaseolorum* f. sp. *meridionalis* and would be a poor choice as a rotational crop. Doublecropping with wheat may not result in disease suppression either. Disease severity was slightly higher in soybean following wheat than for monocropped soybean (Rothrock et al., 1985, 1988).

Planting date has been found to influence the severity and incidence of stem canker (Chambers, 1991). Disease ratings were highest from plantings in early May and were progressively lower as planting date was delayed into late June and early July. Delayed planting appeared to affect yield loss to stem canker in two ways. Plants were able to avoid the initial release of inoculum from crop debris that generally occurs during May, so fewer plants were infected. In those that do become infected, the shorter vegetative and reproductive period induced by late planting allows less time for the disease to develop to the point that plant death occurs before pod fill. The difficulty with using late planting as a disease-control strategy is that yield potential is limited by the shorter photoperiod accompanying the later planting dates.

In some cases (Rhoton, 1989), plant residue left on the soil surface from the previous crop year did not influence disease levels. In other cases, tillage had an impact on stem canker. No-till plots had a greater disease incidence than conventional tillage plots (Rothrock et al., 1985). Burying crop residue helped to reduce disease incidence and severity. In another study (Rothrock et al., 1988), conventional tillage plots of a susceptible variety had 47% higher yield than did no-till plots. Tillage did not have much of a yield effect on a resistant variety.

Anthracnose

Symptoms and losses

Soybean anthracnose is caused by the fungus *Colletotrichum truncatum* (Schw.) Andrus & W.D. Moore, and occurs in all major soybean-producing regions of the world. Soybean plants may be infected in all stages of growth, but symptoms generally appear only during germination and seedling stages, and during pod filling and senescence. Symptoms on emerging seedlings are typically reddish brown, sunken lesions concentrated on the outer surface of the cotyledons. These cankers gradually extend up toward the epicotyl and down to the root. Rot of the epicotyl and hypocotyl is likely to cause pre- and postemergence damping-off of seedlings. If the seedling does survive, the fungus may completely destroy one or both cotyledons or grow from them into the stem tissues. Once on the stem, the fungus produces many small, shallow, elongated, reddish brown lesions or large, deep-seated, dark brown lesions that kill the young plant. Seedlings that avoid the death of critical tissue may survive and after 3 to 4 weeks symptoms usually disappear. These plants, however, often lag in their development compared with plants that show no seedling disease.

Stems, pods, and leaves may be infected during the vegetative stages but symptoms are usually absent. Roots may also be infected. Symptoms reappear when conditions are favorable for spore production. Lower branches, leaves, and pods may be attacked. On older plants, anthracnose reaches its most destructive stage during rainy periods in late summer.

Symptoms on leaves are varied. Necrotic local lesions are centralized on the underside of the leaflet veins. In response to severe aerial attacks of soybean anthracnose, leaves may cup downward and fall prematurely. Local lesions on petioles are nearly rectangular in shape with a dark brown color. Premature defoliation can also occur when the lesions girdle the leaf petiole. Irregularly shaped brown areas develop and coalesce to cover the entire surface of the diseased stems and pods. Numerous black fruiting bodies are formed on the diseased area. As these structures develop, several short, dark spines emerge from each fruiting body, such that each fruiting body looks like a tiny pincushion.

In early stages of pod development, infection results in pod abortion. This early infection of pods can result in either no seeds forming (pod blanking), or the seeds that develop are fewer or smaller than normal. Later infections cause local lesions on the outer surface of the pod. Seeds in diseased pods may be shriveled, dark brown, and moldy, or infected seeds may show no outward sign of disease. Mycelium of the fungus occasionally can completely fill the pod cavity.

Pod and seed damage is the major component of yield loss for soybean infected with anthracnose. Anthracnose can cause significant yield losses in the warm and humid areas of the midsouthern U.S. Anthracnose has been described as the most serious fungal disease in Alabama in terms of yield reduction when compared with brown spot, frogeye leaf spot, and pod and stem blight. In a study conducted in Alabama, natural infection caused yield losses ranging from 16 to 26%, with an average of 20% (Backman et al., 1982). Using artificial inoculation, yield losses ranged from 17 to 30% in Illinois (Khan and Sinclair, 1992). When both are present in the same seed, *C. truncatum* and *Phomopsis longicolla* (the cause of Phomopsis seed decay) have an additive effect on the destruction of seed coat tissues (Sinclair and Backman, 1989).

Disease cycle

C. truncatum has a wide geographic distribution, possibly because it is a good dormant parasite and saprophyte of many diverse plant species. The fungus survives from one growing season to the next in diseased crop residue left in the field. Anthracnose can also overwinter in infected seed, which serves as a source of primary inoculum as well as an agent for spreading the fungi to new locations. The fungus can become established in infected seedlings without symptoms developing until the plants begin to mature. Latent or symptomless infections act as reservoirs of the fungus. A secondary cycle of inoculum production starts as lower leaves die from old age or shading. These late-season infections cause the most damage to yield and seed quality.

Environmental effects

Prolonged periods of high rainfall, humidity, and temperature favor severe disease development. Detailed information on environmental conditions that promote disease development is lacking. *C. truncatum* depends on a film of water to penetrate the soybean epidermis. With free water, spores germinate within 4 to 16 h at 68 to 86°F. Fungal growth is sustained when relative humidities exceed 83%. Infections typically occur if rain, dew, or fog provides free moisture for periods of 12 h or more. The optimum temperature for root infection is when soil temperature is 86°F.

Late-season and full-season weed competition greatly increases soybean anthracnose under humid conditions. Weeds alter the microclimate and improve conditions for disease development. Weeds can also serve as alternate hosts for anthracnose, acting as possible sources of inoculum. *C. truncatum* was isolated from 35% of the cocklebur plants sampled and from 15% of the purple nutsedge plants sampled.

Mineral nutrition can influence the severity of anthracnose. Calcium and potassium deficiency can predispose the plant to infection. In both greenhouse and hydroponic tests, the addition of calcium reduced the incidence of seedling anthracnose. In field tests, disease ratings decreased as potassium rates increased to a maximum of 400 lb/acre. When soybean is planted on the same land the following year, the effect of residual potassium on disease suppression was also evident.

Control

Anthracnose can be managed by (1) use of disease-free planting seed; (2) treating seed with a fungicide (Vitavax CT, Vitavax — M, or Vitavax M DC); (3) applying a foliar fungicide, such as 0.25 lb benomyl (0.5 lb Benlate®) at R3 and R5; (4) plowing under crop residues; and (5) rotating soybean with other crops.

There are no commercially available anthracnose-resistant varieties. Attempts to identify resistant germplasm have been unsuccessful (Bowers, 1984; Manandhar et al., 1988). All material screened was found to be susceptible to some degree. However, there is some evidence of varietal differences in yield loss caused by anthracnose.

Using fungicides to control anthracnose results in yield increases (Hepperly, 1985). Treating infected seed with a fungicide improves seedling stands (Athow, 1987). In Texas, applying 0.5 lb/acre of benomyl at both R3 and R5 reduced disease incidence by as much as 60%, resulting in higher yields (Sij et al., 1985). Similar work in Alabama measured a yield response to benomyl only when conditions were favorable for disease development. Favorable conditions were defined as at least 5 days between bloom and pod fill receiving a minimum of 0.08 in. of rain. Under such conditions, benomyl increased yield of early-maturing varieties by about 15% and of late-maturing varieties by 11%. If the period between bloom and pod fill was dry, yield increases associated with benomyl application were less than 1%. Researchers in Alabama developed a predictive system to reduce the use of a fungicide when no benefit would be realized. This system recommends the application of 0.25 lb/acre benomyl if 2 or 3 favorable days for infection occur in the period between R1 and R3. If it is still dry 1 week after R3, then the use of fungicides should not be considered. A second application would be warranted if 2 more favorable days occur in the period between R3 and R3 plus 10 days. The cost of this or any other control measure must be considered when deciding to implement a disease management strategy. Benlate cost approximately $15/lb, plus application costs.

Pod and stem blight and phomopsis seed decay

Symptoms and losses

These associated diseases are caused by a complex of fungi including *Diaporthe phaseolorum* (Cke. & Ell.) Sacc. f. sp. *sojae* (Lehman) Wehm., *Phomopsis sojae* Lehman, and *Phomopsis longicolla* Hobbs. *Phomopsis* seed decay is caused primarily by *P. longicolla*. The most noticeable symptom of pod and stem blight is the linear rows of dark, speck-sized fruiting structures called pycnidia on stems, petioles, and pods. Under field conditions, no definite leaf or stem lesions are produced. The disease can first be detected in the field as pycnidia on petioles of abscised leaves or on broken branches. Pycnidia develop on the main stem and upper branches after death of the entire plant following maturity or death from other causes. Dead stems may be covered with the pycnidia, or the pycnidia may be limited to small patches, usually near the nodes. Pycnidia are also scattered on dry, poorly developed pods. Not all infected pods produce pycnidia, but mature pods with pycnidia always contain infected seeds.

The most important aspect of this disease complex is its effect on seed. *Phomopsis* is a major cause of poor seed quality in regions where the climate is warm and humid during and after crop maturity. *Phomopsis* destroys the protective value of the seed coat, resulting in a general loss of seed quality. Seed quality losses attributed to "weathering" are due in part to *Phomopsis* infection. Soybean seed germination is closely related to the incidence of *Phomopsis* seed infection. Field emergence declines with an increase of *Phomopsis*-infected seed. Symptomatic seed are smaller in size, lower in density, produce lower-quality oil and flour, and have lower durability than symptomless seed. Severely infected seed crack and shrivel and are often covered with a white to gray mold. However, most infected seed display few visible symptoms. *Phomopsis* can lower seed grade by reducing test weight and by increasing the number of split seed. Flour produced from moldy seed may be unacceptable for human consumption. The oil produced from moldy seed is dark and has a rancid odor.

There is little relationship between seed infection and seed yield (Bisht and Sinclair, 1985). In fact, the effect of either pod and stem blight or *Phomopsis* seed decay on yield is unclear. In one study (Bisht and Sinclair, 1985), inoculation of plants with *P. sojae* did not reduce yield compared with the noninoculated check. Horn and co-workers (1975) reported the variety Dare inoculated with *D. phaseolorum* f. sp. *sojae* yielded 41.2 bu/acre, while the noninoculated control yielded 46.6 bu/acre, and the plots treated with a fungicide yielded 53.4 bu/acre. Pod and stem blight, with or without the development of *Phomopsis* seed decay, was reported to reduce yield in a high-yield environment but not in a low-yield environment. This yield reduction was a result of a shortened seed-filling period caused by premature plant death.

Disease cycle

The pod and stem blight/*Phomopsis* seed decay fungi overwinter in soybean plant debris and infected seed. The overwintering straw contains an abundant number of pycnidia and spores and is a more important inoculum source than seed. Seed-borne *Phomopsis* can be transmitted to the resulting seedling, but dead *Phomopsis*-infested seed do not infect adjacent seedlings in the field (Garzonio and McGee, 1983). There is no relationship between the percentage of *Phomopsis*-infected seed planted and the subsequent severity of symptoms on seedlings or on mature plants. Plant infection can occur at any point in the growing season. This complex of fungi is among the first microorganisms to infect a soybean plant. Generally, young plants are infected by spores (originating from crop

debris) deposited on the plant by splashing rain. Infections of green cotyledons, hypocotyls, stems, petioles, and pods are localized and latent. Initially, the fungi colonize plant
tissue within 1 in. of the point of infection (Sinclair and Backman, 1989). Under the proper
environmental conditions, physiological changes within a senescing or dying plant allow
the fungi to spread.

Pod infections occur independently of stem infections (Kmetz et al., 1979). Infected
stems and petioles do not lead to a systemic spread of the fungus to the seed. Inoculating
plant stems 5 weeks prior to maturity had no effect on seed infection (Athow and Laviolette, 1973). The source of inoculum for late-season infections could be from latent infections on stems and pods, from current crop debris, or from the previous season's straw.
Whatever the source, the maximum production of spores during the season occurs during
pod fill.

Most seed infection occurs during or after the beginning-maturity stage (R7). Colonization of immature seed begins in infected pods after R6. In fact, there is a close relationship
between pod infection at R6 and seed infection at harvest maturity. In one study (McGee,
1986), low levels of pod infection were detected between R4 and R7, at which time the
incidence of detectable pod infection increased markedly. It was suggested that this
increase in detectable infection was due to infection already present and not to new
inoculum produced after R6. There was a further colonization of pods already infected
because pods are not susceptible to infection at R7 or R8. After R7, seed water content
declines from about 55% to about 12 to 14% at harvest maturity. *Phomopsis* grows from
infected pods into seeds during this declining seed moisture period. At 77°F, seed infection
rates were linearly related to seed water content within the range of 19 to 35% moisture
(Rupe and Ferriss, 1986). Seed infection did not occur when seed moisture was below
19%. In other research (Balducchi and McGee, 1987), seed infection was greater in pods
detached at R7 and exposed to 85 to 100% relative humidity for a period of 7 days than
when the pods were detached at R8 when seed moisture was lower. Much seed infection
seems to occur after seed is mature and may be associated with the deterioration of the
pod wall. Physically protecting the pod prior to maturity did not affect seed infection or
germination. Protecting the pod for 6 weeks after maturity significantly increased germination and reduced seed infection.

Environmental effects

Optimum conditions for infection are high levels of rain, temperature, and relative humidity. Moisture (humidity) is the most important factor influencing *Phomopsis* seed infection
during and after maturation. Temperature plays an important, though lesser, role in
infection. High temperature, in combination with high humidity, contributes to greater
fungal growth on seed and lower germination than does lower temperature. There is a
low positive correlation between temperature and seed decay, and a high positive correlation between rainfall during pod fill and seed decay. Both the timing and amount of
rainfall affect disease incidence. Another study found levels of seed infection correlated
with temperature and minimum relative humidity during both R5 to R7 and R7 to R8,
but not with total precipitation or precipitation per day. The correlation between minimum
relative humidity and infection was higher than that with temperature during both reproductive periods. Moisture is critical for infection to occur, whereas temperature mainly
affects the rate of colonization. At 77°F, 3 continuous days of 100% relative humidity are
needed for extensive seed infection to occur at R8, 4 days are needed at 68°F, and 5 days
are needed at 59°F. The optimum temperature for seedling infection is 76°F.

Late-season rainfall or sprinkler irrigation increases the incidence of seed infection.
Extensive seed infection may be expected only when there is precipitation, particularly

during the day, that keeps relative humidity near 100% for prolonged periods. Precipitation can be in the form of rain or fog. The duration rather than the amount of precipitation is important. If this period of high relative humidity is interrupted with 1 to 3 days of low (40 to 60%) relative humidity, then more time at high relative humidity is required for 90% seed infection to occur. The longer the duration of interruption, the longer the subsequent period of high humidity required. Seed produced under low-soil-moisture conditions during the R2 to R8 period had less *Phomopsis* seed decay (18 vs. 58%) and higher germination (66 vs. 47%) than seed produced under high-soil-moisture conditions.

Seed-borne *Phomopsis* reduces emergence and stand establishment most when seed are incubated in dry to marginally moist soil compared with moist soil. Two related factors may cause this: a growth rate advantage of the fungus over that of the host and a prolonged preemergence period. A soil moisture that is too low to promote vigorous seed germination but high enough to allow pathogen activity results in more seedling infection.

Delaying harvest, or allowing mature soybean plants to remain in the field, significantly increases seed infection. A 6-week delay in harvest was reported to result in a fourfold increase in seed-borne *Phomopsis* (Athow and Laviolette, 1973). Delayed harvest affects both seed vigor and seed viability. Declines in seed vigor associated with a delayed harvest occurred 3 to 6 weeks sooner than similar declines in viability (TeKrony et al., 1980). Vigor declined to less than 50% germination within 30 days of harvest maturity. The time required for seed vigor to decline significantly with delayed harvest was closely related to average air temperature, average minimum relative humidity, and precipitation per day. Delaying harvest 4 weeks increased the incidence of seed infection from 3% for harvest at maturity to 20% for delayed harvest. Field emergence of the seed from the delayed harvest was reduced 37%.

Maturity date influences the severity of seed infection. Within a region of adaptation, *Phomopsis* seed decay is greatest on early-maturing varieties and decreases on later-maturing varieties. The higher disease severity for the early-maturing varieties is related to higher temperatures experienced during seed maturation; therefore, seed are more likely to be infected during wet periods (Balducchi and McGee, 1987). Maturity late in the season results in the plants escaping conditions conducive to a high incidence of seed decay. Using flowering date and maturity isolines, it was reported that late flowering and late maturity decreased seed infection 48% and improved seed germination 97% compared with normal isolines. Artificially delaying maturity by removing some of the pods increased seed infection, probably as a result of lengthening the period during which infection and colonization could take place. There was a positive relationship between the length of the period between R6 and R7 and Phomopsis seed decay. The length of the period between R6 and R7 was more important than the period between R7 and R8 or between R8 and harvest, which is probably because the pathogen is more active in the pods and seeds prior to R7 than during senescence.

The location of a seed on a plant is also important regarding the incidence of seed infection. More *Phomopsis*-infected seed are found in pods produced on lower nodes than on upper nodes. This relationship probably occurs because the lower pods on a plant are closer to a major source of inoculum, which is infected plant debris on the soil surface,. Also, lower pods may stay wet slightly longer than pods on upper nodes, creating a better environment for seed infection.

Several studies have found significant relationships between potassium deficiency and increases in moldy seed. High levels of potassium (exceeding that necessary to maximize yield) generally reduced levels of moldy seed. However, germination and yield responses to increased potassium fertilization were less consistent. Even where seed germination was improved by higher rates of potassium fertilization, the incidence of seed infection showed no response to potassium. Their results suggested that, whereas potassium fertilization might not influence the infection of seed directly, it might reduce the severity of infection and limit fungal growth after infection has occurred. In another study, significantly less seed decay occurred as potassium level increased in seed, although the addition of potassium chloride or potassium sulfate did not completely prevent infection. Again, potassium had no significant effect on yield. In a third study, potassium fertilization reduced the incidence of *Phomopsis* in seed harvested at maturity from 35 to 1% and in 4-week delayed-harvest seed from 52 to 28%. Additions of phosphorus or nitrogen fertilizer did not influence seed infection.

Antagonistic interactions between *Phomopsis* and other seed-borne fungi have been reported. Studies have indicated an inverse relationship of seed infection between *Phomopsis* and *Alternaria* (Ross, 1975, 1982a), *Phomopsis* and *Fusarium* (Ross, 1982a), and *Phomopsis* and *Cercospora kikuchii* (Hepperly and Sinclair, 1981). However, the recovery of *Phomopsis* and *C. sojina* from pods, stems, and seeds is not affected by the presence of the other (Bisht and Sinclair, 1985). These two fungi act independently and the effect of their damage is additive.

Control

Pod and stem blight and *Phomopsis* seed decay can be managed by (1) planting only high-quality, disease-free seed; (2) treating seed with a fungicide (Vitavax CT, Vitavax — M, or Vitavax M DC); (3) use of late-maturing varieties; (4) applying a foliar fungicide, such as 0.25 lb benomyl (0.5 lb Benlate) at R3 and R6; (5) harvesting soybean promptly at maturity; (6) plowing under crop residues; and (7) rotation with other crops.

At present, no satisfactory resistance to *Phomopsis* is available in widely used, adapted soybean varieties. Variation in seed infection between genotypes of identical maturity has been reported (TeKrony et al., 1984). However, information on the relative susceptibility of currently grown varieties to *Phomopsis* seed decay is generally unavailable. A MG IV germplasm line with a very high level of resistance to seed infection was identified after examining over 3000 lines. However, this resistance is not yet incorporated into commercial varieties. In the future, breeders might also alter the development of a plant to provide a genetic control for *Phomopsis* seed decay. Lines with a shorter period between R7 and R8 had better seed quality than lines with relatively longer maturation periods.

Fungicides have provided effective control of pod and stem blight and *Phomopsis* seed decay. Seed treatment with captan reduced *Phomopsis* infection of germinating seed. Applying 0.5 lb/acre of benomyl at both R3 and R6 reduced the number of plants with symptoms of pod and stem blight. In another study, application of 0.125, 0.25, or 0.5 lb/acre of benomyl at either R3, R6, or at both R3 and R6 resulted in the reduction of pod and stem blight symptoms and reduced seed infection below that for the nonsprayed control. The authors of that study suggested that a single application between R3 and R6 would provide effective disease control. Generally, applying 0.25 to 0.5 lb/acre of benomyl results in decreased seed infection and increased seed germination. The success of a foliar fungicide depends upon applying it before seed infection occurs. Benomyl does not eradicate seed infection already present at application time, although it may arrest further progress of the disease. The application of 0.5 lb/acre at both R4 and R6 or 1.0 lb/acre at R6 lowered levels of seed infection but waiting until R7 to apply 1.0 lb/acre did not (TeKrony et al., 1985a).

The effect of benomyl on increasing the seed yield of *Phomopsis*-infected plants is less clear. TeKrony and others (1985a) found no yield increase associated with fungicide application. Another study reported the yield of Dare soybean increased from 41.2 to 53.4 bu/acre with the application of benomyl. This difference in results may be explained by the fact that benomyl was reported to increase yield only in an environment with high irrigation or rainfall and not under low irrigation or rainfall conditions (Slater et al., 1991).

Applying a fungicide when it is not needed has unacceptable economic and environmental costs. Knowing if a disease (that shows no symptoms at the time control measures need to be implemented) will be a problem requires some type of predictive measure. Several states, including Alabama, Arkansas, and Mississippi, have developed predictive point systems for scheduling fungicide spraying. These point systems are based on various factors that include cropping history, variety, planting date, rainfall between R2 and R7, yield potential, seed production, and other environmental considerations. Fields receiving scores exceeding a threshold level need to be sprayed. Fields with marginal scores require the use of a pod test (TeKrony et al., 1985b). This pod test is based on the fact that there is a strong correlation between natural pod infection at R6 and later seed infection. Pod infection can be measured and a fungicide applied at R6 without the risk of seed infection having already occurred.

Cropping history is an important factor in disease development, but one should not automatically apply a fungicide because of growing continuous soybean (TeKrony et al., 1985b). Alternatively, it should not be assumed that by rotating with other crops the disease will be completely eliminated. *Phomopsis* seed decay was found to be highest in continuous soybean, less in a corn–soybean rotation, and least following continuous corn (Garzonio and McGee, 1983). However, *Phomopsis* was still detected in seed harvested from soybean grown in a field that had been continuously cropped to corn for 10 years.

Since the main source of disease inoculum is from overwintering soybean debris, it is logical to assume that burying crop residue to hasten decomposition will reduce subsequent disease incidence. However, in the only published study (Jeffers and Schmitthenner, 1981), burying soybean residue by fall or spring plowing did not affect germination or disease incidence when compared with soybean grown without tillage. A secondary source of disease inoculum is infected seed. Using high-quality, disease-free seed is a good general recommendation in soybean production. However, using good-appearing seed does not always ensure a lack of *Phomopsis* infection. Dead, shrunken seed are easily removed by seed conditioning equipment, but viable infected seed rarely show visible symptoms and are normal in size (Garzonio and McGee, 1983).

Rhizoctonia foliar blight

Symptoms and losses

Rhizoctonia foliar blight, also known as aerial blight, is caused by the fungus *Rhizoctonia solani* anastomosis group (AG) 1. This disease was first reported from Louisiana in 1951 and was considered to be epidemic there by 1973. According to Joye and co-workers (1990), *Rhizoctonia* foliar blight affects more than half of the soybean acreage in Louisiana. The disease also occurs in other states along the Gulf of Mexico. *Rhizoctonia* foliar blight is caused by two types of *R. solani* AG-1: intraspecific group A (IA), which causes aerial blight, and intraspecific group B (IB), which causes web blight (Yang et al., 1990a). Fairly large (up to 0.16 in. in diameter) sclerotia are produced by IA isolates, whereas microsclerotia produced by IB isolates resemble sand grains. The IA type of *R. solani* AG1 also causes sheath blight in rice, and rice–soybean doublecropping may contribute to increased disease incidence on both crops. Soybean yield loss in Louisiana was estimated at 3% in 1992, but losses in individual fields as great as 35% have been reported. However, controlled studies to measure yield loss in soybean accurately are lacking.

Seeds and seedlings can be infected by both types of *R. solani* AG-1. These pathogens cause seeds to rot prior to emergence and produce lesions on hypocotyls and shoot apices of surviving seedlings. Lesions initially are red but later turn brown and dry. Mycelial webbing and production of sclerotia and microsclerotia may be observed on diseased seedlings. These symptoms differ from those described previously for seedling disease caused by *Rhizoctonia*. Plants with this disease exhibit water-soaked lesions on roots and stems accompanied by as much as 50% reduction in soybean stand.

Foliar symptoms can be found throughout the soybean canopy later in the season. Lesions on leaves first appear water soaked and grayish green, with mycelium spreading over the leaf surface in advance of the lesion. Mycelium spreads by contact between leaves, resulting in mats of webbed, diseased foliage followed by defoliation. Diseased leaves frequently fall onto lower leaves, stems, or pods, which can be subsequently colonized by the fungi. Considerable effort has been made to determine differences in symptomatology between the IA and IB isolates. In addition to production of larger sclerotia, infections resulting from IA isolates typically exhibit extensive blighting of foliage. These symptoms generally are restricted to within the soybean crop canopy. In contrast, IB infections frequently are associated with distinct leaf spots, which results in a shot-hole appearance when weather is unfavorable for disease development. Microsclerotia are common on such plants. Leaf spot symptoms occur on more-exposed portions of the canopy, which suggests a role for airborne microsclerotia in disease development. Both IA and IB can occur in the same field and even on the same plant. Therefore, symptoms tend to be a combination of both types and frequently are difficult to separate. Routine evaluations of disease severity do not distinguish between these two types but rather combine all symptoms together (Harville et al., 1996).

In addition to foliage, symptoms also can occur on stems and pods. Reddish brown to brown lesions can form on stems and pods after canopy closure. Severe infection can cause blighting and death of young pods. This pod-blighting phase is believed responsible for most yield losses. Seed infection also results from pod infection.

Disease cycle

The epidemiology of *Rhizoctonia* foliar blight has been divided into two distinct phases, one before and one after canopy closure. The fungi overwinter as sclerotia or microsclerotia in soil or diseased plant debris. These inocula infect seedlings or can be splashed onto stems and foliage during rainfall. Infected seedlings may become sources of inoculum and have a significant impact on subsequent disease development. Under very wet soil conditions, mycelia also can reach foliage by growing directly up the outside of stems. Following canopy closure, the pathogens can spread extensively along and across rows. It is at this time that the disease expands most rapidly. Production of sclerotia and microsclerotia is abundant at that time (Yang et al., 1990a). Diseased soybean debris supporting mycelia, sclerotia, and microsclerotia eventually falls to the soil surface. Because it has both soilborne and foliage components, *Rhizoctonia* foliar blight is a highly clustered disease.

Several weed species are hosts for *R. solani* AG-1 IA and IB. Black and co-workers (1996a) reported that 14 weed species common to Louisiana soybean fields were hosts for both types of AG-1. These included species of broadleaf and grass weeds, as well as sedges. They showed further that the fungi can spread from inoculated weed plants to adjacent healthy soybean plants, with weeds thereby acting as "bridge hosts" for the pathogens. Sclerotia and microsclerotia were produced on these weed species, which suggests that infected weeds also may be important for overwintering of inoculum or for survival of the pathogens in the absence of soybean.

Environmental conditions

Development of *Rhizoctonia* foliar blight depends on several environmental conditions. The primary means of pathogen spread in the soybean canopy is through water splashing of inoculum onto foliage or by growth of mycelia on overlapping leaves. Therefore, high temperatures accompanied by extended periods of high humidity, rainfall, and cloudy weather commonly are favorable for disease development. Free moisture on plant surfaces is very important for disease development.

In a field study, Joye and others (1990) reported that disease severity was reduced when row spacings were ≥20 in. but was not affected by within-row plant populations that varied from 3 to 12 plants per foot of row. Although environmental conditions were not measured, wider row spacings presumably allowed greater air circulation within canopies which reduces disease development. Several postemergence herbicides influence severity of *Rhizoctonia* foliar blight under experimental conditions. Notable among these was paraquat, which showed modest reductions in disease severity when applied post-directed (Black et al., 1996b).

Control

Currently, there are no effective controls for *Rhizoctonia* foliar blight. Soybean varieties resistant to *Rhizoctonia* foliar blight have not been identified, although some are less susceptible than others (Harville et al., 1996). Occasional control has been achieved using high rates of benomyl or chlorothalonil applied at R3 and R5.

Brown spot

Symptoms and losses

Brown spot is caused by the fungus *Septoria glycines* Hemmi. This disease is primarily a leaf spot disease, although seeds, stems, and pods of maturing plants can also be infected. The first symptom is the appearance of irregular, dark brown patches on the cotyledons. Next, reddish brown angular spots ranging from tiny specks to areas 0.15 in. in diameter appear on both upper and lower surfaces of unifoliate leaves. These leaves quickly turn yellow and drop. Adjacent lesions frequently merge to form irregularly shaped blotches. Numerous irregular, light brown lesions form on trifoliolate leaves. These lesions gradually darken until they become chocolate brown to blackish brown. The fungus progresses from lower to upper leaves. Late in the growing season, leaves become rusty brown and drop prematurely. Moderate to severe infection can cause premature defoliation.

Irregular brown lesions with indefinite margins form on the main stem, branches, petioles, and pods. These lesions range from tiny specks to areas of several square inches. Symptoms on these plant parts are not sufficiently distinct from those of other diseases to be diagnostic.

Brown spot is considered the most prevalent foliar disease in the midwestern U.S. (Lim, 1979) and sporadically causes yield losses in the southeastern U.S. (Ross, 1982b). Due to the defoliating nature of this disease, infection during early- and mid-reproductive growth stages may decrease yield. Yield reductions ranging from 2 to 34% have been reported under artificially inoculated conditions and from 8 to 10% for natural infections. Reductions in yield are due to smaller seed. Yield and seed weight have been negatively correlated with the percentage diseased leaf area (Young and Ross, 1979).

Disease cycle

Primary inoculum for this disease comes from diseased seeds or from the fungus over-wintering in leaf and stem debris. Spores are spread by wind and splashing rain. Young plants are commonly infected in the spring, with infection occurring first on unifoliate leaves. Lesions can form and sporulate throughout the season. The disease and subsequent defoliation progress from the lower leaves to the upper leaves of the plant. Usually the hot dry weather of midseason arrests disease development, but the disease may become active again near maturity. If conditions favorable for disease development continue into the season or recur before maturity, the resulting defoliation can cause serious yield reductions. Infection rates of artificially inoculated plants were highest at the time pods were still growing or during pod filling.

Generally, most severe brown spot epidemics occur late in the season. The development of a large number of lesions in the fall apparently results from an accumulation of infections that occurred at flowering and remained latent until late reproductive stages. The accumulation of infections accounts for the rapid development of the disease observed in the fall (Young and Ross, 1979).

Even low disease severity leads to premature defoliation. This is important because these dropped leaves act as an inoculum base for epidemic outbreaks following favorable weather conditions for disease development. The amount of brown spot is related to the amount of defoliation, which is associated with yield loss. Premature senescence (defoliation) occurs when there is at least 10% infected leaf area on the lower leaves or at least 25% infected leaf area on the upper leaves. Loss of 25 to 50% of the upper leaves of a plant has resulted in yield reduction.

Environmental effects

Warm, humid, rainy conditions favor increased disease severity and spread. High temperature and wet leaves increase the number of lesions formed. Leaf infection occurs between 60 and 90°F with an optimum temperature range of 77 to 82°F.

Spores do not need free moisture to germinate as is the case for many other foliar diseases. However, leaf wetness is needed for infection to occur. Research done in a growth chamber reported that a minimum of 72 h of leaf wetness is required for infection, with longer periods resulting in accelerated fungal development (Peterson and Edwards, 1982). Other work done in the field found infection occurring after only 6 h of leaf wetness. As the period of leaf wetness increased from 6 to 36 h, disease severity increased. It has been demonstrated that a daily dew in lieu of continuous wetness following inoculation is adequate for infection to occur (Ross, 1982b).

The higher brown spot disease levels observed in the spring and fall are due to more favorable temperatures occurring in the field at those times. In Illinois, less disease has been observed in an early-maturing variety in a year with little rain and frequent temperatures above 80°F in July and August compared with a year having more rain and cooler temperatures. Considering that high temperatures generally occur in late summer in the midsouthern U.S., it is reasonable to expect brown spot to be limited to a late-season problem. Plants maturing in August and early September, such as early-planted MG IV and MG V varieties, should be little affected by brown spot. However, prolonged periods of cloudy, rainy weather in late summer could increase the severity of this disease.

Control

Brown spot can be managed by (1) use of disease-free planting seed; (2) rotation with a non-susceptible crop for at least 1 year; and (3) applying a foliar fungicide, such as 0.25 lb benomyl (0.5 lb Benlate) at R3 and R6.

Although differences in susceptibility exist, no highly resistant varieties are available. Several thousand soybean lines were screened for their reaction to brown spot using artificial inoculation, but no resistant lines were identified (Young and Ross, 1978; Lim, 1979). The best lines still suffered 30% defoliation. Differences in the rate of disease development were observed. Yield losses caused by brown spot have been found to vary by variety. In addition, the maturity of a variety may affect the severity of brown spot symptoms. Brown spot can be severe on both early- and late-maturing varieties when conditions are favorable for disease development. However, brown spot is more severe on late-maturing varieties under less-favorable conditions.

Benomyl reduces the incidence of brown spot and significantly increases yield when conditions are favorable for disease development. Disease control is achieved primarily by slowing the rate of disease development. However, applying 1.0 lb/acre benomyl at R3 or applying 1.0 lb/acre at both R1 and R6 has effectively reduced disease incidence (Pataky and Lim, 1981a). When benomyl was only applied at R6 the epidemic was severe because the disease developed unsuppressed throughout most of the reproductive development of the plant, and when the fungicide application occurred only at R1, brown spot infection rates increased during the pod-filling period, resulting in severe disease.

Rotation is recommended because it has been observed that the severity of brown spot is greatest in fields where soybean are grown consecutively (Lim, 1979). Row spacing was found to have little impact on brown spot severity or incidence. Yield loss caused by this disease was no different between soybean grown in 7 or in 30 in. rows. The canopies of the narrow rows closed 2 to 3 weeks before those of the wide rows and might contribute to a microclimate conducive to the development of brown spot. However, disease severity was not any more severe under narrow-row culture.

Frogeye leaf spot

Symptoms and losses

Frogeye leaf spot, caused by the fungus *Cercospora sojina* Hara, is primarily a disease of the leaves, but stems, pods, and seeds may be infected. Leaf lesions are circular to angular spots varying in size from less than 0.05 to 0.2 in. in diameter. When leaf spots are first visible on the upper surface of a leaf, they are approximately the size of mature lesions. Leaf spots begin as dark, water-soaked areas with or without a lighter center. Young, fully developed lesions have a gray to brown center with a distinct purplish to reddish brown margin. As lesions age, the center becomes tan to nearly white and the margin darkens. Several of these lesions may coalesce to form larger, irregular spots. The spots are darker on the lower surface of leaves. The absence of yellowing around the spot is a distinguishing characteristic. Numerous lesions cause the leaf to become dry and drop.

Stem symptoms appear late in the season, are less common, and are not distinctive. The lesions are two to four times as long as they are wide and may extend up to half the circumference of the stem. Young lesions are deep red with a narrow, dark brown to black margin. As they age, their centers become brown and then turn a pale smoky gray. Stem infections may be latent with symptoms not appearing until after the plant reaches maturity (Bisht and Sinclair, 1985).

Lesions on pods are circular to elongate, slightly sunken, and reddish brown. Older lesions become brown to light gray, usually with a narrow, dark brown border. Often seeds opposite a pod lesion will also become infected. Seeds infected with *C. sojina* develop conspicuous light to dark gray or brown areas which vary from specks to large blotches covering the entire seed coat. Some lesions show alternating bands of light and dark brown. Usually, the seed coat cracks or flakes.

In inoculated plots in Indiana, frogeye leaf spot reduced yield from 17 to 21% (Laviolette et al., 1970). Frogeye leaf spot reduced yield 10%, in similar studies in Illinois, but this reduction was not statistically significant. Similar information is not available for the midsouthern U.S., but environmental conditions for disease development and spread are more conducive in this area than in the midwestern U.S. Therefore, it is not unreasonable to expect disease losses, under favorable conditions, to equal or exceed those reported for the midwestern U.S.

Disease cycle

The fungus survives in infected seed and soybean debris. Germination of infected seeds is slightly reduced. Only severely infected seeds may fail to germinate. Infected seeds that germinate may produce weak, stunted seedlings with lesions on the cotyledons. Sporulation on the cotyledonary lesions provides inoculum for infecting young leaves. On artificially inoculated plants, the first lesions are visible after 9 to 10 days, and the first spores are produced within the next 24 to 48 h (Sinclair and Backman, 1989). In warm, humid weather, sporulation is profuse. The spores are carried short distances by wind and splashing rain. Under favorable conditions, secondary infections of leaves, stems, and pods arise throughout the season. Secondary spread can be rapid and uniform over the plant. Young expanding leaves are extremely susceptible to invasion, while fully expanded ones are more resistant. However, because visible lesions develop nearly 2 weeks after infection, lesions are never seen on young expanding leaves.

Environmental effects

This disease occurs most commonly in the midsouthern U.S. where warm, humid conditions prevail. In other areas, the occurrence of frogeye leaf spot is sporadic, but is more severe in seasons with frequent rainfall. With adequate moisture, new leaves become infected as they develop, until often all the leaves of a plant are infected. Leaves that develop and become fully expanded during dry periods unfavorable for invasion may remain relatively disease-free. As a result, a layered pattern of heavily diseased and very lightly diseased leaves sometimes occurs on the same plant.

Control

Frogeye leaf spot can be managed by (1) planting resistant varieties; (2) use of disease-free planting seed; (3) rotation with a nonleguminous crop; (4) treating seed with a fungicide; and (5) applying a foliar fungicide, such as 0.5 lb benomyl (0.5 lb Benlate) at R3 and R5.

Frogeye leaf spot has been effectively controlled with resistant varieties. Reactions of different varieties vary from immunity to highly susceptible. The more susceptible the variety, the more uniform and larger the spots and the more profuse the sporulation. Varieties having an intermediate reaction have fewer sporulating spots. Three major dominant genes (Rcs_1, Rcs_2, and Rcs_3) confer resistance to frogeye leaf spot.

C. sojina is an extremely variable pathogen, encompassing physiological races that differ in their capacity to cause frogeye leaf spot on a wide range of soybean varieties. Five races of the fungus have been reported and described (Phillips and Boerma, 1982). In the 1960s and 1970s, many of the soybean varieties grown were resistant to races 1 to 4. Race 5 was found in 1978 and many important varieties were found to be susceptible (Phillips and Boerma, 1980). The variety Davis was found to be resistant to race 5 and to all other described races in the U.S. Recently, native isolates found in the midsouthern U.S. did not correspond to any of the previously described races from the U.S. A large number of varieties was screened for their reaction to one of these undescribed isolates; 37% were rated susceptible (Pace et al., 1993). It is possible that the fungus is changing, and our current varieties and sources of resistance may not provide long-lasting protection against this disease.

Foliar fungicides such as benomyl applied from bloom to early pod set have been effective in controlling frogeye leaf spot. Benomyl application increased yield when disease potential was high. Yield increases as high as 73% have been attributed to chemical control of frogeye leaf spot (Horn et al., 1975).

Cercospora leaf blight and purple seed stain

Symptoms and losses

Cercospora leaf blight and purple seed stain are both caused by the fungus *Cercospora kikuchii* (T. Matsu. & Tomoyasu) Gardner. Under field conditions, the first symptoms of the disease are observed during late pod fill and full seed growth stages. Foliar symptoms become most evident during the latter part of the growing season as the plants approach maturity. Upper leaves exposed to the sun have a light purple, leathery appearance. This discoloration can deepen and extend over the entire upper surface of affected leaves, giving the leaves a dark, reddish purple appearance highlighted with bronzing. Reddish purple, angular-to-irregular lesions later occur on both upper and lower leaf surfaces. Lesions vary from pinpoint spots to irregular patches up to 0.5 in. in diameter that may coalesce to form large necrotic areas. Veinal necrosis can also occur. Numerous infections cause rapid chlorosis and necrosis of leaf tissue, resulting in defoliation starting with the young upper leaves. Such defoliation has often been mistaken for early maturity by growers. However, green leaves usually remain below the defoliated area.

Lesions on petioles and stems are slightly sunken, reddish purple areas. Infection of petioles increases defoliation, but the petioles remain attached to the plants. On more susceptible varieties, round reddish purple lesions, which later become purplish black, occur on pods. A general dark reddish purple discoloration similar to that on upper leaves also occurs on upper canopy stems, petioles, and pods exposed to the sun. Heavily infected stems have a dull gray to dark brown appearance and dry up 7 to 10 days prematurely.

The disease is most conspicuous and easily distinguished on seeds. Seed discoloration varies from pink or pale purple to dark purple. The discolored areas range from specks to large, irregular blotches that may cover the entire surface of the seed coat. Seed discoloration is often accompanied by wide cracks in seed coats, usually transversely along the seed. Cotyledons are generally not discolored. Seeds that carry the pathogen may not show symptoms.

It is possible but not typical in the midsouthern U.S. to have significant yield losses due to premature defoliation caused by Cercospora leaf blight. Most of the economic losses caused by *C. kikuchii* are due to the purple seed stain phase of the disease. The incidence of purple seed stain can be as high as 50% of a seed lot. Purple seed stain can reduce seed grade. U.S.DA No. 1 soybean are allowed up to 5% purple-stained seed and No. 3 soybean are allowed up to 20% purple-stained seed.

Disease cycle

C. kikuchii overwinters in diseased leaves, petioles, stems, and seeds. When diseased seeds are planted, the fungus may grow from the seed coat into the cotyledons and from these into the stem of a small percentage of the seedlings. There is little relationship between seed infection by *C. kikuchii* and the results of standard germination tests. However, seedling emergence from purple-stained seeds has been reported to be 5 to 15 percentage points lower than that from seed free of purple stain. This may mean that *C. kikuchii* does not affect viability but has a negative impact upon seed vigor.

Fungal spores in large numbers can be detected in the field beginning in early spring and throughout the growing season. Seedling infection, leaf infection, and pod infection are independent events. Most early-season infections are latent. Only a small, limited number of cells are infected and colonized without any macroscopic signs of disease. These latent infections, along with infected crop residue, are sources of airborne spores for secondary disease spread. Even though foliar infection may be of little economic importance, its role in sustaining the epidemic through production of inoculum for pod infections is important (Orth and Schuh, 1994).

Seed infection is a direct result of pod infection. As the plant begins to mature, the fungus grows from the pod wall and colonizes the seed coat. Inoculating flowers with spores of the fungus does not result in the production of purple-stained seed. Inoculating pods from R3 to R6 results in seed infection and the production of purple-stained seed. Apparently, the fungus is unable to grow directly from other infected plant parts and infect the pods and, hence, the seeds.

After harvest maturity is reached, delaying harvest for up to 6 weeks does not increase the incidence of *C. kikuchii* seed infection as has been found for other seed-borne fungi, such as *Phomopsis* sp. This may be explained by the relationship between *C. kikuchii* and other seed-borne fungi. *C. kikuchii* is antagonistic to *Phomopsis* sp., *Diaporthe* sp., and *Fusarium* sp. As the incidence of *C. kikuchii*-infected seed increases, the incidence of infection of these other fungi decreases and vice versa. Seed infection by *Phomopsis* sp. has a strong tendency to reduce seed germination. In some instances, purple-stained seed had higher germination rates at harvest than unstained seed because of the reduction in seed-borne *Phomopsis* sp. It was also shown that a delay in harvesting reduced the incidence of seed-borne *C. kikuchii* because of an increase in infection by *Phomopsis* sp.

Environmental effects

Weather at the time of inoculation and several days afterward appears to have the most impact on the incidence of purple seed stain. Infection of both leaves and pods is most severe during warm, wet periods. Moisture serves as a threshold determining if infection takes place. Temperature functions quantitatively in determining the speed and range of symptom development. Infection does not occur when it is below 60°F or above 86°F. The optimum temperature for infection is around 75 to 77°F. Splashing rain is not needed for spore dispersal, but sufficient humidity and dew are required for spore germination and subsequent infection to occur (Schuh, 1991). An 18- to 24-h period of surface wetness is required for successful leaf and pod infection. These moisture requirements can be achieved when there are extended periods of dew, fog, or rain. Significant infection can still occur under shorter (8 to 10 h) dew periods, as frequently occurs in the midsouthern U.S., if the dry period is accompanied by relative humidities in excess of 90%.

Control

Cercospora leaf blight and purple seed stain can be managed by (1) use of disease-free planting seed; (2) planting early-maturing varieties late or planting late-maturing varieties; (3) applying a foliar fungicide, such as 0.05 lb benomyl (0.5 lb Benlate) at R3 and R6; and (4) rotation with a nonleguminous crop.

Using disease-resistant varieties is the most cost-effective control method. Differences in varietal susceptibility to both *Cercospora* leaf blight and purple seed stain exist. Most varieties have some level of resistance to both diseases (Walters, 1985). However, complete immunity has not been reported, and the resistance to foliar symptom development is not related to the resistance to purple seed stain (Orth and Schuh, 1994). One could select varieties that may avoid these diseases. It has been noted that disease severity is higher on early-maturing varieties than on later-maturing varieties (Walters, 1980). This difference in disease severity is probably related to the lower temperatures that usually occur during the maturation period of the later-maturing varieties. Development of varieties with resistance to *C. kikuchii* does not receive much emphasis in either public or private soybean breeding programs.

There has been little reported yield increase in response to application of benomyl (TeKrony et al., 1985a). Application of 0.5 lb/acre benomyl at R4 and again at R6 or 1.0 lb/acre at R7 lowered seed infection and the incidence of purple seed stain (TeKrony et al., 1985a).

Spores of *C. kikuchii* can be spread a long distance by wind. Spores can be readily collected in the field in the absence of plant symptoms and previous crop residue (Orth and Walters, 1994). In light of these facts, crop rotation can be expected to reduce early-season infection but not eliminate later (reproductive-phase) infection. Selecting seed free of purple stain may reduce levels of the initial inoculum but will not completely eliminate seed-borne *C. kikuchii*. Seed lacking visible signs of disease can harbor the fungus.

Variety Resistance

Disease development is controlled by the interaction between host plant and pathogen. The basis for this interaction is genetic. Resistant varieties have been used effectively for the control of diseases, and genetic resistance in plants is a major form of biological control of pests. Genetic resistance is probably the best component of a disease management strategy, requiring no additional economic input and imparting no negative environmental impact.

The planting of resistant varieties is the first line of disease control. Resistance may be due to morphological or physiological characteristics of the plant. It is common to find that a plant has more than one resistance mechanism that operates against a pathogen. These mechanisms include immunity, resistance, tolerance, and avoidance.

Immunity

Immunity is freedom from attack and injury by a pathogen. It means complete disease resistance. Strictly speaking, all plant species that a given pathogen is incapable of infecting are immune to that pathogen. Most plants are immune to most pathogens. Immunity plays no role when making variety selection (e.g., all soybean varieties are immune to stem rust of wheat).

Resistance

Resistance refers to the ability of the host to interfere with the normal growth and/or development of the pathogen. Resistance mechanisms are often chemical in nature and may be naturally occurring or induced. Naturally occurring resistance compounds are present in the host tissue prior to its contact with the pathogen, whereas induced compounds occur only after such contact. One type of induced resistance is called hypersensitivity. Hypersensitivity is the sudden localized host response to the presence of a pathogen that involves the quick death of host cells around the pathogen and the restriction and/or death of the pathogen. The host plant "sacrifices" some cells to keep the disease from spreading. Latent infections are related to this concept. A fungus, such as *Cercospora sojina*, infects a plant and the infection is localized to only a few cells. Throughout most of the growing season, symptoms do not develop and there is no additional spread of the fungus within the plant. However, as the plant begins to senesce its defenses weaken and then disease rapidly develops.

Resistance does not mean the complete lack of disease development as immunity does. A resistant plant may support some disease development but to a lesser degree than a susceptible plant. Differences between resistant and susceptible may be on a continuum.

Tolerance

Tolerance is the ability of a plant to perform well even though it exhibits the symptoms of a susceptible host. A tolerant plant lacks the ability to prevent the establishment of a pathogen or to retard its development after establishment. On the basis of a visual assessment of symptoms, a tolerant plant would be rated as susceptible, just like a nontolerant one. Despite the exhibition of disease symptoms, however, the performance of a tolerant plant would be similar to that observed in plants without infection. Generally, the tolerance of a variety against a particular pathogen is not known, or, if it is, the information is not presented to assist in variety selection. However, tolerance is an important factor in final yield and economic return.

Avoidance

Avoidance (escape) is the capacity of a susceptible plant to avoid infectious disease through some character of the plant or other factor(s) that prevent infection. Most mechanisms for avoidance are morphological, such as the presence of leaf hairs, the thickness of the cuticle, or susceptible tissue being present for only a short time during the life cycle of the host. Avoidance is not resistance, which does not mean that avoidance as a control measure is valueless; quite the contrary, any measure that lessens disease, or the chance for disease, is certainly worthwhile.

Avoidance may be accomplished by the planting of an early-maturing variety that escapes infection because it matures before the environment becomes favorable for infection and disease development. This is the case for frogeye leaf spot. Another example is planting a late-maturing variety or the late planting of an early-maturing variety. Plants escape infection by senescing at a time when environmental conditions are less favorable for disease development. *Phomopsis* seed decay is a good example.

Seldom is the mechanism of avoidance so efficient as to prevent disease totally. Rather, avoidance is a tool for lessening the occurrence and severity of disease. Plants possessing the capacity of avoidance are still susceptible. Therefore, if a change of conditions renders the escape mechanism inoperative, disease can occur.

Disease management

Cultural practices can play a role in disease management by reducing initial inoculum or by reducing the rate of disease development. Some cultural practices reduce disease by reducing stress on the host plant. Other practices encourage the destruction of the fungus. Knowing which practice works for a disease is based on a detailed knowledge of the ecology of both the soybean plant and the pathogen. Unfortunately, much of this information is either lacking or incomplete for soybean diseases in the midsouthern U.S. Therefore, a complete understanding of the impact of all cultural practices on the incidence and severity of the diseases discussed in this chapter is not always possible.

Plant debris, soil, and various living plants harbor great quantities of inoculum, and it is against these sources of inoculum that many cultural practices are directed. For the most part, inoculum is not completely eliminated. Generally the goal is to reduce the inoculum to the point where a crop can be economically grown and harvested. It is often possible to utilize two or more procedures to reduce the level of the inoculum.

Tillage

In conventional tillage systems, a crop residue is plowed under sometime after fall harvesting but before spring planting. Such tillage practices result in burying crop residue and fungal-resting structures. This places them in close contact with soil-inhabiting micro-organisms and in an environment more favorable for the decomposition of plant residue in which pathogens may be surviving. Tillage also encourages the decomposition of fungal-resting structures.

Reduced-tillage systems are being adopted to reduce soil erosion, retain soil moisture, and reduce farming operations. Appreciable crop residues are left on the soil surface by reduced tillage. The long-term impact of not burying crop residue on soybean diseases is not yet clear. The crop residue left on the soil surface may contain inoculum for soybean diseases and allow it to overwinter. The problem inherent in leaving exposed crop residue on the soil surface is the almost certain increase of particular plant diseases. However, leaving the soil undisturbed for long periods may help reduce the inoculum for other diseases by allowing the development of a diverse soil microflora, including microorganisms antagonistic to or parasitic on pathogens.

Crop rotation

Crop rotation is routinely recommended as a disease management strategy. The theory is that growing the same crop season after season allows a disease to perpetuate itself. Rotating with a nonhost crop interrupts the disease cycle and allows time for the decomposition of inoculum. In the past, crop rotation has often resulted in the practice being abused, since it will not always reduce the amount of inoculum in the soil. Therefore, for some diseases crop rotation may be useless. Conversely, continued soybean cropping may increase the severity of other diseases, and crop rotation may be valuable for decreasing the inoculum. What is needed is a thorough knowledge of each pathogen. Will the inoculum remain viable for years in the soil without a host crop or not? Will there be enough reduction of inoculum to allow a crop rotation program to be workable?

Pathogens that are soil inhabitants, such as *Rhizoctonia*, may grow on a wide variety of crops and can survive indefinitely in the soil, or at least survive for a long enough time to make crop rotation impractical. On the other hand, soil invaders, such as *Colletotrichum*, are usually killed soon after the crop residue has decomposed. Therefore, a 3- to 4-year crop rotation for control of such diseases is a practical management procedure.

Row spacing/plant population

Temperature, relative humidity, and dew period all influence infection incidence and rate of disease spread and development. Changing row spacing, plant population, or otherwise altering the characteristics of the plant canopy may result in changes in the microclimate within the canopy. However, there is little published evidence to prove that such changes can be used as a method of disease control in soybean.

Planting date

Altering planting date or maturity group has been suggested as a method of disease management. This approach is based on the concept that temporally shifting a vulnerable development stage of the soybean plant to a time less conducive to disease development will reduce disease incidence or severity. For example, delaying planting until mid-June or later is a recommended procedure for the control of stem canker. The drawback to this approach is that planting to avoid disease may force planting to occur at a time less favorable for other reasons. As in the case of stem canker, delaying planting places the plant in shortening day lengths. These shorter photoperiod conditions immediately limit the yield potential of the plant. The negative effect of photoperiod on plant growth and yield probably outweighs any advantage gained in disease control.

Fertility

The application of fertilizer, such as potassium, has been found to reduce the incidence of some soybean diseases. However, overfertilization can promote disease development by promoting excessive plant growth. The best fertility levels for disease management will be the ones resulting in a healthy plant, since this type plant best resists disease.

Foliar fungicides

A foliar fungicide such as benomyl applied at R3 and R6 can effectively reduce the incidence of many diseases (Horn et al., 1975; Backman et al., 1979). However, yield increases resulting from such a reduction in disease only occur in irrigated or otherwise high-yield environments. Fungicides may also result in an economic return when applied to a higher-value crop such as soybean being grown for seed production.

Many growers have failed to realize the benefits that can be obtained by using other control methods in conjunction with fungicides, because the use of chemicals seldom, if ever, provides complete disease control. There is ample evidence that the effectiveness of fungicides increases when there is a decrease of the inoculum potential. A partial reduction of inoculum by any method and the use of resistant varieties and fungicides will provide optimum control.

Conclusions

1. Select adapted, high-yielding, disease-resistant varieties. Emphasis should be placed on stem canker resistance in combination with nematode resistance.
2. Use high-quality seed treated with a fungicide (such as Apron 50W, Apron FS, Apron XL LS, Vitavax CT, Vitavax — M, or Vitavax M DC) . Fungicide seed treatment is an inexpensive method of reducing risks associated with seed-borne fungi and seedling diseases.
3. Tillage practices, row spacing, and seeding rates may all impact the total disease picture. However, the selection of these practices should be based on agronomic considerations.
4. Fertilize based on soil tests and with the objective of optimizing yield. Fertility may also influence disease development. A healthy plant will best resist disease.
5. Planting date, and subsequent maturity date, can influence the chance for the development of several diseases. Planting beyond the optimum planting window may reduce disease pressure. However, planting when photoperiods are too short for optimum growth will reduce the yield potential of a variety.
6. Use a foliar fungicide only when yield potential is high (40 bu/acre or more) or if the objective is to produce seed for planting.
7. Harvest in a timely manner, or as soon as seed are mature.

References

Ammon, V. D., T. D. Wyllie, and M. F. Brown. 1974. An ultrastructural investigation of pathological alterations induced by *Macrophomina phaseolina* Tassi (Goid.) in seedlings of soybean *Glycine max* (L.) Merrill, *Physiol. Plant Pathol.* 4:1–4.
Ammon, V. D., T. D. Wyllie, and M. F. Brown. 1975. Investigation of the infection process of *Macrophomina phaseolina* on the surface of soybean roots using scanning electron microscopy, *Mycopathologica* 55:77–81.
Anahosur, K. H., S. H. Patil, and R. K. Hegde. 1984. Effect of herbicides on *Macrophomina phaseolina* (Tassi) Goid. causing charcoal rot of sorghum, *Pesticides* 18:11–12.

Anderson, T. R. and R. I. Buzzell. 1982. Efficacy of metalaxyl in controlling Phytophthora root and stalk rot of soybean cultivars differing in field tolerance, *Plant Dis.* 66:1144–1145.

Athow, K. L. 1985. Phytophthora root rot of soybean, in R. Shibles, Ed., *World Soybean Research Conference III: Proceedings*, Westview Press, Boulder, CO, 575–581.

Athow, K. L. 1987. Fungal diseases, in J. R. Wilcox, Ed., *Soybeans: Improvement, Production, and Uses*, 2nd ed., ASA, CSSA, and SSSA, Madison, WI, 687–727.

Athow, K. L. and F. A. Laviolette. 1973. Pod protection effects on soybean seed germination and infection with *Diaporthe phaseolorum* var. *sojae* and other microorganisms, *Phytopathology* 63:1021–1023.

Atkins, J. G., Jr. and W. D. Lewis. 1954. Rhizoctonia aerial blight of soybeans in Louisiana, *Phytopathology* 44:215–218.

Ayanru, D. K. G. and R. J. Green, Jr. 1974. Alteration of germination patterns of sclerotia of *Macrophomina phaseolina* on soil surfaces, *Phytopathology* 64:595–601.

Aycock, R. 1966. Stem Rot and Other Diseases Caused by *Sclerotium rolfsii*, North Carolina Agr. Exp. Sta. Tech. Bull. 174, 202 pp.

Backman, P. A. 1984. Effects of timing and rates of application of fungicides for control of stem canker of soybeans, in M. M. Kulick, Ed., *Proc. Conf. on* Diaporthe/Phomopsis, Ft. Walton Beach, FL.

Backman, P. A., R. Rodríquez-Kabana, J. M. Hammond, and D. L. Thurlow. 1979. Cultivar, environment, and fungicide effects on foliar disease losses in soybeans, *Phytopathology* 69:562–564.

Backman, P. A., M. A. Crawford, and J. White. 1981. Soybean stem canker: a serious disease in Alabama, *Highlights Agric. Res.* 28:6.

Backman, P. A., J. C. Williams, and M. A. Crawford. 1982. Yield losses in soybeans from anthracnose caused by *Colletotrichum truncatum*, *Plant Dis.* 66:1032–1034.

Backman, P. A., D. B. Weaver, and G. Morgan-Jones. 1985. Etiology, epidemiology, and control of stem canker, in R. Shibles, Ed., *World Soybean Research Conference III: Proceedings*, Westview Press, Boulder, CO, 589–597.

Baker, D. M., H. C. Minor, M. F. Brown, and E. A. Brown. 1987. Infection of immature soybean pods and seeds by *Phomopsis longicolla*, *Can. J. Microbiol.* 33:797–801.

Balducchi, A. J. and D. C. McGee. 1987. Environmental factors influencing infection of soybean seeds by *Phomopsis* and *Diaporthe* species during seed maturation, *Plant Dis.* 71:209–212.

Berner, D. K., G. T. Berggren, M. E. Pace, E. P. White, J. S. Gershey, J. A. Freedman, and J. P. Snow. 1986. Red crown rot: now a major disease of soybeans, *La. Agr.* 29:4–5, 24.

Berner, D. K., G. T. Berggren, J. P. Snow, and E. P. White. 1988. Distribution and management of red crown rot of soybean in Louisiana, *Appl. Agric. Res.* 3:160–166.

Beute, M. K. and R. Rodríguez-Kabana. 1979. Effect of volatile compounds from remoistened plant tissues on growth and germination of sclerotia of *Sclerotium rolfsii*, *Phytopathology* 69:802–805.

Beute, M. K. and R. Rodríguez-Kabana. 1981. Effects of soil moisture, temperature, and field environment on survival of *Sclerotium rolfsii* in Alabama and North Carolina, *Phytopathology* 71:1293–1296.

Bisht, V. S. and J. B. Sinclair. 1985. Effect of *Cercospora sojina* and *Phomopsis sojae* alone or in combination on seed quality and yield of soybeans, *Plant Dis.* 69:436–439.

Black, B. D., J. L. Griffin, J. S. Russin, and J. P. Snow. 1996a. Weeds as hosts for *Rhizoctonia solani* AG-1, causal agent for Rhizoctonia foliar blight of soybean (*Glycine max*), *Weed Technol.* 10:(in press).

Black, B. D., J. S. Russin, J. L. Griffin, and J. P. Snow. 1996b. Herbicide effects on *Rhizoctonia solani* AG-1 IA and IB *in vitro* and on Rhizoctonia foliar blight of soybean (*Glycine max*), *Weed Sci.* 44:711–716.

Bowen, C. R. and W. T. Schapaugh. 1989. Relationships among charcoal rot infection, yield, and stability estimates in soybean blends, *Crop Sci.* 29:42–46.

Bowers, G. R., Jr. 1984. Resistance to anthracnose, *Soybean Genet. Newslett.* 11:150–151.

Bowers, G. R., Jr., K. Ngeleka, and O. D. Smith. 1993. Inheritance of stem canker resistance in soybean cultivars Crockett and Dowling, *Crop Sci.* 33:67–70.

Brown, E. A., H. C. Minor, and O. H. Calvert. 1987. A soybean genotype resistant to Phomopsis seed decay, *Crop Sci.* 27:895–898.

Canaday, C. H., D. G. Helsel, and T. D. Wyllie. 1986. Effects of herbicide-induced stress on root colonization of soybeans by *Macrophomina phaseolina*, *Plant Dis.* 70:863–866.

Cerkauskas, R. F., O. D. Dhingra, and J. B. Sinclair. 1982. Effects of herbicides on competitive saprophytic ability by *Macrophomina phaseolina* on soybean stems, *Trans. Br. Mycol. Soc.* 79:201–205.

Chambers, A. Y. 1991. Effects of date of planting on severity of soybean stem canker, *Phytopathology* 81:1135.

Cloud, G. L., and J. C. Rupe. 1991. Morphological instability on a chlorate medium of isolates of *Macrophomina phaseolina* from soybean and sorghum, *Phytopathology* 81:892–895.

Cooper, R. L. 1989. Soybean yield response to benomyl fungicide application under maximum yield conditions, *Agron. J.* 81:847–849.

Crawford, M. A. 1984. Seed treatments and tillage practices as they affect spread and control of stem canker, in M. M. Kulick, Ed., *Proc. Conf. on* Diaporthe/Phomopsis, Ft. Walton Beach, FL.

Crittenden, H. W. and L. V. Svec. 1974. Effect of potassium on the incidence of *Diaporthe sojae* in soybean, *Agron. J.* 66:696–697.

Damicone, J. P., G. T. Berggren, and J. P. Snow. 1987. Effect of free moisture on soybean stem canker development, *Phytopathology* 77:1568–1572.

Damicone, J. P., J. P. Snow, and G. T. Berggren. 1990. Spatial and temporal spread of soybean stem canker from an inoculum point source, *Phytopathology* 80:571–578.

Dhingra, O. D. and J. F. da Silva. 1978. Effect of weed control on internally seedborne fungi in soybean seeds, *Plant Dis. Rept.* 62:513–516.

Dhingra, O. D. and J. B. Sinclair. 1975. Survival of *Macrophomina phaseolina* sclerotia in soil: Effects of soil moisture, carbon:nitrogen ratios, carbon sources, and nitrogen concentrations, *Phytopathology* 65:236–240.

Dhingra, O. D. and J. B. Sinclair. 1978. Biology and Pathology of *Macrophomina phaseolina*, Universidade Federal de Viçosa, Viçosa-Minas Gerais-Brazil, Imprensa Universitaria da U.F.V., 166 pp.

Dirks, V. A., T. R. Anderson, and E. F. Bolton. 1980. Effect of fertilizer and drain location on incidence of Phytophthora root rot of soybean, *Can. J. Plant Pathol.* 2:179–183.

Ellis, M. A., M. B. Ilyas, F. D. Tenne, J. B. Sinclair, and H. L. Palm. 1974. Effect of foliar applications of benomyl on internally seedborne fungi and pod and stem blight in soybean, *Plant Dis. Rept.* 58:760–763.

Fernandez, F. A. and J. B. Sinclair. 1990. Colonization of soybean pods and seeds by *Cercospora kikuchii* at different reproductive stages, *Phytopathology* 80:1040.

Ferriss, R. S. 1988. Seedling establishment — an epidemiological perspective, in T. D. Wyllie and D. H. Scott, Eds., *Soybean Diseases of the North Central Region*, APS Press, St. Paul, MN, 7–13.

Filho, E. S. and O. D. Dhingra. 1980a. Effect of herbicides on survival of *Macrophomina phaseolina* in soil, *Trans. Br. Mycol. Soc.* 74:61–64.

Filho, E. S. and O. D. Dhingra. 1980b. Survival of *Macrophomina phaseolina* sclerotia in nitrogen amended soils, *Phytopath. Z.* 97:136–143.

Foor, S. R., F. D. Tenne, and J. B. Sinclair. 1976. Occurrence of seedborne microorganisms and germination in culture for determining seed health in soybeans, *Plant Dis.* 60:970–973.

Gangopadhyay, S., T. D. Wyllie, and V. D. Luedders. 1970. Charcoal rot disease of soybeans transmitted by seeds, *Plant Dis. Rept.* 54:1088–1091.

Gangopadhyay, S., T. D. Wyllie, and W. R. Teague. 1982. Effect of bulk density and moisture content of soil on the survival of *Macrophomina phaseolina*, *Plant Soil* 68:241–247.

Garzonio, D. M. and D. C. McGee. 1983. Comparison of seeds and crop residues as sources of inoculum for pod and stem blight of soybeans, *Plant Dis.* 67:1374–1376.

Gleason, M. L. and R. S. Ferriss. 1985. Influence of soil water potential on performance of soybean seeds infected by *Phomopsis* sp., *Phytopathology* 75:1236–1241.

Gleason, M. L. and R. S. Ferriss. 1987. Effects of soil moisture and temperature on Phomopsis seed decay of soybean in relation to host and pathogen growth rates, *Phytopathology* 77:1152–1157.

Griffin, G. J., D. A. Roth, and N. L. Powell. 1978. Physical factors that influence the recovery of microsclerotium population of *Cylindrocladium crotalariae* from naturally infested soils, *Phytopathology* 68:887–891.

Gupta, J. P., D. C. Erwin, J. W. Eckert, and A. I. Zaki. 1985. Translocation of metalaxyl in soybean plants and control of stem rot caused by *Phytophthora megasperma* f. sp. *glycinea*, *Phytopathology* 75:865–869.

Harville, B. G., G. T. Berggren, J. P. Snow, and H. K. Whitam. 1986. Yield reductions caused by stem canker in soybean, *Crop Sci.* 26:614–616.

Harville, B. G., J. S. Russin, and R. J. Habetz. 1996. Rhizoctonia foliar blight reactions and seed yields in soybean, *Crop Sci.* 36:563–566.

Heatherly, L. G. and G. L. Sciumbato. 1986. Effect of benomyl fungicide and irrigation on soybean yield and yield components, *Crop Sci.* 26:352–355.

Hepperly, P. R. 1985. Soybean anthracnose, in R. Shibles, Ed., *World Soybean Research Conference III: Proceedings*, Westview Press, Boulder, CO, 547–554.

Hepperly, P. R. and J. B. Sinclair. 1978. Quality losses in Phomopsis-infected soybean seeds, *Phytopathology* 68:1684–1687.

Hepperly, P. R. and J. B. Sinclair. 1981. Relationships among *Cercospora kikuchii*, other seed mycoflora, and germination of soybeans in Puerto Rico and Illinois, *Plant Dis.* 65:130–132.

Hepperly, P. R., B. L. Kirkpatrick, and J. B. Sinclair. 1980. *Abutilon theophrasti*: wild host for three fungal parasites of soybean, *Phytopathology* 70:307–310.

Hershman, D. E., J. W. Hendrix, R. E. Stuckey, P. R. Bachi, and G. Henson. 1990. Influence of planting date and cultivar on soybean sudden death syndrome in Kentucky, *Plant Dis.* 74:761–766.

Herzog, D. C., J. W. Thomas, R. L. Jensen, and L. D. Newsom. 1975. Association of sclerotial blight with *Spissistilus festinus* girdling injury on soybean, *Environ. Entomol.* 4:986–988.

Higley, P. M. and H. Tachibana. 1987. Physiologic specialization of *Diaporthe phaseolorum* var. *caulivora* in soybean, *Plant Dis.* 71:815–817.

Horn, N. L., F. N. Lee, and R. B. Carver. 1975. Effects of fungicides and pathogens on yields of soybeans, *Plant Dis. Rept.* 59:724–728.

Jeffers, D. L. and A. F. Schmitthenner. 1981. Germination and disease in soybean seed as affected by rotation, planting time, K fertilization and tillage, *Agron. Abstr.* 73:119.

Jeffers, D. L., A. F. Schmitthenner, and D. L. Reichard. 1982a. Seed-borne fungi, quality, and yield of soybeans treated with benomyl fungicide by various application methods, *Agron. J.* 74:589–592.

Jeffers, D. L., A. F. Schmitthenner, and M. E. Kroetz. 1982b. Potassium fertilization effects on Phomopsis seed infection, seed quality, and yield of soybeans, *Agron. J.* 74:886–890.

Johnston, S. A. and M. K. Beute. 1975. Histopathology of Cylindrocladium black rot of peanut, *Phytopathology* 64:649–653.

Joye, G. F., G. T. Berggren, and D. K. Berner. 1990. Effects of row spacing and within-row population on Rhizoctonia aerial blight of soybean and soybean yield, *Plant Dis.* 74:158–160.

Keeling, B. L. 1982. A seedling test for resistance to soybean stem canker caused by *Diaporthe phaseolorum* var. *caulivora*, *Phytopathology* 72:807–809.

Keeling, B. L. 1984. Evidence for physiologic specialization in *Diaporthe phaseolorum* var. *caulivora*, *J. Miss. Acad. Sci.* Suppl. 29:5.

Keeling, B. L. 1985. Soybean cultivar reactions to soybean stem canker caused by *Diaporthe phaseolorum* var. *caulivora* and pathogenic variation among isolates, *Plant Dis.* 69:132–133.

Keeling, B. L. 1988. Influence of temperature on growth and pathogenicity of geographic isolates of *Diaporthe phaseolorum* var. *caulivora*, *Plant Dis.* 72:220–222.

Keeling, B. L. 1990. Observed spread of soybean stem canker disease, *Phytopathology* 80:1004.

Khan, M. and J. B. Sinclair. 1991. Effect of soil temperature on infection of soybean roots by sclerotia-forming isolates of *Colletotrichum truncatum*, *Plant Dis.* 75:1282–1285.

Khan, M. and J. B. Sinclair. 1992. Pathogenicity of sclerotia- and nonsclerotia-forming isolates of *Colletotrichum truncatum* on soybean plants and roots, *Phytopathology* 82:314–319.

Kilen, T. C. and E. E. Hartwig. 1987. Identification of single genes controlling resistance to stem canker in soybean, *Crop Sci.* 27:863–864.

Kilen, T. C., B. L. Keeling, and E. E. Hartwig. 1985. Inheritance of reaction to stem canker in soybean, *Crop Sci.* 25:50–51.

Kim, K. D. 1994. Susceptibility on Soybean to Red Crown Rot and Characteristics of Virulence in *Calonectria crotalariae*, Ph.D. dissertation, Louisiana State University, Baton Rouge, 119 pp.

Kittle, D. R. and L. E. Gray. 1979. The influence of soil temperature, moisture, porosity, and bulk density on the pathogenicity of *Phytophthora megasperma* var. *sojae*, *Plant Dis. Rept.* 63:231–234.

Kmetz, K. T., A. F. Schmitthenner, and C. W. Ellett. 1978. Soybean seed decay: prevalence of infection and symptom expression caused by *Phomopsis* sp., *Diaporthe phaseolorum* var. *sojae*, and *D. phaseolorum* var. *caulivora*, *Phytopathology* 68:836–840.

Kmetz, K. T., C. W. Ellett, and A.F. Schmitthenner. 1979. Soybean seed decay: sources of inoculum and nature of infection, *Phytopathology* 69:798–801.

Koover, A. T. A. 1954. Some factors affecting the growth of *Rhizoctoni bataticola* in soil, *J. Madras Univ.* 24:47–52.

Krausz, J. P., and B. A. Fortnum. 1983. An epiphytotic of Diaporthe stem canker of soybean in South Carolina, *Plant Dis.* 67:1128–1129.

Kunwar, I. K., T. Singh, C. C. Machado, and J. B. Sinclair. 1986. Histopathology of soybean seed and seedling infection by *Macrophomina phaseolina*, *Phytopathology* 76:532–535.

Kuruppu, P. U. and J. S. Russin. 1995. Effects of planting date and soybean cultivar on soil populations and root colonization by the red crown rot fungus, *Calonectria crotalariae*, *Phytopathology* 85:1193.

Lalitha, B., J. P. Snow, and G. T. Berggren. 1989. Phytotoxin production by *Diaporthe phaseolorum* var. *caulivora*, the causal organism of stem canker of soybean, *Phytopathology* 79:499–504.

Laviolette, F. A., K. L. Athow, A. H. Probst, J. R. Wilcox, and T. S. Abney. 1970. Effect of bacterial pustule and frogeye leafspot on yield of clark soybean, *Crop Sci.* 10:418–419.

Lim, S. M. 1979. Evaluation of soybean for resistance to septoria brown spot, *Plant Dis. Rept.* 63:242–245.

Lim, S. M. 1980. Brown spot severity and yield reduction in soybean, *Phytopathology* 70:974–977.

Manandhar, J. B., G. L. Hartman, and J. B. Sinclair. 1988. Soybean germ plasm evaluation for resistance to *Colletotrichum truncatum*, *Plant Dis.* 72:56–59.

Martin, K. F. and H. J. Walters. 1982. Infection of soybean by *Cercospora kikuchii* as affected by dew temperature and duration of dew periods, *Phytopathology* 72:974.

McGee, D. C. 1986. Prediction of Phomopsis seed decay by measuring soybean pod infection, *Plant Dis.* 70:329–333.

McGee, D. C. 1992. *Soybean Diseases. A Reference Source for Seed Technologists*, APS Press, St. Paul, MN.

McGee, D. C. and J. A. Biddle. 1987. Seedborne *Diaporthe phaseolorum* var. *caulivora* in Iowa and its relationship to soybean stem canker in the southern United States, *Plant Dis.* 71:620–622.

McGee, D. C. and C. L. Brandt. 1979. Effect of foliar application of benomyl on infection of soybean seeds by *Phomopsis* in relation to time of inoculation, *Plant Dis. Rept.* 63:675–677.

Meyer, W. A., J. B. Sinclair, and M. N. Khare. 1974. Factors affecting charcoal rot of soybean seedlings, *Phytopathology* 64:845–849.

Minton, N. A., M. B. Parker, and R. A. Flowers. 1975. Response of soybean cultivars to *Meloidogyne incognita* and to the combined effects of *M. arenaria* and *Sclerotium rolfsii*, *Plant Dis. Rept.* 59:920–923.

Morgan-Jones, G. 1992. The *Diaporthe phaseolorum* complex of soybean, *Fitopatol. Bras.* 17:359–367.

Muchovej, J. J., R. M. C. Muchovej, O. D. Dhingra, and L. A. Maffia. 1980. Suppression of anthracnose of soybeans by calcium, *Plant Dis.* 64:1088–1089.

Mukherjee, B., S. Banerjee, and C. Sen. 1983. Influence of soil pH, temperature, and moisture on the ability of mycelia of *Macrophomina phaseolina* to produce sclerotia in soil, *Indian Phytopathol.* 36:158–160.

Olanya, O. M. and C. L. Campbell. 1988. Effects of tillage on the spatial pettern of microsclerotia of *Macrophomina phaseolina*, *Phytopathology* 78:217–221.

O'Neill, N. R., M. C. Rush, N. L. Horn, and R. B. Carver. 1977. Aerial blight of soybean caused by *Rhizoctonia solani*, *Plant Dis. Rept.* 61:713–717.

Orth, C. E. and W. Schuh. 1994. Resistance of 17 soybean cultivars to foliar, latent, and seed infection by *Cercospora kikuchii*, *Plant Dis.* 78:661–664.

Pace, P. F., D. B. Weaver, and L. D. Ploper. 1993. Additional genes for resistance to frogeye leaf spot race 5 in soybean, *Crop Sci.* 33:1144–1145.

Pacumbaba, R. P. 1995. Effect of fertilizers and their rates of application on the incidence of soybean diseases and crop productivity, *Phytopathology* 85:511.

Paschal, E. H., II and M. A. Ellis. 1978. Variation in seed quality characteristics of tropically grown soybeans, *Crop Sci.* 18:837–840.

Pataky, J. K. and S. M. Lim. 1981a. Efficacy of benomyl for controlling septoria brown spot of soybeans, *Phytopathology* 71:438–442.

Pataky, J. K. and S. M. Lim. 1981b. Effects of row width and plant growth habit on septoria brown spot development and soybean yield, *Phytopathology* 71:1051–1056.

Pearson, C. A. S., F. W. Schwenk, F. J. Crowe, and K. Kelly. 1984. Colonization of soybean roots by *Macrophomina phaseolina, Plant Dis.* 68:1086–1088.

Pearson, C. A. S., J. F. Leslie, and F. W. Schwenk. 1987. Host preference correlated with chlorate resistance in *Macrophomina phaseolina, Plant Dis.* 71:828–831.

Peterson, D. J. and H. H. Edwards. 1982. Effects of temperature and leaf wetness period on brown spot disease of soybeans, *Plant Dis.* 66:995–998.

Phillips, D. V. and H. R. Boerma. 1980. *Cercospora sojina* race 5: A threat to soybeans in the southeastern United States, *Phytopathology* 71:334–336.

Phillips, D. V. and H. R. Boerma. 1982. Two genes for resistance to race 5 of *Cercospora sojina* in soybeans, *Phytopathology* 72:764–766.

Phipps, P. M. and M. K. Beute. 1977. Influence of soil temperature and moisture on the severity of Cylindrocladium black rot in peanut, *Phytopathology* 67:1104–1107.

Ploetz, R. C. and F. M. Shokes. 1985. Soybean stem canker incited by ascospores and conidia of the fungus causing the disease in the southeastern United States, *Plant Dis.* 69:990–992.

Ploetz, R. C. and F. M. Shokes. 1986. Evidence for homothallism and vegetative compatibility in southern *Diaporthe phaseolorum, Can. J. Bot.* 64:2197–2200.

Ploetz, R. C. and F. M. Shokes. 1987. Factors influencing infection of soybean seedlings by southern *Diaporthe phaseolorum, Phytopathology* 77:786–790.

Ploper, L. D. and T. S. Abney. 1985. Effect of late season maturation rate on soybean quality, *Phytopathology* 75:965.

Punja, Z. K., J. D. Carter, G. M. Campbell, and E. L. Rossell. 1986. Effects of calcium and nitrogen fertilizers, fungicides, and tillage practices on incidence of *Sclerotium rolfsii* on processing carrots, *Plant Dis.* 70:819–824.

Quebral, F. C. and D. R. Pua. 1976. Screening of soybeans against *Sclerotium rolfsii, Trop. Grain Leg. Bull.* 6:22–23.

Rhoton, F. E. 1989. Soil properties' role in stem canker infection of soybean, *Agron. J.* 81:431–434.

Rodriquez-Marcano, A. and J. B. Sinclair. 1978. Fruiting structures of *Colletotrichum dematium* var. *truncata* and *Phomopsis sojae* formed in soybean seeds, *Plant Dis. Rept.* 62:873–876.

Ross, J. P. 1968. Additional physiological races of *Cercospora sojina* on soybeans in North Carolina, *Phytopathology* 58:708–709.

Ross, J. P. 1975. Effect of overhead irrigation and benomyl sprays on late-season foliar diseases, seed infection, and yields of soybean, *Plant Dis. Rept.* 59:809–813.

Ross, J. P. 1982a. Preemptive fungal infection of soybean seed, *Phytopathology* 72:974.

Ross, J. P. 1982b. Effect of simulated dew and postinoculation moist periods on infection of soybean by *Septoria glycines, Phytopathology* 72:236–238.

Roth, D. A., G. J. Griffin, and P. J. Graham. 1979. Low temperature induces decreased germinability of *Cylindrocladium* microsclerotia, *Can. J. Microbiol.* 25:157–162.

Rothrock, C. S., T. W. Hobbs, and D. V. Phillips. 1985. Effects of tillage and cropping system on incidence and severity of southern stem canker of soybean, *Phytopathology* 75: 1156–1159.

Rothrock, C. S., D. V. Phillips, and T. W. Hobbs. 1988. Effects of cultivar, tillage, and cropping system on infection of soybean by *Diaporthe phaseolorum* var. *caulivora* and southern stem canker symptom development, *Phytopathology* 78:266–270.

Rowe, R. C., S. A. Johnston, and M. K. Beute. 1974. Formation and dispersal of *Cylindrocladium crotalariae* microsclerotia in infected peanut roots, *Phytopathology* 64:1294–1297.

Roy, K. W. 1982. Seedling diseases caused in soybean by species of *Colletotrichum* and *Glomerella*, *Phytopathology* 72:1093–1096.

Roy, K. W. and T. S. Abney. 1976. Purple seed stain of soybeans, *Phytopathology* 66:1045–1049.

Roy, K. W. and T. S. Abney. 1977. Antagonism between *Cercospora kikuchii* and other seedborne fungi of soybeans, *Phytopathology* 67:1062–1066.

Roy, K. W. and W. A. Miller. 1983. Soybean stem canker incited by isolates of *Diaporthe* and *Phomopsis* spp. from cotton in Mississippi, *Plant Dis.* 67:135–137.

Roy, K. W., G. W. Lawrence, H. H. Hodges, K. S. McLean, and J. F. Killebrew. 1989a. Sudden death syndrome of soybean: *Fusarium solani* as incitant and relation of *Heterodera glycines* to disease severity, *Phytopathology* 79:191–197.

Roy, K. W., K. S. McLean, G. W. Lawrence, M. V. Patel, and W. F. Moore. 1989b. First report of red crown rot on soybean in Mississippi, *Plant Dis.* 73:273.

Rupe, J. C. 1988. Etiology of sudden death syndrome of soybeans, in Pascale, A. J., Ed., *World Soybean Research Conference IV Proceedings*, Orientacion Grafica Editora, Buenos Aires, Argentina, 2085–2090.

Rupe, J. C. 1989. Frequency and pathogenicity of *Fusarium solani* recovered from soybeans with sudden death syndrome, *Plant Dis.* 73:581–584.

Rupe, J. C. 1990. Effect of temperature on the rate of infection of soybean seedlings by *Phomopsis longicolla*, *Phytopathology* 80:1039.

Rupe, J. C. and R. S. Ferriss. 1986. Effects of pod moisture on soybean seed infection by *Phomopsis* sp., *Phytopathology* 76:273–277.

Rupe, J. C. and E. E. Gbur, Jr. 1995. Effect of plant age, maturity group, and the environment on disease progress of sudden death syndrome of soybean, *Plant Dis.* 79:139–143.

Rupe, J. C. and E. A. Sutton. 1994. Effect of temperature on infection of soybeans by the southern biotype of the stem canker pathogen, *Diaporthe phaseolorum* var. *caulivora*, *Phytopathology* 84:1120.

Rupe, J. C., E. E. Gbur, and D. M. Marx. 1991. Cultivar responses to sudden death syndrome of soybean, *Plant Dis.* 75:47–50.

Rupe, J. C., W. E. Sabbe, R. T. Robbins, and E. E. Gbur, Jr. 1993. Soil and plant factors associated with sudden death syndrome of soybean, *J. Prod. Agric.* 6:218–221.

Russin, J. S., N. N. Troxclair, Jr., D. J. Boethel, and E. C. McGawley. 1985. Effect of soybean planting date and soil nutrients on incidence of red crown rot and populations of insects associated with roots, *Phytopathology* 75:1284.

Russin, J. S., D. J. Boethel, G. T. Berggren, and J. P. Snow. 1986. Effects of girdling by the threecornered alfalfa hopper on symptom expression of soybean stem canker and associated soybean yields, *Plant Dis.* 70:759–761.

Russin, J. S., M. B. Layton, D. J. Boethel, E. C. McGawley, J. P. Snow, and G. T. Berggren. 1989a. Severity of soybean stem canker disease affected by insect-induced defoliation, *Plant Dis.* 73:144–147.

Russin, J. S., M. B. Layton, D. J. Boethel, E. C. McGawley, J. P. Snow, and G. T. Berggren. 1989b. Development of *Heterodera glycines* on soybean damaged by soybean looper and stem canker, *J. Nematol.* 21:108–114.

Russin, J. S., C. H. Carter, and J. L. Griffin. 1995. Effects of preemergence herbicides for grain sorghum (*Sorghum bicolor*) on the charcoal rot fungus, *Weed Technol.* 9:343–351.

Schmitthenner, A. F. 1985. Problems and progress in control of Phytophthora root rot in soybean, *Plant Dis.* 69:362–368.

Schmitthenner, A. F. 1988. Phytophthora rot of soybean, in T. D. Wyllie and D. H. Scott, Eds., *Soybean Diseases of the North Central Region*, APS Press, St. Paul, MN, 71–80.

Schmitthenner, A. F. 1989. Phytophthora root rot: detection, ecology and control, in Pascale, A. J., Ed., *World Soybean Research Conference IV Proceedings*, Orientacion Grafica Editora, Buenos Aires, Argentina, 1284–1289.

Schuh, W. 1991. Influence of temperature and leaf wetness period on conidial germination *in vitro* and infection of *Cercospora kikuchii* on soybean, *Phytopathology* 81:1315–1318.

Schuh, W. 1992. Effect of pod development stage, temperature, and pod wetness duration on the incidence of purple seed stain of soybeans, *Phytopathology* 82:446–451.

Schuh, W. 1993. Influence of interrupted dew periods, relative humidity, and light on disease severity and latent infections caused by *Cercospora kikuchii* on soybean, *Phytopathology* 83:109–113.

Schuh, W. and A. Adamowicz. 1993. Influence of assessment time and modeling approach on the relationship between temperature-leaf wetness periods and disease parameters of *Septoria glycines* on soybeans, *Phytopathology* 83:941–948.

Scott, D. H. 1988. Soybean sudden death syndrome, in T. D. Wyllie and D. H. Scott, Eds., *Soybean Diseases of the North Central Region*, APS Press, St. Paul, MN, 67–70.

Sherwin, H. S. and K. W. Kreitlow. 1952. Discoloration of soybean seeds by the frogeye fungus, *Cercospora sojina*, *Phytopathology* 42:568–572.

Short, G. E., T. D. Wyllie, and P. R. Bristow. 1980. Survival of *Macrophomina phaseolina* in soil and in residue of soybean, *Phytopathology* 70:13–17.

Shortt, B. J., A. P. Grybauskas, F. D. Tenne, and J. B. Sinclair. 1981. Epidemiology of Phomopsis seed decay of soybean in Illinois, *Plant Dis.* 65:62–64.

Sij, J. W., F. T. Turner, and N. G. Whitney. 1985. Suppression of anthracnose and phomopsis seed rot on soybean with potassium fertilizer and benomyl, *Agron. J.* 77:639–642.

Sinclair, J. B. and P. A. Backman. 1989. *Compendium of Soybean Diseases*, APS Press, St. Paul, MN.

Slater, G. P., R. W. Elmore, B. L. Doupnik, Jr., and R. B. Ferguson. 1991. Soybean cultivar yield response to benomyl, nitrogen, phosphorus, and irrigation levels, *Agron. J.* 83:804–809.

Smith, E. F. and P. A. Backman. 1989. Epidemiology of soybean stem canker in the southeastern United States: relationship between time of exposure to inoculum and disease severity, *Plant Dis.* 73:464–468.

Spilker, D. A., A. F. Schmitthenner, and C. W. Ellett. 1981. Effects of humidity, temperature, fertility, and cultivar on the reduction of soybean seed quality by *Phomopsis* sp., *Phytopathology* 71:1027–1029.

Stuckey, R. E., W. F. Moore, and J. A. Wrather. 1984. Predictive systems for scheduling foliar fungicides on soybeans, *Plant Dis.* 68:743–744.

Subbarao, K. V., J. P. Snow, G. T. Berggren, J. P. Damicone, and G. B. Padgett. 1992. Analysis of stem canker epidemics in irrigated and nonirrigated conditions on differentially susceptible soybean cultivars, *Phytopathology* 82:1251–1256.

TeKrony, D. M., D. B. Egli, and A. D. Phillips. 1980. Effect of field weathering on the viability and vigor of soybean seed, *Agron. J.* 72:749–753.

TeKrony, D. M., D. B. Egli, R. E. Stuckey, and J. Balles. 1983. Relationship between weather and soybean seed infection by *Phomopsis* sp., *Phytopathology* 73:914–918.

TeKrony, D. M., D. B. Egli, J. Balles, L. Tomes, and R. E. Stuckey. 1984. Effect of date of harvest maturity on soybean seed quality and *Phomopsis* sp. seed infection, *Crop Sci.* 24:189–193.

TeKrony, D. M., D. B. Egli, R. E. Stuckey, and T. M. Loeffler. 1985a. Effect of benomyl applications on soybean seedborne fungi, seed germination, and yield, *Plant Dis.* 69:763–765.

TeKrony, D. M., R. E. Stuckey, D. B. Egli, and L. Tomes. 1985b. Effectiveness of a point system for scheduling foliar fungicides in soybean seed fields, *Plant Dis.* 69:962–965.

Thomison, P. R., D. L. Jeffers, and A. F. Schmitthenner. 1987. Phomopsis seed decay and nutrient accumulation in soybean under two soil moisture levels, *Agron. J.* 79:913–918.

Thomison, P. R., W. J. Kenworthy, and M. S. McIntosh. 1990. Phomopsis seed decay in soybean isolines differing in stem termination, time of flowering, and maturity, *Crop Sci.* 30:183–188.

Vaughan, D. A., R. L. Bernard, and J. B. Sinclair. 1989. Soybean seed quality in relation to days between development stages, *Agron. J.* 81:215–219.

Von Qualen, R. H., T. S. Abney, D. M. Huber, and M. M. Schreiber. 1989. Effects of rotation, tillage, and fumigation on premature dying of soybeans, *Plant Dis.* 73:740–744.

Walters, H. J. 1980. Soybean leaf blight caused by *Cercospora kikuchii*, *Plant Dis.* 64:961–962.

Walters, H. J. 1985. Purple seed stain and cercospora leaf blight, in R. Shibles, Ed., *World Soybean Research Conference III: Proceedings*, Westview Press, Boulder, CO.

Weaver, D. B., B. H. Cosper, P. A. Backman, and M. A. Crawford. 1984. Cultivar resistance to field infestations of soybean stem canker, *Plant Dis.* 68:877–879.

Webster, R. K. and P. S. Gunnell. 1992. *Compendium of Rice Diseases*, APS Press, St. Paul, MN.

Wilcox, J. R. and T. S. Abney. 1973. Effects of *Cercospora kikuchii* on soybeans, *Phytopathology* 63:796–797.

Wilcox, J. R., F. A. Laviolette, and K. L. Athow. 1974. Deterioration of soybean seed quality associated with delayed harvest, *Plant Dis. Rept.* 58:130–133.

Williams, D. J. and R. F. Nyvall. 1980. Leaf infection and yield losses caused by brown spot and bacterial blight diseases of soybean, *Phytopathology* 70:900–902.

Wrather, J. A. and G. L. Sciumbato. 1995. Soybean disease loss estimates for the southern United States during 1992 and 1993, *Plant Dis.* 79:84–85.

Wrather, J. A., A. Y. Chambers, J. A. Fox, W. F. Moore, and G. L. Sciumbato. 1995. Soybean disease loss estimates for the southern United States, 1974 to 1994, *Plant Dis.* 79:1076–1079.

Wyllie, T. D. 1976. *Macrophomina phaseolina* — charcoal rot, in L. D. Hill, Ed., *World Soybean Research, Proceedings of the World Soybean Research Conference*, Interstate, Danville, IL, 482–484.

Wyllie, T. D. 1988. Charcoal rot of soybeans — current status, in T. D. Wyllie and D. H. Scott, Eds., *Soybean Diseases of the North Central Region*, APS Press, St. Paul, MN, 106–113.

Wyllie, T. D., S. Gangopadhyay, W. R. Taegue, and R. W. Blanchar. 1984. Germination and production of *Macrophomina phaseolina* microsclerotia as affected by oxygen and carbon dioxide concentration, *Plant Soil* 81:195–201.

Yang, X. B., G. T. Berggren, and J. P. Snow. 1990a. Types of Rhizoctonia foliar blight on soybean in Louisiana, *Plant Dis.* 74:501–504.

Yang, X. B., G. T. Berggren, and J. P. Snow. 1990b. Seedling infection of soybean by isolates of *Rhizoctonia solani* AG-1, causal agent of aerial blight and web blight of soybean, *Plant Dis.* 74:485–488.

Yang, X. B., G. T. Berggren, and J. P. Snow. 1990c. Effects of free moisture and soybean growth stage on focus expansion of Rhizoctonia foliar blight, *Phytopathology* 80:497–503.

Yang, X. B., J. P. Snow, and G. T. Berggren. 1990d. Analysis of epidemics of Rhizoctonia aerial blight of soybean in Louisiana, *Phytopathology* 80:386–392.

Yang, X. B., J. P. Snow, and G. T. Berggren. 1991. Patterns of Rhizoctonia foliar blight on soybean and effect of aggregation on disease development, *Phytopathology* 81:287–293.

Young, L. D. and J. P. Ross. 1978. Resistance evaluation and inheritance of a nonchlorotic response to brown spot of soybean, *Crop Sci.* 18:1075–1077.

Young, L. D. and J. P. Ross. 1979. Brown spot development and yield response of soybean inoculated with *Septoria glycines* at various growth stages, *Phytopathology* 69:8–11.

Zorrilla, G., A. D. Knapp, and D. C. McGee. 1994. Severity of Phomopsis seed decay, seed quality evaluation, and field performance of soybean, *Crop Sci.* 34:172–177.

GLOSSARY

Acervulus (pl. **acervuli**) — A saucer-shaped structure that produces asexual spores in certain fungi.

Anastomosis — The union of two fungal hyphae that results in fusion of their contents.

Asymptomatic — Without visible symptoms.

Chlorosis — Yellowing of normally green tissue.

Cotyledon — A seed leaf; a primary leaf in the embryo.

Damping-off — Destruction of seedlings near the soil line, resulting in the seedlings falling over on the ground.

Desiccation — Drying.

Epicotyl — The region above the attachment point of cotyledons in seeds or seedlings.

Hypha (pl. **hyphae**) — A single branch of a mycelium.

Hypocotyl — The region below the attachment point of cotyledons in seeds or seedlings.

Inoculum — The portion of a pathogen that comes into contact with the host and causes infection.

Isolines — Plant genotypes that differ in only one trait.

Latent infection — The state in which a host is infected with a pathogen but shows no symptoms.

Microsclerotium (pl. **microsclerotia**) — A small sclerotium.

Mycelium (pl. **mycelia**) — The hyphae or mass of hyphae that make up a fungus.

Necrotic — Dead and discolored.

Oospore — A sexual, resting spore produced by certain fungi.

Perithecium (pl. **perithecia**) — A spherical or flask-shaped structure that produces sexual spores in certain fungi.

Pycnidium (pl. **pycnidia**) — A spherical or flask-shaped structure that produces asexual spores in certain fungi.

Radicle — The rudimentary root of an embryo.

Saprophyte — An organism that uses dead organic material for food.

Sclerotium (pl. **sclerotia**) — A compact mass of hyphae, usually dark, that can survive under unfavorable environmental conditions.

Sporangium (pl. **sporangia**) — A structure that contains asexual spores in certain fungi; can also function as an asexual spore.

Spore — The reproductive unit of fungi, consisting of one or more cells.

Systemic — Spreading internally throughout a plant.

Zoospore — An asexual spore in certain fungi that possesses flagella and hence is motile in water.

chapter fifteen

Soybean insect management

Joe Funderburk, Robert McPherson, and Dave Buntin

Contents

Arthropod communities inhabiting soybean fields

Soybean in the southern U.S. is attacked by numerous occasional arthropod pests (e.g., mites and insects), but only a few pose a persistent economic threat. Injury from arthropod pests can occur during any soybean growth stage, but the greatest losses usually occur during the reproductive stages. Growers must be concerned with both foliage- and soil-inhabiting pests, since economically important species occur in both habitats. Pests of soybean in the southern U.S. include velvetbean caterpillar, soybean looper, corn earworm, bean leaf beetle, lesser cornstalk borer, threecornered alfalfa hopper, and stink bugs. Other pests of less economic importance are whiteflies, armyworms, green cloverworm, Mexican bean beetle, grasshoppers, spider mites, wireworms, grubs, and soybean nodule fly. Table 15.1 gives additional information about these pests.

Table 15.1 Important Soybean Pests in the Southern U.S., Their Scientific Names, and Soybean Plant Parts Injured by Each

Common name	Scientific name	Plant parts injured
Major pests		
Velvetbean caterpillar	*Anticarsia gemmatalis* Hübner	Leaf blades
Soybean looper	*Pseudoplusia includens* (Walker)	Leaf blades
Corn earworm	*Helicoverpa zea* (Boddie)	Leaf blades, pods, seeds
Bean leaf beetle	*Cerotoma trifurcata* (Forster)	Roots, leaf blades, pods, seeds
Lesser cornstalk borer	*Elasmopalpus lignosellus* (Zeller)	Lower stems
Threecornered alfalfa hopper	*Spissistilus festinus* (Say)	Lower stems
Southern green stink bug	*Nezara viridula* (L.)	Pods, seeds
Green stink bug	*Acrosternum hilare* (Say)	Pods, seeds
Brown stink bug	*Euschistus servus* (Say)	Pods, seeds
Less frequently important pests		
Silverleaf whitefly	*Bemisia argentifolii* Bellows	Leaf blades
Banded winged whitefly	*Trialeurodes abutilinea* (Haldeman)	Leaf blades
Fall armyworm	*Spodoptera frugiperda* (J. E. Smith)	Leaf blades
Yellowstriped armyworm	*Spodoptera ornithogalli* (Gueneé)	Leaf blades
Beet armyworm	*Spodoptera exigua* (Hübner)	Leaf blades
Green cloverworm	*Plathypena scabra* (Fabricius)	Leaf blades
Mexican bean beetle	*Epilachna varivestis* Mulsant	Leaf blades
Redlegged grasshopper	*Melanoplus femurrubrum* (De Geer)	Leaf blades, pods, seeds
Differential grasshopper	*Melanoplus differentialis* (Thomas)	Leaf blades, pods, seeds
Twospotted spider mite	*Tetranychus urticae* Koch	Leaf blades
Wireworms	*Melanotus* species, *Limonius* species, *Agriotes mancus* (Say)	Roots
Grubs	*Phyllophaga* species, *Cyclocephala* species	Roots
Soybean nodule fly	*Rivellia quadrifasciata* (Macquart)	Roots, nodules
Potato leafhopper	*Empoasca fabae* (Harris)	Leaf blades and veins

Soybeans are habitat for a wide range of beneficial arthropod species that aid in the suppression of pests. These beneficial insects, mites, and spiders form a complex fauna that occur in great numbers in soybean fields. The natural control agents of soybean pests include ground beetles, bigeyed bugs, damsel bugs, earwigs, the red imported fire ant, and spiders. Additionally, soybean pests are hosts for parasitic wasps and parasitic flies.

There also are diseases associated with soybean arthropod pests, the most common of which include the fungal diseases of lepidopterous pests, spider mites, and whiteflies. Virus and disease organisms of soybean arthropod pests are important. Recognition of these beneficial organisms is important for integrated pest management programs. Table 15.2 gives information about generalist predators inhabiting soybean in the southern U.S. Important parasites and diseases are included later. Producers and their agents (e.g., consultants and scouts) must make reliable and timely identification of arthropods in a soybean field. With practice, important species can be reliably and rapidly identified. Publications are available that contain identification keys and pictures of arthropod pests and predators and are included in the references at the end of this chapter.

Integrated pest management

Highly effective, inexpensive chemical insecticides became widely available shortly after World War II. Purely insecticidal approaches were rapidly adopted to solve pest problems.

Table 15.2 Some Generalist Predators
of Soybean Arthropod Pests

Common name	Scientific name or family
Ground beetles	*Calosoma* species
Bigeyed bugs	*Geocoris* species
Damsel bugs	*Raduviolus* species, *Nabis* species
Earwigs	*Labidura* species
Red imported fire ant	*Solenopsis invicta* Buren
Minute pirate bugs	*Orius* species
Spiders	Araneidae

The practice of spraying at regular intervals without regard to pest density and damage potential was implemented. Two decades of such indiscriminant use resulted in an ecological disaster, as many target and nontarget pests became genetically resistant and some pesticides lost their effectiveness. Contamination of soil and water by long-lived pesticides affected wildlife populations, with some species nearing extinction. Public health was jeopardized by unknown effects of pesticide residues on food. Integrated pest management (IPM) is an ecologically compatible method of pest control that was adopted following this disaster. Other consequences of the pesticide crisis were the passage of stringent laws regarding pesticide use and the establishment of the Environmental Protection Agency (EPA). The EPA is a regulatory agency responsible for ensuring that pesticides do not contaminate the environment or pose an unnecessary health risk.

As the pesticide crisis demonstrated, it is neither possible nor feasible to eliminate all economic losses attributable to arthropod pests in soybean. The focus of soybean IPM programs is to minimize economic losses from arthropod pests. This is done by the careful use of ecologically compatible management tactics. Conservation of natural enemy populations is a primary consideration of IPM programs. Detailed information on pest mortality from individual species or combinations of natural enemies is not available for most situations. There is no doubt that the combined activity of these natural enemies prevents many pests from reaching economically damaging levels and reduces the severity of outbreaks of other pest species.

Most soybean pests are detected by scouting during periods of risk. Scouts monitor crop growth stage, pest development and population density, and occasionally natural enemy development and population density. Management decisions for pest populations in individual fields are based on economic injury levels defined as the lowest population density of each pest likely to cause economic damage. The economic injury level usually changes during the growing season. It is a function of soybean growth stage and growing season environment and reflects changes in the response of the plant to injury from the pest.

Curative control measures should be applied only when pest density exceeds the economic threshold. This threshold is a pest density below the point at which control measures are needed to prevent an increasing pest population from reaching the economic injury level. In soybean, the most common, curative control tactic is the use of insecticides. Effective insecticides are available for most soybean insects. Efforts are made to use insecticides and rates that minimize harmful effects on beneficial predators, parasitoids, and diseases of soybean pests.

Soybean IPM programs integrate chemical control with biological control, cultural control, and plant resistance to reduce dependence on insecticides. This also helps avoid development of resistance in pest populations. Biological control, cultural control, and plant resistance are preventive control tactics employed as production practices without regard to specific pest densities in a field.

The impact of insect injury on yield, plant maturity, and seed quality is fairly well understood with economic injury levels developed for most soybean pests. Usually, management efforts are based on detection and management of individual species. In some cases a group of pests attack the same soybean parts and produce the same types of soybean response to their attack. Such a group is called a guild, and an example for the southern U.S. is defoliating pests such as velvetbean caterpillar, soybean looper, and green cloverworm. When these species occur together in the same field, managing populations as a guild is more efficient than managing each individual species.

A discussion follows of the management programs for the economically important arthropod pests of soybean in the southern U.S. This discussion includes relevant information on economic injury levels, sampling, and preventive and curative control tactics. It covers aspects of biology necessary for effective and efficient management. Problems associated with decision making and control are discussed.

Lepidopterous defoliators

Soybeans are attacked by a complex of lepidopteran caterpillars that consume leaf tissue. The three predominant species are velvetbean caterpillar, soybean looper, and green cloverworm. Fall armyworm, yellowstriped armyworm, and beet armyworm also can be important defoliators in some years. The corn earworm causes defoliation but it causes more serious injury by feeding on flowers, pods, and seeds within pods. Soybean looper also may feed on pods but is primarily a defoliator. The typical phenology of defoliators in southern soybeans is green cloverworm — early to midseason; soybean looper and corn earworm — mid to late season; velvetbean caterpillar — late season; and armyworms — mid to late season. A number of other minor defoliators such as webworms and loopers other than soybean loopers commonly occur on soybean. These species rarely cause serious injury alone but add to the defoliation caused by more important species.

Larvae of the three dominant species are green to pale green in color, although velvetbean caterpillars sometimes are brown to blackish when populations are large. Green cloverworms have a single white longitudinal stripe on each side, whereas velvetbean caterpillars and soybean loopers have several more or less distinct longitudinal stripes. These three species can be easily distinguished by the number of prolegs. In addition to an anal pair of prolegs, loopers have two prolegs, green cloverworms have three prolegs, and velvetbean caterpillars have four prolegs, although the first pair are reduced in small larvae. Armyworms are pale greenish, brown, or black. Fall armyworms generally are brown with alternating broad light and dark longitudinal stripes. Yellowstriped armyworms are blackish with a distinct yellow longitudinal stripe on each side. Beet armyworms are variable in color but have a distinct black dot on each side about ⅕ the larval length behind the head.

The biology of the green cloverworm, soybean looper, and velvetbean caterpillar is similar. Eggs are laid individually and hatch in 3 to 5 days. Larvae usually molt six times and require 14 to 21 days for development. Total developmental time is 25 to 30 days in the summer, but up to 60 days may be needed when temperatures are cooler. All three species have three or four generations each year in the southern U.S. Soybean looper and velvetbean caterpillar do not overwinter in the main soybean production areas in the region. Both species remain active on alternate legume hosts throughout the winter in areas where freezing temperatures rarely occur. Consequently, migrating adults of both species reinfest the southern U.S. each year and continue to migrate northward throughout the summer. Both species overwinter to some extent in southern Florida and Texas, but the Caribbean basin, Mexico, and Central and northern South America probably are the primary source areas for migrating moths. In contrast, the green cloverworm overwinters as an adult in protected areas throughout the southern U.S. Adults are commonly observed during the winter in garages, outbuildings, and sheds. All three species feed on a variety of native and introduced legumes. The three armyworm species have similar life histories as the soybean looper and velvetbean caterpillar, except eggs are laid in masses of up to several hundred eggs and small larvae typically feed in a group before dispersing after one or two molts. These armyworms are general feeders and attack many crops, but the fall armyworm has a natural preference for grasses.

Population levels of these migratory defoliators are determined mostly by the number of migrating adults, weather conditions, and mortality by natural enemies. Weather conditions conducive for adult long-range migration are a high-pressure system in the east and a lower-pressure system with a tailing cold front in the west. Population levels of migrants have been correlated with the number of these weather patterns in the spring and early summer.

Natural enemies have a major role in regulating defoliator populations in soybean. Eggs and larvae of all species are killed by predators, especially bigeyed bugs, damsel bugs, ants, earwigs, and spiders. However, parasitoids and pathogens generally are more important than predators in soybean. Pathogens are key regulators of the green cloverworm, soybean looper, and velvetbean caterpillar. Epizootics of the fungal pathogen *Nomuraea rileyi* (Farlow) Samson typically occur in most years and greatly reduce larval populations of all three species. Epizootics of other fungal pathogens and a nuclear polyhedrosis virus also can decimate defoliator populations. However, unacceptable levels of defoliation may occur before epizootics reduce larval infestations. Parasitoids also cause significant mortality of soybean looper and green cloverworm larvae. Important parasitoids are *Cotesia marginventris* (Cresson), *Meteorus autographae* Meusebeck, *Euplectrus comstockii* Howard, and *Chaetophlepsis plathypenae* Sabrosky. Parasitoids also attack the velvetbean caterpillar, but the incidence of parasitism is very low. A wide variety of natural enemies prevents economic populations of armyworms from developing in most years.

Defoliating larvae affect soybean production by removing leaf tissue. First and second instars cannot cut completely through the leaf; therefore they eat the lower epidermis and mesophyll tissue which creates a "windowpane" effect. Green cloverworms and velvetbean caterpillars usually begin to defoliate the upper canopy before moving to the middle and lower canopy, whereas soybean loopers typically begin to defoliate the interior lower canopy before moving to the upper canopy. Leaf tissue removal reduces the available leaf area for interception of light and subsequent photosynthesis. This injury mostly reduces seed yield by affecting pod and seed numbers, although severe defoliation may reduce seed size. Soybean can tolerate a considerable amount of defoliation before yield is reduced. Research has found that significant yield losses occur when defoliation is >35% before bloom and greater than 15 to 20% after initiation of bloom through seed fill (Hutchins and Funderburk, 1991; Higley and Boethel, 1994).

Soybean defoliating caterpillars are typically managed as a group, but assessment of species prevalence is necessary to select insecticides properly. The green cloverworm occasionally reaches economic levels, whereas the soybean looper and velvetbean caterpillar are much more likely to exceed economic thresholds in the southern U.S. Dry weather encourages outbreaks of these defoliators by preventing epizootics of *N. rileyi* and other naturally occurring pathogens. Furthermore, soybean looper infestations in soybean often are worse when soybeans and cotton are grown in the same area. This is because access to cotton nectar results in a tremendous increase in egg production by female soybean loopers.

Currently, the varieties Lamar and Crockett have some resistance to defoliators. High-yielding varieties with good levels of natural resistance or genetically engineered resistance derived from the Bt (*Bacillus thuringiensis*) toxin may become available in the near future. Recently, early-maturing varieties (maturity groups III and IV) have been developed for the southern U.S. These varieties mature early and may avoid some or most "late-season" defoliation.

Defoliator management in soybean consists of scouting followed by judicious pesticide use. Ideally, soybean fields should be scouted for defoliators beginning in late vegetative stage or early bloom (stage R1) through seed fill (stage R6). Densities can be estimated weekly at five to ten locations per field using a ground cloth or sweep net. Sampling for velvetbean caterpillars may need to be more frequent in late September and early October because this insect can rapidly defoliate soybeans. Economic thresholds for treatment are four to eight larvae (½ in. or longer) per foot of row depending on soybean market value and control costs. Generally, the lower number should be used if defoliation is already near the level causing yield loss.

Numerous conventional insecticides are effective against the green cloverworm and velvetbean caterpillar. However, soybean looper has developed resistance to organophosphate, carbamate, and pyrethroid insecticides. Resistance to insecticides varies across the U.S., but resistance generally is greatest in areas with cotton production. Commercial formulations of Bt are increasingly being used because they control these defoliators, including resistant soybean loopers, without killing predators and parasitoids. Early-season populations of the green cloverworm also may be helpful in building up natural enemy populations that attack soybean looper and other defoliators later in the season.

A preventive approach to velvetbean caterpillar control is gaining popularity in the extreme southern U.S. where the risk of economic infestations is considerable. This treatment consists of a foliar application of an insect growth regulator, diflubenzuron, plus boron fertilizer during August. Diflubenzuron prevents populations from reaching economic levels by killing eggs and newly hatched larvae. The boron does not affect the defoliators, but increases soybean yield by 1 to 2 bu/acre, thereby increasing the cost/benefit of the preventive application, especially when velvetbean caterpillar populations are low. Theoretically, preventive insecticide treatments are undesirable because of the potential for development of insecticide resistance. However, velvetbean caterpillar populations die out each fall, so the development of resistance may be slowed or prevented.

Bean leaf beetle

The bean leaf beetle is a native species found throughout the eastern half of the U.S. Adults are about ¼ in. long and typically are light yellowish with four black spots and a black marginal stripe. Adults also may be crimson in color with spots present but less distinct. Eggs are laid in soil, and larvae, which are white with a black head, feed on soybean roots and nodules.

Bean leaf beetle has three generations. Adults overwinter in clumps of grass or leaf litter in field margins, pastures, and woods. Adults become active on alternative hosts such as forage legumes in the spring before migrating to soybean fields. Adults lay eggs in soil, and the larvae feed on roots. A second adult flight and generation occur during soybean pod and seed filling. Larvae from this later generation feed partly on soybean nodules. Weather conditions influence bean leaf beetle populations, but key factors regulating bean leaf beetle populations are not well known.

Both larvae and adults have chewing mouthparts and injure soybean. Feeding by larvae is considered to be less important than adult injury. The most obvious form of adult injury is defoliation. Adults chew small irregular round holes in leaves. Defoliation usually is of greatest concern during the seedling stage when adults can cause extensive defoliation to small plants, especially along field edges. Defoliation becomes less important as soybean plants increase in size but may be important in combination with defoliation by other defoliators. Chewing injury to pods is the most important type of bean leaf beetle injury. Adults will eat pod wall tissue down to the endocarp that encloses the seed. Seeds beneath lesions become shrunken and discolored, which reduces seed quality. Pod lesions also increase seed exposure to excess moisture and secondary molds. In addition, adult chewing may result in pod clipping and loss. However, seed weight losses occur at a much lower beetle density than seed quality losses.

Bean leaf beetles also are known to transmit bean pod mottle virus, cowpea mosaic virus, cowpea chlorotic mottle virus, and the cowpea strain of southern bean mosaic virus. These viruses generally are of minor importance in the southern U.S. However, bean pod mottle virus can reduce soybean yield when 20 to 40% of plants become infected, and infection rates exceeding this level have been observed in the southern U.S., especially on late-maturing soybean. Infection of soybean by the cowpea viruses usually occurs only when soybeans are grown near cowpeas.

Bean leaf beetles are occasional pests but can reach economic levels in some years. Conventional management programs for bean leaf beetle consist of estimating adult numbers and damage by sampling at regular intervals, and applying insecticides when populations exceed economic thresholds. Adults should be sampled using a sweep net or ground cloth and/or injured-pod counts. Fields should be sampled during the seedling stage and pod-elongation and seed-fill stages. Economic thresholds for seedling plants are six beetles per plant or 50% defoliation. Thresholds for pod injury with soybean market values of $6 to 8/acre and pesticide application costs of $7 to 10/acre are 6 to 11 injured pods per plant, 3 to 5 adults per sweep, or 5 to 8 adults per foot of row. Bean leaf beetles are easily controlled by minimal rates of labeled insecticides.

Trap cropping has been shown to be an effective preventive approach where 5 to 10% of fields in an area are planted 10 to 21 days earlier than the main plantings. Adults are attracted to these early-planted fields in the spring and can be controlled with a foliar-applied insecticide. Colonization of the later plantings is minimized and bean leaf beetle pod injury is prevented over the entire area by treating only a few fields. Weather conditions and cropping practices, especially doublecropping, may not permit such an organized soybean-planting effort.

Threecornered alfalfa hopper

The threecornered alfalfa hopper is an occasional pest of vegetative and reproductive soybeans throughout the southern U.S. Eggs are initially transparent but later turn white. First-stage nymphs are nearly transparent, and second-stage nymphs are buff colored. The third-stage nymphs are light green. The fourth- and fifth-stage nymphs are usually green, but are sometimes brown or mixed brown and green. Nymphs have 12 pairs of projections along the top of the body. Adults are "threecornered" when viewed from the top. They are usually bright green, but occasionally may be brown.

Three to four generations occur per year. Eggs are laid individually or in clusters just beneath the leaf epidermis. The last generation enters reproductive diapause (inactivity) and migrates to overwintering hosts. Overwintering adults are known to reside in mats of pine needles, but come out to sun on pine trees during warm, sunny weather conditions. The first spring generation develops on other hosts, especially legumes. At least two complete generations develop in soybeans. A strepsipteran parasitoid develops on adult threecornered alfalfa hoppers. Little is known about the impact of other parasitoids and disease on populations. Important predators include ants, damsel bugs, minute pirate bugs, and bigeyed bugs.

Nymphs and adults are phloem feeders. Feeding punctures may be random or in a series circumscribing the stem, branch, or leaf petiole. Such girdling disrupts vascular tissue and results in a nutrient sink above the girdle. Adults and nymphs frequently feed near girdles. Plant mortality, plant lodging, and reduced yield can result from main stem girdling. Threecornered alfalfa hopper feeding on peduncles and pedicels may cause shedding of pods and yield reduction.

Curative management approaches are best for threecornered alfalfa hopper. Adults are sampled with the sweep net and nymphs with the ground cloth, vertical beat sheet, or a modification of the sweep net technique. Economic thresholds for vegetative-stage soybean are 50% of plants girdled, or with threecornered alfalfa hoppers present if fewer than four to six ungirdled plants per row-foot remain in a field. Economic thresholds starting at pod-set stage of soybean growth are three nymphs per row-foot or one adult per sweep.

Lesser cornstalk borer

The lesser cornstalk borer is a common pest of seedling soybean in the southern U.S. The pest occurs on over 60 crops in 14 plant families, although grasses and legumes are preferred. Eggs are greenish white and turn pink to red before hatching. They are laid individually or in small groups directly on the stem near the soil surface or on the soil surface. Larvae may be present in fallow fields and feed on weeds or leftover crop residues. Larvae can survive tillage operations and transfer to germinating soybean. Pupae are greenish and later become brown. Adult females are dark brown or black, and males are tan with a dark stripe down the center of the back. The wings are folded together at rest. The lower developmental threshold is 58°F, and developmental time is greatly dependent on temperature conditions. A complete generation from egg, larva, pupa, and adult is possible in 2 weeks during hot, dry weather conditions.

Five or more broadly overlapping generations occur in the extreme southern U.S., with the pest overwintering in all life stages. In other areas of the southern U.S., two to four generations occur per year, and populations do not overwinter in these regions in some years. Numerous species of natural enemies provide biological control of larval populations. Parasitoids include the parasitic wasps *Orgilus elasmopalpi* Meusebeck, *Pristomerus spinator* (F.), and *Chelonus elasmopalpi* McComb, and the tachinid fly *Stomatomyia floridensis* Townsend. There are virus and fungal pathogens, while important predators include red imported fire ants and earwigs.

Larvae of the lesser cornstalk borer are the only injurious life stage. Larval habitat is the upper soil just below the soil surface. They bore into the main stem and tunnel up and down the plant, and construct silken tubes covered with soil particles that are attached to the main stem of the soybean plant. Numerous seedlings can be injured by a single larva. Seedling soybeans usually do not survive larval attack. Older plants sometimes survive tunneling, but may lodge or have reduced yield.

Outbreaks of this pest typically occur during periods of hot, dry weather, because pest development is more rapid and natural enemy populations are adversely affected. Efficient sampling methods and economic thresholds are not available for this concealed soil pest. Soybean is able to withstand stand reductions to a minimum of about four plants per row-foot without yield reduction. Lesser cornstalk borer problems are common in the extreme southern U.S. in doublecropped soybeans following winter wheat in which the crop residues have been burned. Damage potential is high during periods of hot, dry weather, especially in fields with sandy soil and those with previous infestation histories. Outbreaks are rare in other areas of the southern U.S. Irrigation reduces potential damage. Curative management with labeled organophosphate insecticide application is efficacious if soon followed by rainfall or irrigation. For most labeled insecticides, minimum rates are effective. Granular organophosphate insecticide application at planting applied in-furrow is warranted under conditions of high risk.

Corn earworm

The corn earworm is a serious pest widely distributed throughout the entire southern U.S. Population densities on soybean vary greatly from year to year. Years with severe population outbreaks usually are characterized as being abnormally hot and dry. Larvae consume foliage, blooms, pods, and seeds of both full-season and doublecropped soybean. The highest infestations occur when fields are in full bloom during the time of peak moth emergence.

Corn earworm moths are robust and about ¾ in. long with a wingspan of 1 to 1.5 in. The forewings of the male are usually a light yellowish green, while the female has yellowish brown to pinkish forewings. Each forewing has a dark spot near the center. The hindwings are cream colored with a broad, dark-colored, outer band. Eggs are laid individually throughout the soybean plant, but are usually concentrated on the terminal growth and in close proximity to blooms and young pods. The tiny eggs are dome-shaped with a flattened base and prominent ribs. They are creamy white when first deposited but darken and often have a reddish brown band before hatching. Larvae develop through five or six instars before pupating. First instars are around ¹⁄₁₆ in. long, while the last instar larvae are around 1¾ in. Newly hatched larvae are light-colored and nearly translucent, while larger larvae have a considerable variation in color patterns. Midsize larvae are usually reddish brown, and large larvae are either green, yellowish, or dark brown. They have cream to yellowish longitudinal bands and dark spots along the sides, an orangish head capsule, and numerous short hairs on each body segment. All corn earworm larvae have four pairs of fleshy prolegs on the middle body segments and one pair on the anal segment. They tend to curl up when disturbed. Small larvae are usually found inside folded soybean leaflets and around flowers and young pods, but larger larvae are found throughout the entire plant. Pupae are reddish brown and are located in the soil. From egg to adult requires about 4 weeks under ideal conditions. Longer developmental times are required in cooler weather.

The corn earworm overwinters as a pupa in an earthened cell about 2 to 4 in. below the soil surface. As temperatures warm in the spring, hibernation is terminated and moths begin to emerge, usually in April, in the southern states. This overwintering generation infests corn, alfalfa, vegetables, weeds, and other available host plants. First-generation moths emerge during June and are highly attracted to corn that is silking. Second-generation moths emerge during July and August and infest soybeans, corn (especially if it is still green and lush), vegetables, and other host plants. Soybeans that are blooming during this period are highly attractive to egg-laying corn earworms, especially soybeans planted on wide rows that have open spaces between the rows during this moth flight. The third-generation moths emerge in September, but this population is usually much smaller. This is because many of the pupae have already begun to hibernate, plus natural enemies and insecticide applications have reduced the larval populations across the region. Soybean is usually not a suitable host for the fourth-generation corn earworm larvae because the crop is past the blooming stage and is filling pods with seeds during September. In fact, the early-maturing varieties (Group IV and V) are already beginning to dry down at this time. Corn earworm moths are strong fliers and are capable of migrating long distances. Thus, soybean fields may be invaded by moths from distant hosts or locally infested by moths from adjacent fields.

Corn earworm larvae have chewing mouthparts and feed on leaves, stems, flowers, pods, and seeds. All larval stages feed; however, most of the consumption is confined to the last two instars. Small larvae feed on the newly formed leaflets and blooms, whereas large larvae feed throughout the plant. High population densities result in yield losses due to excessive defoliation and removal of flowers and seeds. Yield loss is dependent on many factors, including environmental stress, plant density, and timing of the infestation. Plants already stressed by hot/dry weather, low fertility, nematodes, weeds, and other factors already have a low yield potential. However, these plants are particularly susceptible to corn earworm infestations and further yield reduction, because there is little plant compensation for damage. Lush growing soybeans are able to compensate for much of the corn earworm–induced plant injury through delayed maturity and setting new pods.

Both the ground cloth and sweep net are used to monitor corn earworm populations on soybean. Correlations have been developed for full-season and doublecropped fields and for wide, narrow, and drilled row spacings to determine if economic injury levels are present. The economic threshold is 1 larva/row-foot (3/25 sweeps) in stunted soybeans, 2/ft in normal soybeans, and 3/ft in lush soybeans. Monitoring fields on a regular basis is highly encouraged because corn earworm populations can vary greatly between fields and throughout the season.

Cultural practices can be used to lessen the likelihood of an economically damaging population of corn earworms. Early plantings of early-maturing varieties often escape heavy corn earworm populations, because such fields are already setting pods and are not attractive to ovipositing females migrating from corn fields in July. Practices that encourage a closed canopy, where all the space between rows is filled in with foliage, also discourage earworm infestations and allow plants to compensate for feeding damage that does occur. Planting on time, narrowing the row spacing, and irrigating will help create a closed canopy.

Many arthropod predators and parasites, as well as viral and fungal pathogens, often maintain corn earworm population densities below the economic injury level. Limiting insecticide use only to those fields where it is justified and using less-disruptive insecticides when they are needed help to preserve these natural enemies.

Stink bugs

The stink bug complex is one of the most important insect pest threats to soybean production in the southern U.S. The southern green stink bug, the green stink bug, and the brown stink bug are annual pests of soybean in the southern states. Other species may be present at low population densities. The seasonal abundance of stink bugs varies greatly from year to year on soybeans. However, the years when serious economic losses occur are years when southern green stink bug populations are high.

All these stink bug species complete five nymphal stages before becoming adults. Eggs are laid in masses that are arranged in a series of rows. From egg hatch to adult takes about 3 weeks at the optimal temperature of 86°F, but can take as long as 1 to 2 months at cooler temperatures. First instar southern green stink bugs are black to reddish black and tend to remain clustered on the egg mass. Second instars are also black with white spots on the top of the abdomen. They remain grouped together near the egg mass. Third instars are usually black, although some may be green with pinkish markings, with white spots on the abdomen. They move away from the eggs but usually remain clustered unless disturbed. Fourth and fifth instars disperse widely and are pale green with pink and black markings and white spots. Fifth instars are larger and have pale green wing pads forming. Nymphal behavior of the other stink bug pests on soybean is similar to that of the southern greenstink bug. First instar green stink bugs are reddish brown, and second and third instars are pale green with black and white stripes on their abdomen. Fourth and fifth instars are either green with yellow and black stripes on the abdomen or pale yellow green with black pronotum and a black area in the center of the abdomen. All instars of the brown stink bug are yellow to tan with a brown pronotum and brown spots on the abdomen.

Stink bug species overwinter as adults under leaf litter, bark, woodpiles, barns, and other areas offering protection from harsh winter climate. In the southern U.S., where warmer weather prevails, hibernation is only partial, and adults can be observed feeding on succulent plants during warm periods in the winter months. In early spring, hibernating adults move from winter cover and begin feeding on clover, small grains, vegetables, weeds, and other host plants where the first generation is completed. Four to five generations occur annually during April to October with many vegetables, row crops, and weeds serving as suitable host plants. Many of the third- through fifth-generation stink bugs occur on soybeans in the southern states.

The stink bug mouthparts are modified for piercing and sucking. Both nymphs and adults obtain their food by piercing plant tissue with their beaks and extracting plant fluids. They feed on stems, foliage, blooms, and seeds of the soybean plant, but prefer the developing seeds. Adults and fifth instar nymphs cause the most damage. The feeding punctures that result from the insertion of the beak form minute discolored spots on the plant or seed. The loss of plant fluids, the injection of digestive enzymes, the deformation and abortion of seeds, and the onset of disease and decay organisms at the puncture site are the primary ways that stink bug feeding causes yield and quality losses to soybean. Delayed plant maturation, retention of foliage, and the production of numerous abnormal leaflets and pods close to the main stem also are attributed to stink bug feeding. Stink bug injury during early seed formation can result in shriveled, deformed, and even aborted seeds. Injury to fully formed seeds causes only minor shriveling and spotting. A large amount of stink bug-damaged kernels can result in reduced market values or even prevent the sale of the crop.

Both the ground cloth and sweep net are used to monitor stink bug population densities on soybeans. Correlations have been developed between the two sampling methods for rows that are wide, narrow, or drilled to determine if economically threatening populations are present. The economic threshold is 0.3 bugs/row-foot (2/25 sweeps) from bloom through mid-pod-fill. The threshold is 1 stink bug/row-foot (6/25 sweeps) during the period from pod filling to maturity.

Life histories of the stink bug pests on soybeans are synchronized to the development and growth of the seeds. Because of this, they can often be managed through certain deliberate alternations in crop production. Cultural practices can be used to affect the stink bug pests adversely or to enhance the buildup of natural enemies. Trap cropping is a management practice that is being used to suppress stink bug populations on soybeans. This concept utilizes the fact that stink bugs infest soybeans during the pod-set and pod-fill stages of plant development. Planting soybeans early or planting an early-maturing variety in a small area on the farm where a major planting of soybeans will be made can lure in large populations of stink bugs into the small trap crop. The heavy concentration of egg-laying adults and the subsequent nymphal population can be effectively and economically controlled in the trap crop before the next generation adults move into surrounding fields, and this is the key to the success of this cultural practice. If effective insecticide control is not achieved in the trap crop, this management practice is of no value, and in fact may worsen the stink bug problem in the main soybean crop. Trap cropping is an effective management practice and has been incorporated into many existing cooperative extension service soybean IPM programs in the southern states.

Stink bug populations often can be maintained at low levels by a number of natural control agents, especially arthropod parasites and predators, entomopathogens, and environmental conditions. These natural controls, along with trap cropping, have helped to reduce insecticide use and have helped to avoid insecticide resistance problems. Insecticides are not economically justified on soybeans except during the period from the pod-filling stage until maturity. However, insecticidal control is necessary to protect soybean seeds from serious stink bug injury and even crop failure in many fields throughout the southern states when economic threshold levels are present.

Mexican bean beetles

The Mexican bean beetle is one of the few lady beetles that feeds on the foliage of several legume plants, including lima bean, snap bean, cowpea, and soybean. The adult is oval-shaped, about ⅓ in. long and ¼ in. wide, orange to copper colored, with eight black spots on each wing cover. The larvae are yellowish, soft bodied, and covered with branching spinelike hairs. There are four larval instars ranging in size from ⅒ to ⅓ in. long. Pupae are also yellowish, about ¼ in. long, and usually found attached to the underside of the host plant leaves.

Mated and unmated adults overwinter in areas offering protection from the winter weather, such as leaf litter, tree bark, and woodpiles. With warming spring temperatures, the adults emerge and seek various early-season legumes, such as beans, clover, vetch, and kudzu. The earliest-planted soybeans in the area are the most susceptible to colonization. The yellow orange eggs are laid in clusters of 20 to 60 eggs and are deposited with one end sticking upward. Each female lays several hundred eggs during her lifespan of about 40 days. The eggs hatch in about 1 week and the four larval instars require around 20 days to complete development. There are 3 to 4 generations per year in the southern states. Population outbreaks vary considerably from year to year, but mild/wet seasons favor Mexican bean beetle population buildups.

Both adults and larvae injure soybean by scraping, crushing, and ingesting fluids and tissue between the leaf veins. This feeding produces a lacy appearance to the foliage. As the foliage begins to yellow late in the season, the adults may feed on stems and the outer tissue of the pods. This pod injury rarely penetrates the pod to the developing seed. However, soybean yield reductions are common when Mexican bean beetle populations are high. These yield losses are usually associated with defoliation of the plants and not pod and stem feeding. Although uncommon, Mexican bean beetle populations can cause economic damage early in the season by feeding on the cotyledons and growing terminals. However, most economic damage occurs from blooming to full-seed stages. After leaves and pods begin to yellow, yield losses are not likely.

Two soybean varieties, Lamar and Crockett, have been released that offer moderate resistance to Mexican bean beetle feeding. Natural enemies such as damsel bugs, bigeyed bugs, and minute pirate bugs help reduce beetle populations. A parasite control program was introduced by U.S.D.A.–APHIS in the 1970s in which mass releases of a parasite were made throughout the eastern U.S. This program helped lower the beetle populations throughout the region. Hot, dry weather, which commonly occurs in the southern U.S., is also detrimental to the establishment and buildup of Mexican bean beetle populations on soybeans.

Whiteflies

The silverleaf whitefly (SLWF), formerly referred to as sweet potato whitefly, and the bandedwinged whitefly (BWWF) are the most common whitefly species that infest soybeans in the southern U.S. These pests are currently not a major economic threat to soybean, although they are important pests of other agricultural crops, including ornamental plants. Because whitefly populations and subsequent damage have been on the increase on many crops throughout the south and southwest, these insects have the potential to become a major soybean pest in the future.

The SLWF and BWWF are very similar in appearance, especially the egg and immature stages. Adults of both species are about the size of a pinhead. The BWWF adult has two irregular and curved dark bands on each forewing and the wings are held loosely to the body. The SLWF adult has no dark bands on the forewings and the wings are held tightly against the body. Eggs of both species are laid individually, although they may be touching in rows or groups of two to ten or more if population densities are very heavy. Newly laid eggs are green to yellowish brown, then turn brown before hatching. The first instar (crawler stage) is greenish yellow and oval, with well-developed legs and antennae while mobile (a few hours), then withdrawn. The second instar is also greenish yellow, elongate oval, legless, transparent or semitransparent, and flat. The third instar is greenish yellow, oval or elongate oval, legless, flat or slightly convex, and transparent or semitransparent. The fourth instar or pupa is yellow and oval, with a convex dorsal surface with two red eyespots. The BWWF pupa is dark colored in the middle of the body and pale green to yellowish on the outer edges. The BWWF pupa, but not the SLWF pupa, has waxy spinelike hairs on the outer edges. It takes about 3 weeks for both species to complete the life cycle during the summer months, but considerably longer in cooler temperatures.

During winter months SLWF populations are low; however, immature stages and adults are present on several weed species and cultivated crops produced in the southern U.S. Females lay an average of 160 eggs, primarily on the undersurface of the leaflets. After hatching, the crawler moves to a suitable site to feed and becomes stationary. The rest of the life cycle is spent at this site on the leaflet. Both adults and nymphs have piercing–sucking mouthparts and cause direct damage by extracting plant nutrients. High population densities cause stressed plants to wilt, become stunted, and have discolored leaflets. These pests also secrete honeydew during the feeding process. This sugary solution drips onto the lower leaves, stems, and pods. A secondary fungus, sooty mold, grows on this honeydew and causes the leaflets to turn black. Heavy sooty mold growth interferes with the normal photosynthesis of the plant and causes the loss of leaf vigor and premature leaf drop. In addition to feeding injury, whiteflies are potentially damaging as vectors of several soybean viruses.

Adult whiteflies can fly both short distances within the plant canopy and long distances on air currents. Adults emerge from pupae during the morning and are more active from early morning until midafternoon as temperatures increase. Population densities on soybeans are generally low until August, then steadily rise until the plants begin to mature. SLWF is the most abundant species on soybeans throughout the season, although BWWF is also present at low levels. Perrin and Cook soybean varieties tend to have lower population densities than other varieties. Perrin and Cook have trichomes (leaf hairs) that lay flat on the leaf surface, while the trichomes of most other varieties are erect or at least partially erect.

Natural sources of mortality of whiteflies include predation by arthropod predators such as lacewings and lady beetles, parasitism by certain wasps, desiccation, and diseases. Currently, there are no truly resistant varieties (although Perrin and Cook offer suppression), rotation programs, or other cultural practices that can effectively manage whiteflies on soybeans. Insecticide controls are very expensive but not very effective. Thus, whiteflies, particularly SLWF, have the potential to reach major pest status on soybeans in the near future.

Grubs

The term *grubs* refers to the larvae of certain scarab beetles. The adults are referred to as May or June beetles and masked chafers. These insects attack a wide variety of grasses, grain crops, vegetables, and ornamentals. Damage to soybean is uncommon. The larvae are white, C-shaped, robust, and up to 1½ in. long. They have a brownish head capsule and a dark tail section. The adults are about 1 in. with dark brown or black bodies and long spiny legs.

Eggs are laid in early summer, primarily in grassy soil. Larval development takes from 1 to 4 years, during which time the larvae feed on upper roots in the warm months and overwinter deep in the soil during the winter. The larvae develop through three instars, pupate, and adults emerge in the spring. The second and third instar larvae can cause considerable damage to soybean by pruning the roots, killing young plants, and reducing stands. Plants that are less severely injured may lodge easily because of a weakened root system. Damage to soybeans is usually patchy and is most severe in fields having a 2-year corn then 2-year soybean rotation (Eastman, 1980).

Management of grubs is difficult. No insecticides are registered for their control on soybeans. Crop rotation is not very effective because of their multiyear life cycle. Soybean fields with severe stand loss could be replanted with another crop where a soil insecticide could be applied prior to or at planting. Also, planting soybeans in an area that has historically been infected with grubs should be avoided.

Wireworms

Wireworms are serious seed, seedling, and root pests of several row crops and grasses, although they are only minor pests of soybean. Adults are called click beetles because of the clicking sound they produce when they snap the segment between the thorax and abdomen. Adults are brown, gray, or black and have two rear-facing points. The larvae are cream to dark brown, elongated and cylindrical, and 1¼ to 1¾ in. long when full grown.

In the spring, females lay eggs in the soil, primarily around roots of grasses. Larvae feed on germinating seeds, the underground stems of seedlings, and the roots of older plants. This injury kills the seedling or results in a stunted, weakened plant that is more susceptible to disease and other stresses. Replanting a portion of the crop, after treating with a soil insecticide, may be necessary if plant stand loss is severe. The larval stage lasts 2 to 6 years; thus, subsequent crops are susceptible to feeding injury. Wireworm damage to soybeans is most likely to occur if soybeans are planted on land that has been fallowed or following grasses, clover, or alfalfa. Fields that have low-lying areas or that are poorly drained are more likely to have wireworm infestations.

If there is a previous history of wireworms, the field is poorly drained, or in grass or fallow, then the field should be monitored for wireworm infestations. From five to ten bait stations should be placed in each field in grassy areas and in areas where wireworms were present the previous year. Each station consists of a 4- to 6-in.-deep hole that is 6 to 9 in. wide, containing 4 oz of either corn, wheat, or a mixture of these and covered with soil. The station is then covered with black plastic to enhance seed germination. Each station is dug up and checked for wireworms prior to planting. An average of one wireworm per station is considered an economic threat.

Twospotted spider mite

The twospotted spider mite is distributed worldwide and attacks many different plants. Spider mites are extremely small, wingless, oval in shape, and have eight legs. They are greenish yellow to dull orange with two irregular black spots on each side of the body. Mites usually feed on the underside of leaves and produce webbing that is noticeable on infested leaves.

Spider mites overwinter in sheltered field margins, but females remain partly active during the winter in the southern U.S. Spider mites initially move into a field on the wind or by crawling. Females lay eggs on leaves, and mites develop through three nymphal stages before becoming adults. Immatures resemble adults. Spider mites can complete one generation in 5 to 8 days during the summer and have many generations during a season. Spider mite populations normally are regulated by weather conditions and a fungal pathogen, *Neozygites floridana* Weiser and Muma. The fungus attacks all stages and is host specific. Predaceous mites also attack spider mites, but their role in regulating spider mites is not known.

Spider mites have piercing–sucking mouthparts and feed by inserting their mouthparts into individual leaf cells and removing cell contents. Feeding reduces leaf photosynthetic capacity by removing chlorophyll and also causes loss of leaf water through feeding and leaf wounds. Mite injury appears as yellow stippling and bronzing on the upper leaf surface. Severely injured leaves typically turn brown and drop off the plant prematurely. Yield reductions of 40 to 60% have been documented during outbreaks. Injury also can cause poor soybean harvestablity and reduce seed size and grain quality.

Spider mite outbreaks are infrequent in the South, but typically occur during periods of extended hot and dry weather, when populations can reach enormous levels. Hot, dry weather inhibits the fungal pathogen that normally regulates mite populations. Droughty conditions also may make soybean plants more nutritionally suitable for spider mites. Mite infestations usually start along field margins, but by the time infestations are detected, they often have spread throughout a field.

Currently, resistant varieties are not available, and control is limited to applications of insecticide/acaracide. Well-defined economic thresholds are not available, but treatment should be considered when injury is evident, large populations are present, and dry weather is forecasted. Late-season infestations are difficult to treat because most labeled chemicals have 21- to 28-day harvest intervals. If dry weather persists, populations may rebound after a midseason application and result in the need for multiple applications. However, because spider mite infestations occur during droughty conditions, yield responses to chemical control may be less than expected.

Grasshoppers

The redlegged grasshopper is the most common grasshopper causing damage to soybean in the southern U.S., but the differential grasshopper also occurs in the region. Eggs are laid during summer and fall in the soil in grassy habitats such as field borders and weedy fallow fields. The nymphs hatch the following May and June, and there are five or more instars. Adults appear in July and August. Both redlegged and differential grasshoppers have one generation per year. Redlegged nymphs are tannish yellow to yellowish green. There are strongly contrasting black markings. Adults are bright yellow on the undersurface and reddish brown above. The back part of the hind leg is red. Differential nymphs have a mottled appearance and are green, yellow, and tan. Adults are yellowish brown with contrasting black marks.

Low rainfall increases the number of eggs laid, and egg and nymphal survival. Outbreaks are common after consecutive years of drought. Late-instar nymphs and adults are migratory and drought conditions favor migratory behavior. There are many natural enemies of grasshoppers. Predators include insects, birds, and mammals. A fungal disease is favored by warm, humid weather and is important in reducing outbreak populations. Numerous parasites and other diseases also cause natural mortality.

Adults and nymphs feed on leaves and pods. Pod feeding may predispose remaining seeds to fungal infection. Infestations usually occur first along field edges next to grassy areas and in fields of preferred crops such as small grains, alfalfa, and clover. Populations then spread throughout the soybean field.

Soybean producers should be aware that problems are most likely during growing seasons following 2 or more years of drought. Cultural and insecticidal management tactics are used for grasshoppers. Tillage of small-grain fields after harvest and tillage of fallow fields are effective in preventing egg laying, and can be used to prevent buildup of grasshopper populations. Insecticides are avaliable for suppressing outbreak populations in soybean fields. Economic thresholds are 30 nymphs/yd^2 and 8 adults/yd^2. Usually, economic thresholds will occur in border areas of soybean fields first. Treating these areas with insecticide frequently will suppress populations and prevent further outbreaks for the remainder of the growing season. Additional areas will require treatment under heavy infestations.

Soybean nodule fly

The soybean nodule fly is a very minor pest. Egg, larval, and pupal stages occur in the soil. The larvae or maggots are white. The adult flies have a reddish brown, oval head and four black bands across the wings. The insect overwinters as a larva, and there are one or two generations per year.

Soybean nodule flies injure soybean by larval feeding on the nitrogen-fixing nodules. Soybean compensates for considerable nodule injury without yield loss. Losses from this pest are considered minor.

The larvae live in the soil, and injury is cryptic. Sampling programs for scouting are not available. Because injury is not known to impact yield, there are no management programs for this insect in soybean.

Weeds, soybean diseases, and insect management

Southern soybeans often are infested with many species of weeds. Increasing the presence of weeds has been suggested as a mechanism for reducing pest outbreaks in crops. Weeds either may affect pest colonization by making the crop more difficult to locate or attract more natural enemies resulting in better pest control. Only a few studies have examined the effect of weeds on insect populations, but these have generally found that weeds do not greatly affect insect populations in soybeans. Furthermore, any benefit in pest control from increased weed incidence most likely would not compensate for the adverse effects of the weeds on soybean production. Conversely, flowering weeds such as sicklepod may attract egg-laying moths of the corn earworm or velvetbean caterpillar, thereby increasing numbers of these insects on soybeans.

Pubescence on soybean plants has a dramatic effect on potential injury from potato leafhopper. Most commercial soybean varieties typically grown in the southern U.S. have at least moderate pubescence, and the potato leafhopper is not an economic pest on these varieties. There have been indications that recently released herbicide-ready varieties were injured by the potato leafhopper. Economic thresholds are five leafhoppers per plant for vegetative-stage soybeans and nine leafhoppers per plant for first-bloom-stage soybeans. Several insecticides provide effective control.

Some soybean diseases are vectored by insects. For example, bean pod mosaic virus is transmitted by the bean leaf beetle. Insects also sometimes indirectly influence soybean losses from disease. Insect injury may provide entrances for disease infection, weaken the plant, and predispose it to disease, or alter plant architecture and provide a more-suitable environment for disease.

Although arthropod management activities can influence losses from weeds and diseases, such interactions are complicated, poorly understood, and not a major concern for soybean IPM programs at this time. Growers, however, must be concerned about any interactions between pesticides that may reduce pesticide effectiveness or stunt soybean plants.

Implementing integrated pest management

IPM arose out of the pesticide crisis and originally embraced integrated insecticide use with natural enemies. This remains a core element in IPM programs developed for soybean pests in the southern U.S., but cultural tactics and resistant varieties have been developed and are now commonly used practices.

Soybean IPM has been developed from knowledge of pest biology, soybean response to pests, and the economics of management and production. Important considerations are the economic sustainability of the agricultural production system, sustainability of management tactics, and efficient utilization of natural resources. IPM is recognized as the optimal approach for dealing with soybean arthropod pests. The IPM practices detailed in this chapter are cost-effective and have been widely implemented on soybean in the southern U.S.

An important barrier to wider adoption is the knowledge required in using IPM. Nevertheless, the importance of IPM will continue to grow, not only because of economic and other benefits to soybean producers, but also because IPM serves the public interest. Reductions in the use of broad-spectrum, highly toxic pesticides result from the implementation of IPM, which reduces environmental and food safety concerns about agricultural production by the general public.

Another major benefit to producers is reduced health risk to themselves and other farmworkers by reduced use of highly toxic pesticides. New worker protection standards have recently been enacted that protect soybean producers.

The Worker Protection Standard is a federal regulation designed to protect agricultural workers and pesticide handlers. This program is not only a good idea, but it is the law. The Worker Protection Standard is a guide to use correctly pesticides labeled for use on soybean and other agricultural crops. This information does not take the place of the product label. Always read and follow the current label directions for each pesticide that is applied. Also, check with the local cooperative extension service office to see if any amendments or additions to this law have occurred.

For more-complete information, refer to a current copy of the *Reference Guide for EPA's Worker Protection Standard*. Copies can be obtained from various chemical supply dealers and local cooperative extension service offices.

Interpretive summary

Soybean fields in the southern U.S. are inhabited by numerous pest insects, beneficial insects, and other arthropod species. Important economic pests are velvetbean caterpillar, soybean looper, corn earworm, bean leaf beetle, lesser cornstalk borer, threecornered alfalfa hopper, and stink bugs. Other pests of less economic importance are whiteflies, armyworms, green cloverworm, Mexican bean beetle, grasshoppers, spider mites, wireworms, grubs, and soybean nodule fly. Each pest has predator, parasitoid, and disease natural enemies. Integrated pest management (IPM) programs are designed to minimize economic losses by arthropod pests. Conservation of natural enemy populations is a primary consideration of IPM programs. Most soybean pests are detected by scouting during periods of risk. Management decisions are based on economic injury levels defined as the lowest population density likely to cause economic damage. Curative control with an insecticide is applied only when there is a high probability that the pest density has reached or will reach the economic injury level. IPM programs integrate chemical control with biological control, cultural control, and plant resistance to reduce dependence on insecticides. This also helps avoid development of resistance in pest populations and reduces risk to workers and the environment from insecticide use. Insecticides and rates are available for most pests that minimize harmful effects on natural enemy populations.

References

Boethel, D. J., J. S. Mink, A. T. Weir, J. D. Thomas, B. R. Leonard, and F. Gallardo. 1992. Management of insecticide resistant soybean loopers, *Pseudoplusia includens,* in the southern United States, in L. Copping, M. Green, and R. Rees, Eds., *Pest Management in Soybean,* Elsevier, London, 66–87.

Buntin, G. D., W. L. Hargrove, and D. V. McCracken. 1995. Populations of foliage-inhabiting arthropods on soybean with reduced tillage and herbicide use, *Agron. J.* 87:789–794.

Daingle, C. J., D. J. Boethel, and J. R. Fuxa. 1990. Parasitoids and pathogens of soybean looper and velvetbean caterpillar (Lepidoptera: Noctuidae) in soybean in Louisiana, *Environ. Entomol.* 19:746–752.

Eastman, C. E. 1980. Sampling phytophagous underground soybean arthropods, in M. Kogan and D. C. Herzog, Eds., *Sampling Methods in Soybean Entomology,* Springer-Verlag, New York, 327–354.

Environmental Protection Agency. 1993. Protect Yourself from Pesticides — Guide for Agricultural Workers, U.S. Environmental Protection Agency Publ. 735-B-93-002.

Funderburk, J. E., D. C. Herzog, R. K. Sprenkel, and R. E. Lynch. 1984. Parasitoids and pathogens of larval lesser cornstalk borer (Lepidoptera:Pyralidae) in northern Florida, *Environ. Entomol.* 13:1319–1323.

Funderburk, J. E., L. G. Higley, and G. D. Buntin. 1993. Concepts and directions in arthropod pest management, *Adv. Agron.* 51:125–172.

Hammond, R. B., R. A. Higgins, T. P. Mack, L. P. Pedigo, and E. J. Bechinski. 1991. Soybean pest management, in D. Pimentel, Ed., Handbook of Pest Management in Agriculture, Vol III, CRC Press, Boca Raton, FL, 341–472.

Higley, L. G. and D. J. Boethel, Ed., 1994. *Handbook of Soybean Insect Pests,* Entomological Society of America, Lanham, MD, selected chapters.

Hutchins, S. H. and J. E. Funderburk. 1991. Injury guilds: a practical approach for managing pest losses to soybean, *Agric. Zool. Rev.* 4:1–21.

Klubertanz, T. H., L. P. Pedigo, and R. E. Carlson. 1991. Impact of fungal epizootics on the biology and management of the twospotted spider mite (Acari: Tetranychidae) in soybean, *Environ. Entomol.* 20:731–735.

Kogan, M., G. P. Waldbauer, G. Bioteau, and C. E. Eastman. 1980. Sampling bean leaf beetles on soybean, in M. Kogan and D. C. Herzog, Eds., *Sampling Methods in Soybean Entomology,* Springer, New York, 201–236.

McPherson, R. M., G. K. Douce, and R. D. Hudson. 1993. Annual variation in stink bug (Heteroptera: Pentatomidae) seasonal abundance and species composition in Georgia soybean and its impact on yield and quality, *J. Entomol. Sci.* 28:61–72.

McPherson, R. M. and A. L. Lambert. 1995. Abundance of two whitefly species (Homoptera: Aleyrodidae) on Georgia soybeans, *J. Entomol. Sci.* 30:527–533.

Nolting, S. P. and C. R. Edwards. 1988. Yield response of soybean to defoliation by the Mexican bean beetle (Coleoptera: Coccinellidae), *J. Econ. Entomol.* 82:1212–1218.

Pedigo, L. P., E. J. Bechinski, and R. A. Higgins. 1983. Partial life tables of the green cloverworm in soybean and a hypothesis of population dynamics in Iowa, *Environ. Entomol.* 12:186–195.

Smelser, R. B. and L. P. Pedigo. 1992. Soybean seed yield and quality reduction by bean leaf beetle (Coleoptera: Chrysomelidae) pod injury, *J. Econ. Entomol.* 85:2399–2404.

Sparks, A. N., Jr. 1986. The Threecornered Alfalfa Hopper on Soybean: Determination of Damaging Stages, Development of Sampling Methodology, Refinement of the Action Threshold, and Evaluation of Insecticidal Control, Ph.D. dissertation, Louisiana State University, Baton Rouge.

Stinner, R. E., J. R. Bradley, Jr., and J. Van Duyn. 1980. Sampling *Heliothis* spp. on soybean, in M. Kogan and D. C. Herzog, Eds., *Sampling Methods in Soybean Entomology,* Springer-Verlag, New York, 407–421.

Tippins, H. H., Ed., 1982. A Review of Information on the Lesser Cornstalk Borer *Elasmopalpus lignosellus* (Zeller), Georgia Agric. Exp. Sta. Spec. Publ. 17.

chapter sixteen

Plant-parasitic nematode pests of soybean

G. W. Lawrence and K. S. McLean

Contents

0-8493-2301-0/99/$0.00+$.50
© 1999 by CRC Press LLC

Introduction

Some of the most serious pests associated with soybean production in practically every state where the crop has been produced are plant-parasitic nematodes. Plant-parasitic nematodes are multicellular, microscopic, wormlike animals that feed primarily on the root system of plants. These animals use a specialized mouthpart called a stylet to puncture the cells of the root and withdraw the cellular contents.

Typically, plant-parasitic nematodes feed either from the outside of the root as an ectoparasite or inside the root tissue as an endoparasite. The ectoparasitic nematodes damage the plant by feeding on the epidermal root cells. The endoparasitic nematodes enter the root and establish permanent residency by altering the physiology of the plant. The nematode will induce the plant to produce specialized nurse cells, which may be associated with enlarged roots or root galls. The nurse cells act as a sink to direct the natural flow of nutrients produced by the plant to the feeding site of the nematode. The reduction of available nutrients for plant growth and development generally results in plants that appear stunted and have chlorotic (yellow) leaves. Infected root systems may have galls, be reduced in size, and inefficient in nutrient uptake.

More than 100 species of plant-parasitic nematodes have been reported associated in some way with the soybean plant (Sinclair and Backman, 1989). However, only a few species are considered to be of economic importance. These include the soybean cyst, reniform, root-knot, and lance nematodes.

The soybean cyst nematode is the most serious nematode pest of soybeans in the U.S. (Riggs, 1977; 1982; Riggs and Schmitt, 1988). This nematode has been reported in 26 states with infestation in some areas so severe that soybean production is not economically possible without control measures (Sinclair and Backman, 1989). In Mississippi, the soybean cyst nematode has been identified in every county that grows soybeans and is estimated to affect over 1 million soybean acres across that state (Patel, 1995).

The reniform nematode recently has been determined to result in economic losses to soybean production in Mississippi. This nematode has been identified in 53 counties across Mississippi and infects both cotton and soybean. Reniform nematode population numbers greater than 30,000/pt of soil may be recovered at planting in Mississippi following cotton. An at-planting nematode population of 2500 nematodes/pt of soil may reduce the yield of a susceptible soybean variety 5 bu/acre (Cornelius and Lawrence, 1993a, b) (Table 16.1).

Table 16.1   Effect of Reniform Nematode Initial Inoculation Numbers on the Growth and Yield of Coker 156 Soybean

Reniform/pt of soil		Plant height (in.)	Nodes/plants	Pods/plant	Bu/A
1	0	30.6	13.8	37.7	30.1
2	500	29.4	13.4	37.5	30.5
3	1000	28.9	13.6	41.0	30.4
4	2500	29.9	14.1	32.2	25.1
5	5000	26.4	13.5	26.0	23.3
6	7500	27.4	13.2	31.9	23.0
7	10000	27.5	12.9	36.6	20.8
LSD ($P = 0.05$)		3.1	0.8	10.3	8.5

Reniform nematodes were incorporated into the top 4 in. of a Stough fine sandy loam soil (68.4% sand, 11.6% clay, 20.0% silt, 0.3% om, 8.0 CEC, pH 6.4) in field microplots. All soil was fumigated with 1.5 lb of methyl bromide per 100 ft^2 of soil 30 days prior to nematode infestation.

Seed yields will decline progressively with time if a susceptible variety is planted more than 1 year in a reniform-infested field (Cornelius and Lawrence, 1994a).

The root-knot nematode is one of the most common species of plant-parasitic nematodes found in Mississippi and in the world. The most common species of root-knot associated with soybean production include *Meloidogyne incognita*, *M. arenaria*, and *M. javanica* (Sinclair and Backman, 1989). *M. incognita* is the most economically important species on soybean; however, in warm climates, *M. arenaria* and *M. javanica* are increasing in importance. Soybean yield losses as high as 90% have been attributed to *M. incognita* in Florida (Schmitt and Noel, 1984).

The lance nematodes are probably the least-understood species in Mississippi. *Hoplolaimus columbus*, *H. galeatus* (Kinloch, 1980), and *H. magnistylus* have all been associated with soybean production in the southeastern U.S. *H. columbus* is the most pathogenic species of lance nematode on soybean (Fassuliotis et al., 1968). This nematode has been reported to reduce soybean yield in South Carolina by 70% when present in the field (Mueller and Sanders, 1987; Mueller et al., 1988). *H. galeatus* and *H. magnistylus* are the most frequently identified species in Mississippi and have been identified in 36 counties across Mississippi.

The life-stage development of most plant-parasitic nematodes is fairly simple; however, development of specific species varies in specific details. In general, juveniles hatch from eggs produced by the mature females. Eggs may be deposited in plant root tissue (lance nematodes), in a gelatinous egg mass (root-knot and reniform), in the cadaver of the female (soybean cyst), or free in the soil. The juvenile stage usually resembles the adults and develops through four growth stages or molts, reaching maturity after the final molt. Mature nematodes, depending on the specific species, either reproduce sexually (requiring both males and females) or parthenogenetically where females reproduce without the presence of males, as with the root-knot nematodes. Regardless of the method of reproduction, the females produce eggs to start the next generation. Under optimal conditions most nematode species on soybean complete their life cycle in 3 to 4 weeks. This allows three to four generations to be produced in one soybean growing season. Each female nematode potentially can produce over 100 eggs each generation, thus allowing the nematode population to increase dramatically.

Although plant-parasitic nematodes have the potential to reduce soybean growth and yield, soybeans can be successfully cultivated in their presence with acceptable returns to the producer. A management plan must be developed to reduce the nematode population numbers in a given field. Soybean varieties vary in their sensitivity to a specific nematode species as well as to races within a single nematode species (Schmitt and Noel, 1984). As an example, the variety Young is reported as susceptible to the reniform (Cornelius and Lawrence, 1994b, d), root-knot, and soybean cyst nematode races 3 and 14. However, Sharkey is susceptible only to race 14 of the soybean cyst nematode. A knowledge of the nematode species, race, and population numbers prior to planting allows the soybean producer to select a management technique or combination of tactics that will suit the particular needs at each location. This helps to ensure the production of a crop profitable to the producer.

Nematode sampling

The only accurate way to determine if potentially damaging nematode infestations are present in a field is through a nematode analysis. A nematode analysis will provide the soybean producer with the specific species of plant-parasitic nematode that may be present in a field, the specific race of the nematodes present, and provide an estimate of the

Table 16.2 Relative Critical Densities of Plant-Parasitic Nematodes to Soybean Related to Sample Date and Soil Types

Nematode	Sample date and soil type					
	Spring			Fall		
	Sand	Silt	Clay	Sand	Silt	Clay
Soybean cyst	1	1	1	1	1	1
Reniform	100	100	100	1000	1000	1000
Root-knot	50	75	100	200	250	300
Lance	400	500	750	600	700	1000

Numbers represents the critical density, nematodes/pt of soil that would warrant the implementation of a nematode management technique.

Patel, M. V. et al., Mississippi Cooperative Extension Service, 1994.

Table 16.3 State and Private Laboratories That Provide Nematode Assays

State	Location	Phone number
Alabama	Auburn University	334/844 5507
Arkansas	University of Arkansas	501/777 9702
Florida	Florida State University	352/392 1994
Georgia	University of Georgia	706/542 2685
	Water's Labs	912/336 7216
	Hickey Agri-Service Labs	912/336 0105
	MicroMacro Interactive	706/548 4557
Louisiana	Louisiana State University	504/388 2186
	Plant Diagnostics	318/342 1778
Mississippi	Mississippi State University	601/325 2146
North Carolina	North Carolina State University	919/515 2826
South Carolina	Clemson University	864/656 2292
Tennessee	University of Tennessee	423/832 6802
	A&L Labs	901/527 2780

Please contact a specific laboratory prior to submission of samples for their charges for a routine nematode analysis.

population densities of individual species (Moore et al., 1984). These results will help determine if parasitic nematodes are present and if the population has reached critical levels (density thresholds) injurious to the crop. The critical levels differ among nematode species and will vary with the particular soil texture in which they are found (Table 16.2).

Ideally, the best time to collect a soil sample is near or immediately after harvest. In most cases the nematode populations will be near their highest reproductive level. The field should be divided into 10- to 20-acre blocks that are uniform in both soil texture and cropping history to provide an accurate representation of the field (Moore et al., 1984). Soil should be collected with a soil sample probe or shovel to a depth of 6 to 8 in. in the root zone of the crop. Each sample should consist of at least 15 to 20 probes collected in a systematic zigzag fashion across the sample area. Soil cores should then be thoroughly mixed and approximately 1 to 2 pt placed in a plastic bag, sealed to prevent drying, and correctly labeled for field identification and location. Soil samples should be kept out of direct sunlight and in a cool location. Samples should be shipped at the earliest possible date to a state or private nematode diagnostic laboratory (Table 16.3). Along with the sample, include a letter stating the type of information needed (routine extraction, race identification, etc.) and provide information about the cropping history of each location.

Costs of the specific analysis needed and time required to process the samples will vary. The results can then be compared with established threshold numbers to help determine the specific management strategy to include in a production system.

Soybean cyst nematode

History and development

The soybean cyst nematode (SCN) is the most serious nematode pest of soybean in the U.S. (Riggs, 1977). Some locations are so heavily infested that soybean production is not possible without some type of control measure. Losses attributed to this nematode were estimated to be 17.5 million bu in the southern soybean-producing states in 1996 (Pratt, 1997).

The soybean cyst nematode (*Heterodera glycines*) was first observed on soybean in Japan in 1881 (Schmitt and Noel, 1984). Circular patches of stunted yellow soybeans were found to be infected with a cyst-producing nematode first believed to be the sugar beet nematode. In 1954 the soybean cyst nematode was identified in three soybean-producing counties of North Carolina. Symptoms were identical to those described on soybean in Japan. Since its first introduction in North Carolina, this nematode has now been identified in 26 central and southeastern soybean-producing states in the U.S. (Sinclair and Backman, 1989; Baldwin and Mundo-Ocampo, 1991).

SCN was first identified in Mississippi in 1957. It is believed that the cyst stages of the nematode may have survived in contaminated bags shipped from Japan into the Mississippi Delta (Schmitt and Noel, 1984). In Mississippi, SCN has been reported in all soybean-producing counties, and infests over 1 million acres of farmland (Patel, 1995). A minimum yield loss of at least 5 bu/acre may be expected from infection by this nematode. Losses of as high as $3 million have been attributed to SCN in Mississippi. In 1994, 49% of all soil samples assayed from soybean fields by the Mississippi Cooperative Extension Service were reported to be infested with SCN (Patel, 1995). In addition, nine of the known biotypes or physiological races 2, 3, 4, 5, 6, 7, 8, 9, and 14 have currently been identified in soybean fields across the state.

Soil texture

Soil texture is recognized to affect both crop productivity and to influence nematode movement and subsequent damage potential to the plant (Koenning et al., 1988; Heatherly and Young, 1991). It is generally recognized that the major damage from plant-parasitic nematodes to most agricultural crops occurs when the crop is grown in sandy-textured soils. This is partially due to the low soil moisture and nutrient capacity of sandy-textured soils compared with fine-textured or clayey soils. Combined with the additional stress imposed by the nematode results in a greater damage potential to the crop.

Soil texture may also have an important influence on SCN reproduction and development. It recognized that soil texture restricts the movement of some plant-parasitic nematodes due to the nematodes size in relation to the soil particle and pore size. Research has shown that SCN development is positively related to the increased sand content in a soil (Dropkin et al., 1976). However, SCN will also increase in population numbers in soils that are fine textured or clayey (Young, 1992b). In greenhouse studies, SCN was demonstrated to increase to similar numbers in soil textures ranging from a Ruston loamy sand to a Sharkey clay (Young, 1992b). However, in a separate study it was reported that SCN was unable to maintain a population in a Sharkey clay under greenhouse conditions (Heatherly and Young, 1991).

Under field conditions, SCN is capable of reproducing quite readily in fine-textured to clayey soils. Infected plants growing in these soils generally do not exhibit the symptoms that are characteristic of plants growing in sandy-textured soils. The only symptoms that may result from SCN in clayey soils is a reduction in expected yields.

It is extremely important for producers that grow soybeans on fine-textured or clayey soils to obtain a nematode analysis to determine if SCN is present. The continual growth of soybeans in these nematode-infested fields may result in a race shift of the nematode. This is extremely important if the variety that is grown has genes for resistance to specific races of SCN. The continued planting of the resistant variety could lead to the development of an SCN race that cannot be managed by currently available resistant varieties.

Genetic variation

The soybean cyst nematode possesses a high degree of genetic variability. This has resulted in the occurrence of biotypes or races to develop in a field. The ability of SCN to reproduce differentially on four soybean varieties and breeding lines has resulted in a classification of 16 races of soybean cyst nematodes (Riggs and Schmitt, 1988). This race scheme has been useful because varieties can be bred for resistance to selected races more rapidly than to all races. The resistance of a soybean variety will favor the increase in density of one race while decreasing the density of another race. When an SCN race 3–resistant soybean variety is grown for several years in a field infested with race 3, there is a shift in the predominant SCN population through natural selection to another race such as race 5 (Lawrence and McGuire, 1987). Race shifts also occur in field populations of SCN through interbreeding. This further enhances the chances of an SCN population to parasitize a resistant variety successfully. Currently, breeding lines have been found in Missouri that are resistant to all known races of SCN. This resistance has been incorporated into the Hartwig variety. However, if this variety is grown continuously in an SCN-infested field, a race may develop that is capable of reproducing on this highly resistant variety.

Symptoms

Foliar symptoms on soybean plants vary from stunting to severe chlorosis and possible plant death. Infected plants generally exhibit poor growth compared with healthy, non-infected plants. Symptoms generally are easily confused with nutrient deficiencies and other soybean disorders; therefore, diagnosis of the disease must be based on the presence of white to yellow females which erupt from the roots and on a soil analysis.

Causal organism

The SCN (*Heterodera glycines*) is a sedentary endoparasitic species of nematode (Baldwin and Mundo-Ocampo, 1991). The females are lemon-shaped and cream white, yellow, or light brown in color. The females are easily observed attached to the roots of infected plants without the use of a hand lens. The cyst stage is produced at the time of death of the female, and the cyst serves as a protective mechanism for egg survival. Eggs may survive more than 8 years inside the cysts. The life cycle of SCN requires 21 to 24 days on soybean (Moore et al., 1984). Starting with an egg inside the cyst, embryonic development within the egg results in a first-stage juvenile. The juvenile will molt in the egg and forms a second-stage juvenile which is the infective stage.

The second-stage juvenile will hatch from the egg after it receives a stimulus from a host plant, usually in the form of a root exudate. The second-stage juvenile swims in the soil moisture film that surrounds soil particles and searches for a soybean root. When the root is located, the juvenile penetrates the developing root behind the zone of cell differentiation near the root tip. It then migrates through the cortical tissue and establishes a feeding site in the vascular system of the root. At this point the nematode becomes sedentary and begins to feed. The feeding activity causes the root cells to enlarge and forms syncytia or specialized nurse cells, which serve to divert nutrients produced for plant growth to the developing juvenile. The second-stage juvenile matures through a series of molts and becomes an adult.

As the females develop they enlarge and rupture through the root epidermis and protrude from the roots. The females head is still located near the nurse cells so it can continue to feed. Males develop faster than the females and leave the root to inseminate the protruding females. Once inseminated the females start producing eggs, which are extruded from her body in a gelatinous egg mass. When conditions for soybean growth become unfavorable or the plant begins to senesce, eggs will be retained inside the female's body.

Upon the death of the female, the cuticle that surrounds her body tans, becomes tough and leathery, and serves as a protective cyst. A single female may retain from 200 to 600 eggs inside her body at the time of her death. The protective nature of the cyst makes this stage an effective dispersal unit. Cysts may be spread by birds, water, machinery, soil peds in seed, or by any means in which soil is moved.

Management

Crop rotation is an effective means of managing SCN (Wrather et al., 1992). The nonhost crops of corn, cotton, and grain sorghum have been successfully incorporated into rotation programs for SCN management. Growing of a nonhost crop for 2 to 3 years on a nematode-infested field is required before a susceptible soybean cultivar can be grown with full yield potential expectations (Moore et al., 1984). The inclusion of a race-specific resistant soybean variety for a nonhost crop is also an economical method for nematode management (Lawrence and McGuire, 1987; Wrather et al., 1992). However, the continuous or frequent use of a resistant variety may result in a race shift that eventually renders the resistant variety useless (Young, 1994). The individuals of the nematode population that are able to feed and reproduce on the resistant variety will increase with each generation and thus build up in population numbers. These individuals eventually become the majority of the population. It is therefore important to know the race that is present in the field and the genetics of the previously planted soybean varieties when considering planting a resistant soybean variety in a crop rotation system.

Early planting of soybean may be beneficial in a field infested with SCN. Generally, the soybean cyst nematode will not hatch and become active until soil temperatures reach 68°F (Ross, 1963). Therefore, in the spring, nematode numbers in the soil are relatively low due to cool soil temperatures and possible overwinter mortality. This should allow the soybean plant time to become established with a minimum of nematode pressure. Reduced nematode numbers at planting is the goal of all nematode management practices and is required for production of high yields. In addition it has been shown that SCN numbers are lower at harvest on early-maturing varieties compared with the later-maturing varieties (Wrather et al., 1992).

The use of nematicides has been found effective in reducing SCN populations in certain soil types; however, nematicide use is limited and generally based on soybean prices (Epps and Young, 1981; Hartwig et al., 1987; Sinclair and Backman, 1989; Rodriguez-Kabana, 1992). Nematicides may be beneficial when used in fields that are planted with

Table 16.4 Nematicides Registered for Use on Soybean

Product and formulation	Manufacturer	Nematicide class	Rate and application method
Telone II[a]	Dow Elanco	Fumigant	Apply 52–106 oz/1000 ft of row; apply as a broadcast or row treatment with a fumigant shank. Shank trace should be sealed with soil to present loss of fumigant. Leave soil undisturbed for 7 days after application.
Vapam	Zeneca	Fumigant	50–100 gal/acre; Inject 4 in. deep into a well-prepared soil. Roll soil to smooth and compact; allow 21 days before planting
Temik 15G[a]	Rhone-Poulenc Ag	Nonfumigant	10–20 lb/acre; apply in a 6–8-in. band and immediately work into soil
Vydate L	DuPont	Nonfumigant	2–4 pt in 10–20 gal of water/acre as in-furrow band treatment (36" row spacing) 2–4 pt in 10–20 gal of water/acre as a broadcast treatment (1 week before planting); incorporate 2–4 in. into soil in the seed zone 1–2 gal in 10–20 gal of water/arce as a broadcast treatment (1 week before planting); incorporate 2–4 in. into soil in the seed zone
Furadan 4F	FMC	Nonfumigant	3.75–5 fl oz/1000 linear feet of row (40-in. row spacing). Apply at planting on 12-inch surface band.

[a] Telone II and Temik 15G are the only two products recommended for use in Mississippi by MCES (Patel and Fox, 1996).

Please consult actual nematicide label for a more-detailed description of application rates and methods.

the more-SCN-tolerant varieties or where there is more than one population of plant-parasitic nematodes in the field. The option to include a nematicide in an SCN management system should not be overlooked. The level of SCN resistance or tolerance is not broad enough to eliminate the use of nematicides in soybean production (Rodriguez-Kabana, 1992).

The decision to use a nematicide should be based on the results of soil assays, the degree of resistance in the variety to be planted, cost vs. expected return, and available application equipment. Long-term SCN management requires an integration of crop rotation, resistant varieties, nematicides, and good land management (Sinclair and Backman, 1989). In some problem fields, rotation with nonhost crops and resistant and susceptible varieties can be effectively used without the inclusion of a nematicide (Young, 1992a, 1994). This requires a susceptible soybean variety to be grown no more than every third year. Also the infrequent rotation of a resistant soybean variety will help minimize the genetic race shift in the nematode.

Nematicides that are currently recommended for use in soybean production include Telone II, Vapam, Temik 15G, Vydate L and Furadan 4F (Table 16.4). However, only Telone II and Temik 15G are recommended for use in Mississippi by the Cooperative Extension Service (Patel and Fox, 1994).

Reniform nematode

History and distribution

The reniform nematode (*Rotylenchulus reniformis* Linford and Olivera) was first observed in 1930 on cowpea in Hawaii (Schmitt and Noel, 1984; Cornelius and Lawrence, 1993a). In 1940, Linford and Olivera established the genus *Rotylenchulus* and described the species *Rotylenchulus reniformis* (Linford and Olivera, 1940).

The reniform nematode was first identified as a parasite on soybean in 1956 in the Gold Coast (West Africa) (Sinclair and Backman, 1989). In 1967, it was identified on soybean in South Carolina. Currently, the reniform nematode has been reported in all Gulf Coast states, plus Arkansas, Missouri, Tennessee, Georgia, North Carolina, and South Carolina. The reniform nematode was first identified in Mississippi in 1968 on centipede grass. It was not reported again until 1980 when it was found on soybean in five counties. The reniform nematode has currently been reported in 54 Mississippi counties (Patel, 1995).

The distribution of the reniform nematode is not limited by soil types. No consistent relationships have been determined between the presence of the reniform nematode and soil texture, soil pH, or the amount of rainfall. In general the finer-textured silt or clayey soils seem to support larger nematode populations than the coarser-textured sandy soils (Heald and Robinson, 1990).

Symptoms

The reniform nematode significantly affects both the growth of the soybean plant and subsequent yields. Symptoms of this nematode on soybean include chlorosis of the foliage, stunting of the plant, root decay, small seed size, empty pods, and reduced yields (Table 16.1). In field research conducted at Mississippi State University, yield reductions as high as 33% have been attributed to the reniform nematode (Cornelius and Lawrence, 1993a, c, 1994a). Research has also demonstrated that soybean yields decreased significantly with increasing reniform nematode numbers. Compared with control plots, which received no reniform nematodes, soybean yields were reduced by 5.0, 6.8, 7.1, and 9.3 bu/acre when planted in field plots infested with the at-planting nematode numbers of 2500, 5000, 7500 and 10,000 nematodes/pt of soil, respectively (Cornelius and Lawrence, 1993a, c). Reniform nematode numbers used in this test are similar to the numbers that are found in some Mississippi soils at planting. It is not uncommon to recover reniform population numbers as high as 40,000 to 60,000/pt of soil after cotton. However, reniform populations may average 10,000 to 20,000/pt of soil in many soil types.

Causal organism

The reniform nematode (*Rotylenchulus reniformis*) is a semiendoparasitic species of nematode. The anterior portion of the females body is embedded in the root, with the nematode head located near the vascular system. The posterior portions of the body remain outside the root. This nematode is often overlooked when roots are visually examined because of the subtle symptoms that are produced on the root systems.

The life cycle of the reniform nematode is unique in that the juvenile stages do not generally feed on a host. On soybean, the reniform nematodes life cycle is completed within 17 to 23 days when soil temperatures are approximately 80°F. The female lays eggs in a small gelatinous egg mass on the surface of the root. Within 4 days, a fully developed first-stage juvenile matures within the egg. Within 24 h, the juvenile will molt to a second-stage juvenile while still inside the egg. The second-stage juvenile is active and forces its way out of the egg shell. Therefore, in 6 to 7 days after the egg was laid, a second-stage juvenile is free in the soil. Within 3 to 4 additional days, the juvenile undergoes three successive molts to form young immature adults, which are the infective stage of this nematode. This adult stage, which is often observed within the cuticle of the previous molt, may serve as a resistant stage to adverse environmental conditions. It is believed that this stage may serve as the resistant overwintering stage in Mississippi. This is supported by the recovery of this developmental stage in high numbers throughout the winter.

After soybeans are planted, the immature adult female locates a root and inserts the anterior one third of her body into the root and begins to feed near the phloem. Within 24 to 48 h, the female begins to enlarge in the region of the vulva, and within 4 to 5 days assumes the characteristic reniform or kidney shape. When the females are fully developed, males can be found near the female or coiled around the female's body. The female, inseminated by the male, then secrete a gelatinous matrix that surrounds her body and provides protection to her and the eggs that she produces. Egg production generally occurs in 7 to 8 days after root infection. Females lay an average of 75 eggs per egg mass. This nematode is often overlooked in the field because root galls are not produced. Soil particles also adhere to the gelatinous egg mass, which makes detection difficult.

Management

The reniform nematode is capable of reducing yields when susceptible soybeans are planted in infested fields (Table 16.1). Therefore, steps to manage and maintain low nematode numbers are important. The most important management tool for this nematode is the use of resistant soybean varieties. An extensive screening program is in progress at Mississippi State University to provide soybean producers with the latest information on soybean varietal reaction (Cornelius and Lawrence, 1993b, 1994a, c). Over 130 soybean varieties in maturity groups IV to VII have currently been examined and 63 have thus far been identified with resistance to the reniform nematode (Tables 16.5 through 16.8).

Crop rotations are also effective in reducing reniform nematode populations. The production of corn or grain sorghum for 1 year reduces reniform nematode numbers to levels that allow the production of a soybean crop (Lawrence et al., 1990, 1991, 1992). Chemical nematicides are also useful in reducing nematode numbers at planting, but the cost of a nematicide and the current price of soybean may not justify their inclusion as a managment option in a soybean production system. Nematicides currently available for use on soybean are included in Table 16.4.

Root-knot nematodes

History and distribution

The root-knot nematodes (*Meloidogyne* spp.) are the most widely distributed group of plant-parasitic nematodes. The southern root-knot (*Meloidogyne incognita*) and javanese root-knot (*M. javanica*) are considered the most important species on soybean. The peanut root-knot (*M. arenaria*) has also been identified on soybean in Mississippi and is becoming increasingly important in the warmer southern climates (Riggs, 1982).

Soybean yield losses due to the root-knot nematode vary from inconsequential to complete depending on the species present, the number of nematodes in the soil, and susceptibility of the variety. The greatest yield losses occur where soybeans are produced in sandy, light-textured soils (Schmitt and Noel, 1984; Sinclair and Backman, 1989).

Symptoms

The main symptom resulting from root-knot nematode infection is characteristic galls or knots that are produced on infected roots. Root galls vary in size and number, depending on the intensity of infection and the specific root-knot nematode that is present.

Aboveground symptoms on the soybean plant may vary from a slight stunting to severely suppressed growth and chlorosis of the foliage. Early foliage senescence is also associated with root-knot infections. Yield losses attributed to the southern root-knot have been as great as 90% in Florida (Kinloch, 1974). The severity of damage is dependent both on the initial population numbers present in the soil at planting and on the environmental conditions the soybeans are subjected to during the growing season.

Causal organism

There are four common species of root-knot nematodes (*Meloidogyne* spp.) known to parasitize soybeans. These include the southern root-knot, *M. incognita*, the peanut root-knot, *M. arenaria*, the javanese root-knot, *M. javanica*, and the northern root-knot, *M. hapla*. All root-knot nematodes are considered sedentary endoparasitic species. The females are embedded within the root tissue for the duration of their lives. The javanese root-knot is considered the most-damaging species on soybean in the most southerly areas of the U.S. (Schmitt and Noel, 1984); however, the extent of this nematode distribution in Mississippi has not been determined.

The southern root-knot and peanut root-knot are the most common species of the root-knot nematodes associated with soybean production (Sinclair and Backman, 1989). Damaging populations are commonly found in the soybean-producing areas of the southern U.S., including Alabama, Florida, Georgia, and South Carolina, as well as the sandy-textured soils of Louisiana and Mississippi (Sinclair and Backman, 1989).

The life cycle of the root-knot nematode requires approximately 30 to 39 days on soybean and begins with the nematode egg. The egg undergoes embryonic development forming a first-stage juvenile in the egg, which molts to form a second-stage juvenile inside the egg. If environmental conditions are favorable for plant growth, the second-stage juvenile hatches from the egg and swims in the moisture film that surrounds the soil particles in search of a soybean root. The second-stage juvenile is the only life stage that can penetrate the root cells, and once a root is located it penetrates just behind the root tip. It then migrates through the root and begins feeding in the region that develops into the phloem. Giant cells or nurse cells are initiated by the feeding activity of the nematode, and the second-stage juvenile then begins to swell and enlarge. Within a few days, the nematode stops feeding and undergoes three rapid molts to form an adult pyriform or pear-shaped female. The adult female resumes feeding in the nurse cells. The female produces a gelatinous matrix that is deposited on the outside of the root and lays eggs into the matrix to produce an egg mass. Each female is capable of producing 500 to 1000 eggs during her lifetime. Males that are produced usually will emerge from the roots after the final molt and generally do not feed, and they are not normally required in reproduction.

Management

Root-knot nematode management is complicated by a wide host range and by mixed populations of the nematode species and races occurring in the same field. Therefore, the accurate identification of the root-knot species and race that is present in any given location

Table 16.5   Disease Reaction of Selected Soybean Varieties in Maturity
Group IV to the Soybean Cyst and Reniform Nematodes

Variety	Brand	Soybean cyst nematode		Reniform nematode
		Race 3	Race 14	
A-4138	Asgrow	R	MR	R
A-4341	Asgrow	S	S	R
A-4415	Asgrow	S	S	S
A-4539	Asgrow	MR	MR	U
A-4595	Asgrow	S	S	R
A-4715	Asgrow	R	R	U
A-4922	Asgrow	R	R	R
Avery	Missouri Agr. Expt. Sta.	R	R	R
BU-44	Buckshot	S	S	S
Cx-458	Dekalb Genetic Corp.	S	S	R
CX-469c	Dekalb Genetic Corp.	MR	MS	R
CX-478	Dekalb Genetic Corp.	MS	S	S
CX-499c	Dekalb Genetic Corp.	R	R	R
Delsoy-4710	Missouri Agr. Expt. Sta.	R	R	U
Delsoy-4500	Missouri Agr. Expt. Sta.	R	S	R
Delsoy-4900	Missouri Agr. Expt. Sta	R	S	R
DG-3495	Dyna-Gro	S	S	R
Dixie 478	Dixie	MS	S	R
DK-4875	Delta King	R	MR	S
DP-3456	Deltapine	S	S	R
DP-3478	Deltapine	S	S	S
DP-3499	Deltapine	S	S	U
FFR-464	Stoneville	S	S	S
FFR-499	Stoneville	S	S	R
H-4464	Hartz	R	MS	R
H-4994	Hartz	R	R	S
HBK-49	Hornbeck	S	S	U
Hill	Mississippi Agr. & Forestry Expt. Sta./USDA-ARS	S	S	S
HY-498	HyPerformer	S	S	U
Manokin	Maryland Agr. Expt. Sta./USDA-ARS	R	S	R
P-9442	Pioneer	S	S	R
P-9444	Pioneer	R	—	R
P-9451	Pioneer	MR	MR	R
P-9452	Pioneer	S	—	R
P-9472	Pioneer	R	—	R
P-9481	Pioneer	R	R	S
P-9501	Pioneer	S	S	R
PHAROAH	Illinois Agr. Expt. Sta./USDA-ARS	R	S	R
RA-451	Northrup King	S	S	S
RA-452	Northrup King	M	S	R
RVS-499	Riverside	S	S	R
S46-44	Northrup King	R	MR	R
S46-84	Northrup King	MR	MR	R
SPRY	Illinois Agr. Expt. Sta./USDA-ARS	S	S	S
STAFFORD	—	S	S	R
Stressland	Ohio Agr. Expt. Sta	S	S	S
TN4-86	Tennessee Agr. Expt. Sta.	R	R	R
TN4-94	Tennessee Agr. Expt. Sta.	MR	MR	R
TV-4452	Terral	MR	S	S

Table 16.5 (continued)   Disease Reaction of Selected Soybean Varieties in Maturity
Group IV to the Soybean Cyst and Reniform Nematodes

Variety	Brand	Soybean cyst nematode		Reniform nematode
		Race 3	Race 14	
TV-4990	Terral	R	R	R
TVX-4596	Terral	MS	R	R
TVX-4479	Terral	R	R	R
Wicomico	Maryland Agr. Expt. Sta. /USDA-ARS	R	S	R

Reniform nematode disease ratings are based on one year's data. Disease ratings and interpretation of data presented may change with additional experimentation and specific nematode populations.

Soybean cyst nematode ratings as reported by Askew et al., 1997.

All nematode disease reactions are based on evaluations conducted in a greenhouse.

R = resistant; MR = moderately resistant; MS = moderately susceptible; S = susceptible; U = undetermined (mixed reaction).

is important. This can be accomplished by submitting a soil sample for nematode analysis to either a state or private nematode diagnostic laboratory (Table 16.3).

Crop rotations are difficult to develop for root-knot nematode management because these nematodes have a wide host range (Schmitt and Noel, 1984). Many grass crops, however, appear to be less susceptible to the various root-knot species and have been shown to reduce the subsequent residual populations (Schmitt and Noel, 1984). In addition, the field should be maintained free of weeds since many serve as hosts for the nematode. Planting and cultivating equipment should be washed free of contaminated soil to prevent transporting this nematode to noninfested fields.

Public and private soybean-breeding programs have been developing many varieties resistant to the southern root-knot nematode (*M. incognita*) (Riggs et al., 1988). Soybean varieties in maturity groups V through VIII have been developed with resistance and tolerance to this nematode (Riggs, 1982; Sinclair and Backman, 1989). Resistant varieties generally outperform susceptible varieties even when a nematicide is used to protect the susceptible variety. However, physiological races specific to geographic areas are known to occur; therefore, a specific variety may respond differently when exposed to these isolates. This geographic response indicates that varietal screening must be conducted in the specific state where the crop will be grown. Centralized screening programs may lead to erroneous results with management and production failures.

Several nematicides are effective for controlling root-knot nematodes (Kinloch, 1980). In general, fumigants are more effective than nonfumigants (Table 16.4). The contact and systemic chemicals are also more effective against only certain species of *Meloidogyne* rather than all species. Nematicides are rarely cost-effective for controlling root-knot nematodes in soybean, but may have to be used in situations where resistance to a specific species or race is not available or mixed populations are found in the field.

Lance nematodes

History and distribution

Three species of lance nematodes have been described on soybean. These include *Hoplolaimus columbus* (Columbian lance), *H. galeatus*, and, more recently, *H. magnistylus*.

Table 16.6 Disease Reaction of Selected Soybean Varieties in Maturity Group V to the Soybean Cyst and Reniform Nematodes

Variety	Brand	Soybean cyst nematode		Reniform nematode
		Race 3	Race 14	
A-5403	Asgrow	R	R	S
A-5547	Asgrow	MR	R	R
A-5560	Asgrow	R	MR	S
A-5885	Asgrow	MR	MR	R
A-5979	Asgrow	R	MR	S
AT-2555	Agra Tech	R	MS	S
AT-2665	Agra Tech	S	S	S
AT-555	Agra Tech	R	MR	R
BU-55	Buckshot	R	MS	S
Clifford	North Carolina Agr. Expt. Sta./USDA-ARS	S	S	S
DG3576	Dyna-Gro	MS	S	R
Dixie 544	Dixie	S	S	S
Dixie 579	Dixie	S	S	S
DK-5850	Delta King	MR	MR	S
DP-3588	Deltapine	R	S	S
DP-3519	Deltapine	MR	MS	S
DP-3589	Deltapine	R	S	R
DP-105	Deltapine	M	M	R
DP-415	Deltapine	R	S	R
FFR-595	Stoneville	R	R	S
H-5050	Hartz	MS	MS	R
H-5070	Hartz	S	S	S
H-5088	Hartz	R	S	S
H-5164	Hartz	R	R	R
H-5350	Hartz	R	MR	R
H-5454	Hartz	MS	MS	R
H-5810	Hartz	R	MS	S
HARTWIG	Missouri Agr. Expt. Sta.	R	R	R
HBK-58	Hornbeck	S	S	S
HSC-557	Hyperformer	R	R	R
HSC-591	Hyperformer	R	R	S
HUTCHENSON	Virginia Agr. Expt. Sta.	S	S	S
Hy-574	HyPerformer	MR	S	R
P-9551	Pioneer	R	MS	S
P-9593	Pioneer	R	S	U
P-9501	Pioneer	S	—	R
R90-515	Arkansas Agr. Expt. Sta.	R	—	R
RHODES	Missouri Agr. Expt. Sta.	R	R	R
RVS-549	Riverside	S	S	S
RVS-577	Riverside	R	S	S
RVS-77	Riverside	S	—	S
Robin-5	Riverside	R	S	R
RVS-77	Riverside	S	—	S
S59-60	Northrup King	R	R	S
S59-95	Northrup King	R	—	S
TV-515	Terral	R	S	R
TV-5452	Terral	R	S	S
TV-5693	Terral	S	S	S
TV-5555	Terral	R	R	S

Table 16.6 (continued)   Disease Reaction of Selected Soybean Varieties in Maturity Group V to the Soybean Cyst and Reniform Nematodes

		Soybean cyst nematode		Reniform
Variety	Brand	Race 3	Race 14	nematode
UW-509A	Underwood	MR	S	S
WALTERS	Arkansas Agr. Expt. Sta.	R	S	S

Reniform nematode disease ratings are based on one year's data. Disease ratings and interpretation of data presented may change with additional experimentation and specific nematode populations.

Soybean cyst nematode ratings as reported by Askew et al., 1997

All nematode disease reactions are based on evaluations conducted in a greenhouse.

R = resistant; MR = moderately resistant; MS = moderately susceptible; S = susceptible; U = undetermined (mixed reaction).

The Columbian lance is reported to be the most pathogenic lance nematode on soybean (Fassuliotis, 1974; Sinclair and Backman, 1989). The first record of infection and subsequent yield reduction of soybean due to *H. columbus* was in 1963 in South Carolina (Sinclair and Backman, 1989). Although this nematode has not been identified in Mississippi, it is now found in Alabama, Georgia, Louisiana, and North and South Carolina. Soybean yield losses up to 70% have been attributed to this nematode in South Carolina (Mueller et al., 1988). Although some soybean varieties appear to be tolerant to the Columbian lance, yield losses still result with nematode infection. The greatest damage from this nematode is reported to occur on plants growing in sandy-textured soils.

H. galeatus occurs in all southern states, but damage to soybean growth and yield is limited. However, this nematode has an extensive host range and can cause injury to corn and cotton (Rhoades, 1987). A third lance species, *H. magnistylus*, was identified from soybean in 1980 (Robbins, 1982). Although no tests have been conducted on soybean to determine yield losses, this species will significantly reduce the yield of cotton. In 1990, *H. magnistylus* was isolated from chlorotic and severely stunted soybean plants near Charleston, MS. All life stages of the nematode were isolated from the soil and soybean roots. An average population density of 175 nematodes/g of root and 500/pt of soil were recovered.

Symptoms

Lance nematodes are capable of producing extensive damage to soybean roots. These nematodes feed from both the outside of the roots as ectoparasites and the inside the root tissue as endoparasites. The migration of the nematode within the root during feeding results in roots with ruptured cells and subsequent cell death. Infected plants are generally stunted, chlorotic, and have reduced pod production (Sinclair and Backman, 1989). Stunting and low yields are more severe in sandy-textured soils, which are subject to hardpans, especially during dry seasons (Sinclair and Backman, 1989).

Causal organism

The lance nematode is considered an ectoparasitic species of plant-parasitic nematodes. Although lance nematodes tunnel extensively in the soybean root system, they are exposed to the soil environment during some point in their life-stage development. Most of the damage produced by these nematodes is a result of the nematode migrating through the

Table 16.7 Disease Reaction of Selected Soybean Varieties in Maturity
Group VI to the Soybean Cyst and Reniform Nematodes

		Soybean cyst nematode		Reniform nematode
Variety	Brand	Race 3	Race 14	
Leflore	Mississippi Agr. & Forestry Expt. Sta./USDA-ARS	R	R	S
P-9641	Pioneer	S	S	S
RA-606	Northrup King	R	MS	S
RVS-699	Riverside	MS	MS	S
RVSL-91-42	Riverside	MR	MS	R
RVSL-91-85	Riverside	R	MS	S
SAMPSON	Agripro	S	S	S
S64-23	Northrup King	R	R	S
S62-66	Northrup King	R	R	S
SHARKEY	Mississippi Agr. & Forestry Expt. Sta./USDA-ARS	R	S	R
TN6-90	Tennessee Agr. Expt. Sta.	R	R	S
TRACY-M	Mississippi Agr. & Forestry Expt. Sta./USDA-ARS	S	S	S
TV-616	Terral	R	S	S
TV-6653	Terral	S	S	S
VERNAL	Mississippi Agr. & Forestry Expt. Sta./USDA-ARS	S	S	S
YOUNG	North Carolina Agr. Expt. Sta. and USDA-ARS	S	S	S
A-6297	Asgrow	R	R	S
A-6785	Asgrow	S	S	R
A-6961	Asgrow	R	R	R
AT-695	Agra Tech	R	R	S
BRYAN	Georgia Agr. Expt. Sta.	S	S	R
BU-62	Buckshot	MS	S	R
BU-66	Buckshot	R	S	S
BU-67	Buckshot	S	S	R
BU-68	Buckshot	S	S	S
Cajun	Riverside	MR	MS	S
DP-3627	Deltapine	MR	MS	S
DP-3682	Deltapine	R	S	R
FFR-646	Stoneville	R	R	S
FFR-671	Stoneville	S	S	S
H-6500	Hartz	MS	S	R
H-6200	Hartz	R	S	R
H-6686	Hartz	S	S	S
HBK-65	Hornbeck	MS	S	S
HSC-B2J	Hyperformer	R	S	R
HSC-623	Hyperformer	R	MS	R
LAMAR	Mississippi Agr. & Forestry Expt. Sta./USDA-ARS	S	S	S
DAVIS	Arkansas Agr. Expt. Sta.	S	S	S

Reniform nematode disease ratings are based on one year's data. Disease ratings and interpretation of data presented may change with additional experimentation and specific nematode populations.

Soybean cyst nematode ratings as reported by Askew et al., 1997.

All nematode disease reactions are based on evaluations conducted in a greenhouse.

R = resistant; MR = moderately resistant; MS = moderately susceptible; S = susceptible; U = undetermined (mixed reaction).

Table 16.8 Disease Reaction of Selected Soybean Varieties in Maturity Group VII and VIII to the Soybean Cyst and Reniform Nematodes

Variety	Brand	Soybean cyst nematode		Reniform nematode
		Race 3	Race 14	
Braxton	Univ. of Florida Agr. Expt. Sta./USDA-ARS	S	S	R
BU-723	Buckshot	S	S	S
Cook	Georgia Agr. Expt. Sta.	S	S	S
C-6847	Northrup King	R	S	R
DP-3773	Deltapine	R	MR	R
DP-3776	Deltapine	S	S	R
HAGOOD	South Carolina Agr. Expt. Sta.	R	S	S
H-7190	Hartz	S	S	R
MAXCY	South Carolina Agr. Expt. Sta.	R	S	R
PERRIN	South Carolina Agr. Expt. Sta.	S	S	S
P-9761	Pioneer	R	S	R
P-9791	Pioneer	S	S	S
P-9831	Pioneer	S	S	R
RV-757	Riverside	R	S	R
STONEWALL	Alabama Agr. Expt. Sta.	R	S	R
TV-727	Terra-Vig	R	MS	R
UW-701 P	Underwood	R	MR	S

Reniform nematode disease ratings are based on one year's data. Disease ratings and interpretation of data presented may change with additional experimentation and specific nematode population.

Soybean cyst nematode ratings as reported by Askew et al., 1997.

All nematode disease reactions are based on evaluations conducted in a greenhouse.

R = resistant; MR = moderately resistant; MS = moderately susceptible; S = susceptible; U = undetermined (mixed reaction).

root tissues. The average life cycle for the lance nematodes is 28 days under favorable environmental conditions. Females produce at least 15 eggs; however, fecundity is difficult to establish because of nematode migration. Eggs usually hatch in 9 to 15 days after oviposition (Schmitt and Noel, 1984). All juvenile stages of lance nematode feed on the soybean roots.

Management

Management of the lance nematode is difficult because of a wide host range (Rhoades, 1987). Rotation crops such as cotton, corn, and grain sorghum are also susceptible to the lance nematode and, therefore, do not reduce nematode numbers. Currently, several soybean lines or varieties have been identified as tolerant to the lance nematode; however, none are considered resistant (Mueller et al., 1988).

Early-planted soybeans have resulted in higher yields than late-planted soybeans in South Carolina (Mueller and Sanders, 1987). This was attributed to greater nematode activity at higher soil temperatures (Nyczepir and Lewis, 1979). The early-planted soybeans were able to become established before optimal temperatures for nematode infection and reproduction (Fassuliotis, 1974). Subsoiling has also been an effective cultural means of nematode management (Bird et al., 1974; Mueller et al., 1988). Breaking the hardpan allows deeper root penetration. Although this does not directly reduce nematode numbers, it allows plants to develop a larger root system for better growth and yields. The use of nematicides has been effective in controlling *H. columbus* and improving soybean yields (Schmitt and Noel, 1984; Sinclair and Backman, 1989). However, the decision to use chemical treatments should be based on soybean prices and may not always be economical.

Summary

The association of plant-parasitic nematodes and their effects on soybean production in Mississippi is described. Over 100 species of plant-parasitic nematodes have been reported associated with soybean; however, only a few are considered economically important in Mississippi. The soybean cyst nematode (SCN), reniform nematode, root-knot nematode, and lance nematode have been shown to reduce the economic yield potential of soybean. The basic life-stage development of each species is described. Management techniques unique to each individual nematode species are also discussed.

Disclaimer

The interpretation of data presented in this section may change with additional experimentation. Information is not to be construed as a recommendation for use or as an endorsement of a specific product.

References

Askew, J. E., Jr., A. Blaine, F. Boykin, R. Dobbs, F. Handcock, D. Ingram, R. Martin, D. Reginelli, A. Smith, T. R. Vaughan, and M. Young. 1997. Soybean Variety Trials — 1996, Mississippi Agricultural and Forestry Experiment Station, Information Bulletin 315, 124 pp.

Baldwin, J. G. and M. Mundo-Ocampo. 1991. Heteroderinae, cyst and non-cyst forming nematodes. in W. R. Nickle, Ed., *Manual of Agricultural Nematology*, Marcel Dekker, New York, 275–362.

Bird, G. W., O. L. Brook, C. E. Perry, J. G. Futral, T. D. Canadany, and F. C. Boswell. 1974. Influence of subsoiling and soil fumigation on cotton stunt disease complex, *Hoplolaimus columbus* and *Meloidogyne incognita*, *Plant Dis. Rep.* 58:541–544.

Cornelius, J. J. and G. W. Lawrence. 1993a. The effect of increasing initial inoculum levels of *Rotylenchulus reniformis* on soybean, *Proc. Southern Soybean Dis. Workers* 20:14.

Cornelius, J. J. and G. W. Lawrence. 1993b. Evaluation of soybean cultivars for resistance to reniform nematode, *Phytopathology* 83:1417.

Cornelius, J. J. and G. W. Lawrence. 1993c. Population dynamics of (*Rotylenchulus reniformis*) on soybean, *Proc. Mississippi Assoc. Plant Pathol. Nematol.* 12:5.

Cornelius, J. J. and G. W. Lawrence. 1994a. The effect of the reniform nematode on monoculture soybean production, *Phytopathology* 84:1129.

Cornelius, J. J. and G. W. Lawrence. 1994b. Resistance of soybean cultivars in maturity groups IV–VIII to *Rotylenchulus reniformis*, *Proc. Southern Soybean Dis. Workers* 21:7.

Cornelius, J. J. and G. W. Lawrence. 1994c. Availability of soybean cultivars in maturity group IV–VIII resistant to *Rotylenchulus reniformis*, *Proc. Mississippi Assoc. Plant Pathol. Nematol.* 13:19.

Dropkin, V. H., C. N. Baldwin, T. Gaither, and W. Nace. 1976. Growth of *Heterodera glycines* in soybeans in the field, *Plant Dis. Rep.* 60:977–980.

Epps, J. M. and L. D. Young. 1981. Evaluation of nematicides and resistant cultivars for control of soybean cyst nematode race 4, *Plant Dis.* 65:665–666.

Fassuliotis, G. 1974. Tolerance of *Hoplolaimus columbus* to high osmotic pressures, dessication, and high soil temperature, *J. Nematol.* 3:309–310.

Fassuliotis, G., G. J. Rau, and F. H. Smith. 1968. *Hoplolaimus columbus*, a nematode parasite associated with cotton and soybeans in South Carolina, *Plant Dis. Rep.* 52:571–572.

Hartwig, E. E., L. D. Young, and N. Buehring. 1987. Effects of morocropping resistant and susceptible soybean cultivars on cyst nematode infested soil, *Crop Sci.* 27:576–579.

Heald, C. M. and A. F. Robinson.1990. Survey of current distribution of *Rotylenchulus reniformis* in the United States, *J. Nematol.* 22:695–699.

Heatherly, L. G. and L. D. Young. 1991. Soybean and soybean cyst nematode response to soil water content in loam and clay soils, *Crop Sci.* 31:191–196.

Kinloch, R. A. 1974. Response of soybean cultivars to nematicidal treatments of soil infested with *Meloidogyne incognita*, *J. Nematol.* 6:7–11.

Kinloch, R. A. 1980. The control of nematodes injurious to soybean, *Nematropica* 10:141–153.

Koenning, S. R., S. C. Anand, and J. A. Wrather. 1988. Effect of within-field variation in soil texture on *Heterodera glycines* and soybean yield, *J. Nematol.* 20:373–380.

Lawrence, G. W. and J. M. McGuire. 1987. Influence of soybean cultivar rotation sequences on race development of *Heterodera glycines*, race 3, *Phytopathology* 77:1714.

Lawrence, G. W., G. L. Windham, K. S. McLean, W. E. Baston, Jr., and J. C. Borbon. 1990. Corn–cotton rotations for the management of the reniform nematode, *Phytopathology* 80: 1047.

Lawrence, G. W., G. L. Windham, and K. S. McLean. 1991. Reniform nematode management in a corn-cotton rotation system, *Phytopathology* 81:1211.

Lawrence, G. W., G. L. Windham, and K. S. McLean. 1992. Corn as a non-host for reniform nematode management in cotton production, *J. Nematol.* 24:604.

Linford, M. B. and J. M. Oliveira. 1940. *Rotylenchulus reniformis*, nov. gen., n. sp., a parasite of roots, *Proc. Helmithol. Soc. Washington* 7:35–42.

Moore, W. F., S. C. Bost, F. L. Brewer, R. A. Dunn, B. Y. Endo, C. R. Gray, L. L. Hardman, B. J. Jacobson, R. Leffel, M. A. Newman, R. F. Nywall, C. Overstreet, and C. L. Parrs. 1984. Soybean Cyst Nematode, Mississippi Cooperative Extension Service, Mississippi Agricultural and Forestry Experiment Station, 23 pp.

Mueller, J. D. and G. B. Sanders. 1987. Control of *Hoplolaimus columbus* on late-planted soybean with aldicarb, *Ann. Appl. Nematol.* 1:123–126.

Mueller, J. D., D. P. Schmitt, G. C. Weiser, E. R. Shipe, and H. L. Musen. 1988. Performance of soybean cultivars in *Hoplolaimus columbus* infested fields, *Ann. Appl. Nematol.* 2:65–69.

Nyczepir, A. P. and S. A. Lewis. 1979. Relative tolerance of selected soybean cultivars to *Hoplolaimus columbus* and possible effects of soil temperature, *J. Nematol.* 11:27–31.

Patel, M. V. 1995. Plant Pathology Laboratory Report 1995, Mississippi Cooperative Extension Service, 27 pp.

Patel, M. V. and J. A. Fox. 1996. Soybean Cyst Nematode, Mississippi Cooperative Extension Service, Publ. 1293, 4 pp.

Patel, M. V., F. Killebrew and J. A. Fox. 1994. Population Densities (Threshold Level) of Plant-Parasitic Nematodes, Mississippi Cooperative Extension Service, 22 pp.

Pratt, P. W. 1997. Southern United States soybean disease loss estimate for 1996, *Proc. Southern Soybean Dis. Workers* 24:1–7.

Rhoades, H. L. 1987. Effect of *Hoplolaimus galeatus* on ten vegetable crops in Florida, *Nematropica* 17:213–218.

Riggs, R. D. 1977. World wide distribution of soybean-cyst nematode and its economic importance, *J. Nematol.* 9:34–39.

Riggs, R. D. 1982. Nematology in the Southern Region of the United States, Southern Cooperative Series Bull. 276.

Riggs, R. D. and D. P. Schmitt. 1988. Complete characterization of the race scheme for *Heterodera glycines*, *J. Nematol.* 20:392–395.

Riggs, R. D. M. L. Hamblen, and L. Rakes. 1988. Resistance in commercial soybean cultivars to six races of *Heterodera glycines* and *Meloidogyne incognita*, *J. Nematol.* 2:70–76.

Robbins, R. T. 1982. Description of *Hoplolaimus magnistylus* n. sp. (Nematoda: Hoplolaimidae), *J. Nematol.* 14:500–506.

Rodriguez-Kabana, R. 1992. Chemical control, in Robert D. Riggs and J. Allen Wrather, Eds., *Biology and Management of the Soybean Cyst Nematode*, American Phytopathological Society Press, St. Paul, MN, 115–123.

Ross, J. P. 1963. Seasonal variation of larval emergence from cysts of the soybean cyst nematode, *Heterodera glycines, Phytopathology* 53:608–609.

Schmitt, D. P. and G. R. Noel. 1984. Nematode parasites of soybean, in W. R. Nickle, Ed., *Plant and Insect Nematodes*, Marcel Dekker, New York, 13–59.

Sinclair, J. B. and P. A. Backman. 1989. *Compendium of Soybean Diseases*, American Phytopathological Society, St. Paul, MN, 106 pp.

Wrather, J. A., S. C. Anand, and S. R. Koenning. 1992. Management by cultural practices, in R. D. Riggs and J. A. Wrather, Eds., *Biology and Management of the Soybean Cyst Nematode*, American Phytopathological Society Press, St. Paul, MN, 125–131.

Young, L. D. 1992a. Cropping sequences effects on soybean and *Heterodera glycines, Plant Dis.* 76:78–81.

Young, L. D. 1992b. Epiphytology and life cycle, in R. D. Riggs and J. A. Wrather, Eds., *Biology and Management of the Soybean Cyst Nematode*, American Phytopathological Society Press, St. Paul, MN, 27–36.

Young, L. D. 1994. Changes in the *Heterodera glycines* female index as affected by ten-year cropping sequences, *J. Nematol.* 26:505–510.

chapter seventeen

Sampling tips and analytical techniques for soybean production

J. L. Willers, G.W. Hergert, and P. D. Gerard

Contents

Overview

Sampling is an extensively discussed topic in the culture of many crops including soybean. Thus, most sampling methods applicable to soybean production are already well described. Therefore, the primary focus of this chapter is to provide concepts that help implement a sampling plan and analyze and interpret its sample data. We believe teaching how to interpret sample data is the chief shortcoming of the sampling literature available to soybean producers. Our goal is to help the reader (i.e., a consultant, farm manager, or producer) better understand the sampling process and give guidance (through a few "hands-on" examples) on what to do with sample data acquired from soybean fields. To aid in this process, several sections provide descriptions of pencil and paper methods that can be applied to sample information. Applying these simple methods will build confidence and proficiency in sampling and interpreting skills. These skills will also help one to better read and understand a greater portion of the sampling literature available. The use of these analysis techniques may result in more insight (by identifying patterns) that could lead to the development of improved production methods. Several sets of data will be analyzed to illustrate key concepts about sampling. Some discussion is provided for a few of the more recent sampling concepts and techniques, in particular, those pertaining to site-specific management (SSM), otherwise known as precision agriculture.

The emphasis on analysis and the description of several hands-on methods is what makes this chapter unique. We have not, however, been able to avoid repeating elementary statistical concepts and some rudiments of sampling theory. We also introduce a modification of the line-intercept sampling (LIS) method, a procedure currently underutilized in row crop agriculture. The use of a graph, called an interaction (or effects) plot, is also described.

Sampling is an activity of obtaining information. The next major breakthrough in agricultural production will likely result from better applications of information. Sampling will be one of several processes at work to provide information that leads to better productivity. In summary, our hope is that this chapter will help the reader be more aware of the need to sample, become more skilled in the analysis of information, and be better informed to select and properly apply soybean sampling techniques.

Introduction

Sampling correctly is a difficult task to learn and one that can be time-consuming and monotonous. Another hindrance is the difficulty of mastering methods of analysis necessary for interpreting sample data. Nevertheless, the reader is invited to travel on a journey into the world of sampling and expected to have a learning experience. You will have to do some work, but once you have acquainted yourself with some key concepts, skill in the use of a valuable tool will have been acquired. Eventually, the time required to learn new sampling concepts will begin to decrease dramatically and improvements in the skills of getting better information for making production decisions will follow. It is hoped the practice of sampling will become more of a faithful companion rather than a distant land better heard about than traveled.

Proper applications of sampling techniques can help increase profits and improve management of the crop. On the other hand, poor sampling techniques that produce "bad" data can lead to erroneous conclusions and be expensive wastes of time and resources. Sampling correctly is a skill that must be mastered. Without learning how to sample, one may never discover improvements that produce better crops. This chapter provides information to identify the important literature on traditional sampling techniques useful for soybean production. These excellent sources are quite complete. To merely repeat their information here would be redundant. The chapter bibliography can be used to help locate these materials from the county extension office, local or university library, or bookstore by special order.

Basic concepts and principles of sampling

In this and following sections, words that have unique meaning appear in bold font. The concept or idea of that word, if it is not immediately described, should become clearer as the discussion proceeds. However, it is hoped that by calling them to your attention a mental model of the concepts and the relationships between them will begin to form in your mind.

Sampling is a specialized discipline of the field of **statistics**. Statistics has as its principal purpose the goal of summarizing collections of data into smaller sets of data (or information) so that relationships (as similarities or differences) between (or among) **populations** can be identified. Thus, the word *statistic*(s) can refer (1) to an academic field of study or (2) to a collection of one or more numbers (i.e., the smaller sets of data) used to summarize (or describe) the characteristics of a population.

Sampling is the application of a prescribed method to a population of individuals (insects, soybean plants, fields, farms, people, etc.) for the purpose of obtaining information (data) to describe that population. A particular method of sampling is called a **sampling design**. The basic task of sampling is to collect, without prejudice, a small sample of individuals from a much larger population of interest and measure from them one or more characteristics of interest. These results are then generalized to the larger population. Often, a mathematical equation, called a **probability density function** or pdf for short, can be used to describe different population traits that are sampled. Estimates are made from the sample to determine **parameters** that "fit" the pdf function to that particular population and not to some other. It is important to keep in mind the distinction between a **population parameter** and an **estimate** (i.e., statistic) of that parameter. The parameter is the true value of a **variable** (i.e., characteristic) when all individuals of a population are measured; that is, it is the value that actually exists if it could be measured. Since it is impossible (in many cases) to measure all individuals of a population, one must sample the population and estimate the value of a parameter from the sample. Later, examples are provided that illustrate fitting the **normal pdf** to different populations of soybean yields and plant heights.

The normal pdf is a common distribution used in many sampling applications. It is not, however, the only one that can be used; for example, the **lognormal** pdf is another common distribution. The binomial distribution is often used in sampling designs where the sample response is a "yes" or a "no" (i.e., does the selected plant in the sample have aphids or not?). More than a dozen other distributions can be applied in sampling designs.

Therefore, sampling involves the use of many concepts, including (1) how to select **sample units**, (2) determining the size (**sample unit size**) and (3) the number of sample units (or **sample number size**), (4) the specification of how often samples will be collected, (5) who will collect the samples, (6) the nature of the data collected, (7) how the data will be analyzed, and (8) to whom the data and analysis will be presented. A sampling design, or plan, describes how these traits are related to each other.

A sampling design should accomplish clear and well-defined objectives (Buntin, 1993). These objectives should be established before any data are collected. It is a good practice to enlist the aid of a statistician (or other expert) when developing a novel sampling plan to provide advice and guidance in its development. On the other hand, if you are the user of an already established sampling plan, you should take steps to ensure that the plan is correctly implemented. In short, learn and understand the assumptions of the sampling plan you choose to employ. The authors of that sampling plan should have provided this information in their description of that plan.

Different sampling plans are best suited to specific tasks; therefore, it is necessary to use a sampling plan that matches a specific need. This judgment can only be made when you know the strengths and limitations of alternative sampling plans. The *assumptions* that lie behind any particular sampling plan should be clear to you before you trust its results.

To help facilitate an understanding of the sampling literature, we next present concepts, definitions, and a few commonly used algebraic expressions. We will use several

Table 17.1 Selected Yields (bu/acre) for Maturity Group V Soybeans
in the 1994 Mississippi Soybean Variety Trials

Variety	Brand	Location				Average ± s
		Clarksdale	Rolling Fork	Hernado	Verona	
A5560	Asgrow	32.9	50.8	77.7	48.6	52.50 ± 18.59
A5843	Asgrow	43.4	58.5	77.7	44.0	55.90 ± 16.12
A5885	Asgrow	40.9	58.2	75.8	53.5	57.10 ± 16.80
A5979	Asgrow	54.0	56.6	78.2	45.2	58.50 ± 14.01
Average ± s		42.80 ± 8.71	56.03 ± 3.58	77.35 ± 1.06	47.83 ± 4.26	
DPL 105	Deltapine	46.2	51.1	78.6	43.3	54.80 ± 16.19
DPL 415	Deltapine	48.4	54.6	73.5	48.1	56.15 ± 11.95
DP 3589	Deltapine	42.1	62.6	69.5	58.4	58.15 ± 11.64
Average ± s		45.57 ± 3.20	56.10 ± 3.89	73.87 ± 4.56	49.93 ± 7.71	
H5164	Hartz	44.6	49.3	75.5	46.3	53.92 ± 14.51
H5088	Hartz	45.7	55.4	72.8	51.8	56.42 ± 11.63
H5218	Hartz	46.7	49.1	79.4	47.5	55.67 ± 15.85
H5350	Hartz	44.0	48.9	77.1	37.8	51.95 ± 17.37
H5454	Hartz	49.3	57.0	77.9	51.1	58.82 ± 13.13
H5545	Hartz	54.2	52.3	72.6	43.1	55.55 ± 12.36
H5566	Hartz	43.6	48.2	74.9	42.0	52.17 ± 15.38
H5810	Hartz	44.2	42.0	67.4	42.9	49.12 ± 12.22
Average ± s		46.54 ± 3.61	50.27 ± 4.67	74.70 ± 3.78	45.31 ± 4.77	
9501	Pioneer	36.8	53.1	66.6	37.5	48.50 ± 14.22
9551	Pioneer	37.1	49.4	68.1	38.9	48.37 ± 14.22
9584	Pioneer	44.9	53.8	81.0	50.9	57.65 ± 16.00
9592	Pioneer	52.9	53.1	84.7	51.1	60.45 ± 16.19
9593	Pioneer	50.5	52.9	77.0	51.7	58.02 ± 12.69
Average ± s		44.44 ± 7.43	52.46 ± 1.74	75.48 ± 7.92	46.02 ± 7.16	
Overall average ± s		45.12 ± 5.62	52.84 ± 4.58	75.30 ± 4.71	46.68 ± 5.58	

sample data sets to illustrate the application of these principles. The comment of Cochran (1956) should be kept before you at all times. He stated, "…any sampling plan contains two parts: a rule for drawing the sample and a rule for making the estimates from the results from the sample." The preceding quote is a good summary of the concepts presented so far.

Basic definitions and tools

The first concept to focus on is what is meant by the word **population**. From a sampling perspective, a population is a collection of individuals (or items), such as a field of soybean plants. These individuals will have one or more characteristics, labels, or other criteria that exist and are shared by all individuals. The definition of what is meant by the word *population* can change as different sampling objectives are established. In some instances, the same individual can belong to more than one population depending upon the goals of a sampling activity.

For example, there are several populations that can be constructed from the data in Table 17.1. We could consider the population of each variety by itself (20 populations having four individuals or locations each), or the populations of varieties within brands (4 populations with 16, 12, 32, or 20 individuals each). Other populations that could be defined are the soybean varieties at any of the four locations (4 populations with 20 individuals each). This exercise should make clear that the population of interest should be clearly defined as determined by the question(s) or objectives of interest.

Table 17.1A Frequency Distribution of the Yield Variates
Presented in Table 17.1

Interval	Interval midpoint	Class boundaries	Tally
30–34	32	29.5–34.5	/
35–39	37	34.5–39.5	/////
40–44	42	39.5–44.5	///// ///// //
45–49	47	44.5–49.5	///// ///// ///// //
50–54	52	49.5–54.5	///// ///// ///// //
55–59	57	54.5–59.5	///// //
60–64	62	59.5–64.5	/
65–69	67	64.5–69.5	///
70–74	72	69.5–74.5	////
75–79	77	74.5–79.5	///// ///// /
80–84	82	79.5–84.5	//

Sometimes several different populations or **strata** (smaller subpopulations) might be used simultaneously. Equally important is the fact that when a population is defined, other populations are excluded. For example, populations involving soybean trials from other states, or even other soybean maturity groups, or even other Group V varieties available in Mississippi are not listed in Table 17.1. Always have a clear idea what population is being targeted when sampling.

Replication is another idea that needs to be discussed. When more than one individual of a population is uniquely subjected to a treatment in an experiment, it is said that treatment has been **replicated**. Often, replication is not easy to achieve and is abused in many experiments. Consider a trivial example. Suppose a food chemist wanted to determine the chocolate content of a candy bar. He realizes that he should replicate his study. So, he divides a single candy bar into four parts and calls each part a replicate and measures the chocolate content of each part. Has replication really been accomplished? As an answer consider the same experiment, but now the food chemist spends considerable *effort* to locate four candy bars from *different batches* produced by the candy maker. Each bar is now called a replicate, and the chocolate content of each bar is determined. It should be obvious that the second approach is much better than the first. A good rule of thumb (Milliken and Johnson, 1984, p. 50) to remember is that if replication is achieved by dividing (or splitting) a larger part into smaller parts then one has not properly replicated.

Replication is an important corollary concept to sampling because individuals, even though they come from the same parent population, are different. Each one will respond differently to the same treatment or differ from one another in some character that is being measured. Similarly, not all individuals in a sampled population are exactly the same. For example, the population of all 20-year-old males in Mississippi will not have the same height and/or weight. There will be **variation** in these characters. A statistic, called the sample **variance**, is a measure of the spread in how individuals in the population differ. The **average** response of sampled individuals is used to estimate (or **infer**) the center of the **distribution** of a population trait.

Two additional points are important for understanding how to sample properly. The first point is the fair or equal (**random**) opportunity for any individual in a population to be selected for measurement. This issue implies that there is no bias or prejudice in selecting or not selecting an individual during sampling. If you are uncomfortable with the word *random*, substitute the words *fair, representative,* or *unprejudiced* and you will have a good idea what statisticians have in mind when they use the word *random*. Always know the rules used to select the individuals included in the sample. You should know these even if someone else reports to you the results of a sample. The key point is that for every sample design there are rules that provide a basis for deciding which individuals are included in a sample. For example, if soybean yields for a 100-acre field are estimated using only a 1-acre tract that is the best (or worst) acre in the field, the answer obtained is not going to be of any real use. In this case, there was no rule established that gave the sampler a random sample (apart from a poor rule that said only to sample the same 1-acre tract of the field).

The second point is that the response or sample behavior of individuals can be predicted (within reasonable limits) again and again once that response has been observed for a particular set of conditions. As a result, smaller sample sizes from many similar populations can be pooled or collected together. It is not necessary to use large sample sizes with each population. The chief point is that large sample sizes alone do not result in an increased understanding of population behavior; the proper allocation of sampling effort is the principal key to estimating population traits successfully.

Sample data can be arranged into **one-** or **two-way tables**. A one-way table is a collection of data in which only one classification criterion exists. A two-way table is a collection of data with two classification criteria present. A classification criterion is any definition that can be used to identify a population. Classification variables are also called **independent variables**. The traits that are measured are not part of the classification criteria. The items measured are collectively called **dependent**, or **response variables**. To illustrate, Table 17.1 is a two-way table (ignore for the moment the average and standard deviation statistics) where *Variety* and *Location* are the two classification criteria for the response variable, yield. A one-way table of the data in Table 17.1 can be quickly made if variety is the only classification criterion and yield is again the response variable. Many other examples of one- or two-way tables could be given. Higher levels of data tables can also be constructed. The take-home message is that classification criteria define populations of interest.

It has already been mentioned that a good sampling plan establishes rules that specify (1) which population will be sampled and (2) the chances (i.e., **probability**) that a particular individual from that population will or will not be included in a sample (Thompson, 1992). Different sampling plans have different rules, which is the reason there are so many different sampling schemes. Specifically, the rules that determine if an individual is or is not included in a sample comprise the **randomization scheme** of a sampling plan. There are many different ways to create a randomization scheme. Often, this procedure is that part which gives a particular sampling plan its name. A few common names for different plans are simple random, stratified, sequential, cluster, two-stage, or adaptive sampling. Later, we will contrast some of the strengths and weaknesses of a few of these schemes.

The **sample units** that a particular sampling plan employs should be distinct, non-overlapping, and comprise the fundamental unit that is collected during sampling. Often, the sample unit will be a single individual plant or counts of individual insect pests at a specific stage of development. In this instance, the sample unit is defined by a natural or biological characteristic. Sometimes, the sampling unit will not be so discrete or well defined and will have to be defined by artificial characteristics established by the sampler. In these instances, the sampling unit may be arbitrary, such as a square yard or an acre of land. Even in this case, clear guidelines must exist to define the sample unit. Be wary of sampling plans that are vague in spelling out what is the basic sample unit. An important point raised by Ludwig and Reynolds (1988) is that a sample is a collection of sample units. It is incorrect to call a sample unit the sample.

Most sampling plans will have only one size of sample unit, but some can have more than one size. In some other sampling designs (e.g., belt transects or two-stage sampling designs), the sample unit size can vary. The **sample unit size** is related to questions or objectives established by the sampling plan. Some authors deal with this issue by labeling some sample units the primary units; other sample units used in the design are labeled secondary sample units.

The **sample number** (or **sample number size**) is how many sample units are collected or counted when a sampling plan is implemented. This number can be different for different fields or different sample dates. The sampler should always have in mind well-defined stopping rules that determine when enough sample units have been collected from a population.

Occasionally, a population is small enough that is easy to count completely — e.g., the number of coins of different denominations in the ashtray of a pickup. It would only take a few seconds to determine what was there and how much. Here, one can **exhaustively sample** the population. There is no need to estimate any population parameters for any population that has been completely counted. The answer is known without doubt and the sample size is the same as the number of members in the population.

More often, however, a population is large but countable if given enough time (e.g., the number of soybeans in a bushel). More likely, it is so large that it would be impossible or too expensive to count all the individuals (e.g., the exact number of soybeans stored in a grain bin). In agriculture, most populations of interest are of the latter type. Since this is true, the reason for developing and using a sampling plan should be obvious. Ruesink (1980) stated, "A 'sample' consists of a small collection drawn from a larger 'population' about which information is desired. It is the sample that is observed, but it is the population that is studied." Therefore, to develop a sampling design that can be used to study a large population, general principles have to be employed. When describing the principles that give a sampling plan its distinctiveness, most texts introduce mathematical notation. The use of notation is what makes the reading of sampling plans so difficult for many people. The style of mathematical notation used by different authors is often similar, but frequently different styles of notation exist. Therefore, it is easy to become confused by this unfortunate practice. We shall try to follow as much as possible the notation style of Little and Hills (1978). However, before this notation can be described, it is necessary to first introduce additional definitions.

Unique characteristics of interest measured from individuals sampled from a population without favoritism (or bias) are called **variables**. The individual observations that are recorded for a particular variable are called **variate**s (Little and Hills, 1978). Examples of variables important to soybean production are yield/acre, number of plants/acre, average plant height, percent defoliation by soybean loopers, and so on.

Soon, data from Table 17.1 will be used to create artificially several different sets of samples drawn from different populations. These data are the yields of several (subjectively selected for no particular reason) Group V soybean varieties across several Mississippi locations during 1994 (Mississippi Agricultural and Forestry Experiment Station, MAFES, Information Bulletin 276). Calculations will be performed to estimate several statistics of interest from the variates that result for each "made-up" population. The examples that follow closely model a sampling plan known as **simple random sampling**.

In simple random sampling, each individual of the population has an equal chance of being included in the sample. This does not imply that every individual in the population *will be* sampled. Other sampling plans have different probabilities (or chances) than equal chances of including individuals. Different sampling plans offer different strengths and advantages. In fact, the data of Table 17.1 will eventually illustrate the need to use a stratified sampling plan rather than simple random sampling.

The most commonly used statistic is called the average (or **mean**) and is often represented by one of several symbols: $\bar{x}$ or $\bar{y}$ or $\hat{\mu}$. The average is one of three measures of central tendency. As the examples are worked through, the concept of what is meant by central tendency should become more obvious. For now, consider the following illustration. If two children of equal weight are simultaneously positioned on both sides of a teeter-totter, the two will balance and hang in midair. In this case, the distribution of weight across the board is symmetrical. Similarly, two children of different, but not dramatically different weights can be balanced by moving the position of each child to different distances from the middle of the teeter-totter. The average is the "point" on the teeter-totter board where the opposing forces are equal across the support bar.

Two other statistics of central tendency are known as the **mode** and the **median** and often are not reported. This is not to say that they are unimportant. For precision agriculture applications, they are more useful than the average because the variable of interest is frequently skewed toward large or small values.

Later, the concept of a histogram will be introduced. Soon thereafter, it will be seen that the average is the center of the histogram of a data set. If the histogram is perfectly symmetrical, the mode, median, and average (or mean) will be identical. If the distribution is **multimodal** (i.e, having multiple "peaks"), or **skewed** (i.e., **asymmetrical**), the three measures of central tendency for a set of sample data will differ considerably. Examples and graphs will be given later to illustrate different kinds of histograms and a statistical model will be fit to these distributions.

A second important statistic is called the sample **variance** (s^2). Another major statistic is the sample **standard deviation** (s) and is the square root of the sample variance. These last two statistics measure the spread of sample variates about the average.

Let us now examine what is meant by the terms *average*, *variance*, and *standard deviation* by pretending that the data contained in Table 17.1 can be used to represent different samples collected from several large imaginary populations. To do this exercise, several simple equations are introduced to summarize these sets of fabricated sample data using yield as the **response variable**. In the sampling literature, these equations are used to obtain estimates of the population average and variance.

The estimate of the population **average** for a **variable** is the sum of the **variates** (y_i) divided by the number of individuals sampled, or

$$\text{Average} = \frac{y_1 + y_2 + y_3 + \cdots + y_n}{n} = \frac{\Sigma y_i}{n}. \tag{17.1}$$

The symbol Σ means to "sum over" or, in this instance, take the sum over all the variates in the sample. This sum is next divided by the total number of observations (n). The subscripts ($i = 1, 2, 3, \ldots, n$) represent the first, second, third observations, along with other individuals included in the sample.

The estimate of the **variance** is a little more difficult but can (as the average) be performed with the help of a simple calculator. (See your instruction manual for the calculator or software package to find out if these capabilities are available.) Several formulas are available, but the one selected here is the following:

$$\text{Variance} = \frac{\Sigma y_i^2 - \frac{\left(\Sigma y_i\right)^2}{n}}{n-1}. \tag{17.2}$$

To illustrate what Equations 17.1 and 17.2 involve, the average and the variance will be estimated across four locations for the Asgrow variety A5560 found in Table 17.1. Each step that follows corresponds to the different terms found in Equations 17.1 and 17.2. The average will be estimated first. To begin, correctly identify and list the variates for yield: 32.9, 50.8, 77.7, and 48.6 bu/acre. Next, obtain the **sum** of these four yields:

Step 1: Σy_i = sum = 32.9 + 50.8 + 77.7 + 48.6 = 210.0 bu/acre.

Next, divide the sum by number of variates (n = 4) to obtain the average:

Step 2: $\Sigma y_i/n$ = average = 210.0 ÷ 4 = 52.5 bu/acre.

To estimate the variance of this sample, multiply each variate by itself (known as taking the square of the observation) and obtain the sum of these values:

Step 3: Σy_i^2 = 12,062.3 = (32.9 × 32.9) + (50.8 × 50.8) + (77.7 × 77.7) + (48.6 × 48.6).

The sum obtained in step 1 is then multiplied by itself:

Step 4: $(\Sigma y_i)^2$ = sum squared = (210.0 × 210.0) = 44,100.

Divide the squared sum by the number of samples (n) and subtract this from the result obtained in step 3. You have now obtained the **numerator** of Equation 17.2 as shown below:

Step 5: $\Sigma y_i^2 - (\Sigma y_i)^2/n$ = numerator = 12,062.3 − (44,100 ÷ 4) = 1037.3.

Finally, divide the numerator by the **denominator** that is one less than the total number of variates ($n - 1 = 4 - 1$) or

Step 6: $[\Sigma y_i^2 - (\Sigma y_i)^2/n]/n - 1$ = variance = s^2 = 1037.3 ÷ (4 − 1) = 345.8.

Having now obtained the estimate of the sample variance, it is possible to determine the estimate (s) of the standard deviation, σ. Using a pocket calculator, take the **square root** of the variance:

Step 7: standard deviation = $s = \sqrt{345.8}$ = 18.6 bu/acre.

The units of measurement for the standard deviation are the same as that of the original observations (variates). For the variance, we have not associated units of measure or dimension. For example, with the data being used here, the units would be bushels squared per acre squared, or $bu^2/acre^2$. These are difficult units of measure to interpret and often the variance is reported as a dimensionless number. Remember that these values are estimates of parameters and are not the true but unknown population values (Freese, 1967).

The smaller the variance (or standard deviation), the closer all variates are to the average. Later, several histograms (e.g., Figures 17.8 and 17.9) will be pictured for sets of samples obtained from different populations. Examine these graphs closely to see how different values of variates influence the estimates of the average and variance for the different populations. With practice, you will gain an intuitive understanding for what these statistics are telling you for sets of data acquired from your farm.

There is a shortcut method (Miller and Freund, 1977) that exists to estimate the standard deviation for small-sized samples (two to ten observations). The method has one assumption. It assumes that samples are drawn from a population whose character of interest follows a **normal distribution**. The probability plotting method discussed later in

Table 17.2   Sample Size and Constants Used to Estimate the Standard
Deviation for Small Sample Sizes

n	2	3	4	5	6	7	8	9	10
d_2	1.128	1.693	2.059	2.326	2.534	2.704	2.847	2.970	3.078

this chapter can be used to confirm if a set of data is from a normally distributed population. The normal distribution is a bell-shaped curve that is symmetrical about the average. For symmetrical distributions, the average, mode, and median are equivalent, whereas for asymmetrical (skewed) distributions this is not the case.

This shortcut method uses the **range** (R) of the sample to estimate the standard deviation. The range is the difference between the smallest and largest variates included in a sample. The range is divided by a special number, labeled in this discussion as d_2 (found in Table 17.2) to estimate the sample standard deviation, or $s = R/d_2$. The first row of Table 17.2 depicts the sample size from two to ten sampling units, whereas the second row presents the constant (d_2) used for that size of sample.

Using the data of Table 17.1, several examples now follow to illustrate the use of Table 17.2 to estimate the standard deviation as compared with taking the square root of the variance obtained in Equation 17.2. (Note that the estimates reported in Table 17.1 for the standard deviation were obtained with a software package that used Equation 17.2.) For practice, you may wish to calculate and compare additional values. The shortcut method to estimate the standard deviation (s) for the Asgrow brands (A5560, A5843, A5885, and A5979) at the Clarksdale location uses the information of both Tables 17.1 and 17.2 as follows:

$$s = (54.0 - 32.9)/d_2 = 21.1/2.059 = 10.2 \text{ bu/acre}.$$

The standard deviation for variety A5885 across the four locations (Clarksdale, Rolling Fork, Hernando, and Verona) is estimated as follows:

$$s = (75.8 - 40.9)/d_2 = 34.9/2.059 = 16.9 \text{ bu/acre}.$$

The previous examples all involved sample sizes of $n = 4$. One last example will use the Hartz brand varieties ($n = 8$) at the Rolling Fork location. Here the estimate of the standard deviation is as follows:

$$s = (57.0 - 42.0)/d_2 = 15.0/2.847 = 5.3 \text{ bu/acre}.$$

Additional examples could be given throughout the table, but with the few that have been done it is easy to see that the agreement is within 1 to 3 bu/acre at most, and that the estimates were obtained with less effort. We think this is a very practical way to estimate the standard deviation for most samples having two to ten observations. If more observations are present then one should use a calculator or software package to obtain the estimate.

The examples just completed involved small numbers of samples. In actual practice, an important question to consider is how many sample units (i.e., the sample size, n) are needed to acquire suitable information to make a management decision? The largest sample size that time and effort allows is not always the best number to collect. A balance must be reached between improving the precision of the sample and increasing the cost of sampling while experiencing a diminishing return on learning new information as the sample size increases. But, if the sample size is too small, then poorer or even useless estimates of population traits result. Generally, the more variation about the average, the larger the sample should be to characterize the population, or the sample should be stratified if rules to develop strata can be devised. Presently, other rules will be given as we focus our attention on determining the proper number of samples.

One important consideration that influences the sample size (n) is the distribution of the character being sampled. The shape or pattern the sample data portray can be inferred from a **histogram** (see below). It has been found that agricultural data, when represented by adequate sample sizes, commonly display a histogram that can be described by a special equation known as the **normal frequency distribution**. At times, the histogram may suggest another distribution to use that is not the normal distribution (D'Agostino and Stephens, 1986). When samples are not normally distributed, one has to either (1) find a sampling plan that accommodates the non-normal behavior of the sample data or (2) make modifications to the normal distribution so that it can still be used. The details of how to do either of these alternatives are beyond the scope of this chapter. Other texts like Thompson (1992) should be consulted. The message is that sample data can follow different sampling distributions. The chief concern is to be aware that data may be symmetrical or asymmetrical about the estimate of the average. We provide here several tools to identify key behaviors of the sample data.

The equation that models or approximates the shape (or histogram) of the normal frequency distribution has the following form (Little and Hills, 1978):

$$\text{Normal Frequency Distribution} = f(y) = \frac{N}{\sigma\sqrt{(6.291)}}\, e^{-(y-\mu)^2/(2\sigma^2)}. \qquad (17.3)$$

There are several parts to Equation 17.3 that should be pointed out. First, the value N is the total number of individuals in the sample for a specific character. The number N can be very large, but it is a finite number that is countable. The values μ and σ are the parameters for the average and the standard deviation (where if it were possible to count all individuals in the population they would be numbers whose values are known without doubt). The estimate of μ is the sample average ($\bar{x}$) and σ is estimated by the sample standard deviation s, which is the square root of the estimated sample variance, s^2. The value 6.291 (or 2 * 3.1456 = 2 * pi, or that famous value from the geometry of a circle) is a special constant determined to be necessary by experience.

The formula (Equation 17.3) can also help understand why certain requirements and assumptions are specified for a sampling plan. The parameter σ shown in Equation 17.3 has a large influence upon deciding the sample size or number of samples to take. Generally, as σ gets larger, the sample size must be larger to characterize a population at a given level of precision. There are exceptions to this rule of thumb. For example, using a sampling plan that stratifies the population into smaller strata where the variance is different for each strata can help preserve the use of small samples. Examine closely the material you may acquire about a particular sampling plan to see how it addresses the issues about sample size.

Another thought of interest about Equation 17.3 is to consider how multimodal distributions influence the estimate of the average and standard deviation. Sometimes the population to be sampled will have a bimodal or perhaps a multimodal distribution. A bimodal distribution has two peaks (called **maxima**); a multimodal distribution has several peaks. These peaks represent variates that are most common in a set of sample data. The peaks do not necessarily have to be of the same size, or if more than one, the same distance apart from one another. Naturally, a unimodal distribution will have only one peak.

Table 17.3   Example Sample Data of Soybean Plant Heights
(in.) Categorized into Irrigated (n = 50) and Nonirrigated
(n = 50) Portions of a Field (See text for further discussion)

Irrigated plant heights		Nonirrigated plant heights	
20.4	16.0	20.4	13.8
19.6	20.1	16.2	16.3
19.9	25.2	14.0	14.8
21.5	17.5	12.7	14.9
17.4	22.6	14.1	9.5
20.0	17.1	7.7	17.4
21.2	15.1	17.6	16.5
22.4	22.4	14.0	14.5
18.7	20.4	13.5	11.0
19.1	15.1	14.8	13.7
20.5	24.4	12.8	13.8
22.4	18.9	9.2	18.8
21.8	21.5	15.1	12.5
20.1	18.7	14.4	13.5
18.2	19.4	17.0	12.3
18.0	21.4	13.9	13.8
20.8	22.7	15.7	13.5
19.3	21.7	19.9	12.9
13.8	17.5	15.7	11.3
20.6	20.6	12.4	15.2
21.0	19.7	19.5	15.2
20.3	19.4	16.6	15.1
24.2	19.3	15.0	16.4
18.9	16.7	17.6	14.2
19.5	21.5	20.5	15.5

How does the occurrence of multiple peaks in a collection of sample data influence the use of a sampling plan? Several answers are possible, but one important possibility can be illustrated by a simple example. Suppose the objective is to estimate the height of soybean plants from a particular field where only 80% of the acreage is irrigated. Lately, it has been dry so irrigations have been applied. The sampler collected 100 samples (Table 17.3), with half of the samples selected from the irrigated portion and half from the nonirrigated portion. The frequency distribution of plant heights for this population of soybeans can be expected to follow a bimodal distribution due to the influence of irrigation upon plant growth.

Our first question to the reader is what has the sampler overlooked, or forgotten to do? A second question to the reader is what will be the characteristics of the estimates of the average and the variance? Make a guess to both questions before reading further and, if you wish, apply Equations 17.1 and 17.2, using the variates found in Table 17.3.

Without stratifying (How many populations are being sampled?) the sample between irrigated and nonirrigated plants, the estimate of the population average and variance will behave as if the data were taken from a unimodal population when in fact the actual sample data are bimodal (See Figure 17.1). In this instance, the estimate of the average for the 100 plants is 17.3 in. with a standard deviation of 3.6 in. The curve drawn behind the bars is the shape of the **normal probability density function** (see below) using the above estimates for the population parameters μ and σ.

The sampler overlooked the fact that irrigation can result in the plants having different heights in the two parts of the field. The sampler in this instance made the mistake of not

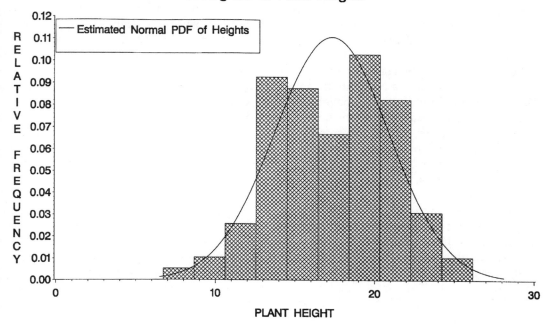

Sample Mean = 17.3 Sample Deviation = 3.6

Figure 17.1 Histogram of the soybean heights found in Table 17.3. The histogram clearly shows the bimodal behavior of these data. The line drawn behind the bars is the fit of the normal pdf to data and ignores the bimodal trait of the data. The frequencies of each class (the bars) are expressed as relative (not absolute, see text) values on the y-axis to the right. The base of each bar is 2 units (in.) of plant height; therefore, sampled plants that are within 2 in. of each other in height are placed into the same class.

stratifying the sample population. The sampler should have divided (i.e., have used a stratified sampling plan) the field into two populations (irrigated and nonirrigated) and then sampled and measured the height of randomly selected plants in each part. If this had been done, the following results would have obtained. The average and standard deviation for the 50 plant heights of the irrigated part of the field is 19.9 in. and 2.4 in. The nonirrigated field plant height average was 14.8 in. with an estimated standard deviation of 2.7 in. Notice that precision was improved by stratification since both estimates of the standard deviation were smaller than the estimate obtained without stratification. The consequence of treating two separate populations as one depends upon the goals established before the samples were collected. At times, serious implications could result; at other times, the effect will be moot. You should keep in mind that sampling plans (e.g., stratified) exist to handle bi- or multimodal distributions of sampled characteristics.

A chief feature of a **stratified sampling** plan is that the sampler is making use of knowledge about the population to do a better job of sampling the population. Stratified sampling divides the population into subpopulations, or strata, that are less variable than the original population (Cochran, 1956). The strata do not need to be the same size in terms of area or number of individuals. When a population has been stratified, different parts of the population can be sampled at different rates (i.e., the number of samples can be different for different strata). This means that sampling effort and resources can be efficiently allocated to those areas where more samples are necessary to characterize the population. The sample data are collected and the parameters are estimated for each strata. The estimates of the strata can later be combined (appropriately) to estimate population parameters with better precision than a simple random sampling plan. The appeal of a stratified sampling scheme is to reduce the **error of the sample**. The previous example of soybean plant heights demonstrated this fact. The error (or **bias**) of a sample is the difference between the estimate and the true (parametric) population value.

The ability to tell if a population should be stratified into different parts can be a good skill to acquire. Practically, whenever a population is stratified, it is an admission that one part of the crop in a field might be managed differently than other parts. In fact, this concept is what lies behind the recent emphasis upon precision agriculture. More comments to this issue from the perspective of sampling for soil characteristics will be given later in this chapter. The failure to use a stratified sampling plan when necessary can lead to wasted resources.

Another common sampling plan is one called **systematic sampling** (or systematic sampling with a random start). Concepts of this sampling plan can be incorporated into other sampling designs. The chief advantage of this plan is reducing the time in selecting sample units. If all samples are randomly drawn, considerable time can be involved. A simple example follows. Suppose one wanted to estimate the average number of soybean pods per plant in a uniform field of 35 acres. The basic sample unit is three successive individual plants. The sample character to be measured is the number of pods per plant. The desired sample size is that at least 35 sets of three plants are to be sampled throughout the entire field (at one set per acre). The sampler has determined that she can walk into a new acre about every 156 paces along a series of parallel lines placed 37 rows apart. Starting at the Northeast corner of the field, she chooses from a random number table two numbers (random number tables can be found in statistics books or generated by a calculator or software package on a computer). The first is a number less than 37, say, 17, and the second is a number less than 156, say, 97. These two numbers define the placement of the first line and the paces on this line to locate the first sample of three plants. For a different sampling date, she will select two new numbers.

With these numbers, she moves to the 17th row from the corner and then walks 97 paces into the field. She will count the number of pods on three successive plants closest to her left foot when she reaches the 97th pace. Thereafter, she will collect samples every 156 paces and over 37 rows until she reaches the opposite side of the field. Paces left uncounted at the end of a line will be resumed on the next line until 156 paces have been walked (idealistically, the line is one long unbroken line). The pattern is repeated until the end of the field is reached. The starting point was selected at random, but the samples thereafter use the same distances of 37 rows between lines and 156 paces apart on each line. The characteristic of systematic sample plans is that the choice of the first member of the sample determines the whole sample (Cochran, 1956). Another advantage is that, with planning, the samples can be more evenly distributed over the defined population. The chief disadvantage is that the samples may follow (by chance) some periodic variation present in the population (Cochran, 1956) and introduce a source of bias. Thus, it may be necessary to stratify the field and use a systematic sampling plan in each strata.

Some sample variates, as mentioned earlier, follow a **skewed** distribution (See Figure 17.13) and are not symmetrically distributed about the mode. For skewed data, the median, mode, and average will be more dissimilar. These distributions can be skewed to the left or to the right. It is suspected that many soil characteristics, when considered on a field-level basis, will follow a skewed distribution. The best sampling plan to use for skewed variates is a modified form of stratified sampling in which the number of samples allocated to a strata is **disproportionate** to the size of the strata (Cochran, 1956). Returning to our plant height example, one could expect the height of plants to vary more in the nonirrigated part than in the irrigated part. So, perhaps it should be considered to have a larger sample size in the nonirrigated than the irrigated portion. Thus, with a stratified sampling plan, the requirement that each individual have an equal chance of being included in the sample refers only to sample units within a stratum and not to the fact that sample units in every strata have the same chance of being sampled as units found in another stratum. The challenging task is to determine the boundaries of the strata followed next by determining the appropriate sample size to use in each stratum. Several years of effort and information from several sources may ultimately have to be used to define the boundaries of strata properly. Thus, in the first year of some sample studies, use the greatest effort and resources on only a small part of a field. Learn a little about a small part, but at a high level of quality. As more is learned, expand the scope of the sampling effort.

At times the distinctions between multimodal and skewed distributions of population attributes become blurred and have little impact. On other occasions, these possible distributions are quite different from one another and can introduce considerable error into sample estimates if not taken into account. The key point is to be aware of these possibilities for any set of sample data. If these features are present, be practical in taking them into account when developing management policy about the crop. The technique of **probability plotting** will be discussed later as a tool to determine if samples are multimodal or heavily skewed.

Sample data can be **discrete** or **continuous**. Discrete data are usually whole numbers. Continuous data are fractional or integer values that occur continuously between specified limits (Fogiel, 1985). Special equations called **probability functions** mathematically describe different behaviors of sample data (Gelman et al., 1995). If the sample data are continuous, the distribution is called a **probability density function**. If the data are discrete, the distribution is called a **probability mass function**. The graph of discrete data looks like a series of small stair steps. Probability functions possess two main features. The area under the curve sums to one and the observed probability values associated with variates of a variable are between 0 and 1 (or 0 and 100% if probability is expressed as percentages). Therefore, one can never have a probability of 2 (or 200%) or –1 (or –100%). However, one can observe a 100% or 200% increase or change in the rate at which some process proceeds. Do not confuse rates of change with the probability of occurrence of some action or item of interest.

The normal frequency distribution has been described in Equation 17.3. The **normal probability density function** is

$$f(y) = \frac{1}{\sigma\sqrt{(2\pi)}}\, e^{-(y-\mu)^2/(2\sigma^2)}. \tag{17.4}$$

The form of Equation 17.4 is very similar to Equation 17.3. The difference between the two is that N has been replaced by 1 with the result that Equation 17.4 now describes areas under a symmetrical bell-shaped curve whose total area is 1. Before, Equation 17.3 modeled the frequency of sampled individuals in different class intervals. Equation 17.4 meets the requirement that the probabilities of occurrence for the different variates (y) in the sample will be between 0 and 1 and that the sum of all the probabilities will be exactly 1. Sometimes the normal probability density function is standardized to have a mean of 0 and a variance of 1. If this is the case, it is called the **standard normal density function**. Sample data that are normally distributed can be transformed to the standard normal function by subtracting the sample average from each of the variates and dividing each difference by the estimate of the standard deviation. These values are called z-scores or standardized scores (Gonick and Smith, 1993) and range between –2 and 2 for 95% of the time. A few z-score values will fall in the intervals –4 and –2 and 2 and 4. These scores are not probabilities and do not have to be restricted between the values 0 and 1.

Probability is a difficult concept for many people. One simple idea to keep in mind is that probability can be a measure of the strength or value of a proposition or occurrence of an event. A probability of 0 means that the event will never occur or that a proposition is not very believable. A probability of 1 means that the event will most definitely occur or that the proposal is very believable. A probability of 0.5 means that an event will occur about one half of the time or that the degree of belief about a proposition is near neutral. Another concept of what is meant by probability is to consider that it is the long-term relative frequency of the occurrence of an event or outcome of an experiment. For example, a fair coin when tossed will turn up as "tails" about 500 times in 1000 tosses for a probability of 0.5.

A good book to help you assimilate and improve your skills is the text by Gonick and Smith (1993). This inexpensive book uses small, related cartoons and simple diagrams to discuss data manipulation, statistics, and sampling. It can be read through in a week or so with 1 or 2 hours of effort per night. Those readers who are interested in precision agriculture, but are unfamiliar with the mathematical concepts that buttress this technology, should consider reading this text.

Determining the size of a sample

Determining the appropriate number of units to sample is challenging. Far too often it is never done. The challenge of modern sampling plans is to balance the **precision** of the sample (or how close the estimate is to the true but unknown population parameter) against the **cost** of taking the sample. For example, if the sample is too small, the estimate will be too inaccurate to be useful (Cochran, 1956). On the other hand, if the sample is too large, it will cost too much in time and labor to collect the samples. Plainly stated, "Samples cost money. So do errors. The aim in planning a (sample) should be to take enough observations to obtain the desired precision — no more, no less" (Freese, 1967). Thus, the proper sample size is a balance between the forces of having the sample be informative and not being too expensive.

The rule of thumb or formula we present to determine sample size is the "confidence limit" approach described by Cochran (1956). Assume that the sampler is willing to take a 5% chance that the allowable error will exceed some value, say L. The following formula to estimate the size of sample n can be derived as

$$n = \frac{4\sigma^2}{L^2}. \tag{17.5}$$

To use Equation 17.5 one must have an estimate of σ from previous samples or knowledge. A simple example is to return to the data of plant heights in Table 17.3. For irrigated soybean plants, how many plants should be sampled if it is desired to estimate the true plant height within 1 in. with a 5% risk that the error will exceed 1 in.? Earlier, it was estimated that the standard deviation for irrigated soybean plant heights in this field was $s = 2.37$. By making use of this estimate in Equation 17.5, the following estimate of sample size is obtained:

$$n = \frac{4 \cdot (2.37)^2}{(1)^2} = 22.47 \approx 22 \text{ plants.}$$

Thus, if 50 samples were actually taken (as shown in Table 17.3), then 28 more plants than necessary were sampled. Cost, in this case, could be considerably reduced by taking only 22 samples instead of the 50 plants illustrated in the example.

For samples that have large standard deviations, or when the sampling objective demands high quality, the estimates of the sample size can be quite large. For example, to be within ½ in. of estimating plant heights for the previous example, a sample size of about 90 plants is required. Here, at this level of precision not enough plants are included in a possible sample. Clearly, sample size is influenced by variability and the degree of precision required with the estimate.

If the sampled attribute is a **binomial proportion**, the following formula should be applied:

$$n = \frac{4pq}{L^2}.$$ (17.6)

A binomial proportion is a ratio between two quantities — the number of individuals classed "yes" and the total number of individuals examined for a specific character (Cochran, 1956). Typically, in row crop agriculture the characteristic of interest is the proportion, or percent, of plants infested with insects or disease. The values p and q are related by the expression $p + q = 1$ if they are proportions, or $p + q = 100$ if they are percentages. The values for p and q in Equation 17.6 should be expressed in the same units.

If Equations 17.5 or 17.6 return estimates that are more than 10% of the total population size, the following correction can be applied:

$$n' = \frac{n}{1+\phi}.$$ (17.7)

The quantity ϕ is the ratio between the total number of individuals that could be sampled and an initial estimate of the sample size that was too large to be practical. To illustrate the application of Equations 17.6 and 17.7, we present the following example of Cochran (1956). A cursory inspection of 480 seedlings suggests that about 15% are diseased. What size of sample is needed to determine p, the percent that are diseased, to within ±5%? By Equation 17.6, the following result is obtained:

$$n = \frac{4\,(15)\,(85)}{(5)^2} = 204 \text{ seedlings}.$$

This estimate of sample size is almost half of the total, and with a sample this large one may just as well examine the whole batch. However, a revised sample number can be obtained from Equation 17.7 that would help avoid the need to examine the whole group. The revised number is

$$n' = \frac{204}{1+\dfrac{204}{480}} = 143 \text{ seedlings}.$$

Remember, we applied Equation 17.7 since the initial estimate of n was more than 10% of the total of the original population. Other authors present other formulas, but the expressions provided above by Cochran (1956) are not as cumbersome to apply. The expressions provided here apply only to simple random sampling plans. More elaborate sampling plans should report applicable formulas (e.g., see Thompson, 1992), but the rules of thumb provided here are good starting points for other sampling plans.

Graphical techniques

Graphical techniques for analyzing data have been available for centuries. However, the advent of computers with greater speed, smaller size, and affordable pricing has caused dramatic changes in graphical methods. Numerous methods of representing data exist or are being developed (e.g., Gazey and Staley, 1986; Friendly, 1991; Fortner, 1995). In this section, we do not describe in great detail any particular graphing technique. However, several pencil and paper methods applicable to small-sized data sets are described in some detail. One needs to understand and appreciate these simpler methods to interpret properly and avoid abusing the power of modern graphical software packages.

Interesting studies exist that demonstrate how one could abuse computer or data management methods (e.g., Fortner, 1995; Calvert and Ma, 1996). Fortner (1995) provides an excellent discussion on how to organize and process data, and avoid common pitfalls that occur while working with technical data (many of the points that are raised apply to sample data as well). Any producer or consultant who works with data and computers should read this book.

Frequency tallies

A **frequency tally**, or frequency distribution table, is a simple but useful tool that summarizes sample data. It is a preparatory step to building a graph. The first step in constructing a frequency tally is to determine suitable **class intervals**. A class interval is the numerical distance, or span, between two whole-number end points that are called the **class limits**. The size of a class, or the class interval, is the difference between its upper and lower class limits. These intervals should be of equal length with midpoints that are convenient whole numbers (Gonick and Smith, 1993). Thus, a class will be a collection of sampled individuals who share similar values for a sample attribute.

The establishment of the appropriate number of classes for a set of sample data is important. Doing this is not an exact science and requires judgment on creating a balance between having too many classes and too few. The number of classes to use generally depends on the number of observations and their range (Little and Hills, 1978). A good general rule is never to use fewer than 5 or more than 15 or 20 classes (Miller and Freund, 1977). Also, a histogram should be constructed from at least 20 or more variates. Other important attributes of a frequency distribution are that its classes will not overlap and be the same size, it will accommodate all the data, and that its class limits will have the same number of decimal placeholders as the original data (Miller and Freund, 1977).

The **class boundaries** can be obtained by adding the upper limit of one class interval to the lower limit of the next higher class interval and dividing by 2 (Spiegel, 1962). If a variate has the same value as a class boundary, a general rule to follow is always to place these variates into the next higher class. The number of variates belonging to each class is called the **class frequency** (Spiegel, 1962).

The data of Table 17.1 will be used to illustrate the making of a frequency tally. It is easier to tally the data if they are first sorted in order from the smallest to the largest value. (The data of Table 17.1 have been sorted in Table 17.4 for use in building a probability plot, a technique that will be discussed shortly.) The smallest variate for yield is the variety A5560 grown at the Clarksdale location (32.9 bu/acre). The largest variate is for

Table 17.4 Rank Order Numbers and Probability Plotting Positions of Soybean Yields
(sorted here in ascending order, found in Table 17.1)

Rank-order number	Variety	Yield	Plotting postion	%	Rank-order number	Variety	Yield	Plotting postion	%
1	A5560	32.9	0.123	1.23	41	H5454	51.1	0.506	50.62
2	9501	36.8	0.025	2.46	42	9593	51.7	0.518	51.85
3	9551	37.1	0.037	3.70	43	H5088	51.8	0.531	53.08
4	9501	37.5	0.049	4.94	44	H5545	52.3	0.543	54.32
5	H5350	37.8	0.062	6.17	45	9592	52.9	0.555	55.55
6	9551	38.9	0.074	7.41	46	9593	52.9	0.567	56.79
7	A5885	40.9	0.086	8.64	47	9501	53.1	0.580	58.02
8	H5566	42.0	0.099	9.88	48	9592	53.1	0.593	59.26
9	H5810	42.0	0.111	11.11	49	A5885	53.5	0.605	60.49
10	DP3589	42.1	0.123	12.34	50	9584	53.8	0.617	61.72
11	H5810	42.9	0.135	13.58	51	A5979	54.0	0.630	62.96
12	H5545	43.1	0.148	14.81	52	H5545	54.2	0.642	64.20
13	DPL105	43.3	0.160	16.05	53	DPL415	54.6	0.654	65.43
14	A5843	43.4	0.173	17.28	54	H5088	55.4	0.667	66.67
15	H5566	43.6	0.185	18.52	55	A5979	56.6	0.679	67.90
16	A5843	44.0	0.197	19.75	56	H5454	57.0	0.691	69.13
17	H5350	44.0	0.209	20.99	57	A5885	58.2	0.704	70.37
18	H5810	44.2	0.222	22.22	58	DP3589	58.4	0.716	71.60
19	Hartz516	44.6	0.234	23.46	59	A5843	58.5	0.728	72.83
20	9584	44.9	0.247	24.69	60	DP3589	62.6	0.740	74.07
21	A5979	45.2	0.259	25.93	61	9501	66.6	0.753	75.30
22	H5088	45.7	0.272	27.16	62	H5810	67.4	0.765	76.54
23	DPL105	46.2	0.284	28.36	63	9551	68.1	0.777	77.78
24	Hartz516	46.3	0.296	29.63	64	DP3589	69.5	0.790	79.01
25	H5218	46.7	0.308	30.86	65	H5545	72.6	0.802	80.24
26	H5218	47.5	0.321	32.10	66	H5088	72.8	0.815	81.48
27	DPL415	48.1	0.333	33.33	67	DPL415	73.5	0.827	82.72
28	H5566	48.2	0.345	34.57	68	H5566	74.9	0.839	83.95
29	DPL415	48.4	0.358	35.80	69	Hartz516	75.5	0.851	85.18
30	A5560	48.6	0.370	37.04	70	A5885	75.8	0.864	86.42
31	H5350	48.9	0.383	38.27	71	9593	77.0	0.876	87.65
32	H5218	49.1	0.395	39.51	72	H5350	77.1	0.889	88.89
33	H5454	49.3	0.407	40.74	73	A5560	77.7	0.901	90.12
34	Hartz516	49.3	0.419	41.97	74	A5843	77.7	0.914	91.36
35	9551	49.4	0.432	43.21	75	H5454	77.9	0.926	92.59
36	9593	50.5	0.444	44.44	76	A5979	78.2	0.938	93.83
37	A5560	50.8	0.457	45.67	77	DPL105	78.6	0.951	95.06
38	9584	50.9	0.469	46.91	78	H5218	79.4	0.963	96.30
39	9592	51.1	0.481	48.15	79	9584	81.0	0.975	97.53
40	DPL105	51.1	0.494	49.38	80	9592	84.7	0.988	98.76

The yields are plotted on the y-axis and the plotting position as percents are plotted along the x-axis of Figure 17.10.

variety 9592 grown at the Hernando location (84.7 bu/acre). The difference between the smallest and largest value (the **range**) is 51.8 bu/acre. By using the range, and a decision to declare that yields that differ by more than 5 bu/acre are different, it was determined that 11 classes should be constructed. This number was reached by merely dividing the range by 5 bu/acre and rounding up to the next whole number. The lowest interval was chosen to begin at 30 bu/acre. Table 17.1A shows the intervals, interval mid-points, class boundaries, and tally marks of each class. The interpretation of these data will be provided as the example is discussed.

Notice first of all that the tally marks create a crude histogram (turn Table 17.1A on its side by rotating it to the left). In brief, the data are bimodal (i.e., there are two peaks), with one peak occurring near the two midpoints of 47 and 52 bu/acre, and another smaller peak occurring at the class midpoint of 77 bu/acre. These facts suggest that Table 17.1 reflects several different patterns or processes at work that determine the yield response of these 20 varieties across the four locations. We next learn how to construct a histogram.

Histograms

Constructing a bar graph known as a histogram uses ideas similar to those used to build a frequency distribution. However, now the goal is to draw a more formal picture of the data. The histograms presented here were drawn using a graphical plotting package (Proc GCHART, SAS Institute, 1990). Many spreadsheet packages can also draw histograms.

Unlike the crude histogram shown in Table 17.1A, histograms can have the feature that the base of the bar multiplied by its height results in an "area" that is the class frequency. The base of these bars are centered on the **midpoint** (Gonick and Smith, 1993). Notice that in Figure 17.1, the base of each bar corresponds to a class interval of size 2. In Figures 17.3 through 17.7 the class interval is 5, but for Figure 17.2 the base of the bar is idealistically one and each bar is centered over a midpoint. Note that the frequencies on the y-axes (or vertical scale) are expressed as **relative frequencies** (Gonick and Smith, 1993) in Figures 17.1 and 17.3 through 17.7. In Figures 17.2, 17.8, and 17.9 the absolute frequencies are used; therefore, the area of each bar is the class frequency. For histograms that graph relative frequencies (like Figures 17.1 and 17.3 through 17.7), the area of each bar must be multiplied by the sample size (the total number of observations used in the sample) to get the original frequency of each class.

By comparing Figure 17.2 and Figure 17.7, it can be seen that either graph gives a similar picture of the data even though the two figures use different units on the y-axis. By selecting different graphing options, both figures could be drawn with the same labels on the horizontal scale or x-axis. A careful look shows that both Figures 17.2 and 17.7 have the same midpoints. This comparison illustrates that you should be alert when examining a histogram and notice which form (absolute vs. relative) is used to present the data on the vertical scale.

Examining the histograms of different samples provides a better understanding of how the average, standard deviation, and variance of different populations compare. For example, Figures 17.3 through 17.6 present both the group (dotted line behind the bars) and the subgroup (solid line behind the bars) "fits" of the normal probability density function (pdf). The "group" fit is the yields of all 20 varieties over all locations, while the "subgroup" fit is the yield of the 20 varieties at each location. Overall, only the Rolling Fork data is the most "normal," while the Clarksdale and Hernando locations are "okay," but slightly skewed. The Verona population appears to be the "poorest" fit to a normal curve because its largest frequency class lies to the right of the fitted mode.

Similarly, Figure 17.7 compares the fit of the normal pdf (solid line behind the bars) of the entire group, but overlays the normal density function fit to only the Hernando location (dashed line). Here, it is seen that the fit of the normal pdf to the entire set of yields is poor because it does not capture the bimodal pattern revealed by the bars of the histogram. Overlaying on the same plot the trace of the normal probability function fit to only the Hernando location (compare Figure 17.5) brings further emphasis to the fact that more than one statistical model is needed to describe these data properly. The dashed line clearly shows which bars of the combined data contain the yields from the Hernando

1994 Yields For All Four Locations

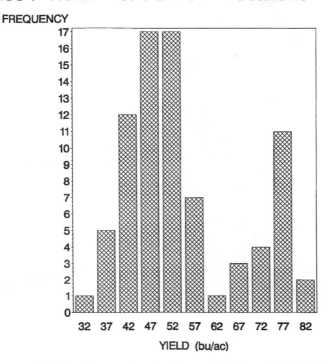

Figure 17.2   Histogram of absolute frequencies of soybean yields from Table 17.1 in 11 classes having class widths of 5 bu/acre. The *y*-axis of this histogram displays the actual (or absolute), not the relative, frequencies of individuals in each class.

variety trial. Interestingly, the Clarksdale, Rolling Fork, and Verona yield data collectively form a population that can be modeled quite well by the normal density function as judged by the shape of the remaining bars that lie to the left of the Hernando trace (Figure 17.7).

The consequences of a poor "fit" depend upon the sampling objectives. You, the decision maker, have the responsibility for determining whether or not these consequences matter for sample data under your control. For example, several histograms for the data of Table 17.1 strongly suggest that the Hernando data should be considered as a distinct population or strata.

Two, three-panel histograms have also been drawn to illustrate further how estimates of an average and a variance relate to each other in describing sample data. In Figure 17.8, the variance is the same for each panel, but the estimate of the average (mean) differs by 10 for values between 30 and 50. The shape of each distribution is the same, but its location on the *x*-axis is different. In Figure 17.9, the means (or the center of the distributions) of three populations are the same at a value of 50, but the variances differ considerably over values of 5, 25, and 50. Here, the location of the distribution is the same on the *x*-axis, but the shape varies from narrow to broad. In Figures 17.8 and 17.9, unlike Figures 17.1, 17.6, 17.7, and 17.13, the distributions are symmetrical; thus, estimates of the average, median, and mode will be equivalent statistics. As you work with more data sets, you will learn of other behaviors. The one pattern you should keep in mind is the fact that sample data expressed as a percentage or proportion will become more skewed (to the right) as the mode gets closer to zero.

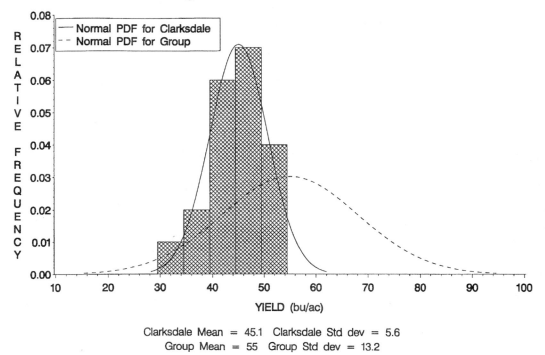

Clarksdale Mean = 45.1   Clarksdale Std dev = 5.6
Group Mean = 55   Group Std dev = 13.2

Figure 17.3 Histogram of relative frequencies of soybean yields for 20 varieties in 1994 at the Clarksdale location. The subgroup mean and standard deviation are for only the yields obtained at the Clarksdale location; the group mean and standard deviation pertain to all the yields found in Table 17.1.

Probability plots

The following discussion is based on information from King (1980) and D'Agostino and Stephens (1986). A probability plot is constructed on **probability paper** by graphing the calculated plotting positions (x_i) of the sample variates (y_i) arranged in **rank order**. Data are rank-ordered when they are sorted from the smallest to the largest value and each variate is assigned a rank.

Probability paper is a special type of graph paper. For the soybean data of Table 17.1, yields comprise the ranked y-values on the **ordinate** or vertical axis. The x-values (or the probability plotting positions) are arranged by increasing value on the **abscissa** or horizontal axis. The plotting positions and ranked values of the variates comprise **ordered pairs** (x,y). Probability graph paper can be obtained from TEAM (Box 25, Tamworth, NH 03886; telephone 603-323-8843). Request a catalog, or ask for pricing and quantities on No. 3111 graph paper for sample data sets having 100 or fewer points, or No. 3211 graph paper for data sets having more than 100, but less than 10,000 values. Software can also be carefully programmed (e.g., a spreadsheet) to accomplish the same task as using special graphing paper.

The value for x (or probability plotting position) is given by a simple formula, or rule. The ith plotting position can be determined by $x_i = i/(n + 1)$, where i is the rank number and n is the total number of values in the set of data. The plotting positions (expressed

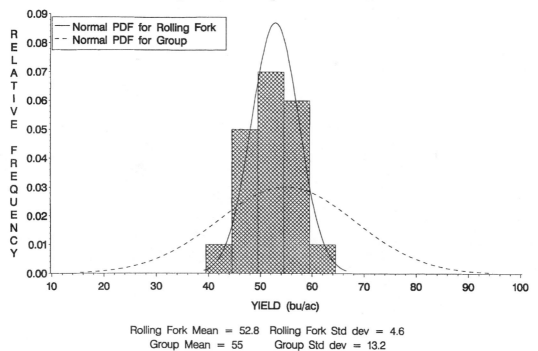

Rolling Fork Mean = 52.8   Rolling Fork Std dev = 4.6
Group Mean = 55   Group Std dev = 13.2

Figure 17.4 Histogram of relative frequencies of soybean yields for 20 varieties in 1994 at the Rolling Fork location. The subgroup mean and standard deviation are for only the yields obtained at this location; the group mean and standard deviation pertain to all the yields found in Table 17.1.

as a percent) using this rule for the example soybean yields are given in Table 17.4 and plotted, by hand, on Figure 17.10 using normal probability paper.

An important modification is necessary for data sets having fewer than 20 points. In these cases, the proper formula (King, 1980) for obtaining the plotting positions is the expression $x_i = (i - 0.375)/(n + 0.25)$. Small-sized samples can introduce artifacts not related to the nature of the data unless corrections are made.

Probability plotting is a graphical technique that can be used to help gain insight about a set of data. The technique can help identify if it is reasonable to assume that a set of sample data is from the same population or not. Probability plotting (1) is easy to use and versatile, (2) is best suited to data that are expensive to obtain and limited in availability or amount, (3) provides for the easy measurement of variability, and (4) helps users understand the implications of that variability in making decisions (King, 1980). The tool should be applicable to data that are obtained from situations where crop heterogeneity exists in a field, or soil fertility, water, and/or insect abundance differ across the field. Differences in the skill of field scouts that cause field-to-field differences in measured responses of interest could also be revealed (be careful here as fields could really be different). Also, these plots can communicate the results of a statistical analysis to interested parties who do not have, or do not desire, the ability to be statistically proficient but still wish to excel in management (King, 1980).

One useful feature of these plots is the ability to discover situations having **contamination** of the data. Contamination of data means that data values occur that do not represent either the original state of nature, or represents the occurrence to two different

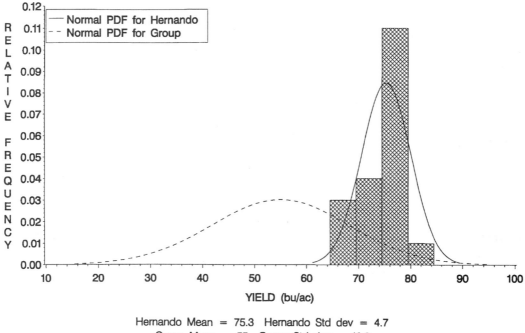

Hernando Mean = 75.3 Hernando Std dev = 4.7
Group Mean = 55 Group Std dev = 13.2

Figure 17.5 Histogram of relative frequencies of soybean yields for 20 varieties in 1994 at the Hernando location. The subgroup mean and standard deviation are for only the yields obtained at this location; the group mean and standard deviation pertain to all the yields found in Table 17.1. Yields are the highest here of all locations.

processes, situations, or events at the same time. Sometimes, contamination represents "bad" or carelessly obtained data. Whenever contamination is observed in a set of data, careful analysis must follow to determine the cause of the discrepancies. Detecting these irregularities is one advantage for drawing a probability plot for a set of sample data. When the cause of the irregularity is identified, valuable insight into system behavior has been obtained. The data of Tables 17.1 and 17.4 show in Figure 17.10 the existence of at least two populations. Yields above 60 bu/acre begin to plot along a different trend. All but one of these yields are from the Hernando location. The higher yields for the Hernando location come from another population, and therefore are governed by different "processes" than those processes that are at work at the other three locations. Earlier, the histograms of all yields (Figures 17.2 or 17.7) showed something unique about the 20 varieties at this location, but these plots could not tell if a difference was due to a different process or due merely to being larger yields. So, while a casual examination of Table 17.1 and the histograms quickly set apart the Hernando yields, it is the probability plot that conclusively shows that these yields are distinct for reasons more than just simply being larger.

   The investigators who conducted the original study would have to explain probable causes. Determining these causes would be instructive. It is beyond the ability of this chapter to determine why, because not all necessary information is available. If a probability plot or histogram shows evidence of "mixed" populations in the sample for data collected, it is hoped that enough background is locally available to allow the identification of possible reasons.

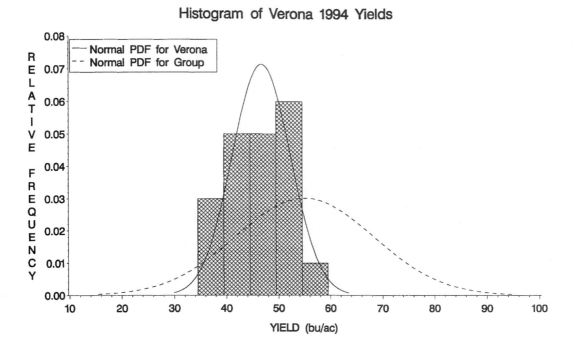

Figure 17.6 Histogram of relative frequencies of soybean yields for 20 varieties in 1994 at the Verona location. The subgroup mean and standard deviation are for only the yields obtained at this location; the group mean and standard deviation pertain to all the yields found in Table 17.1.

Interaction plots

There are two types of interaction plots. Of these, only one can be drawn without using advanced statistical techniques. The one that can be drawn without special skills, called a Type I interaction plot (Milliken and Johnson, 1989), is demonstrated here. Interaction plots are best used with data that can be arranged into a two-way table. A two-way table should be balanced, meaning that all row and column combinations (i.e., cells) have an entry. The entries in each cell should be numerical values and have like units. For example, do not include in the same table entries for some cells that are bu/acre and others plants/acre. The factors of the table can be different numerical levels (e.g., rates of nitrogen fertilizer) or different qualitative labels of the same thing (e.g., location or varieties). Interaction plots permit one to determine quickly if the observed response at different settings of one treatment interact (i.e., nonparallel responses) with different levels of a second treatment. The graphs that are shown were drawn using a computer, but the reader should note that any of these can be drawn by hand using regular graph paper, a ruler, and a pencil.

To draw a Type I plot, label and scale the *y*-axis to the range of the response data. Graph on the *x*-axis different levels (coded as indexes, 1, 2, 3, etc.) of one of the two factors. The second factor of the two-way table is used to define the different lines that are drawn on the plot. Another plot of the same kind can be drawn by reversing the order and scale of the factors on the *x*-axis. If too many lines on the same plot make it difficult to interpret the plot, then select a smaller set of lines and draw only those. For example, an interaction

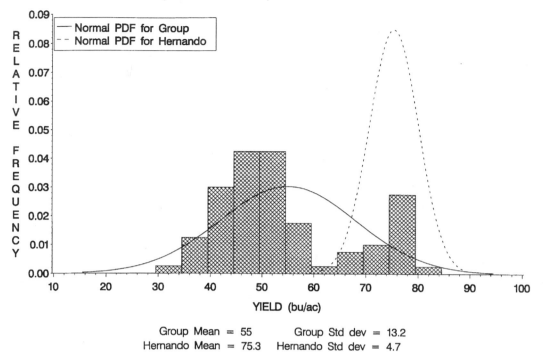

Figure 17.7 is shown below the image.

Figure 17.7 Histogram of relative frequencies of soybean yields for 20 varieties in 1994 at all locations. The group mean and standard deviation are for the yields at all locations. Shown for comparison (dotted line) is the fit of the normal density function for data from the Hernando location. Compare to Figure 17.2, which is a histogram of absolute frequencies.

plot of all 20 varieties over the four locations is too cluttered to be useful. The plots are examined for occurrences, or the lack thereof, of parallel line segments.

To keep the example plots simple, only the yields of three varieties (DP3589, H5454, and H5545) at four locations will be used (see Table 17.1). In Figure 17.11, these yields are graphed across indexes that represent the four locations. The graph indicates that the yield response of location 1 (Clarksdale) shows an increasing trend in yield. Locations 2 (Rolling Fork) and 4 (Verona) show decreasing trends, and location 3 (Hernando) shows the highest, but flattest yield response.

In Figure 17.12, another Type I plot is drawn, but here the yields across the four locations are graphed by indexes that represent the three varieties. The varieties DP3589 (line 1) and H5454 (line 2) yield similarly between Clarksdale and Hernando and differ from H5545 (line 3). But, DP3589 differs from H5454 and H5545 among the Hernando, Rolling Fork, and Verona locations.

The interaction plots very clearly illustrate the challenge of how to select the best variety for a location. A variety that performs poorly in one location may excel at another. The problem of variety selection is even more difficult if different years are considered. Advanced techniques do exist to sift the performance of varieties among locations and different years (see Gauch, 1992). These methods are not discussed in this paper.

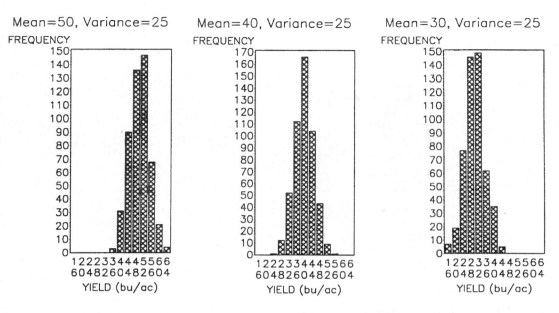

Figure 17.8 Histograms of soybean yields from three imaginary fields, each 500 acres in size, that have the same variance but different means. Reported are the yields from each acre in the field since the sum of the absolute frequencies equals 500. Notice that each histogram has a similar shape, but that as the mean increases, the center of the distribution moves to the right along the *x*-axis.

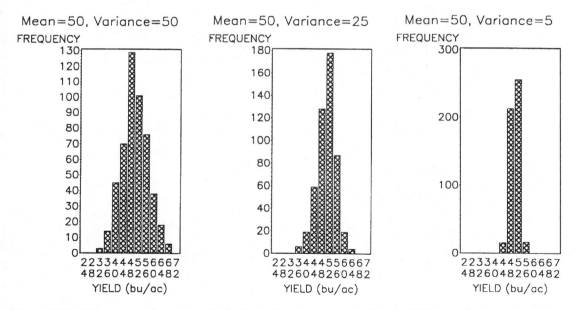

Figure 17.9 Histograms of soybean yields from three imaginary fields, each 500 acres in size, that have the same mean (or average) but different variances. Reported are the yields from each acre in the field since the sum of the absolute frequencies equals 500. Notice that each histogram has a different shape, but each distribution is centered at 50 bu/acre. Each distribution becomes broader as the variance increases.

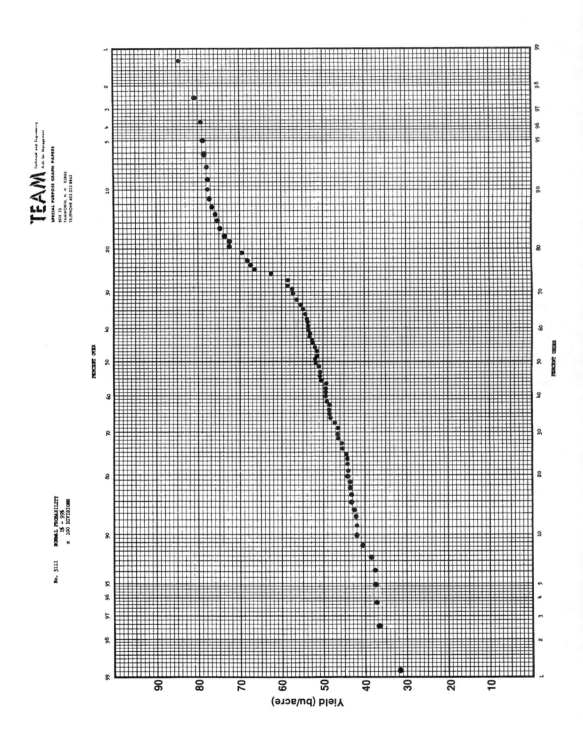

Sampling for soil characteristics

Concepts useful to sampling soil attributes are discussed next. Again, the discussion is somewhat general and interested readers are referred to selected references for details (e.g., Wollenhaupt et al., 1997).

Effect of composite samples

It is often relatively easy to collect soil samples, but the cost of analysis of numerous samples collected on a small scale is expensive. To help keep costs down, a technique known as composite (bulk) sampling is used to estimate the average value of a soil property. In composite sampling, a fixed number of samples are collected by either a random sampling or a systematic sampling scheme. A random sampling scheme simply identifies the points on the field where soil samples are taken in a random manner. A systematic sample uses some sort of predefined grid to determine the sample points. Theoretical studies have shown that unless the value of interest follows a cyclic trend, a systematic sample can result in a more precise estimation. This has also been supported by field and laboratory studies. Once the samples have been collected, they are combined or "composited" and thoroughly mixed. The mixture is then subsampled and those samples analyzed and averaged to obtain an estimate of the average value of interest.

There are several advantages and disadvantages to composite sampling. First, the cost of analysis is typically much lower than if each sample were measured separately. If the act of creating a composite sample yields a homogeneous mixture, then the estimate is unbiased. However, if the mixing is not thorough, a biased estimate of the desired average can result. Composite sampling results in an estimate of the average only; no estimate of the variation of the property in the field can be obtained. Variability among the samples analyzed can be quantified, but this measures the thoroughness of the mixing procedure and the accuracy of the test procedures rather than the variation in the field itself. Since values in the field can vary substantially even for samples taken near each other, reliance on the estimated average value without consideration of the variation in the individual samples can lead to overestimation of the quantity of interest in some areas of the field and underestimation in others. When these estimates are used to provide guidelines for nutrient requirements, significant over- or underapplication can result. Research has also suggested that the act of compositing the samples can change the physical composition of the soil in such a way as to alter the values of some soil properties (Giesler and Lundström, 1993).

Historical overview of soil sampling and interpolation methods

Soil testing has been used since the late 1940s to identify soils that may require lime and fertilizer inputs for optimizing crop production (Bray, 1929; Truog, 1930; Morgan, 1932). For plant nutrients, it involves rapid chemical analyses in addition to interpretation, evaluation, and fertilizer recommendations (Peck and Soltanpour, 1990). With the increasing awareness of fertilizer effects on environmental and soil quality, soil tests can also be used to determine where fertilizers or manure should not be applied as well as where to apply them.

Figure 17.10   Example graph of normal probability plotting using special-purpose graph paper. The data are the yields of Table 17.1 after being arranged in rank order (see Table 17.4). The plot is a reproduction of one drawn by hand.

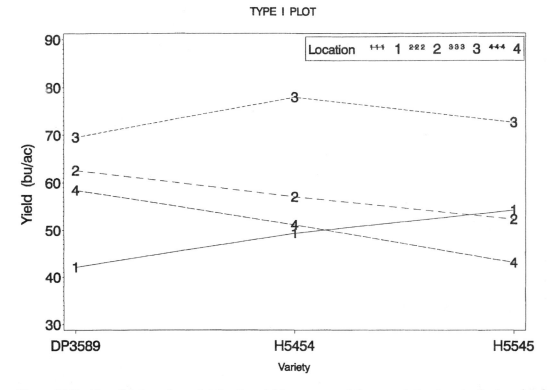

Figure 17.11 Type I interaction plot for the yield response of three varieties (see text) at each of four locations. Parallel line segments indicate where the yield response of a variety is similar to another between locations. Line segments that are not parallel indicate an interaction in yield response of the varieties over locations. Line 1 is Clarksdale, line 2 is Rolling Fork, line 3 is Hernando, and line 4 is Verona.

Soil testing has historically focused on determining the average soil test value for a field or area. It assumed that each observation was independent from other observations, and based on that assumption many chapters and articles have appeared in the literature. Cline (1944) presented general principles of soil sampling that were expanded by Peterson and Calvin (1982) and James and Wells (1990).

Most soil sampling efforts focused on determining an adequate number of samples to provide a reliable estimate of the mean, the most efficient sampling plan, and some measure of spatial variability. Peterson and Calvin (1965) defined the best sampling plan as one that gave the lowest sampling error at a given cost or the lowest cost at a given sampling error. However, past sampling research has shown that grid sampling almost always increases precision compared with random sampling due to the spatial correlation of values (Peck and Melsted, 1967; Sabbe and Marx, 1987). Most producers and agribusinesses have done composite (or bulk) sampling to determine field averages.

The underlying probability distribution functions of many soil parameters are usually not normally distributed but are log normal (Reuss et al., 1977; Parkin et al., 1988; Hergert et al., 1997). If data are distributed log normally, more than 50% of the values are less than the mean (Figure 17.13). A few high testing values can skew the mean and cause an overestimation of the central tendency. This fact probably has led to more confusion and questioning of soil test credibility than any other factor. This was not a major consideration in past soil test correlation/calibration research because plot sizes were generally small

TYPE I PLOT

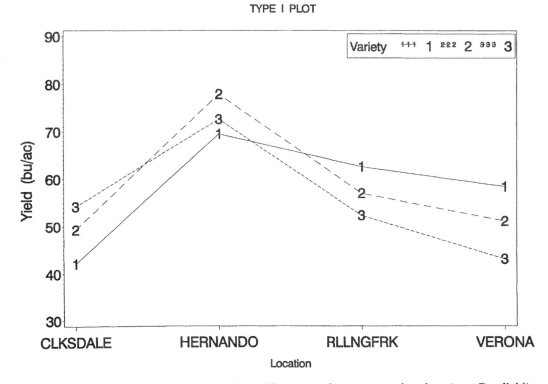

Figure 17.12 Type I interaction plot for the yield response by variety at four locations. Parallel line segments indicate where the yield response of a location is similar to another location for the selected varieties. Line segments that are not parallel indicate an interaction in yield response for different locations. Line 1 is the variety DP3581, line 2 is H5454, and line 3 is H5545.

and homogeneous so the critical levels that were developed were not the problem. The problem was one of scale when small-plot information was extended to the field scale, which encompassed much wider variability (Hergert et al., 1997).

Advances in the theory of regionalized variables (geostatistics) enables estimation of the spatial dependence of soil properties regardless of the underlying distribution (Matheron, 1971). Geostatistical analysis provides a method to develop "statistically correct" contour maps of the soil parameter being measured regardless of the underlying frequency distribution. Many additional samples are required, however, compared with classical sampling methods. The advent of site-specific management (SSM) and variable-rate application (VRA) has caused many producers to change their soil sampling methods.

The dilemma is that quantifying the variability of a soil test parameter requires soil sampling at an intensity that will allow the variability to be mapped spatially with some degree of confidence. This is nothing new, as Reed and Rigney (1947) concluded that field variation was much greater than laboratory variation. Each soil property has a unique variation in a specific field, and the specific soil property having the greatest variation could not be anticipated. Soil properties show large differences in spatial and temporal variability. In general, soil properties with a larger variability require more-intensive sampling to quantify the pattern of variability. Wollenhaupt et al., (1997) lists common differences between soil test parameters and their spatial range.

Sample properties can also be classified according to temporal variability. Soil properties that do not change appreciably over many years include organic matter content,

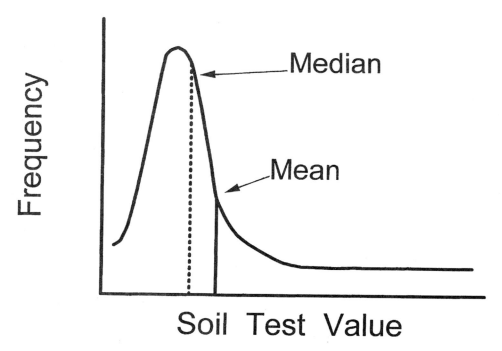

Figure 17.13 Diagram showing the general distribution of many soil characteristics. Notice that the curve is not a normal distribution since it is not symmetrical about the mean and that these data are skewed to the right.

texture, and cation exchange capacity. Soil pH changes slowly over periods of 5 to 10 years. Soil P and K may show dramatic changes over 3 to 5 years, whereas soil nitrate or sulfate can show large changes within one growing season or year. Sampling time and frequency need to be guided by the dynamics of the soil property being measured. Proper timing of sampling is critical for accuracy when measuring dynamic vs. static parameters.

Soil and fertilizer management for crop production has increased field variation in soil nutrients and pH. Most past soil sampling guidelines called for sampling the depth of tillage, normally 6 to 8 in. (Peck and Melsted, 1967). However, fertilizer application methods (band vs. broadcast, manure application, differential crop removal) and changing tillage practices have increased heterogeneity and complicated both the sampling and interpretation process (Randall, 1982; James and Wells, 1990; Kitchen et al., 1990). Additional research will be required to determine future suitable sampling techniques to deal with nutrient and pH stratification.

Fertilization according to VRA and SSM provides additional challenge and opportunity for improved soil sampling techniques. The assumption of a VRA is that it will more closely match productivity, input efficiency, and profitability compared with uniform application (Sawyer, 1994). The concept assumes that within-field variability exists, that it influences crop yields, that the variation can be identified, measured, and mapped, that precise crop-response models are available to determine optimum economic inputs, and that data processing and application equipment are available that can effectively manage and apply the inputs.

This new methodology suggests reevaluating past research and assumptions about soil sampling. We know from past research that uniform fertilizer management across an entire field can result in overfertilization of some areas and underfertilization of others. In either case, this is an economic loss to the producer. Overfertilization increases the probability of nutrient loss from erosion or leaching and may accentuate environmental problems, whereas underfertilization limits yield and product quality.

The focus of this section is a discussion of soil sampling techniques from two perspectives — classical sampling to determine average soil properties for use in whole field management (WFM), and sampling and interpolation techniques that can be used to develop precise maps for SSM.

Sampling frequency

Sampling frequency should be based on expected changes in the soil parameter as discussed earlier. For many properties or parameters, periods of 4 to 5 years will be adequate, whereas mobile nutrients like NO_3–N and SO_4–S may require yearly sampling to detect changes. This recommendation would apply whether sampling is based on WFM or SSM. The challenge in both systems will be to determine the impact and interaction of other factors including uniformity or lack of it, sampling pattern, number of samples, and sampling depth as influenced by past or current tillage.

Sampling methods

The degree of variability for the different scales the producer plans to manage should guide sampling. Soil variability can be classified into three categories of variability, ranging from the smallest to largest: micro-, meso- and macrovariability (see James and Wells, 1990). With WFM, producers are managing above the macroscale by incorporating micro-, meso-, and macrovariability into a bulk (composite) sample that represents the field average. This may not be a major limitation if the field is fairly uniform. SSM is more concerned with variability on the meso- and macroscales, and is most helpful where macrovariation is large.

Sampling uniform fields

Uniform fields have small macro- and mesovariability (James and Wells, 1990). Similarities in slope, elevation changes and drainage, aspect and management history including fertilizer and lime application, and cropping would be considerations to determine if a field may be categorized as "uniform." For these types of fields, WFM and classical random composite soil sampling is still a reliable tool. The value of information gained from extensive and expensive grid soil sampling, the possible savings on fertilizer or lime, and increases in productivity probably would not offset the additional time and money required for a more-intensive sampling (Franzen and Peck, 1995a; Gotway et al., 1997).

Sampling guidelines recommending areas no larger than 40 to 50 acres per sample with a minimum of 25 to 30 cores per area apply for these types of fields for most nutrients and soil properties (Sabbe and Marx, 1987; James and Wells, 1990). A simple random sampling pattern is favored by most agronomists, although the literature suggests other systems including zigzag patterns (Sabbe and Marx, 1987). The unknown is the influence of tillage on required depth of sampling. Changing tillage and fertilizer application techniques have complicated sampling by stratifying nutrients and pH. Variability with depth exists but what has not been determined is how plants respond to these highly variable conditions. Do they simply adapt to the "average" that would be expressed by 0 to 8 or 12 in. samples or is there an effect of stratification on the crop?

Current guidelines suggest shallower sampling depths of 2 to 4 in. can replace the traditional sampling depth of 8 to 12 in. (James and Wells, 1990). This applies for no-till and ridge-till fields for monitoring surface pH and the buildup of immobile nutrients.

Sampling for site-specific management

There is some uncertainty about the exact sampling frequency, method, and pattern required to develop maps for SSM that show significant variation at a reasonable cost compared with classic bulk soil sampling. Information comes at a cost. Producers cannot expect to spend the same amount of money for SSM as for WFM. However, the value of the information should offset the increased soil sampling cost. Because of the intensity that will be required for SSM soil sampling, farmers must first look at sampling frequency. For many properties, one sample every 4 to 5 years will be sufficient. This allows yearly sampling of 20 to 25% of the land area per year.

Another factor that has changed soil sampling is who does the sampling. When operations were smaller, many farmers were thoroughly aware of their fields. They have a distinct advantage in that they have prior knowledge about productivity, problem areas, etc. that can help them direct some of their sampling. As farms have grown larger, much of the soil sampling has been done by agricultural consultants and fertilizer/agricultural chemical company personnel. These individuals may not always have this prior knowledge. Knowing the production history of a field can help guide sampling. This is part of the reason for the excitement of generating yield maps. The exact relationships of productivity as a basis for soil sampling pattern, however, still needs to be established by research. Sharing this information between the producer and an agricultural consultant or a fertilizer dealer would be helpful in developing sampling plans.

A number of recent papers have looked at sampling intensity (Hergert et al., 1995a,b; Franzen and Peck, 1995b; Wollenhaupt et al., 1997). Although there is no single recommended intensity, there is some agreement that sampling units at spacings above 196 to 230 ft causes a loss of information. Simply stated, the more points used to make a map, the better the map. If there is limited background information about a field, some type of systematic sampling is suggested. There are a number of different options, including aligned or unaligned grids. Variations include taking one large core from the square that represents the area, or possibly taking four to eight cores in a random pattern within the square.

Another factor to consider in the sampling pattern is the interpolation technique that will be used to produce a map from the data. Numerous interpolation techniques are available (Wollenhaupt et al., 1997). Most interpolation techniques accommodate data that are not equally spaced. This may be an advantage for some soil properties where there is some idea of the variability. For example, soil organic matter can be mapped based on a bare soil from using aerial photography. Intensive sampling of a corresponding location (Gotway et al., 1997) developed an excellent map with close correspondence to soil organic matter. However, using the bare soil photograph as a guide, a much smaller directed sampling could have been used to determine the same information. In the case of soil organic matter, taking soil samples close to boundaries provides additional information that when interpolated describes changes between soil zones.

For many other soil properties (P, K, pH), there is usually little background information about where to sample. In these situations the best sampling plan is to use some type of regular grid making sure that there are sufficient points to develop a good map. As mentioned previously, the sampling plan can influence the interpolation technique. Most interpolation techniques can handle non-uniformity spaced data.

As more sampling is done for SSM, the selection of which interpolation techniques to use becomes a factor. The selection of the interpolation technique probably will be much less important than taking a sufficient number of sample locations to produce a good map. Sampling intensities that require at least one sample point for each 1.25 acre would be ideal, although samples for areas up to 2.5 acre may be adequate. There is considerable debate concerning these issues, but the bottom line is that information does come at a cost. Therefore, to do a better job of managing a crop, it must be remembered that if you can not measure it, you can not manage it.

Line-intercept sampling for stand analysis

The LIS method for obtaining information about stands was adapted from long-established techniques used in forestry and wildlife biology (Kaiser, 1983). This method has been little used in row crop agriculture, but recently its use has been advocated for stand analysis and scouting for insect pests in early-season cotton (Willers et al., 1992; Williams et al., 1995). A closely related application has been the estimation of crop residue for erosion control. Samples are drawn from row segments of equal length on each row crossed by the transect line (Williams et al., 1995). If seed are broadcast, the method cannot be used as described here, because there will be no parallel rows. (For broadcast fields, a technique can be developed. If this is a need for any fields on your farm, contact the lead author, JLW, and a protocol will be developed). In this discussion, the method estimates only one attribute — the number of soybean plants per acre. Other attributes can be estimated, but the technique has not been modified for other uses in soybean production.

The first step in using LIS to estimate the stand of a soybean field is to divide the field into subunits (strata) where crop phenology is similar by applying concepts analogous to the use of stratified sampling plans. Here, these smaller divisions of a field are called management units. Depending upon the situation, a good range in size for a management unit is between 50 and 100 acre. It is recommended that at least one sample line be used per management unit and, if time is available, up to four lines per management unit. Generally, each line should be at least as long as the width of one planter pass and no longer than the width of four planter passes. As the row spacing of the crop gets narrower, it becomes more convenient to use only one line per pass. The other remaining requirement is that the sample lines should not overlap.

A starting point for a single sample line is chosen at random in the management unit. In soybeans, it is best if the starting point be midway between the first or last drill of one pass of the drilling machine (seeder or planter) and the first or last drill of the next adjacent pass. Sampling is conducted along the transect line across consecutive rows. The total number of plants in a fixed length of row is counted on each row crossed by the transect line. Typically, data are collected from 1- to 3-ft sections of row. If the number of plants per foot is large and there are not too many gaps in the drill, the 1-ft length is sufficient. If the number of plants per row is variable, then 3 ft should be used. If the stand is extremely sparse, one could even use a 5-ft sample length per row. The process of moving across rows gives this method its strength by capturing the variability in the crop due to planting irregularities that occur among planter boxes and other causes that vary the number of plants in each row.

The estimate of plants per acre depends on the variability of the stand in each row crossed by the transect line, the length of row sampled, and number of transect lines used per field. Do not change the length of row sampled on each row for any lines in a management unit; that is, do not sometimes use 1 ft of row and another time use 3 ft and another time use a 5-ft sample on the same line. Also, make sure the transect line is as straight as possible. Do not let the line become crooked or vary too far from being perpendicular to the row direction. Some people actually stretch a small cord or rope between two stakes and obtain a very straight line. Then, they lay one end of the yardstick or ruler against the rope on each row and make their count.

Table 17.5 LISs for Estimating the Number of Soybean Plants per Acre

Row	Sample variates	Sorted variates	Cumulative sum	Cumulative %
1	6	1	1	0.005
2	8	3	4	0.022
3	12	3	7	0.038
4	10	3	10	0.054
5	1	4	14	0.076
6	9	5	19	0.103
7	10	5	24	0.130
8	5	6	30	0.163
9	6	6	36	0.196
10	11	8	44	0.239
11	3	8	52	0.283
12	8	8	60	0.326
13	10	9	69	0.375
14	10	9	78	0.424
15	5	9	87	0.473
16	9	10	97	0.527
17	4	10	107	0.582
18	12	10	117	0.636
19	10	10	127	0.690
20	3	10	137	0.745
21	12	11	148	0.804
22	9	12	160	0.870
23	3	12	172	0.935
24	8	12	184	1.000

Average/3 ft = 7.67   Standard deviation/3 ft = 3.25   Range/3 ft = 1-12 plants
Average/1 ft = 2.56   Number row-feet/acre = 37,336   Estimated plants/acre = 95,580

Reported are the number of plants per row for each of 24 drills in one pass of the seeder, at 14-in. row widths. The fixed row length per sample on each drill was 3 ft. In practice, at least four drill passes per managment unit should be sampled. See text for further explanation and compare with Figure 17.14, which graphs the empirical distribution function (ECDF) of these data.

Calculations

If the rows are parallel, the only number needed is the total number of linear row-feet per acre for that row spacing. This value must be known in order to calculate a per-acre estimate of the number of soybean plants using the line-intercept method. Mississippi Cooperative Extension Service Bulletin Publication 883 reports the number of linear row per acre for several different row widths from 6 to 40 in.

Presented in Table 17.5 is an example data set of a 3-ft sample across 24 rows of a single pass of the seeder from a field of soybeans in the Mississippi Delta during 1996. The row spacing was 14 in., which means, in round figures, there are 37,336 linear row-feet per acre. In this 24-row sample, a total of 184 soybean plants were counted for this transect line. The total number of row-feet examined in the sample is 72 (24 rows × 3 ft/row). Dividing 184 plants by the total number of row-feet sampled gives an estimate of 2.56 plants per foot of row. This value when multiplied by the number of linear row-feet per acre for this row spacing provides an estimate of 95,580 plants per acre.

Table 17.5 also reports the final result and presents several descriptive statistics based upon the 3-ft sample units along the transect line. The table also shows the sample sorted

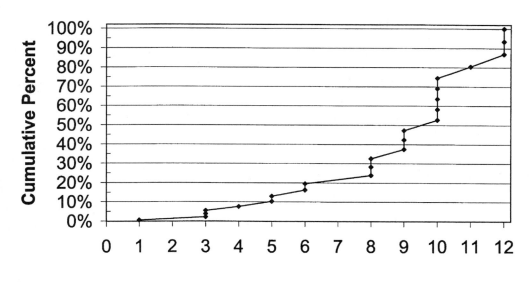

Figure 17.14 Diagrammatic distribution of stand variability for narrow-drill soybeans. Counts were obtained from one 24-row-long transect line where the number of plants was counted in each 3-ft length for each row. If the stand where uniform, the plot would be a single vertical line at one of the values for plants per 3 ft of row.

into ascending order, the cumulative sum of the values, and the **cumulative percentage**. The cumulative percentage is obtained by dividing the cumulative sum column by the total. The cumulative percentages can be graphed against the sorted observations (Figure 17.14). This graph represents the within-row variability in stand for the field. The median number of plants per 3 ft of row is the value on the x-axis where the cumulative percentage is 50. For these data (Table 17.5) the estimated median is 9.5 plants. The mode is 10 plants per 3 ft of row which can be seen easily in Figure 17.14 since this value has the most points. Note from Table 17.5 that the estimated average is 7.67 plants per 3 ft of row. When the mean is smaller than the median and the mode, the distribution of the sampled attribute is skewed left.

If several lines are collected in a management unit, the average for that unit is the average of the averages estimated for each line. For example, if four transect lines were sampled in one management unit and provided estimates (rounded to the nearest whole plant) of 88,165, 86,958, 91,355, and 89,753, the average using Equation 17.1 would be 89,058 soybean plants per acre. The estimated standard deviation after taking the square root of the result of Equation 17.2 is 1912. By the use of the information in Table 17.2, the estimate of the standard deviation is (91,355 − 86,958)/2.059 = 2135. Again, as before, the short-cut approach agrees favorably with the results given by the equation. Armed with these results, the farm manager can make appropriate management policy for any decision that involves knowledge of the stand. The method described here to obtain the stand estimate of soybeans is not difficult to employ. It is a simple process and does not require tremendous amounts of time to collect the data from one sample line whose length is the same as the width of one planter pass.

Compendium of soybean sampling literature

Several reference texts describing entomological sampling methods are available for soybeans. Two major reference books by Kogan and Herzog (1980) and Pedigo and Buntin (1993) are extremely thorough. McDonald and Manly (1989) have also written a technical text on sampling. A recent handbook edited by Higley and Boethel (1994) has appeared that includes a black-and-white pictorial key to insects and the damage they cause at different stages of soybean growth. Numerous color plates showing the adult and most immature life stages of soybean insects (including beneficial species) are displayed. The handbook also provides information on the life cycle and biology of soybean pests including range maps showing the distribution of the species in North America.

Sequential sampling plans for several insects that attack soybeans have been described. These plans include ones for green cloverworm (Hammond and Pedigo, 1976), velvetbean caterpillar (Strayer et al., 1977), and other defoliators (Bellinger and Dively, 1978). Waddill et al., (1974) describes a sequential plan for beneficial insects, including several species of the genera *Nabis* and *Geocoris*. Additional details about sequential sampling can be found in several chapters of Pedigo and Buntin (1993) and Kogan and Herzog (1980). Two bibliographies of sequential sampling plans for insects on several crops, including soybeans, are Fowler and Lynch (1987) and Pieters (1978).

From time to time, research groups put together bulletins on specialized topics. Long-standing examples are the various bulletins in the Southern Cooperative Series that are edited and authored by researchers affiliated with the agricultural experiments stations of different states. One handbook discussing the sampling of Heliothine moth pests on soybeans (along with cotton, corn, and other crops) is Southern Cooperative Series Bulletin No. 231 (1979). Another, Southern Cooperative Series Bulletin No. 377 (1994), describes soil sampling procedures for the southern region of the U.S. Peck and Melsted (1973) and Wollenhaupt et al., (1997) are other useful references on soil sampling.

Other publications such as state extension service information sheets and publications on numerous sampling issues are also available. The title and number of these small publications produced by the Mississippi Cooperative Extension Service (along with a few from other states) are listed at the end of the reference section.

Many sampling plans described from a general biology or wildlife point of view have been described in an excellent text by Thompson (1992). Discussed are basic concepts and simple random, stratified, cluster, survey, line-transect, and capture–recapture sampling plans. Spatial sampling plans (including a technique known as kriging) are also discussed. Also, several commercial companies are making information sheets that contain useful information. For example, The Potash & Phosphate Institute (Suite 110, 655 Engineering Drive, Norcross, GA 30092-2837; Telephone: 770-447-0335) has a small flyer entitled "Site-Specific Nutrient Management Systems for the 1990s" (Item #01-1180) that presents an overview and generalized recommendations for soil sampling for precision agriculture.

We have pointed out several times how computers are bringing dramatic changes in data collection, manipulation, and analysis. Similar changes are at work in the area of sampling. One method of sampling, as well as an analysis technique, is called the "Bootstrap." It is briefly discussed by Efron and Tibshirani (1991). This method, also known as resampling statistics, is popularized in a text by Simon and Bruce (1993). Resampling methods may prove to be extremely useful to those who do not have specialized training in statistics.

Other innovative sampling techniques are computer intensive (Schmitt, 1969; Plant and Wilson, 1985; Nyrop et al., 1986; Binns and Bostanian, 1990; Willers et al., 1990a,b;

Binns et al., 1996). Recently, adaptive sampling techniques applicable to rare populations have been described (Thompson, 1992). The chief advantage of many of these methods is that they utilize smaller-sized samples, or are more efficient at allocating sampling effort than many traditional methods but at the cost of a greater computational burden. Traditional methods use larger sample sizes, but often can be analyzed with a pencil and paper, or pocket calculator, hence, their greater popularity for the present time.

We anticipate that as speech recognition (Schindler, 1996), sensor development, computer advances, and other technologies continue to decrease in cost (McKinion, 1992), sampling techniques that are accurate, timely, and cost-efficient with small sample sizes will be developed. The trend toward the invention, adoption, and application of innovative sampling techniques will be pushed by increasing labor and chemical control costs.

Concluding remarks

We have described several concepts that require the use of little more than a pencil, ruler, graph paper, and handheld calculator. This information is what we hope you will begin to use. Therefore, you are encouraged to use some of these methods with data obtained from fields you manage.

Without a doubt, computers are going to be used more and more in soybean production in the years ahead. The development of better methods of retaining farm production records should also follow. Historical records that are complete, well documented, and retrievable are also sources that can be "sampled or mined" for information to produce better crops. However, to use best most material discussing the analysis of sample data, one should seek additional training and expert advice, and invest in more-sophisticated software packages. In the future, we anticipate that many techniques will be managed by "expert systems" that place the technical burden on the software and not the user. Users of such software (often called a decision support system) can instead focus their efforts on what is important — the production of an excellent crop at the lowest cost and in an environmentally acceptable manner.

There are three major points that we hope you never will forget. First, when embarking on a sampling effort, have a well-defined goal or question in mind. Understand the advantages and limitations of the sampling design used to answer your question. If you do not perform the sampling yourself, ask the person who collects the sample to tell you these things. If they do not know, ask that they find out and then let you know. The costs of production today are too high and profit margins too thin to let shoddy sample data be used to make management decisions. With respect to the sampling goal, establish well-defined sampling units and sample sizes and clearly know what attribute of the population is being sampled. In a sense, wed the biology of the attribute of interest to the appropriate sample plan. Be clear in your mind what population is being sampled. Second, be aware of variability. Do not combine samples carelessly, especially if the sampled attribute is being measured on distinct populations. Do not take an average of different attributes unless you can interpret what that average means. Third, learn and remember the distinction between the mode, median, and mean or average. If the sampled data are heavily skewed, which one of these three measures do you wish to use and which one should you use? By being wise in using and investing your sample resources, perhaps you may discover some hidden treasure in your own data that will help you better manage your soybean crop.

References

Bellinger, R. G. and G. P. Dively. 1978. Development of sequential sampling plans for insect defoliation on soybeans, *J. N.Y. Entomol. Soc.* 86:278–279.

Binns, M. R. and N. J. Bostanian. 1990. Robust binomial decision rules for integrated pest management based on the negative binomial distribution, *Am. Entomol.* 36:50–54.

Binns, M. R., J. P. Nyrop, and W. Van der Werf. 1996. Monitoring pest abundance by cascading density classification, *Am. Entomol.* 113–121.

Bray, R. H. 1929. A Field Test for Available Phosphorus in Soils. Ill, Agric. Exp. Stn. Bull. 337:589–602.

Buntin, G. D. 1993. Developing a primary sampling program, in *Handbook of Sampling Methods for Arthropods in Agriculture*, Pedigo, L. P. and G. D. Buntin, Eds., CRC Press, Ann Arbor, MI, 99–115.

Calvert, W. S. and J. M. Ma. 1996. *Concepts and Case Studies in Data Management*, SAS Institute, Inc., Cary, NC.

Cline, M. G. 1944. Principles of soil sampling, *Soil Sci.* 58:275–288.

Cochran, W. G. 1956. Design and analysis of sampling, in *Statistical Methods*, 5th ed., G. W. Snedcor, Ed., Iowa State College Press, Ames, 489–523.

D'Agostino, R. B. and M. A. Stephens. 1986. *Goodness-of-Fit Techniques*, Marcel Dekker, New York.

Efron, B. and R. Tibshirani. 1991. Statistical data analysis in the computer age, *Science* 253:390–395.

Fogiel, M., Ed., 1985. *The Statistics Problem Solver*, Research and Education Association, New York.

Fortner, B. 1995. *The Data Handbook. A Guide to Understanding the Organization and Visualization of Technical Data*, 2nd ed., Springer-Verlag, New York.

Fowler, G. W. and A. M. Lynch. 1987. Bibliography of sequential sampling plans in insect pest management based on Wald's sequential probability ratio test, *Great Lakes Entomol.* 20(3):165–171.

Franzen, D. W. and T. R. Peck. 1995a. Field soil sampling density for variable rate fertilization, *J. Prod. Agric.* 8:568–574.

Franzen, D. W. and T. R. Peck. 1995b. Sampling for site-specific application, in P. C. Robert, R. H. Rust, and W. E. Larson, Ed., *Site-Specific Management for Agricultural Systems*, 2nd Intl. Conf., ASA, CSSA, SSSA, Madison, WI, 535–551.

Freese, F. 1967. Elementary statistical methods for foresters, *Agric. Handb.* 317, U.S. Department of Agriculture Forest Service, Burgess Publishing Company, Minneapolis, MN.

Friendly, M. 1991. *SAS System for Statistical Graphics*, 1st ed., SAS Series in Statistical Applications, SAS Institute, Cary, NC.

Gauch, H. G. 1992. *Statistical Analysis of Regional Yield Trials. AMMI Analysis of Factorial Designs*, Elsevier, Amsterdam.

Gazey, W. J. and M.J. Staley. 1986. Population estimation from mark-recapture experiments using a sequential Bayes algorithm, *Ecology* 67:941–951.

Gelman, A., J. B. Carlin, H. S. Stern, and D. B. Rubin. 1995. *Bayesian Data Analysis*, Chapman & Hall, London.

Giesler, R. and U. Lundström. 1993. Soil solution chemistry: effects of bulking soil samples, *Soil Sci. Soc. Am. J.* 57:1283–1288.

Gonick, L. and W. Smith. 1993. *The Cartoon Guide to Statistics*, HarperCollins, New York.

Gotway, C. A., R. B. Ferguson, G. W. Hergert, and T. A. Peterson. 1996. Comparison of kriging and inverse-distance methods for mapping soil parameters, *Soil Sci. Soc. Am. J.* 60:1237–1247.

Gotway, C. A., R. B. Ferguson, and G. W. Hergert. 1997. The effects of mapping and scale on variable rate fertilizer recommendations for corn, in P. C. Robert and W. E. Larson, Eds., *Site-Specific Management for Agricultural Systems, Third Intl. Conf.*, ASA, CSSA, SSSA, Madison, WI.

Hammond, R. B. and L. P. Pedigo. 1976. Sequential sampling plans for the green cloverworm in Iowa soybeans, *J. Econ. Entomol.* 69:181–185.

Hergert, G. W., R. B. Ferguson, and C. A. Shapiro. 1995a. Fertilizer Suggestions for Corn, University Nebraska NebGuide G74–174 (Revised).

Hergert, G. W., R. B. Ferguson, C. A. Shapiro, E. J. Penas, and F. B. Anderson. 1995b. Classical statistical and geostatistical analysis of soil nitrate-N spatial variability, in P. C. Robert, R. H. Rust, and W. E. Larson, Eds., *Site-Specific Management for Agricultural Systems, Second Intl. Conf.*, ASA, CSSA, SSSA, Madison, WI, 175–186.

Hergert, G. W., W. L. Pan, D. R. Huggins, J. H. Grove, and T. R. Peck. 1997. The adequacy of current fertilizer recommendations for site specific management, in F. J. Pierce and E. J. Sadler, Eds., *The state of Site-Specific Management for Agriculture*, ASA, CSSA, SSSA, Madison, WI, 283–300.

Higley, L. G. and D. J. Boethel. 1994. *Handbook of Soybean Insect Pests*, Entomological Society of America, Lanham, MD.

James, D. W. and K. L. Wells. 1990. *Soil sample collection and handling: technique based on source and degree of field variability*, in R. L. Westerman, Ed., *Soil Testing and Plant Analysis*, 3rd ed., SSSA, Madison, WI, 25–44.

Kaiser, L. 1983. Unbiased estimation in line-intercept sampling, *Biometrics* 39:965–976.

Kitchen, N. R., J. L. Havlin, and D. G. Westfall. 1990. Soil sampling under no-till banded phosphorus, *Soil Sci. Soc. Am. J.* 54:1661–1665.

King, J. R. 1980. Frugal Sampling Schemes, Technical and Engineering Aids for Management, Tamworth, NH.

Kogan, M. and D. C. Herzog, Eds., 1980. *Sampling Methods in Soybean Entomology*, Springer-Verlag, New York.

Little, T. M. and F. J. Hills. 1978. *Agricultural Experimentation. Design and Analysis*, John Wiley and Sons, New York.

Ludwig, J. A. and J. F. Reynolds. 1988. *Statistical Ecology. A Primer on Methods and Computing*, Wiley Interscience, New York.

Matheron, G. 1971. The theory of regionalized variables and its application, Cah. Cent. Morphol. Math. Fontainebleau 5. Centre de Geostatistique.

McKinion, J. M. 1992. Getting started: basics of modeling strategies, in J. L. Goodenough and J. M. McKinion, Eds., *Basics of Insect Modeling*, ASAE Monograph No. 10., 1–8.

McDonald, L. L. and B. F. J. Manly. 1989. Calibration of biased sampling procedures, in L. McDonald, B. Manly, J. Lockwood, and J. Logan, Eds., *Estimation and Analysis of Insect Populations*, Springer-Verlag, Berlin.

Miller, I. and J. E. Freund. 1977. *Probability and Statistics for Engineers*, 2nd ed., Prentice-Hall, Englewood Cliffs, NJ.

Milliken, G. A. and D. E. Johnson. 1984. *Analysis of Messy Data*, Vol. 1, *Designed Experiments*, Van Nostrand Reinhold, New York.

Milliken, G. A. and D. E. Johnson. 1989. *Analysis of Messy Data*, Vol. 2, *Nonreplicated Experiments*, Van Nostrand Reinhold, New York.

Morgan, M. F. 1932. Microchemical Soil Tests, Connecticut Agric. Exp. Stn. Bull. 333.

Nyrop, J. P., R. E. Foster, and D. Onstad. 1986. Value of sample information in pest control decision making, *J. Econ. Entomol.* 79:1421–1429.

Parkin, T. B., J. J. Meisinger, S. T. Chester, J. L. Starr, and J. A. Robinson. 1988. Evaluation of statistical estimation methods for log normally distributed variables, *Soil Sci. Soc. Am. J.* 52:323–329.

Peck, T. R. and S. W. Melsted. 1967. Field sampling for soil testing, in M. Stelly, Ed., *Soil Testing and Plant Analysis,* Part 1: Soil Testing, SSSA, Madison, WI, 25–35.

Peck, T. R. and S. W. Melsted. 1973. Field sampling for soil testing, in L. M. Walsh and J. D. Beaton, Eds., *Soil Testing and Plant Analysis*, SSSA, Madison, WI, 67–75.

Peck, T. R. and P. M. Soltanpour. 1990. Principles of soil testing, in R. L. Westerman, Ed., *Soil Testing and Plant Analysis*, 3rd ed., SSSA, Madison, WI, 3–9.

Pedigo, L. P. and G. D. Buntin, Eds., 1993. *Handbook of Sampling Methods for Arthropods in Agriculture*, CRC Press. Boca Raton, FL.

Peterson, R. G. and L. D. Calvin. 1982. Sampling, in A. Klute, Ed., *Methods of Soil Analysis*, Part 1: Agronomy, 2nd ed., 9:33–51.

Pieters, E. P. 1978. Bibliography of Sequential Plans for Insects, *Bull. Entomol. Soc. Am.* 24(3):372–374.

Plant, R. E. and L. T. Wilson. 1985. A Bayesian method for sequential sampling and forecasting in agricultural pest management, *Biometrics* 41:203–214.

Randall, G. W. 1982. Strip tillage systems-fertilizer management, in *Farm Agric. Resources Management Conf. on Conservation Tillage*, Iowa State University Est. Publ. CE-1755.

Reed, J. F. and J. A. Rigney. 1947. Soil sampling from fields of uniform and non-uniform appearance and soil types, *J. Am. Soc. Agron.* 39:26–40.

Reuss, J. O., P. N. Soltapour, and A. E. Ludwick. 1977. Sampling distributions of nitrates in irrigated fields. *Agron. J.* 69:588–592.

Ruesink, W. G. 1980. Introduction to sampling theory, in M. Kogan and D. C. Herzog, Eds., *Sampling Methods in Soybean Entomology*, Springer-Verlag, New York.

Sabbe, W. E. and D. B. Marx. 1987. Soil sampling: spatial and temporal variability, in *Soil Testing: Sampling, Correlation, Calibration and Interpretation*, SSSA Spec. Publ. 21. SSSA, Madison, WI, 1–14.

Sawyer, J. E. 1994. Concepts of variable rate technology with considerations for fertilizer application, *J. Prod. Agric.* 7:195–201.

Schindler, E. 1996. *The Computer Speech Book*, Academic Press, Boston.

Schmitt, S. A. 1969. *Measuring Uncertainty: An Elementary Introduction to Bayesian Statistics*, Addison-Wesley, Reading, MA, 400 pp.

Simon, J. L. and P. C. Bruce. 1993. *The New Biostatistics of Resampling*, Duxbury Press,

Spiegel, M. R. 1962. *Statistics. Schaum's Outline Series*, McGraw-Hill, New York.

Strayer, J., M. Shepard, and S. G. Turnipseed. 1977. Sequential sampling for management decisions on the velvetbean caterpillar on soybeans, *J. Ga. Entomol. Soc.* 12:220–227.

Thompson, S. K. 1992. *Sampling*, Wiley-Interscience, John Wiley and Sons, New York, 343 pp.

Truog, E. 1930. The determination of the readily available phosphorus in soils, *J. Am. Soc. Agron.* 22:874–882.

Waddill, V. H., B. M. Shepard, S. G. Turnipseed, and C. R. Carner. 1974. Sequential sampling plans for *Nabis* spp. and *Geocoris* spp. on soybeans, *Environ. Entomol.* 3:415–419.

Willers, J. L., D. L. Boykin, J. M. Hardin, T. L. Wagner, R. L. Olsen, and M. R. Williams. 1990a. A simulation study on the relationship between the abundance and spatial distribution of insects and selected sampling schemes, in *Proceedings, Applied Statistics in Agriculture*, Kansas State University, Manhattan, KS, 35–45.

Willers, J. L., R. L. Olson, M. R. Williams, and T. L. Wagner. 1990b. Developing a Bayesian approach for estimating the proportion of cotton plants at risk to insect attack, in *Proceedings, Beltwide Cotton Production Research Conferences*, Las Vegas, NV, 246.

Willers, J. L., S. R. Yatham, M. R. Williams, and D. C. Akins. 1992. Utilization of the line-intercept method to estimate the coverage, density, and average length of row skips in cotton and other row crops, in *Proceedings, Applied Statistics in Agriculture*, Kansas State University, Manhattan, KS, 48–59.

Williams, M. R., T. L. Wagner, and J. L. Willers. 1995. Revised Protocol for Scouting Arthropod Pests of Cotton in the Midsouth, Tech. Bull. 206, Mississippi Agricultural and Forestry Experiment Station, Mississippi State.

Wollenhaupt, N. C., D. J. Mulla, and C. A. Gotway Crawford. 1997. Soil sampling and interpolation techniques for mapping spatial variability of soil properties, in F. J. Pierce and E. J. Sadler, Eds., *The State of Site-Specific Management for Agriculture*, ASA, CSSA, SSSA, Madison, WI, 19–54.

Extension Service and Other Agricultural Publications

Lime Needs — Illustrated. Lime Increases Yield & Profit. Mississippi Cooperative Extension Service, Publ. 720.

Monitoring Soybeans for Insect Pests. Mississippi Cooperative Extension Service, Publ. 1498.

Soil Testing for the Farmer, Mississippi Cooperative Extension Service, Inf. Sheet 346.

Soybean Cyst Nematode, Mississippi Cooperative Extension Service, Publ. 1293.

Soybean Insect Control, Mississippi Cooperative Extension Service, Publ. 883.

Soybean Looper: Biology and Approaches for Improved Management, Mississippi Cooperative Extension Service, Inf. Sheet 1400.

Soybean Seedling Diseases, Mississippi Cooperative Extension Service, Inf. Sheet 1167.

Soybeans: Doublecropping Soybeans after Wheat in Mississippi, Mississippi Cooperative Extension Service, Publ. 1380.

Soybeans: Efficient Production Practices, Mississippi Cooperative Extension Service, Publ. 1559.

Soybeans: Plant Populations and Seeding Rates, Mississippi Cooperative Extension Service, Publ. 1194.

Soybeans: Planting Guidelines for Mississippi, Mississippi Cooperative Extension Service, Publ. 1289.

Stem Canker of Soybean, Mississippi Cooperative Extension Service, Publ. 1827.

Sterling, W. L., Ed., 1979. Economic Thresholds and Sampling of *Heliothis* Species on Cotton, Corn, Soybeans and Other Host Plants, Southern Cooperative Series Bull. 231. Department of Agricultural Communications, Texas A & M University, College Station, TX 77843.

Thom, W. O. and W. Sabbe, Eds., 1994. Soil Sampling Procedures for the Southern Region of the United States. Southern Cooperative Series Bull. 377, Kentucky Agricultural Experiment Station, Department of Agricultural Communications, University of Kentucky, Lexington, KY 40546.

Weeds of the Southern United States, Mississippi Cooperative Extension Service, No. 2500-8-75.

chapter eighteen

Dandy plumbing designs

Harry F. Hodges and K. Raja Reddy

Contents

Introduction

Knowledge of movement of water and nutrients in plants is vital to understanding the way plants grow in response to the environment. The complexity of the processes and forces that cause movement of different materials throughout the plant has been a mystery until recent times, and even now some aspects of the processes are still not understood. The integral way plant cell structures function with several plant processes is intriguing and only recently has become explainable as related to plant growth and development.

      The engineering requirements for plants to move materials are very complex and demand an impressive plumbing system to meet the needs of various organs. Consider the roots (water- and mineral-gathering system) and the extensive network of veins in the leaves. Actually, growing plants have a plumbing system that is continually under construction, while at the same time serving as the vital linkage between the roots, leaves, and other growing organs. Leopold and Kriedeman (1975) observed that if one were to design the translocation system for a plant, it would have to have the following characteristics.

1. It must be capable of moving carbohydrates up to the top of the stem, where growth is most active, and down to the lower stem and roots. In short, it must be bidirectional.
2. It must be able to move (translocate) carbohydrates to growing sites in both the shoot and the root while moving water, minerals, and hormones from the roots to the leaves. In short, these systems must be separated from one another.
3. It must be capable of moving highly concentrated solutions, which are difficult for living cells to handle. The concentration of carbohydrates in the translocation stream is so high that it will cause most plant cells to plasmolyze (shrink) if in contact with them.
4. The translocation system utilizes pressure to cause the flow of water and nutrients in the plant. Any break could destroy this internal pressure with lethal consequences unless there is a means of plugging the leak. Therefore, it must be capable of mending breaks in the system so that removal of a leaf, or part of a stem, does not destroy the whole system.
5. It must be sufficiently flexible to tolerate movement caused by wind.

There are two major translocating systems in plants. The **xylem** moves water, minerals, and hormones from the roots to the leaves. The **phloem** moves both organic and inorganic nutrients in either direction. The movement of solutions in the plant plumbing system is governed by two factors: the driving force (pressure) and the conductance of the flow path.

The water conduit

The driving force for movement of water in any system originates from differences in solute (dissolved substances) concentration and pressure. The differences in solute concentration in the xylem are minor and inconsequential; therefore, the primary driving force controlling water movement in the xylem is differences in pressure. Solid surfaces, principally cell walls, exert forces on water that express themselves as local tensions. These tensions arise from the effects of surface tension at air–water interfaces and attractive forces at solid–water interfaces. At the water contents found in cell walls, these surface effects can be described as negative pressures (Boyer, 1985). Measurements of water potentials are always compared with the reference water, which is pure free water at atmospheric pressure and a water potential of zero. One may visualize the concept of negative pressures associated with these systems by thinking of what happens to a drop of water on a napkin. The water is associated with the cellulose fibers and would require considerable force to remove it from those fibers. However, if the napkin was placed in a powerful centrifuge and force applied, the water could be removed similarly to the way free water is removed by the spin cycle in a home washing machine.

The path of water conductance

Leaves

Plant leaf surfaces appear relatively simple at a glance. However, one may be impressed if leaf surface details and the complexity of the processes that occur beneath those surfaces are considered. First of all, the shapes and arrangement of leaves are designed to capture solar radiation with a minimum use of plant energy. The large flat or nearly flat surfaces are ideal for intercepting light, and yet they cost relatively little energy to make. The surfaces are covered with cutin, and a thin waxy layer that acts as an interface between the leaf and its environment. The cuticle and waxy layer gives the leaves the necessary support, acts as an outer skeleton, and retards water loss. The leaf surface is made up of many regularly spaced cells in which small openings called *stomata* are interspersed (Figure 18.1A). These stomata have guard cells on either side that expand and contract in

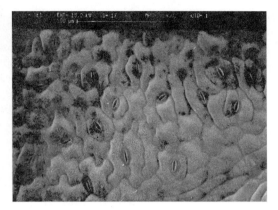

a

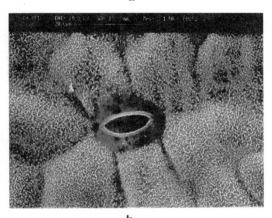

b

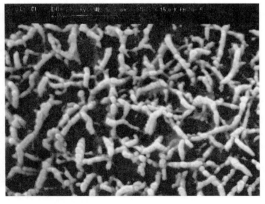

c

Figure 18.1 Different magnifications of the lower surface of a soybean leaf. A, 400× magnification of lower leaf surface showing several open stomata and guard cells. B, Lower surface of soybean leaf showing one stomata and surrounding cells. Magnification 1500×. (Stomates open in light, but may close due to low turgor. Atmospheric carbon dioxide enters the leaf via the open stomata and water vapor escapes through the same stomata.) Surface texture of area surrounding the stomata is wax. C, Higher magnification of leaf surface (10,000×) showing details of wax rods. Waxy layer reduces unregulated water loss from the leaf. Water loss through the wax and cuticle layer is 10 to 20%. Many herbicides, fungicides, and insecticides have to make their way through the wax and cuticle layer into the leaf to exert an action.

response to environmental conditions and water content. There are approximately 22,000 stomata per square inch of soybean leaf surface, with about three times more openings on the bottom than on the top (Carlson, 1973). These stomata work in a coordinated fashion to permit water loss to the atmosphere (evaporative cooling of the plant) and simultaneously allow uptake of atmospheric carbon dioxide (CO_2), the raw material used in photosynthesis for food production. The water lost through stomata is replaced by water removed from the soil. This process induces a water flow from the soil to the atmosphere that carries essential minerals throughout the plant.

Plants do not have muscles as animals do. Therefore, chemically induced pressure must be used to accomplish movements that are similar to those animals might do with contractions. The guard cells on either side of a stomata cause the stomata to open in light and to close in darkness or in water-deficit (drought stress) situations (Figure 18.1B). In daylight, solar energy activates a potassium pump in the guard cells, which causes potassium to move from surrounding cells into the guard cells. This process increases the salt concentration in guard cells and a simultaneous increase in water content that results in turgid guard cells and open stomata. The guard cells have cross-wall-like structures that cause them to become shaped somewhat like a kidney bean when turgid (not visible in Figure 18.1). Such a shape opens a pore (stomata) to the inside of the leaf. In darkness, the potassium pump stops, both potassium and water diffuse to equilibrium, the guard cells lose pressure, and the stomata close.

Open stomata allow air and moisture to move in and out of the leaf. Carbon dioxide from the air is absorbed on the inner surfaces of the leaf and diffuses to chloroplasts in nearby cells. In the chloroplasts, the carbon dioxide is reduced to sugars by energy obtained from the sun. As carbon dioxide is diffusing into the leaf from a high concentration in the atmosphere to low concentrations at the chloroplasts, water vapor diffuses out of the leaf.

The flow of water vapor through the stomata into the atmosphere is due to a concentration of essentially 100% saturated air inside the leaf, caused by the wet cell surfaces, to the much lower water vapor concentration that occurs in the bulk air outside the leaf. The actual rate of water vapor flow in well-watered plants depends primarily on the leaf temperature, relative humidity of the outside air (atmospheric demand), and the wind. On high-water-use days, nearly *0.5 in. of water per acre* may flow through the leaves of soybeans. How that flow of water gets from the soil through the whole plant system and out into the atmosphere depends on several unique properties of water and the plant structural characteristics. About 90% of the water loss is through the stomata. The rest is lost through the cutin and waxy layer associated with other epidermal cells (Figure 18.1B).

Wax is formed below the leaf surface and is pushed through pores in the cutin, thus forming tiny rodlike structures on the leaf surface (Figure 18.1C and 18.2B). The structure of these wax rods and their density on the leaf varies widely among plant species. Their density is also affected by environmental conditions, but relatively little is known about plant waxes, their chemistry, and the physiological conditions that cause their production. They effectively limit water loss from nonstomatal portions of leaf surfaces, and may play an important role in the way herbicides and other chemicals are absorbed.

One of the unique physical properties of water is its strong cohesiveness, caused by the tendency of water molecules to hold together. Water molecules are strongly attracted to each other giving it unusually high tensile strength. This unusual property of water allows it to be drawn (pulled) along the walls of cells located between the stomata and the xylem (Figure 18.2A). The water also adheres to the cellulose along the path. The flow of water from the xylem to the atmosphere is much like water in a sponge that has one side in water while the other side is exposed to the sun and wind. Water moves through

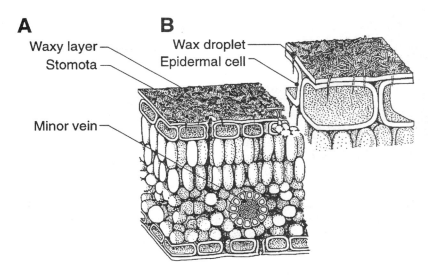

Figure 18.2 A, A diagrammatic sketch of a leaf in cross section showing the cutin and waxy layer on the surface. A stomata in both upper and lower surface, and the internal cell organization including a cross section of a minor vein. Water moves by capillary flow in the cell walls between the xylem cells in the minor vein and the stomata. Sugars produced in the loosely organized cells diffuse to the minor vein where they can be pumped into the phloem sieve element for translocation to other parts. B, An enlarged diagram of the outer cells showing cutin and waxy layer. Wax droplets are formed in the epidermal cells and pushed to the outside as tiny rods. See Figure 18.1B and C.

the fibrous cell walls, keeping the space between cells and under the stomata at near 100% relative humidity. The water vapor escapes through the stomata to the outside atmosphere and is replaced by water pulled into the cell walls because of the attraction of the water molecules to each other. This also resembles the use of a drinking straw.

The water is supplied to the leaf by the xylem. So, the water is pulled from the xylem along the maze of cellulose walls to evaporating surfaces. In a thin, confined column of water, such as that found in the xylem vessels of plants, the high tensile strength allows water to be pulled to the leaf surfaces.

The flow of water through the plant system accomplishes three major functions. First, the flow of water from the soil to the leaves provides vital minerals for plant growth. Second, the plant cells are kept full of water, which allows the plant to keep its shape. Third, water causes cooling due to evaporation. The evaporation of 1 g of water releases 580 cal of energy. Leaves that are wilted or nearly wilted on a hot summer day may be several degrees warmer than the surrounding air temperature, while leaves of well-watered plants typically are 5 to 10°F cooler than air. The cooling effect is caused by evaporation of water from the leaf surface. This process is somewhat similar to the sweating process that occurs in animals.

Stems

The xylem vessels through which water rises are made up of small-diameter, thick-walled dead cells that are resistant to collapsing even under high tension (Figure 18.3). The protoplasm of these cells breaks down and disappears before they become functional. These cells have small holes in the side walls and are spaced end to end with much of the end walls degraded, or open, so that they effectively form tubes with sievelike plates between cells. The resistance to water flow in xylem cells is related to their diameter, but

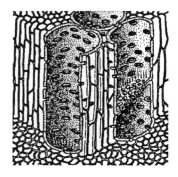

Figure 18.3   A diagrammatic sketch of the two xylem (water-conducting) cells. They are thick-walled cells that resist collapsing when under tension. These cells are aligned so that continuous tubes are formed between the roots and leaves. The side walls have tiny perforations so that if one tube becomes blocked, water may bypass the blockage.

resistance in the xylem is small relative to the resistance across the cells between the root epidermis and the xylem. In other words, it takes more tension to pull water from the soybean root epidermis (root outer surface) to the xylem than it takes to pull the water the rest of the way through the plant.

The stem of a soybean seedling has developing vascular cells in strands around a central pith tissue (Figure 18.4B). These strands develop with the xylem closer to the center and the phloem cells closer to the outside. As the plant matures, dividing cells between the xylem and phloem keep adding new vascular cells. Near the growing tip, these cells have walls that are subject to expansion, but as they mature the walls become thickened and pitted. The vascular strands are so numerous that they nearly form a ring of structures near the outer surface of the stem. Mixed among these vascular cells are fiber cells that have thick heavy walls, which provide strength to withstand wind and gravitational forces.

In hot dry conditions, plants may develop such severe water tensions in the xylem that *embolisms* (or air bubbles) occur in the xylem cells. An air bubble is elastic and stretches under tension, so such embolisms render the xylem ineffective. Tyree and co-workers (Tyree and Sperry, 1989) have demonstrated that such interruptions of the transpiration stream do occur. They even found that by placing an acoustical listening device on the stem of water-stressed plants during a high-evaporative-demand day, snapping sounds could be detected. These snapping sounds were assumed to be caused by the formation of air bubbles inside the xylem.

The formation of such an embolism may result from (1) a coalescing of microbubbles due to extreme tension on the water column or (2) an invasion of an air bubble through a tiny pore in the side wall of the xylem cell. The tiny pore will not allow air through under normal pressures, but will allow air to pass under conditions that cause extreme negative pressure or tension. This is because the size of the air bubble shrinks as the pressure decreases. If a cavitation occurs and the resulting embolism renders a xylem vessel ineffective, the scope of the damage is usually limited to one or only a few cells due to the perforated end plates between cells. Water flow and continuity of water may be only marginally reduced because of redundancy of the vascular system and the hydrated cell walls. Plants can recover from a small number of such embolisms when tension is removed, as occurs in a humid overnight environment or if the plants are watered. Recovery is accomplished through absorption of the air bubble by water from the surrounding cells when tensions are low.

In separate studies, Boyer (1971) found that the recovery was slow and incomplete when sunflower plants were subjected to water stress that resulted in a leaf water tension of about −20 bars, and then rewatered. In that experiment, he was probably inducing air embolisms in the transport system due to insufficient water, and the plants were unable to recover. This phenomenon has been observed in several plant species, including soybeans.

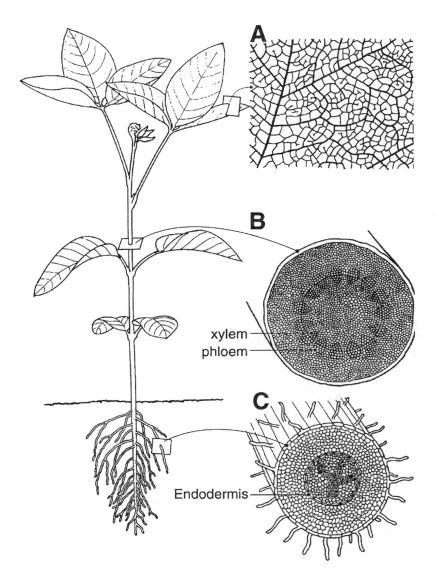

Figure 18.4 A diagrammatic sketch of an immature soybean plant. A, A sketch of a soybean leaflet illustrating the distribution of the plumbing system in the leaves. It is estimated that each leaf cell is within three to five cells of a minor vein. B, A diagram of a soybean stem in cross section. The xylem and phloem cells are the major parts of the vascular bundles. Packed throughout the vascular bundle are thick-walled fiber cells that provide strength. C, A diagrammatic sketch of a young root in cross section. Root hairs are an extension of the epidermal cells. Below the outer layer of cells in a zone of loosely spaced cells followed by a ring of highly specialized cells called an *endodermis* or *inner skin*. The endodermis forms the most important barrier that water and nutrients must overcome to reach the vascular cells. Thick-walled, dark-appearing cells in the center are the xylem, and interspersed near the edges of the xylem are phloem cells.

Roots

The vascular system in roots is surrounded by a layer or ring of cells called the *endodermis* (meaning the inside skin) that has a rather thick layer of water-flow-resistant suberin in its walls (Figure 18.4C). That ring of cells is further surrounded by a zone of loosely packed nondifferentiated cells called a cortical region and finally by an epidermal cell layer. Water and nutrients are thought to move readily both in the cell walls and from cell to cell through the living membranes through the outer several layers of cells. At the endodermis, flow through the cell walls is interrupted. All water and nutrients must go through the living cytoplasm of that layer of cells. From there they go to the xylem, and water may continue to move through the porous cell walls as well as through the living membranes. Once water gets to the xylem, the open tubes provide a clear passage for bulk flow toward the site with the least pressure.

Flow through the living portion of the cells at the endodermis provides a regulatory point that is sensitive to a number of external factors. Water and nutrient uptake by roots is sensitive to temperature, oxygen, salinity or poisons, and *mycorrhiza* (fungus often associated with roots and that facilitates mineral uptake). It is at the endodermis that these external factors affect the root effectiveness. It is the endodermis that provides the major source of resistance between the leaf stomata and the soil water.

Root growth appears to be highly sensitive to both environmental conditions and genetically controlled factors. By genetic control, we are referring to differences among species and varieties, and a tendency of plants to allocate different amounts of their food supply to roots at different times. In general, root growth reflects aboveground growth. Conditions favorable for aboveground growth also result in root growth. Young plants allocate a large percentage of their dry matter production to root formation. As plants age, a lower percentage of the dry weight goes to roots. During the period of rapid seed growth, little new root growth is produced.

Young roots are where the uptake activity occurs. It is believed that most of the water and nutrient uptake occurs in roots that are only hours or days old. Uptake activity of roots 1 week or more old may be severely limited compared with that of younger roots. Growth and exploration of new soil regions by the fine roots and root hairs growing between soil particles occurs continuously.

Whole plant growth is regulated in part by the trade-off relationship between the roots and aboveground parts. Root growth is normally limited by available carbohydrate or energy. Fine-tuning the balance of growth among stems, leaves, and roots may be considered as a trade-off of water (provided by roots) for carbohydrate (provided by the leaves) and is described in more detail in a later section.

The driving force

The key to understanding water movement through plants is to have an appreciation of the great capacity of dry air to hold water vapor. As the relative humidity of air drops below 100%, its affinity for water increases dramatically. At 100% relative humidity, the water potential (suction) at 68°F is 0; at 98% relative humidity, the water vapor pressure drops to −27.5 bars (an equivalent pressure would support a water column 843 ft high). At 50% relative humidity and 68°F, the water potential of the atmosphere is −944 bars. An equivalent pressure would support a water column over 2800 ft high. Thus, air does not have to be very dry to establish an extremely steep water potential gradient. It is this water potential gradient or demand of dry air for water that pulls water from the soil through the plant plumbing system.

The plant complex includes the space between leaf cells (substomatal cavities), leaf cell walls between the stomatal cavities, the extensive vascular network of the leaf, the petiole, and stem xylem (the vessels extending all the way to the roots), and the root cells between the vessels and the root epidermis. To move water through this complex, a pressure gradient, which determines water transport and use, between the atmosphere and soil solution is established. When the soil is well watered, only a small suction from the roots is required to pull the water into the plant. As it becomes drier, the remaining soil water is held with increasingly greater forces so that a greater suction from the atmosphere is required for water extraction. For water to move via the cohesion mechanism described earlier in this chapter, the driving force or the gradient in water suction from the atmosphere to the soil through the plant must exceed the pressure difference required to lift and move the water from the soil to the atmosphere. If cohesion is operating, tension must exist in the stem water.

Measuring tension

If a tension exists in the vascular system of plants, there must be a way to measure it. In 1969, scientists described an elegantly simple technique that has become known as the pressure-bomb method. Removing a leaf from the plant interrupts the xylem water. Such an interruption results in the water column snapping back into the leaf petiole. The pressure-bomb technique requires removing a leaf or leaflet from the plant and placing the leaf in a pressure chamber with the petiole extending into the outside atmosphere. A pressure-tight seal can be made around the petiole with a compression washer without damaging the vessels. As pressure is slowly added to the chamber, the leaf cells are squeezed and water in the vessels is gradually forced back toward the cut end of the petiole. The pressure required to bring the water to the point that it was at the time of the cut is considered to be equivalent to the tension on the water in the vessel at the time the leaf was severed. This has been done repeatedly in a large number of environmental conditions and on many plant species. The evidence that these general concepts are valid is overwhelming, and the relationship of several physiological processes appears to be intimately associated with leaf water potential (Salisbury and Ross, 1978; Boyer, 1985).

Leaf water potential and effects on physiological processes

Crop growth and productivity are greatly dependent on water status. A young, well-watered soybean plant typically has a leaf water potential of about –2 bar at dawn. As the sun rises, the leaf water potential decreases to –12 to –15 bar at solar noon (even when well watered). As solar radiation decreases, the leaf water potential returns toward 0 until it reaches about –2 bar the next morning. Boyer (1970) related photosynthesis, transpiration (water use), and leaf growth to water potential of soybean plants grown in a controlled environmental chamber, and found that the most responsive of those processes was leaf growth. Photosynthesis and transpiration appeared to be related to stomatal closure and were at approximately 50% of maximum values when the leaf water potential was about –12 bar. Leaf growth was 50% of maximum values when the leaf water potential values were only about –3 to –4 bar. Physiological processes of field-grown plants responded similarly to plants grown in controlled environmental chambers. Stem elongation and plant height responded similarly to leaf growth.

This means that leaf and stem growth stops much more quickly in a dry-down cycle caused by water deficits than do either photosynthesis or transpiration. Photosynthesis and transpiration were reduced as the stomata closed. Stomata closed because of low water pressure in the guard cells. Since leaf growth slows earlier because of water deficits, there is a sparing of sugars that might otherwise be used for leaf growth. Some of the spared sugar is translocated to the roots so that in marginally dry conditions more root growth occurs. That is, a reprioritization of resources occurs in marginally dry conditions causing production of more roots. This allows roots to explore more soil and alleviate or delay the effect of drought. The remainder of the spared sugar accumulates in the tissues as plants undergo moderate to severe drought stress. In fact, a common characteristic of drought-stressed plants is higher than normal concentrations of sugars in the leaves. As the time from the last rain or irrigation becomes longer, leaf water potential values become more negative, the plants close their stomata, and wilt. Timely irrigation alleviates these water-limited growth processes.

Leaf growth and water deficits

Leaf growth is obviously closely linked with canopy development and efficient light interception. The obvious goal of soybean producers is to develop a healthy, vigorous leaf canopy quickly that captures as much solar radiation as possible, since conversion of solar energy is the source of crop yield. Canopy development occurs by adding new leaves to both the mainstem and branches. Each of these leaf types has its own rate of development, but their growth also depends on water potential, nutrients, and temperature. Even though the leaf water potential declines every day as the sun rises and the atmospheric humidity drops, in well-watered conditions the leaf continues to grow. Midday growth may be slowed for a short period when the water potential is at its lowest. Leaf growth over an extended time period is able to compensate for mild water deficit conditions and show only small reductions. If soil moisture is limited, however, the water potentials become too negative to allow leaf growth. When soil moisture is marginally limited, much of the leaf expansion occurs at night when the atmospheric evaporative demand is less and cell pressure is sufficient to stretch the walls of young leaves. As soil water becomes severely depleted, even nighttime expansion of leaves is stopped.

Cell expansion

Cell growth occurs as a combination of several forces acts on the young cell. The protoplasm of the cells of a young plant is encased in a thin matrix of cellulose somewhat like a cardboard box. As the protoplasm absorbs minerals, sugars, and other organic compounds, the combined effect of these chemicals attracts an inflow of water. The additional inflow of water increases the size of the protoplasm much as increased air inflates a balloon. Increased pressure on the interior walls of the cellulose matrix causes those fibers to slip, stretch, and expand the size of the box (cell). The cellulose fibers are stretchy and plastic for only a limited period, after which they become rigid and set so that no further expansion can occur. Secondary wall growth may occur after expansion has ceased so that the walls become thicker, heavier, and more rigid. Expansion of a soybean leaf occurs over a 14- to 16-day period in optimum conditions, but may take longer in less than optimum conditions. Therefore, several leaves on a plant grow simultaneously. Usually the top three or four leaves on the mainstem of vegetatively growing plants are expanding, in addition to some young branch leaves.

Solute conduit

Moving energy and building blocks

Clearly, the engineering requirements for the water translocation system of plants are demanding, but movement of solutes (substances that are produced in leaves dissolved in water) also demands a sophisticated design. How the design meets the plant and environmental requirements and accomplishes the complicated tasks is fascinating and impressive. The material being transported is largely sugar, but amino acids (the building blocks for proteins), hormones, and other substances, including some herbicides, are also

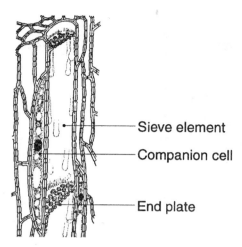

Sieve element

Companion cell

End plate

Figure 18.5 A diagrammatic sketch of a longitudinal section through phloem cells. The end walls of the sieve element are open when functional. The sieve element is a nonliving cell when functional. Cells are aligned so that sugars can be transported from the leaves to other plant organs. The adjacent living, companion cells pump sugars and other vital nutrients into the sieve elements.

transported in the phloem. Large molecules such as proteins, nucleic acids, starch, and lipids are not transported from cell to cell, but must be synthesized in or near the cells in which they are used.

The transport of organic solutes from the leaf to sites where they can be used is accomplished through several steps via some highly specialized plant structures. Sugars produced in the leaf through the process of photosynthesis diffuse from the cells in which they are produced (Figure 18.2A). There is a single layer of unique specialized cells in soybean leaves. This layer is several layers of cells below the upper surface and appears to facilitate the transport of sugars to the phloem from the cells in which they are produced. This layer of cells may also function as a temporary site for the storage of nitrogen-containing compounds before they can be used in building seed proteins.

The phloem structure is made up of more than one kind of cell (Figure 18.5). The *sieve element* is a small cell that has a relatively large diameter and appears to be in a state of disintegration at the time it is functional. That is, the internal structure normally found in actively metabolizing cells appears to be breaking down and becoming typical of old and dying cells. The protoplasm, with its internal parts, appears disorganized and broken up into small globs that are stuck to the cell walls. Other parts of these cells may be missing. Bodies of slime or poorly defined structures are also present. The ends of the sieve elements are perforated plates that join to other sieve elements aligned to form tubes, with many sieve plates marking the ends of old cells. Closely associated with the sieve elements are smaller *companion cells*. These cells are physically adjacent to the sieve elements but are quite the opposite from the sieve elements in physical condition. They have prominent nuclei, dense, highly structured protoplasm, and many mitochondria. In short, they appear to be young, actively metabolizing cells.

There is now considerable evidence that sugars are captured by a protein associated with the companion cells, moved across the membranes, and "pumped" into the phloem sieve element. This work uses some metabolic energy, but results in high concentrations of organic materials in the old disintegrating sieve elements. The rate of translocation is probably controlled by the amount of organic solutes pumped into the sieve elements, and that is likely controlled by the available materials that diffuse from the actively

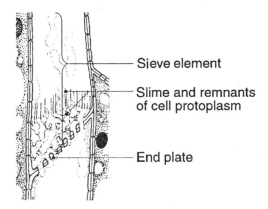

Figure 18.6 A diagrammatic sketch of a phloem sieve element end plate that is plugged with slime and remnants of cell protoplasm. Such end plate blockage is caused by a surge of contents when tissue is cut or broken. The surge, caused by the break or cut, rips away the cell wall slime, sealing the end, and preventing bleeding.

producing cells. As materials accumulate in the sieve elements, the increasing concentrations of solutes attract water. As water pressure increases in the sieve elements, the contents move through the open sieve ends, carrying the organic materials to other plant parts.

Discovering a functional solute transport system

Until about 25 years ago, the mechanism by which plants translocated sugars and other compounds in the phloem was a mystery. Numerous studies had shown those cells just described to be present, and it was widely assumed that they functioned as a part of the translocation system. However, no one had been able to demonstrate that was indeed their function, because the situation was clouded by the fact that the ends of the sieve elements nearly always appeared to be plugged. Thus, the openings in the sieve plates between cells looked to be nonfunctional. Fisher (1975) demonstrated that one could show certain phloem cells of intact soybean petioles were functional. He fed radioactive carbon dioxide to a leaf and then detected the radioactive sugars in the phloem cells. He froze those cells very quickly with an ultracold solution (approximately −300°F) to prevent any movement of cell contents as the petioles were being sampled and handled. The frozen solution in the cells was slowly dehydrated without thawing, and the tissue was examined microscopically. It showed the sieve plates were open and functional. Previous experimenters, without taking such extraordinary measures as Fisher, had sampled tissues and found the sieve plates plugged. Under normal sampling procedures, a surge of phloem contents occurs when the tissue is sampled, causing the sieve plates to be plugged with the clumps of slime and degraded protoplasm. The column of thin-walled tubular cells is normally under pressure. A break results in a surge that strips the slime from the inner surfaces of the degenerating cell. The slime plugs holes in the sieve plates thus preventing bleeding (Figure 18.6).

Energy sinks

Any storage site or growing or metabolizing plant part that is not producing its own energy is an energy sink. Roots are an energy sink and may use energy for taking up nutrients for growth (both increasing length and diameter), for active metabolism that keeps them alive, and for supporting the nitrogen-fixing activity that occurs in their closely

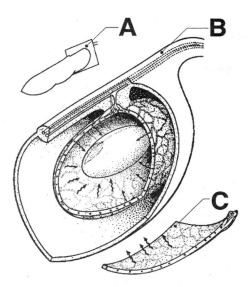

Figure 18.7 A, A diagrammatic sketch of a soybean pod. B, A cross section through a pod showing immature seed and veins that transport sugar and other nutrients from the leaves. The veins stop in the seed coat, and arrows show the nutrient pathway to the developing embryo and cotyledons without vascular connections. C, Sketch of one portion of seed coat showing crude cross sections of veins. The veins form a fine network in the seed coat that delivers the nutrients destined for the immature seed.

associated nodules. Stems serve as supporting structures for the plant and as conduits of water from the roots to the leaves, and conduits of sugars from the leaves to roots and seed. However, they are also energy sinks since the cellulose, lignin, and other compounds that make up their mass must be produced from the phloem-translocated materials. Young, growing leaves are also energy sinks since they must import essentially 100% of their needs until they begin exporting materials. They begin exporting materials when they are about 50% of full size. For a brief period, sugars and amino acids may be both imported into and exported from a growing leaf. The direction of flow, whether into or out of developing leaves, appears to depend on the developmental status of the phloem cells. Sugar is not only used as the source of energy to drive the cell and tissue building process, but it also provides most of the carbon structures for the cells. About 50% of the sugar undergoes several chemical transformations and ultimately becomes a part of cellular structure.

Unloading solutes at the seed

Solute flow into the developing seed is accomplished via the phloem leading through the pods and a seed stalk that attaches the seed to the pod. Seeds are primary sinks for photosynthates. The ability of soybean seed to assimilate and store the products of the transport stream into a valuable combination of proteins, lipids, and other materials is what makes the crop so valuable.

The transport stream comes to the seed coat, where it branches and extends around the circumference of the seed (Figure 18.7). From these two veins, the seed coat is filled with a reticulate venation much like minor veins found in leaves. The phloem bundles are embedded in thin-walled cells near the inner surface of the seed coat. All of the vascularization responsible for assimilate import into the seed stops within the maternal seed coat. There are no vascular connections to the embryo or cotyledons of the next-generation developing seed. The mechanism of phloem unloading and movement of assimilates into the next-generation structures is not well understood.

There is probably an energy-requiring step that moves the assimilates either out of the phloem tissues or out of the interior layer of cells in the seed coat. The result is the translocated materials diffuse across an open space between the seed coat and the embryonic seed. Evidence that the phloem-unloading step is energy requiring was shown by Thorne (1982) and Thorne and Rainbird (1983). They found the process limited by low temperature, the absence of oxygen, or respiratory poisons that prevent metabolic energy from being produced.

Sugar is not broken down and resynthesized during transport to the cotyledons in soybean seeds as it is in some species. However, considerable transformation of nitrogenous compounds occurs near the end of the translocation process. Complex nitrogen-containing molecules found in the transport system, pod walls, and seed coat do not arrive at the next-generation seed structures in the same chemical form they were in just prior to leaving the phloem. Transformations of those molecules therefore occur somewhere en route between the phloem and the developing seed. The exact location and mechanism of these biochemical transformations are not known, but what happens in this process is a conversion of nitrogen-containing molecules into chemical forms that can be readily used to produce proteins and other nitrogen-containing compounds in the developing seed.

The direction of solute flow is probably determined by the concentration gradient established at the sink or growing organ. The control of flow toward competing growing leaves, new roots, or seeds is not understood. How one organ gets the available nutrients at the expense of the others has been a difficult process to determine. Proximity to the source is surely not the answer; otherwise roots would not survive. Raper and Barber (1970) and others have found that soybean and corn roots continue to be produced throughout much of the seed-filling period, but the rate of root growth slows considerably during seed fill. Reduced root-growth rate during seed filling is probably due to competition for the limited sugar supply. Isotope-labeling studies have clearly shown that radioactive CO_2 fed to the plant is distributed throughout the plant with the greatest concentrations ending up in the growing structures. It seems reasonable to assume that the amount of sugar that arrives at any particular sink is determined by the concentration of unused sugar of any sink relative to other competing sinks, and the total amount available. Thus, if such a mechanism is controlling the flow, a seed that can remove nearly all the sugar from a phloem terminus will get a higher percentage of the available sugar than a root or young, growing leaf that is less effective at removing sugar to such a low concentration. Thus, the concentration of sugar maintained at the sink end of the phloem terminus may be the mechanism the plant uses to establish priority for the tissue getting the most nutrients. The sugar concentration loaded into the phloem at the leaf appears to control flow rate (how rapidly sugar and other nutrients leave the leaf), while the degree of utilization at the sink end of the phloem determines the direction of flow. There is little evidence that the transport process itself is limiting.

Summary

The translocation of water and minerals occurs in the xylem cells that form a continuous vascular system throughout the plant. These thick-walled cells are specialized and do not collapse even under negative pressure. The evaporation of water from the leaves and the strong cohesive forces that hold water molecules together pull the water through tiny vessels from the soil and roots by suction or negative pressure. Minerals are carried in the water stream after being "pumped" across a selectively impermeable barrier located in the roots.

Sugars and other complex molecules diffuse from the cells in which they are produced to the leaf phloem cells where they are "pumped" with an energy-requiring process into the sieve elements. Minerals that accumulate in the leaf from the water stream also diffuse to the phloem cells. The concentration of these products attracts water, increasing the pressure in these cells, and causes flow toward tissues with lower nutrient concentrations. The sugar and other nutrients are removed from the translocation stream and used for growth processes in the various tissues. If a break occurs, the resulting surge of pressurized phloem contents causes slime-like material from the sieve element walls to plug the leaks and prevent bleeding.

Acknowledgments

Part of the research was funded by the USDoE National Institute for Global Environment Change through the South Central Regional Center at Tulane University (DoE cooperative agreement no. DE-FC03-90ER 61010).

References

Boyer, J. S. 1970. Differing sensitivities of photosynthesis to low leaf water potentials in corn and soybeans, *Plant Physiol.* 46:236–239.

Boyer, J. S. 1971. Recovery of photosynthesis in sunflower after a period of low leaf water potential, *Plant Physiol.* 47:816–820.

Boyer, J. S. 1985. Water transport, *Annu. Rev. Plant Physiol.* 36:473–516.

Carlson, J. B. 1973. Morphology, in *Soybeans: Improvement, Production, and Uses*, B. E. Caldwell, Ed., Agronomy Monograph 16, American Society of Agronomy, Madison, WI, 17–95.

Fisher, D. B. 1975. Structure of functional soybean sieve elements, *Plant Physiol.* 56:555–569.

Franceschi, V. R. and R. T. Giaquinta. 1983. The paraveinal mesophyll of soybean leaves in relation to assimilate transfer and compartmentation. I. Ultra structure and histochemistry during vegetative development, *Planta* 157:411–421.

Leopold, A. C. and P. E. Kriedeman. 1975. *Plant Growth and Development*, McGraw-Hill, New York.

Raper, C. D., Jr. and S. A. Barber. 1970. Rooting systems of soybeans. I. Differences in root morphology among varieties, *Agron. J.* 62:580–584.

Salisbury, F. B. and C. W. Ross. 1978. *Plant Physiology*, 2nd ed., Wadsworth Publishing, Belmont, CA.

Thorne, J. H. 1982. Temperature and oxygen effects on 14C-photosynthate unloading and accumulation in developing soybean seeds, *Plant Physiol.* 69:48–53.

Thorne, J. and R. M. Rainbird. 1983. An *in vitro* technique for the study of phloem unloading in seed-coats of developing soybean seeds, *Plant Physiol.* 72:268–271.

Tyree, M. T. and J. S. Sperry. 1989. Vulnerability of xylem to cavitation and embolism, *Annu. Rev. Plant Physiol. Mol. Biol.* 40:19–38.

chapter nineteen

Use of crop simulation models and decision support systems in soybean production

Frank D. Whisler, K. Raja Reddy, and Harry F. Hodges

Contents

Introduction

One of the unusual features of agricultural production is the uniqueness of every production season. Each year is unique in the timing of rain, temperature regimes, etc., and when the uniqueness of weather is combined with the individuality of cultural practices, soils, and variety characteristics, the crop production manager has more variables to consider than the human mind can reasonably organize. Complicating these many variables, the producer must also deal with economic and institutional (governmental or societal) constraints. Crop managers, therefore, do not want to be confronted by a huge body of research data, but instead want the best information available, applied directly and specifically to their fields and perhaps even to specific soils within their fields.

0-8493-2301-0/99/$0.00+$.50
© 1999 by CRC Press LLC

Definition of some terms and concepts

We are being told in the popular press almost daily of the wonders of computing power available for the price of the latest personal computer. But how is that affecting soybean producers? Many record-keeping, marketing aids, and accounting packages have become available and are being used to varying degrees by producers. Since the mid-1970s, agricultural research and extension workers have been developing software programs designed to assist in making various agricultural production decisions. These programs usually addressed relatively narrow, production-related topics such as a decision relative to the control of crop-damaging pests. During the mid to late 1980s, software for developing expert systems became available, and facilitated developing computer programs for agricultural problems that provided additional decision aids. Basically, "expert systems" allow the computer user to apply certain rules that can be applied repetitively to a situation to arrive at a "best" alternative selection; e.g., one can select a variety that meets several criteria. The variety must have resistance to certain races of nematodes, disease x and y, and yield in the top 50% of the variety test at certain locations.

Simulation as used herein means using mathematical equations written in computer code to predict how a crop, in this case, soybeans, will grow on a particular soil at a specific location given specific weather for that location, crop management inputs, and specific growth parameters for a soybean variety (Whisler et al., 1986). A mechanistic model attempts to represent causability between variables explicitly, such as the rate of plant development in height and numbers of nodes being "caused" or driven by air temperature, time (age of organs), water status of the plant and soil, etc. This is in contrast to empirical models where plant height might be a function solely of time or time and temperature. All crop simulation models have a range of mechanistic and empirical equations.

Model users

We have been asked by the editors to speculate on who the future users of a model such as GLYCIM/GUICS might be. Since GLYCIM has only been used on the farm in a closely monitored research mode, this is only speculation. Based upon our experience with the cotton model (McKinion et al., 1989), individual growers, consultants, extension agents, and grain elevator cooperatives are all possibilities.

Crop simulation models

Another approach to developing a computer aid for crop production decision making is to develop a highly technical simulator of crop responses to environment. Ideally, such a program would incorporate the important physiological processes involved in crop growth and appropriately addresses the various physical processes that concern crop production into one package that would correctly predict growth, development, and yield. The physical processes would include the movement and availability of water and nutrients in the soil. The plant processes would include modeling light interception, photosynthesis, respiration, growth, and developmental rates such as number of nodes, branches, and flowering sites, and the setting of flowers and pods. Also included must be appropriate responses of the crop to stresses and the timing of maturation.

Considerable progress along these lines was made in developing a simulator for cotton known as GOSSYM (Baker et al., 1983). Almost immediately after GOSSYM was tested on the farm (McKinion et al., 1989), producers indicated that the information provided was interesting and useful, but too complicated and technically involved for practical use. Therefore, an expert system was developed to facilitate information input, to provide simplified reports that could more readily be read and interpreted, and to suggest alternative management practices. The expert system was called COMAX, derived from the words CrOp MAnagement eXpert. The COMAX is basically a rule-based expert system that takes information obtained from the crop simulator, GOSSYM, and applies that information as an extension specialist would to a production recommendation, such as when to apply PIX, irrigate, or apply N. Such a recommendation would be based on a particular field and its crop growth history, the crop condition that is controlled by planting date and other cultural practices, as well as the past weather received by that crop. Future weather, required to complete the simulation of the season, may be last year's weather, a typical wet or dry year, or a whole file of weather records of many years.

Ideally, the model would simulate the soil moisture and root distribution as well as the crop growth and other aspects of development. It should account for movement of nitrogen and water and the growth of roots into various soil layers. There has been considerable testing and fault-finding with different aspects of this attempt at crop simulation and its applications to production decision making. However, it seems apparent that such a model is the first of a kind, and with improvements and refinements subsequent models will more accurately simulate crop responses to the physical environment. As more accurate models become available, they will be more useful for predicting responses to alternative cultural practices and for aiding production decisions.

Since the development and field testing of GOSSYM/COMAX (Baker et al., 1983; Hodges et al., 1997; Reddy et al., 1997a), a similar model called GLYCIM was developed for soybeans (Acock et al., 1985). Some of the user input/output characteristics of GLYCIM have been obtained from the commercially available Windows 95 program. The resulting soybean model is called GUICS (Reddy et al., 1997b).

The following description of GLYCIM was extracted from Acock and Trent (1991). It is a dynamic (continuous) simulator of soybean crop growth. The model runs on hourly time steps, which means that the processes are calculated hourly throughout the day and night using actual weather conditions (temperature and sun intensity), as well as soil moisture. GLYCIM consists of a collection of subroutines called modules coded in C computer language.

The modules describe related sets of physical or physiological processes. The model describes the growth of an average plant in a uniform crop that is free of pests and diseases. (Balances of materials such as mass or dry weight and nitrogen content of individual leaves and petioles on the plant are kept to allow subsequent interfacing with pest and disease models.) Balances of materials are calculated daily for other plant organs by type (i.e., stems, flowers, seeds, roots) and for the soil by cells assigned according to horizontal and vertical location in the soil profile. These hypothetical cells are rectangular blocks of soil with a thickness of almost 0.5 in. in the direction of the plant row and a width that is a fraction of row spacing. Cell depth must be specified, but is routinely assumed to be 4 in. per cell down to a depth of 3 ft. Root activity and soil processes are assumed to be symmetrical about the plant row, so movement of materials are calculated only for a representative set of cells from the row to the row middle and to the depth specified. Movement of water, heat, nitrate, and oxygen are simulated for the soil, whereas movement of carbon, nitrogen, and other structural dry matter are simulated for the plant.

The model was designed to simulate the growth of any soybean maturity group on any soil and planted at any location and time of year. A maturity group is a group of varieties with a similar response to day length resulting in control of time of flowering and maturity. Maturity group number has proved to be an inadequate description of differences between varieties, and individual varieties are now being characterized. All soil processes in the model are mechanistic (use cause and effect relationships to predict a result), and soil characteristics by horizon are required for all locations simulated. For example, a combination of gravity and soil texture determines the water-holding capacity of a soil. Movement of water in the soil is predicted by mathematical equations that relate to those factors.

Simulations are initiated at the cotyledon stage with appropriate data on the number, size, and weight of organs on the plant. Thereafter, the only information required is daily weather data of the type collected by class-A stations and daily total solar radiation. The growth in size of the plant and its advances through the growth stages are all predicted by the model. In the course of the simulation, the model provides predicted values for many of the physiological variables that can be readily measured. This feature is intended to facilitate the improvement of the model.

Depending on the nitrogen supply/demand ratio, carbon compounds produced in the leaves may be sent to the nodules, where they are used for energy to fix nitrogen, grow nodules, or initiate new nodules. The inert atmospheric nitrogen can be converted in the nodules by bacteria to biologically usable forms that may be transported to other plant parts and used to make proteins and other structural compounds. If the shoot is not turgid, the carbon that was allocated for shoot growth instead goes to grow roots. If the plant is in the reproductive stage, both carbon and nitrogen compounds are moved to the developing pods and seeds to support their growth.

To simulate mechanistically the daytime reduction in shoot growth and the midday stomatal closure that occurs during water stress, the model runs in time steps of 1 h. For each of the hourly calculation periods, process rates are calculated for the midpoint of the current period based on states at the end of the previous period. For most processes, these rates are assumed to be constant during that hour.

Usefulness of computer decision aids

The use of computerized management decision aids has increased rapidly over the last two decades and will probably continue even more rapidly into the foreseeable future. This is due to the decrease in price of computers and/or the increase in computer power and capability, and the need for soybean producers to have as much information as possible at their fingertips for making production decisions. Some of these decision aids are what the computer software developers call expert systems; examples are HERB (Wilkerson et al., 1991), a weed growth and weed-control computer program for soybeans, and WHIMS (Wagner et al., 1992), an insect-control program for cotton. These expert system programs ask for information from the grower about the number and kinds (species) of weeds or insects, their size and the crop size and stage of development, and some information about the weather and soil. Using this information, these expert systems then give recommendation alternatives as to what chemicals to use, and how much. They do not try to simulate how the crop will grow with or without treatment from the day of observation until harvest or what the next stages of crop development will be. HERB has been adapted for Mississippi conditions (Rankins et al., 1995), and it also provides estimates of the crop loss with and without weed control.

Another class of decision aids are empirical models of plant development, such as the DD50 programs for rice (Slaton et al., 1993). These models require the user to provide inputs of planting or emergence dates, variety, and daily air temperatures. The model then predicts the crop developmental stages, i.e., the number of mainstem leaves, when the panicle is initiated, time of floral initiation, etc., based upon actual temperatures since emergence and expected future temperatures from long-term weather records. Recommendations are then made about fertilizer, herbicide, and insecticide applications based on the current crop growth stage. Since rice is not usually water stressed and if well fertilized not usually nitrogen stressed, such models are very useful.

Types of management decisions

Soybean producers make many kinds of production decisions, but we might classify them into two groups: preseason, or strategic decisions, and within-season or tactical decisions. Many of the decisions are interrelated, and thus the use of a model is most helpful. For example, some varieties perform and yield better on certain soils and at certain plant populations and even row spacings with certain methods of irrigation while other varieties are better in other systems and locations. With a model such as GLYCIM, growers can make many "runs" or computer simulations for their conditions or expected conditions to relate to a decision such as leasing/purchasing of a new field, or installing a new irrigation system on a field where they have farming experience. They can make the simulations during December and January. One producer told us that he usually makes 400 runs during this period and thus has planned the variety, row spacing, and seeding rate he will use on each field before he ever puts the first seed in the ground. This saves him time and money at planting time.

Other strategic decisions might involve soil fertility, seed treatment, and weed control. By taking soil fertility samples early, producers know their fertility needs and also the soil organic matter and pH levels. Also, by knowing the disease and weed infestation levels in specific fields from the past growing season, the producer can use an aid such as HERB for next year's seed treatment and herbicide purchases.

During the growing season, producers make tactical decisions about when to cultivate and apply herbicides, when to irrigate, and when to harvest. By using an aid like HERB, it forces users to identify their specific weed species, measure plant height, and consider soil moisture. The program will then provide specific herbicide recommendations and alternatives. By using GLYCIM, the program will forecast when irrigation is needed for each soil in a field. Thus, the grower will know when to start irrigation by using the soil series that requires water the earliest. By making simulations with the other soils and the same irrigation schedule, it is hoped the user will not overwater the other soils in that field. One grower-cooperator using GLYCIM in 1992 found the model recommending an irrigation on a Sharkey clay soil much earlier than he normally irrigated that field (Remy, 1994). From his previous experience, he knew that it took him 4 to 5 days to get the water across this deeply cracked field. By starting earlier it took only 36 h to get the job done and this increased his irrigation efficiency by 400%. It also took less water at each irrigation and lowered pumping costs. On the other hand, another cooperator farming a field of Dubbs sandy loam found the model recommending a delay in his normal irrigation timing. In 1993, using the model allowed him to reduce the number of irrigations by one on that field without sacrificing any yield compared with another field of the same soil type and variety.

Who should use computer decision aids

The question is often asked, "Who will or should use these decision aids?" Some of our more experienced users have told us, "Any producer who is in the soybean business for a living should use these aids." Almost any PC computer user could install HERB on his or her machine. It would, however, take a machine with Windows 95 capability to run GLYCIM. More about the hardware requirements later. One grower-cooperator has GLY-CIM on his laptop computer that he carries in his pickup. As he goes from field to field, he checks the model predictions with what he finds in the field. Such an exercise forces him to evaluate the crop status more critically and be a better manager. However, many growers do not feel that they have the time to run these aids on their own PCs. There are several crop consultants and some extension agents who are equipped to perform this task for a fee. It takes close communication, however, between the grower and consultant to keep track of all the specific operations in a field to make the outcome as useful as possible. A wrong herbicide concentration or chemical or a missed irrigation can void the recommendations of any decision aid. Other persons, such as researchers, lending agents, co-op operators, etc., can also make use of these aids to answer "what if" types of questions and/or evaluate yield and economic returns, or potentials for new land leases or purchases.

Costs of computer decision aids

The costs of using these decision aids are also varied. As we stated earlier, almost any recent-model PC (386 processor or higher) can be used to run HERB. A Windows 95–compatible PC system (486 processor or higher) is usually a little more expensive, but should be in the $1000 to $4000 range depending upon how many extras, such as sound cards, CD players, and games, etc., are added. Additionally, one needs at least 8 megabits of random access memory (RAM), 250 meg of disk, and a color monitor. In 1997, WINDOWS 95 is the required operating system. An automated weather station would be needed if your field(s) is more than 5 or 6 miles from an existing station. Such a weather station costs $4000 to $5000 plus a telephone line of $30 to $50/month. A $10 to $20 rain gauge is needed for each field. If the soil (that is, the soil series and texture) that is in the field is not one for which there is a measured hydrology file, it would also need to be sampled and those factors measured. Arrangements for that can be made through the local extension agent or the Mississippi State University Department of Plant and Soil Sciences.

Seemingly one of the greatest costs that some producers have complained about is the time required to make the runs and assimilate the results. The GLYCIM is perhaps the biggest time user of the aids mentioned herein. It takes 1 to 3 min, depending upon the computer, to run one soil and one year's simulation. The most time-consuming items, however, are to set up the initial files and keep the weather data up-to-date for each field. That requires counting the plants at emergence on several consecutive days and then averaging that data to initialize the crop file in the machine. It also requires someone to read the rain gauges on a daily basis and update the weather file before making a computer run. The plant population counting is done rapidly but needs to be checked two or three times after emergence and during the season. The weather updating should only take a minute or two for each field. The building of the initial files should take about 5 min per field at the start of the season, but will have to be changed if the population count changes during the course of the season. For example, at emergence there may be 6.5 plants per foot of row, but later due to seedling diseases, hail storms, or whatever (deer browsing), the average population might be 2.2 plants per foot of row.

Additional benefits of crop simulation models

There are also some intangible values associated with using these management decision aids that are impossible to express in dollars. Most growers who have used the crop simulation models have said after their first season of use, "I learned more about growing soybeans this year than I ever knew before!" or "It made me get out of my truck and into the field more than ever." Most producers approach the use of crop simulation models with a healthy skepticism, which we encourage. They say or think "No computer is going to raise a better crop than I can!" so they get out of their pickups and measure plant height, count the number of nodes, and check the fruiting stage at several places in a field and compare it to the computer predictions. Meanwhile, they see the weed and/or insect pressures or disease symptoms or nutrient deficiency symptoms and make plans to correct

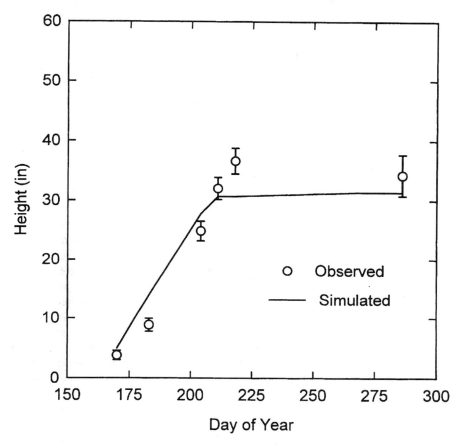

Figure 19.1   Plant height vs. day of the year for Hutcheson soybeans grown on an irrigated Decatur silt loam soil by Mr. Bragg in 1997. The solid line is the GLYCIM simulated results. The open circles are observed average values of five measurements with 1 standard deviation about the average.

these problems. That makes them better producers, but it is impossible to put dollar values on that education or improvement. However, a word of warning should be added. If problems such as insect damage, herbicide damage, disease, etc. are present to a large extent in a field, the usefulness of the model will be lost. At present, the model assumes that all of these items are being properly managed. Its predictions could or should only be viewed under such adverse circumstances as being related to the *potential* performance of the crop. The actual crop performance will be much less depending upon the severity of the problem.

Comparison of simulated and measured plant responses

Some examples of GLYCIM simulations compared with measured plant data from a producer's field should serve to show the precision (or lack thereof) of the model's performance. Figures 19.1 through 19.3 show changes in plant height and vegetative and reproductive growth stages (Fehr and Caviness, 1977) over the growing season for a field of Hutcheson soybeans grown by Mr. Dennis Bragg on a Decatur silt loam in Madison County, Alabama in 1997. The solid lines are the model predictions, and the circles with the vertical error bars are the averages of five sites, plus or minus the statistical standard deviations (i.e., ⅔ of all the observations at one date are within the error bars). In this case

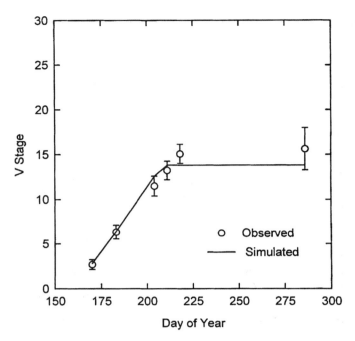

Figure 19.2 Vegetative growth stages (nodes) vs. day of the year for Hutcheson soybeans grown on an irrigated Decatur silt loam soil by Mr. Bragg in 1997. The solid line is the GLYCIM simulated results. The open circles are observed average values of five measurements with 1 standard deviation about the average.

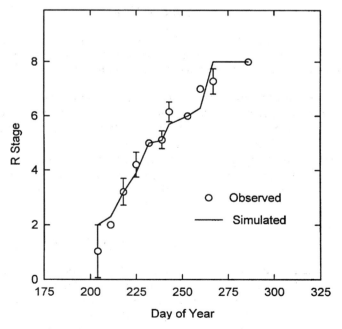

Figure 19.3 Reproductive stages vs. day of the year for Hutcheson soybeans grown on an irrigated Decatur silt loam soil by Mr. Bragg in 1997. The solid line is the GLYCIM simulated results. The open circles are observed average values of five measurements with 1 standard deviation about the average.

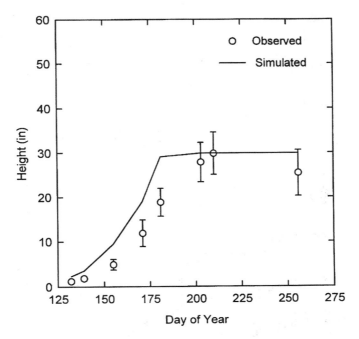

Figure 19.4   Plant height vs. day of the year for Asgrow 4922 soybean grown on an irrigated Dundee sandy loam soil by Mr. McCain in 1997. The solid line is the GLYCIM simulated results. The open circles are observed average values of five measurements with 1 standard deviation about the average. There were no variety parameters available for Asgrow 4922, so Pioneer 9501 parameters were used in the model.

we had a calibration file for this variety based upon 5 prior years data collected on other soils. The model predicted the plant height, vegetative growth stages and reproductive growth stages relatively well.

On a Dundee sandy loam soil, Mr. William McCain in St. Francis County, Arkansas planted Asgrow 4922. We had not had any prior experience in growing or monitoring this variety and therefore had to use another variety file available for the model. From the 1994 Mississippi Soybean Variety Trial report (Askew et al., 1995) and our own research plot results, we selected Pioneer 9501 variety file because it was from the same maturity group and yielded about the same in variety trials. We had the data for Pioneer 9501 from earlier years. The results are shown in Figure 19.4 through 19.6. The heights and vegetative stages were overestimated by the model, but the reproductive stages were simulated very well. These results illustrate the need to have specific files for each variety, and also show that the model did not provide useful vegetative developmental results for that Asgrow 4922 variety. As we test the crop simulators under a wide range of conditions and cultural practices, we parameterize the variety-specific characteristics so the model can be used in a wider array of conditions. There does not appear to be any way to avoid developing variety-specific parameters.

Crop simulation and precision agriculture

There has been a lot of publicity over the last 2 to 3 years about precision agriculture geographic information systems (GIS), global positioning systems (GPS), and yield monitors. Over the last two to three decades, efforts were made to utilize remote-sensing technology for helping make management decisions, but so far there have not been any

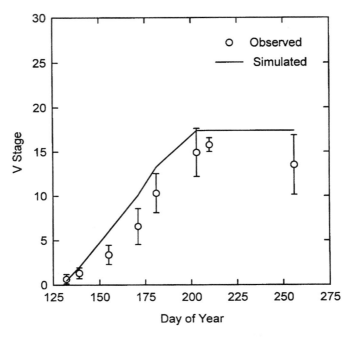

Figure 19.5 Vegetative growth stages (nodes) vs. day of the year for Asgrow 4922 soybeans grown on an irrigated Dubbs sandy loam soil by Mr. McCain in 1997. The solid line is the GLYCIM simulated results. The open circles are observed average values of five measurements with 1 standard deviation about the average. There were no variety parameters available for Asgrow 4922, so Pioneer 9501 parameters were used in the model.

dramatic successes. Precision agriculture demands a rational process for understanding the spacial distribution of yield, so that one can manage the field to obtain the potential growth and yield from all sites in the field. To do that, the farm manager has to know which of the many factors might limit crop yield on any day. Past experience and knowledge testify that no single measurement will determine the optimum decision. Instead, the field drainage, pH, organic matter, fertility, weeds and other pests, water-holding capacity, weather, crop cultural practices, and variety, etc. each play a role in determining yield and yield distribution over the field. Some sites in the field may be limited by one or more factors, while other sites are limited by additional factors. By improving our records and methods of analysis of causes for yield reductions, we should be able to eliminate some of the yield-reducing causes and improve overall yields.

Crop yield monitors, variable-rate applicators, and associated computer technology may provide tools essential for diagnosing causes of yield reduction. More-sophisticated crop models and methods to link the information provided by models with that provided by remote-sensing and other site-specific information may allow the type of analysis needed. This merger of technologies will challenge even the most diligent producers, but the tools to do it are becoming available and it appears almost feasible. Sampling techniques for use with precision agriculture technologies are discussed in the Chapter 17 by Willers et al., in this book.

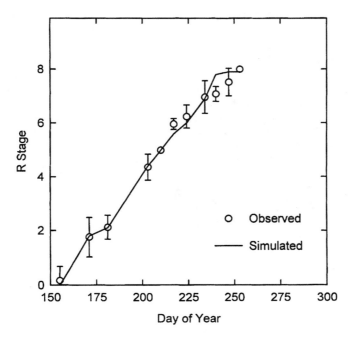

Figure 19.6 Reproductive stages vs. day of the year for Asgrow 4922 soybeans grown on an Dubbs sandy loam soil by Mr. McCain in 1997. The solid line is the GLYCIM simulated results. The open circles are observed average values of five measurements with 1 standard deviation about the average. There were no variety parameters available for Asgrow 4922, so Pioneer 9501 parameters were used in the model.

References

Acock, B. and A. Trent. 1991. The soybean crop simulator GLYCIM: documentation for the modular version 91. Response of vegetation to carbon dioxide, No. 017, Joint Program of the USDA and the USDE.

Acock, V., V. R. Reddy, F. D. Whisler, D. N. Baker, J. M. McKinion, H. F. Hodges, and K. J. Boote. 1985. The soybean crop simulator GLYCIM. Model documentation 1982, PB85171163/AS, U.S. Department of Agriculture, Washington, D.C.

Askew, J. E., Jr., A. Blaine, J. Coccaro, C. Estess, G. Jones, F. Hancock, D. Ingram, D. Reginelli, A. Smith, and T. R. Vaughan. 1995. Soybean 1994 Variety Trials, MAFES Inf. Bull. 276.

Baker, D. N., J. R. Lambert, and J. M. McKinion. 1983. GOSSYM: A Simulator of Cotton Growth and Yield, S.C. Agric. Expt. Sta. Bull. Tech. Bull. No. 1089.

Fehr, W. R. and C. E. Caviness. 1977. Stages of soybean development, Coop. Ext. Ser. Iowa State U. Spec. Rep. 80.

Hodges, H. F., F. D. Whisler, S. M. Bridges, K. R. Reddy, and J. M. McKinion. 1997. Simulation in crop management — GOSSYM/COMAX, in *Agricultural Systems Modeling and Simulation*, R. M. Peart and R. B. Curry, Eds., Marcel Dekker, New York, Chap. 8, 235–282.

McKinion, J. M., D. N. Baker, F. D. Whisler, and J. R. Lambert. 1989. Application of the GOSSYM/COMAX system to cotton crop management, *Agric. Syst.* 31:55–65.

Rankins, A., Jr., D. R. Shaw, J. D. Byrd, Jr., and W. C. Elkins. 1995. Modification for HERB for postemergence weed control recommendations in soybean, *Proc. Southern Weed Sci. Soc.* 48:197.

Reddy, K. R., H. F. Hodges, and J. M. McKinion. 1997a. Crop modeling and applications: a cotton example, in *Advances in Agronomy*, Vol. 59, D. L. Sparks, Ed., Academic Press, San Diego, CA, 225–290.

Reddy, V. R., Ya. A. Pachesky, F. D. Whisler, and B. Acock. 1997b. A survey to develop a decision-support system for soybean crop management, ACSM/ASPRS, Annual Convention of Exposition, Tech. Papers, 4:11–18.

Remy, K. 1994. GLYCIM soybean model proves its worth, in Research Highlights, MAFES Publ. 57, Mississippi State University.

Slaton, N. A., R. S. Helms, C. E. Wilson, Jr., and B. R. Wells. 1993. DD50 Computerized Rice Management Program, Computer Tech. Ser. Ark. Coop. Ext. Service.

Wagner, T. L., J. L. Willers, M. R. Williams, and R. L. Olson. 1992. Field application of rbWHIMS — an expert system for cotton pest management, in *Proc. Belt. Cotton Conf.*, 790–792.

Whisler, F. D., B. Acock, D. N. Baker, R. E. Fye, H. F. Hodges, J. R. Lambert, H. E. Lemon, J. M. McKinion, and V. R. Reddy. 1986. Crop simulation models in agronomic systems, *Adv. Agron.* 40:141–208.

Wilkerson, G. G., S. A. Modena, and H. D. Coble. 1991. HERB: decision model for postemergence weed control in soybean, *Agron. J.* 83:413–417.

Index